Advanced Topics on Semilinear Evolution Equations

SERIES ON CONCRETE AND APPLICABLE MATHEMATICS

ISSN: 1793-1142

Series Editor: Professor George A. Anastassiou
Department of Mathematical Sciences
University of Memphis
Memphis, TN 38152, USA

Published*

*To view the complete list of the published volumes in the series, please visit:
http://www.worldscientific/series/scam

Series on Concrete and Applicable Mathematics – Vol. 25

Advanced Topics on Semilinear Evolution Equations

Mouffak Benchohra
Djillali Liabès University of Sidi Bel-Abbès, Algeria

Gaston M. N'Guérékata
Morgan State University, USA

Abdelkrim Salim
Hassiba Benbouali University of Chlef, Algeria

World Scientific

NEW JERSEY · LONDON · SINGAPORE · BEIJING · SHANGHAI · HONG KONG · TAIPEI · CHENNAI

Published by

World Scientific Publishing Co. Pte. Ltd.

5 Toh Tuck Link, Singapore 596224

USA office: 27 Warren Street, Suite 401-402, Hackensack, NJ 07601

UK office: 57 Shelton Street, Covent Garden, London WC2H 9HE

Library of Congress Control Number: 2024046669

British Library Cataloguing-in-Publication Data
A catalogue record for this book is available from the British Library.

Series on Concrete and Applicable Mathematics — Vol. 25
ADVANCED TOPICS ON SEMILINEAR EVOLUTION EQUATIONS

ISBN 978-981-98-0318-7 (hardcover)
ISBN 978-981-98-0319-4 (ebook for institutions)
ISBN 978-981-98-0320-0 (ebook for individuals)

For any available supplementary material, please visit
https://www.worldscientific.com/worldscibooks/10.1142/14092#t=suppl

Desk Editors: Murali Appadurai/Lai Fun Kwong

Typeset by Stallion Press
Email: enquiries@stallionpress.com

We dedicate this book to our family members. In particular, Mouffak Benchohra makes his dedication to the memory of his wife K. Bencherif; Gaston M. N'Guérékata makes his dedication to his wife Béatrice N'Guérékata; Abdelkrim Salim makes his dedication to the memory of his father Hamza.

Preface

Differential evolution equations serve as mathematical representations that capture the progression or transformation of functions or systems as time passes. Currently, differential equations continue to be an active and thriving area of study, with continuous advancements in mathematical methodologies and their practical applications spanning diverse fields such as physics, engineering, and economics. In the late 20th century, the notion of "Differential Evolution Equations" emerged as a distinct field applied to optimization and machine learning challenges. Evolution equations hold immense importance in numerous realms of applied mathematics and have experienced notable prominence in recent times. This book delves into the study of several classes of equations, aiming to investigate the existence of mild and periodic mild solutions and their properties such as approximate controllability, complete controllability and attractivity, under various conditions. By examining diverse problems involving second-order semilinear evolution equations, differential and integro-differential equations with state-dependent delay, random effects, and functional differential equations with delay and random effects, we hope to contribute to the advancement of mathematical knowledge and provide researchers, academicians, and students with a solid foundation for further exploration in this field. Throughout this book, we explore different mathematical frameworks, employing Fréchet spaces and Banach spaces to provide a comprehensive analysis. Our investigation extends beyond traditional solutions, encompassing the study of asymptotically almost automorphic mild solutions, periodic mild solutions, and impulsive integro-differential equations. These topics shed light on the behavior of equations in both bounded and unbounded domains, offering valuable insights into the dynamics of functional evolution equations. The rigorous mathematical analysis and the diverse range of equations considered in this book make it a valuable resource for anyone seeking a deeper understanding of the existence and controllability of solutions in nonlinear systems.

Throughout the book, the authors provide numerous examples as well as a dedicated applications chapter to help readers understand and apply the concepts presented. The book is intended for advanced undergraduate and graduate students, as well as researchers and professionals in the field of mathematics science and engineering.

We are grateful to S. Abbas, A. Benaissa, A. Bensalem, S. M. Bouguima, J. Graef, K. A. Kada, E. Karapinar, A. Moussaoui, J. J. Nieto, N. Rezoug, C. Tunc and Y. Zhou for their contributions to research on the problems covered in this book.

M. Benchohra
G. M. N'Guérékata
A. Salim

Contents

Chapter 1

Introduction

Differential evolution equations are mathematical descriptions of the evolution or change of functions or systems over time. The study of these equations dates back to the 17th century when mathematicians like Isaac Newton and Gottfried Leibniz introduced calculus and techniques for solving differential equations. Throughout the 18th and 19th centuries, notable mathematicians such as Leonhard Euler, Joseph-Louis Lagrange, Augustin-Louis Cauchy, Henri Poincaré, and George Green made significant contributions, advancing the study to include ordinary and partial differential equations. In the 20th century, the focus expanded to nonlinear differential equations, thanks to mathematicians like Andrey Kolmogorov, Jürgen Moser, and Stephen Smale. Today, differential equations remain a vibrant field of research with ongoing developments in mathematical techniques and applications across various disciplines, including physics, engineering, and economics. In the late 20th century, the concept of "Differential Evolution Equations" emerged as a specific field used in optimization and machine learning problems. This stochastic optimization technique utilizes the principles of differential equations to guide the optimization process. Evolution equations play a significant role in various domains of applied mathematics [1, 2]. Notably, there have been numerous advancements in understanding second-order differential equations, as evidenced by multiple works [3–9]. In particular, the field of semilinear functional evolution equations has witnessed substantial progress in recent times. Several papers [10, 11] and additional references provide valuable insights into this area. Noteworthy achievements include the establishment of global existence results for functional evolution equations and inclusions on Fréchet spaces, employing various approaches [12, 13]. It is worth noting that earlier studies often imposed certain assumptions, such as Lipschitz conditions and the compactness of the semigroup.

Periodic solutions are solutions to specific types of partial differential equations that exhibit periodic behavior over time. These solutions are particularly valuable in studying systems that demonstrate periodic tendencies, commonly encountered in physics and engineering. An essential characteristic of periodic solutions is their validity over an extended period rather than just at a single time point. This enables the exploration of a system's long-term behavior rather than focusing solely on its short-term dynamics. The applications of periodic solutions span various fields, including physics, engineering, and economics. In physics, these solutions serve as models for systems like oscillators and wave propagation. In engineering, they offer insights into mechanical and electrical system behaviors. In economics, they facilitate the understanding of market dynamics and population phenomena. Furthermore, periodic solutions find utility in the realm of optimization, particularly in addressing optimization problems featuring periodic constraints. For more details, see the following publications [14–20, 22–26].

Attractivity, in the context of differential equations, describes the behavior of solutions as they approach a specific point or set of points in the system's phase space. An attractive point or set is one where solutions, starting sufficiently close, gradually converge towards it over time. Different types of attractivity exist, depending on the characteristics of solutions and the corresponding points or sets in the phase space. The most common types are: asymptotic attractivity, global attractivity, exponential attractivity and Lyapunov attractivity. The concept of attractivity holds great significance in the study of dynamical systems, shedding light on the long-term behavior of differential equation solutions. Attractivity finds applications in various fields, including physics, engineering, economics, and biology, providing valuable insights into system behavior under diverse conditions [27–30].

Controllability, within the realm of differential equations, refers to the capacity to govern the behavior of a system described by a set of such equations. A system is deemed controllable if it can be propelled from any initial state to any desired final state through a suitable control input, a time-dependent function applied to modify the system's behavior. Controllability assumes diverse forms, contingent upon the system's nature and the control input involved. The most prevalent types encompass: complete controllability and approximate controllability. Controllability theory is crucial for understanding the dynamics of abstract control systems. Exact controllability refers to the ability to steer the system precisely to the end state, while approximation controllability allows for directing the system to a neighborhood of the final state, however small. Approximate controllability is particularly relevant for real-world systems that often exhibit uncertainty or imprecision. Numerous researchers have studied approximate or complete controllability of control systems, leading to various published papers (see references [31–46] and the related works). Generalizations of controllability to infinite-dimensional systems in Banach spaces with bounded operators have been made by many researchers (see [47,48,350]). Some authors, such as the ones in [49–51], have investigated the controllability problem by transforming it into a fixed point problem. Semilinear ordinary differential equations have been explored by Balachandran and Dauer in [52], while Benchohra *et al.* have studied various classes of functional differential equations and inclusions using fixed point arguments and provided controllability results in [53–57]. Controllability also holds great significance in control theory and engineering as it relates to pole assignment, quadratic optimal control, observer design, and structural decomposition. Several authors, including Benchohra *et al.* [58–61], Balachandran *et al.* [62], and Mophou *et al.* [63], have extensively contributed to the study of controllability for different types of differential equations, such as neutral functional differential equations, integrodifferential equations, differential inclusions, and impulsive differential inclusions in Banach spaces. The idea of controllability has been expanded to infinite-dimensional Banach space systems with unbounded operators, as explored in the monographs [40, 64–66].

Asymptotically automorphic solutions of evolution equations refer to solutions of specific partial differential equations that exhibit a periodic behavior over time, with the period of the solution growing indefinitely as time progresses. These solutions are particularly valuable when studying systems that display asymptotic periodicity, which is characterized by an increasing period as time goes on. Understanding the long-term behavior and stability of these solutions is the central objective in the field of asymptotically automorphic solutions of evolution equations, which is a relatively new and actively researched area. The concept of almost asymptotic automorphy is closely related to asymptotic periodicity, as both pertain to the long-term behavior of a system. By studying the properties of solutions that possess asymptotic automorphy, we gain insights into the system's stability and how it evolves over extended periods. The concept of automorphic functions, introduced by Bochner [67], has found applications in various domains, including ordinary and partial differential equations, abstract differential equations, functional differential equations, and integral equations [68–72]. For a comprehensive understanding of almost automorphic functions and their applications, readers can refer to the monographs by N'Guérékata [25, 73]. Building upon this notion, the concept of asymptotically almost automorphic functions was introduced by N'Guérékata [74], leading to significant developments and applications in subsequent works [75–78]. While the investigation of almost periodic, almost automorphic, and weighted pseudo almost periodic solutions to various evolution equations has been undertaken by multiple authors [75, 79–81], the idea of S-asymptotically ω-periodic functions has proven valuable in addressing problems related to functional differential equations, integro-differential equations, fractional differential equations, and fractional integro-differential equations. The concept of S-asymptotically ω-periodic functions was initially introduced by Henriquez *et al.* [82,83], and it has since garnered considerable attention in deterministic settings [84–91]. The existence of S-asymptotically almost periodic solutions holds great significance in the qualitative theory of differential equations, captivating mathematicians, due to its mathematical curiosity and wide-ranging applications in physics, mathematical biology, and other fields [92–95]. Almost automorphic functions, an extension of almost periodic functions introduced by S. Bochner [96], have been extensively studied, with fundamental results and applications to deterministic differential systems documented in [97–99]. Moreover, investigations on almost automorphic functions and solutions for differential systems have been pursued by numerous mathematicians [100, 101]. The pseudo nearly periodic function has attracted the interest of several researchers [102], and the study of almost periodic solutions for stochastic differential equations has witnessed progressive research [103]. Notably, Tudor [104] explored nearly periodic solutions for affine and stochastic evolution equations, while Mokkedem and Fu [105] introduced the concept of square-mean almost automorphic process and studied square-mean almost automorphic mild solutions for stochastic differential equations, expanding the realm of almost automorphic theory from deterministic to stochastic settings.

Differential equations with delay are a class of mathematical equations used to describe systems that exhibit a time delay in their response to changes. This delay is incorporated into the equations by introducing a term that represents the time between an event and the system's corresponding reaction. Different types of delays can be considered, such as constant delay, state-dependent delay, and functional delay. The field of studying differential equations with delay is highly active and holds wide-ranging applications in physics, engineering, biology, and economics. The choice of the delay type depends on the characteristics of the system and the specific problem under investigation. Each type of delay has its own distinct features, advantages, and drawbacks. In the modeling of physical processes, for instance, state-dependent delay is often employed. Unlike a fixed delay, the delay in these systems depends on the state of the system itself. The study of state-dependent delay systems is a rapidly growing field with numerous recent advancements. For readers interested in delving deeper into the latest developments in this area, we recommend several publications that offer comprehensive investigations of state-dependent delay systems. These papers, including [32, 39, 41, 106–113], provide valuable insights and a thorough overview of the current research in the realm of state-dependent delay systems.

Differential equations with infinite delay are a type of mathematical equations that describe the behavior of systems that have an infinite delay in their response to changes in the system. The delay can be modeled by introducing an infinite delay term in the equation, which represents the time between the occurrence of an event and the system's response to that event, when it approaches infinity. The study of differential equations with infinite delay is an active area of research and it has a wide range of applications in fields such as physics, engineering, biology, and economics. In physics, they can be used to model the behavior of systems such as oscillating systems and wave propagation. In engineering, they can be used to model the behavior of systems such as mechanical and electrical systems. In biology, they can be used to model the behavior of systems such as population dynamics and neural networks. In economics, they can be used to model the behavior of systems such as markets and populations. In contrast to the finite delay differential equations, the infinite delay ones have some specific properties, such as the existence of equilibrium points that are not globally asymptotically stable, the existence of multiple periodic solutions, and the possibility of the system to exhibit phenomena such as oscillations and chaos. Many researchers have expressed interest in the study of differential equations with state-dependent delay since they are fundamental in applications and qualitative theory, and they describe many physical, chemical, and biological problems, see [24, 114, 115]. State-dependent delay differential equations are used to provide more realistic modeling in systems where delay changes depending on the system's internal effects. In a basic population dynamics model, for example, the time to maturity is assumed to be a constant delay, see [116]. In [117], it was discovered that the length of time to maturity of Antarctic whales and seals varies depending on population status, and this was investigated in [118] using a mathematical model using state-dependent delay differential equations.

Random differential equations are a form of mathematical equations used to describe the behavior of systems influenced by random or uncertain factors. These equations incorporate one or more random variables or processes, representing the unpredictable or random influences acting on the system. Random differential equations offer several advantages, including their ability to model the impact of randomness on a system, which is crucial in various real-world applications. Moreover, they provide a means to model the behavior of systems that are challenging to predict or control, such as those characterized by significant uncertainty or subject to random influences. Random differential equations arise naturally as extensions of deterministic differential equations and find widespread application across various domains. Interested readers can refer to the papers [119–123] for more detailed information. The nature of a dynamic system depends on the accuracy of our knowledge regarding its characteristics. A deterministic dynamical system arises when we possess precise information about the system. However, the identification and assessment of dynamic system characteristics involve dealing with erroneous, unclear, or ambiguous data. In other words, the process of determining the parameters of a dynamical system is inherently uncertain. When we have statistical knowledge about a dynamic system's characteristics, indicating a probabilistic understanding, the standard approach in mathematical modeling is to utilize random or stochastic differential equations.

Fixed point theory is a field of mathematics that focuses on studying the existence and properties of solutions to equations in the form $x = f(x)$, where f is a given function. These solutions, known as fixed points, have been subject to investigation for centuries and find applications in various areas, including functional analysis, differential equations, and nonlinear dynamics. The origins of fixed point theory can be traced back to ancient Greece, where mathematicians like Euclid and Archimedes studied the existence and properties of fixed points. However, it was during the 19th century that fixed point theory began to receive systematic attention. An important early contribution was made by the French mathematician Cauchy in 1821, who established the existence of fixed points for certain types of functions known as contraction mappings. Presently, fixed point theory is a vibrant and rapidly advancing field, with applications across mathematics, including functional analysis, differential equations, and nonlinear dynamics. It also extends its reach to other disciplines such as physics, engineering, economics, and computer science. Fixed point theory has been utilized to study properties of chaotic systems, fractals, and to address problems in optimization, control theory, and image processing. In 1964, Perov extended the classical Banach contraction principle to contractive maps on vector-valued metrics in the work cited as [124]. Additionally, R. Graef *et al.* presented vector versions of certain fixed point theorems, including Schaefer's fixed point theorem in a vector Banach space, in [125]. In recent years, many researchers have explored the existence of solutions for systems of ordinary differential, integral, and semilinear differential equations using vector versions of fixed point theorems. Relevant studies and references can be found in [38, 125–145] and related works. For some details on other used methods then the fixed point theory, see [146–153].

Impulsive differential equations provide suitable models for physical phenomena that experience short-term disturbances caused by external interventions during their evolution. In the literature, two types of impulses, namely instantaneous impulses and non-instantaneous impulses, are commonly discussed. Instantaneous impulses have negligible durations, while non-instantaneous impulses exhibit abrupt changes that remain active over finite initially given time intervals. The interpretation of solutions for impulsive fractional differential equations typically involves two approaches: one maintains the lower bound of the fractional derivative at the fixed initial time, while the other switches the lower limit of the fractional derivative at the impulsive points. The nature of the fractional derivative significantly affects the formulation of the problem. Fractional derivatives share certain properties with ordinary derivatives, such as the derivative of a constant, resulting in similar initial value problems and impulsive conditions, whether instantaneous or non-instantaneous. Considerable research has been devoted to the study of fractional differential equations with abrupt and instantaneous impulses, exploring various aspects of the existence and qualitative properties of solutions (see references such as [154–156]). However, when it comes to pharmacotherapy, instantaneous impulses fail to capture the dynamics of specific evolutionary processes. For instance, in analyzing an individual's hemodynamic equilibrium, the introduction of drugs into the bloodstream and their subsequent absorption by the body occur gradually and continuously. The literature contains numerous studies on different types of initial value problems and boundary value problems for fractional differential equations with both instantaneous and non-instantaneous impulses (see, for example, [154, 157–168]).

Both Instantaneous Impulsive Differential Equations (IIDEs) and Non-instantaneous Impulsive Differential Equations (NIIDEs) possess distinct characteristics, advantages, and disadvantages. IIDEs are more mathematically manageable and offer a clearer physical interpretation. However, in certain scenarios where the duration of the impulse cannot be ignored, NIIDEs are better suited for modeling system behavior. Impulsive differential equations have gained increasing importance in recent years, particularly in mathematical models of real phenomena, for instance in the fields of biology, medicine, and control theory. Relevant literature on the topic includes monographs such as [169–171], as well as papers like [172–174]. Domains such as population dynamics and optimal control naturally involve impulsive integral equations, impulsive integro-differential equations, and impulsive differential equations, as explored in monographs like [43, 154, 175–177, 350]. The earliest discussions on impulsive systems can be traced back to the work of Krylov and Bogolyubov [178]. Notably, the authors of [179–182] have investigated various aspects of existence and uniqueness of mild and classical solutions for different differential problems. It has been observed that models utilizing instantaneous impulses are insufficient in explaining the specific dynamics of evolutionary processes in pharmacotherapy. For instance, when analyzing the hemodynamic equilibrium of an individual, drug entry into the bloodstream and subsequent absorption by the body occur gradually and continuously. Hernandez and O'Regan [177], as well as Pierri *et al.* [183], have explored Cauchy problems for first-order evolution equations involving both instantaneous and non-instantaneous impulses. Works such as [20, 21, 184–187] and their references provide current results concerning evolution equations with non-instantaneous impulses.

The measure of non-compactness, a fundamental tool in nonlinear analysis, originated from the pioneering articles by Alv'arez [188] and Mönch [7]. It has since been further developed by Banaş and Goebel [189], as well as numerous researchers in the field. The measure of non-compactness finds applications in various branches of applied mathematics, including the theory of differential equations (see [190, 191] and related references). Recently, authors such as [189, 192, 193] have employed the measure of non-compactness in studying certain classes of differential equations in Banach spaces. More details are found in Aissani and Benchohra [194], Akhmerov *et al.* [195], Alv́ares [196], Banaś and Goebel [197],

Guo *et al.* [23], Olszowy and Wędrychowicz [198], Olszowy [199], and the references therein.

Nonlocal conditions were introduced by Byszewski [200] to establish the existence and uniqueness of mild and classical solutions for nonlocal Cauchy problems. These nonlocal conditions offer a more effective description of certain physical phenomena compared to standard initial conditions. The study of fractional differential equations with nonlocal conditions can be found in [201–203] and related references. Nonlocal conditions frequently arise in applications and can be viewed as feedback controls in mathematical models, ensuring that a qualitative feature or magnitude of the solution aligns with its original state during its evolution. Various physical problems have motivated the consideration of nonlocal problems, as they outperform standard Cauchy problems in applications. They have been employed in constructing mathematical models for the evolution of phenomena such as nonlocal neutral networks, nonlocal pharmacokinetics, nonlocal pollution, and nonlocal combustion. The nonlocal Cauchy problem was initially studied by Byszewski in 1991 (see [204–206]). Balachandran and collaborators have also investigated different classes of nonlinear integrodifferential systems [207]. Numerous authors have explored qualitative aspects of differential equations with nonlocal conditions, including existence, uniqueness, and stability (see [208–210]). The study of abstract nonlocal Cauchy problems was initiated by Byszewski in 1991 [211]. Evolution equations with nonlocal initial conditions have been motivated by physical problems, and it has been demonstrated that they yield superior results in various applications compared to classical Cauchy problems. Examples include their use in mathematical models for the evolution of phenomena such as nonlocal neural networks, nonlocal pharmacokinetics, nonlocal pollution, and nonlocal combustion [212,213]. Due to the wide range of applications of nonlocal problems in real-world scenarios, evolution equations with nonlocal initial conditions have been extensively studied by many authors.

In the following we give an outline of this monograph organization, which consists of 11 chapters defining the contributed work.

In **Chapter 2**, we will provide the necessary notations and preliminary results for the study at hand. This includes descriptions, theorems, and other auxiliary findings that are essential for our research. Our exploration will focus on the significant properties of Banach spaces, phase spaces, evolution systems, and semigroups. Moreover, we will delve into the fundamental aspects of the measure of non-compactness and fixed point theorems, as they are vital in our examination of fractional differential equations.

To establish a solid foundation, we will revisit fundamental definitions and results related to almost automorphic, asymptotically almost automorphic, and S-asymptotically ω-periodic functions. Subsequently, we will present the required lemmas, theorems, and properties.

Chapter 3 deals with some existence of mild solutions and asymptotically almost automorphic mild solutions for certain classes of second-order semilinear evolution equations. We will utilize the measure of non-compactness techniques to conduct our investigation. The analysis will be founded upon the Kuratowski measure of non-compactness and various fixed point techniques. The chapter is divided into four sections. We start by Section 3.1, which provides an introduction and some motivations, then finish the chapter with Section 3.4, which contains some remarks and suggestions. The main results of the chapter begin with Section 3.2 where we provide the existence of asymptotically almost automorphic mild solution for second differential equations. More precisely, we will consider the following problem:

$$
\begin{cases}
y''(t) - A(t)y(t) = f(t, y(t)), \ t \in \mathbb{R}^+ := [0, +\infty), \\
\\
y(0) = y_0, \quad y'(0) = y_1,
\end{cases}
$$

where $\{A(t)\}_{t \in \mathbb{R}^+}$ is a family of linear closed operators from E into E that generates an evolution system of linear bounded operators $\{\mathcal{U}(t, s)\}_{(t,s) \in \mathbb{R}^+ \times \mathbb{R}^+}$ for $0 \leq s \leq t < +\infty$, $f : \mathbb{R}^+ \times E \rightarrow E$ is a Carathéodory function, and $(E, \|\cdot\|)$ is a real Banach space. The investigation is based on a new fixed point result which is a generalization of the well known Darbo's fixed point theorem. Finally examples are given to illustrate the analytical findings. In Section 3.3, we discuss the non-autonomous second order semilinear evolution equation of the type:

$$
\begin{cases}
y''(t) - A(t)y(t) = f\left(t, y(t), \int_0^t K(t, s, y(s))ds\right), \ t \in J := [0, T], \\
\\
y(0) = y_0, \quad y'(0) = y_1,
\end{cases}
$$

where $\{A(t)\}_{0 \leq t < +\infty}$ is a family of linear closed operators from E into E, that generates an evolution system of linear bounded operators $\{\mathcal{U}(t, s)\}_{(t,s) \in \Delta}$, $f : J \times E \times E \rightarrow E$ is a Carathéodory function, $K : \Delta \times E \rightarrow E$ is a continuous function, $\Delta := \{(t, s) \in J \times J : s \leq t\}$.

The results are obtained by using the fixed point technique and Kuratowski measure of non-compactness. We give an example for illustration.

The aim of **Chapter 4** is to study the existence of mild solutions and approximate controllability for a general class of abstract nonlinear differential and integro-differential equations with state-dependent delay. To construct our criterion, we use a fixed point theorem such as the nonlinear alternative of Granas-Frigon theorem for contraction and a generalization of the classical Darbo fixed point theorem in Fréchet spaces, in conjunction with measures of non-compactness. Examples are included to show the applicability of our results for each section. The first result is provided in Section 4.2 where we consider the existence and uniqueness of mild solutions defined on unbounded interval for semilinear integro-differential equations of fractional order of the form:

$$\begin{cases} y'(t) - \int_0^t \dfrac{(t-s)^{\zeta-2}}{\Gamma(\zeta-1)} Ay(s)ds = f(t, y_{\varrho(t,y_t)}), \quad \text{a.e. } t \in \mathbb{R}_+ := [0, +\infty), \\[4mm] y_0 = \Psi(\sigma(y), y) \in C([-r,0], E), \end{cases}$$

where $1 < \zeta < 2$, $A : D(A) \subset E \to E$ is a closed linear operator, and $(E, \| \cdot \|)$ is a Banach space. $f : \mathbb{R}_+ \times C([-r,0], E) \to E$, $\sigma : C([-r,+\infty), E) \to \mathbb{R}_+$, $\Psi : \mathbb{R}_+ \times C([-r,+\infty), E) \to E$ and $\varrho : \mathbb{R}_+ \times C([-r,0], E) \to \mathbb{R}$, are suitable functions. If $y \in C([-r,+\infty), E)$, then for any $t \in \mathbb{R}_+$, define y_t by $y_t(\varkappa) = y(t + \varkappa)$ for $\varkappa \in [-r,0]$. In this section, we establish sufficient conditions to ensure the existence of a unique mild solution for fractional integro-differential equations with state-dependent nonlocal conditions in Fréchet spaces. Our study uses a nonlinear alternative of Leray-Schauder type for contraction maps in Fréchet spaces, which is attributed to Frigon-Granas. To exemplify the obtained result, we provide an illustrative example. In Section 4.3, we discuss the existence of mild solutions defined on unbounded interval for general class of abstract second-order differential equations with state dependent delay of the form:

$$\begin{cases} y''(t) = Ay(t) + f(t, y_{\sigma_1(t,y_t)}, y'_{\sigma_2(t,y_t)}), \quad \text{a.e. } t \in \mathbb{R}_+ := [0, +\infty), \\[4mm] y_0 = \Psi \in C([-r,0], E), \\[4mm] y'(0) = \Psi'(0) \in E, \end{cases}$$

where A is the generator of a strongly continuous cosine family of bounded linear operator $(\Phi(t))_{t\in\mathbb{R}}$ on a Banach space $(E, \|\cdot\|)$. $f(\cdot), \sigma_i(\cdot)$ are given functions. In this work we examine some existence results for abstract second order differential equations with state dependent delay, our method relies on the technique of generalization of the classical fixed point theorem for Fréchet spaces associated with the concept of measures of non-compactness. In Section 4.4, we discuss the approximate controllability and complete controllability for second-order integro-differential equations with state-dependent delay described in the form:

$$\begin{cases} y''(t) = A(t)y(t) + f(t, y_{\rho(t,y_t)}, (\sigma y)(t)) + \int_0^t \Upsilon(t,s)y(s)ds + \mathcal{P}u(t), \ t \in J, \\ \\ y'(0) = \zeta_0 \in E, \quad y(t) = \Psi(t), \ \text{if } t \in \mathbb{R}_-, \end{cases}$$

where $J = [0, T]$, $A(t) : D(A(t)) \subset E \to E$, $\Upsilon(t,s)$ are closed linear operators on E, with dense domain $D(A(t))$, which is independent of t, and $D(A(s)) \subset D(\Upsilon(t,s))$, for any $t \in \mathbb{R}_+$, define y_t by $y_t(\varkappa) = y(t+\varkappa)$ for $\varkappa \in \mathbb{R}_-$. The operator σ is defined by

$$(\sigma y)(t) = \int_0^T \Xi(t,s,y(s))ds, \quad a > 0,$$

the nonlinear terms $\Xi : J \times J \times E \to E$, $f : J \times \mathcal{B} \times E \to E$, $\Psi : \mathbb{R}_- \to E$, $\rho : J \times \mathcal{B} \to (-\infty, \infty)$, are a given functions, the control function u is give function in $L^2(J, U)$ Banach space of admissible control with U as a Banach space. $\mathcal{P}$ is a bounded linear operator from U into E, and $(E, \|\cdot\|)$ is a Banach space. To construct our criterion, we use a fixed point theorem in conjunction with measures of non-compactness. A practical example is used to illustrate the obtained results.

In **Chapter 5**, we focus on investigating the existence and controllability of a specific class of semilinear differential evolution equations, both first and second order, which incorporate random effects. Our analysis is conducted within the framework of Fréchet spaces. To establish the existence of random mild solutions, we use a technique that extends the original Darbo fixed point theorem for Fréchet spaces. This approach is supplemented by the concept of measure of non-compactness. Throughout the chapter, we provide illustrative examples to showcase the practical relevance and applicability of our obtained results for each section. After the introduction section, in Section 5.2, we study the following problem:

$$\begin{cases} y'(t,\delta) = Ay(t,\delta) + f(t,y(t,\delta),\delta); & \text{if } t \in \mathbb{R}_+ := [0,\infty), \ \delta \in \Omega, \\ \\ y(0,\delta) = y_0(\delta) \in E; \ \delta \in \Omega, \end{cases}$$

where (Ω, F, P) is a complete probability space, $y_0 : \Omega \to E$ is a given function, $f : \mathbb{R}_+ \times E \times \Omega \to E$ is a given function, $(E, \| \cdot \|)$ is a Banach space, and $A : D(A) \subset E \to E$ is the infinitesimal generator of a C_0-semigroup $S(t)$, $t \geq 0$. Section 5.3 deals with the controllability of the functional differential equation with random effect:

$$\begin{cases} y''(t,\delta) = A_1 y(t,\delta) + f(t,y(t,\delta),\delta) + A_2 g(t,\delta), & \text{a.e. } t \in \mathbb{R}_+ := [0,\infty), \\ y(0,\delta) = \varpi_1(\delta), \ y'(0,\delta) = \varpi_2(\delta), \ \delta \in \Omega, \end{cases}$$

where (Ω, F, P) is a complete probability space, $f : \mathbb{R}_+ \times E \times \Omega \to E$ is a given function, $\varpi_1, \varpi_2 : \Omega \to E$ are given measurable functions, $A_1 : \Phi(A_1) \subset E \to E$ is the infinitesimal generator of of a strongly continuous cosine family of bounded linear operators $(S_1(t))_{t \in \mathbb{R}}$ on E, and $(E, \| \cdot \|)$ is a real Banach space. $g(\cdot, \delta)$ is the control function defined in $L^2(\mathbb{R}_+, U)$, a Banach space of admissible control functions with U as a Banach space, and A_2 is a bounded linear operator from U into E. The main result is based upon a generalization of the classical Darbo fixed point theorem,and the concept of measure of non-compactness combined with the family of cosine operators. Examples are included to show the applicability of our results.

The focus of **Chapter 6** is to explore the existence and controllability of specific classes of functional differential equations with delay and random effects. Our analysis relies on the utilization of a relevant random fixed point theorem with a stochastic domain and the integrated semigroup, along with a generalization of the classical Darbo fixed point theorem for Fréchet spaces, incorporating the concept of measures of non-compactness. To validate the theoretical findings, we provide illustrative examples. Section 6.2 deals with the integral solutions to the following problem:

$$\begin{cases} y'(t,\delta) = Ay(t,\delta) + f(t,y_t(\cdot,\delta),\delta), & \text{a.e. } t \in J := [0,T], \\ \\ y(t,\delta) = \Psi(t,\delta), & t \in (-\infty, 0], \end{cases}$$

where (Ω, G, P) is a complete probability space, δ is a random variable in Ω, $f : J \times \mathcal{B} \times \Omega \to E$, $\Psi \in \mathcal{B} \times \Omega$ are given random functions which represent random nonlinear of the system, $A : D(A) \subset E \to E$ is a nondensely defined closed linear operator on E, $\mathcal{B}$ is the phase space and $(E, | \cdot |)$ is a real Banach space. For any function y defined on $(-\infty, T] \times \Omega$ and any $t \in J$ we denote by $y_t(\cdot, \delta)$ the element of $\mathcal{B} \times \Omega$ defined by $y_t(\varepsilon, \delta) = y(t+\varepsilon, \delta)$, $\varepsilon \in (-\infty, 0]$. Some notations will be given later. In Section 6.3, we discuss the existence of mild solutions defined on unbounded interval for general class of abstract second-order differential equations with state dependent delay of the form:

$$
\begin{cases}
y''(t, \delta) = Ay(t, \delta) + f(t, y_{\sigma_1(t, y_t, \delta)}, y'_{\sigma_2(t, y_t, \delta)}, \delta), \ t \in \mathbb{R}_+, \ \delta \in \Omega, \\
y(t, \delta) = \Psi(t, \delta), \ t \in [-r, 0], \\
y'(0, \delta) = \Psi'(0, \delta),
\end{cases}
$$

where $A : D(A) \subset E \to E$ is generator of a strongly continuous cosine family of bounded linear operator $(\Phi(t))_{t \in \mathbb{R}}$ on a Banach space $(E, \| \cdot \|)$, and $f : \mathbb{R}_+ \times C([-r, 0], E) \times C([-r, 0], E) \times \Omega \to E$, $\sigma_i : \mathbb{R}_+ \times C([-r, 0], E) \times \Omega \to E$ and $\Psi \in C([-r, 0], E) \times \Omega$ are suitable functions. In Section 6.4, we consider the following functional differential equation with delay and random effect:

$$
\begin{cases}
y''(t, \delta) = A_1 y(t, \delta) + f(t, y_t(\cdot, \delta), \delta) + A_2 g(t, \delta), \ \text{a.e.} \ t \in J := [0, T], \\
y(t, \delta) = \varpi_1(t, \delta); \ t \in (-\infty, 0], \\
y'(0, \delta) = \varpi_2(\delta),
\end{cases}
$$

where (Ω, F, P) is a complete probability space, $f : J \times \mathcal{B} \times \Omega \to E$, $\varpi_1 \in \mathcal{B} \times \Omega$ are given functions, $A_1 : D(A_1) \subset E \to E$ is the infinitesimal generator of of a strongly continuous cosine family of bounded linear operators $(S_1(t))_{t \in \mathbb{R}}$ on E, $\mathcal{B}$ is the phase space, and $(E, | \cdot |)$ is a real Banach space. The control function $g(\cdot, \delta)$ is given in $L^2(J, U)$, a Banach space of admissible control functions with U as a Banach space, and A_2 is a bounded linear operator from U into E. For a function y defined on $(-\infty, T] \times \Omega$ and each $t \in J$, we denote by $y_t(\cdot, \delta)$ the element of $\mathcal{B} \times \Omega$ given by $y_t(\iota, \delta) = y(t + \iota, \delta), \iota \in (-\infty, 0]$. Here $y_t(\cdot, \delta)$ represents the history of the state from time $-\infty$, up to the present time t. We assume that the histories $y_t(\cdot, \delta)$ belong to some abstract phase spaces $\mathcal{B}$. Next, we consider the following random problem:

$$
\begin{cases}
y^{''}(t, \delta) = A_1 y(t, \delta) + f(t, y_{\sigma(t, y_t)}(\cdot, \delta), \delta) + A_2 g(t, \delta); \quad \text{a.e. } t \in J, \\
y(t, \delta) = \varpi_1(t, \delta); \quad t \in (-\infty, 0], \\
y^{'}(0, \delta) = \varpi_2(\delta),
\end{cases}
$$

where $f : J \times \mathcal{B} \times \Omega \to E$, $\varpi_1 \in \mathcal{B} \times \Omega$ are given random functions, $A_1 : D(A_1) \subset E \to E$, $\mathcal{B}$ is the phase space, $\sigma : J \times \mathcal{B} \to (-\infty, T]$, and $(E, |\cdot|)$ is a real Banach space. We base our arguments for the main results on Schauder's fixed theorem [214] and random fixed point theorem combined with the family of cosine operators.

Chapter 7 focuses on investigating the existence of S-asymptotically ω-periodic mild and almost automorphic solutions for certain classes of second-order semilinear evolution equations in Banach spaces, random functional evolution equations with infinite delay, and semilinear integro-differential systems with nonlocal conditions. The investigation is supported by the utilization of specific fixed point theorems. Finally, we provide illustrative examples to showcase the main findings derived in this study. In Section 7.2, we investigate the existence of S-asymptotically ω-periodic mild solution for second differential equations. More precisely, we will consider the following problem:

$$
\begin{cases}
y''(t) - A(t)y(t) = f(t, y(t)), \ t \in \mathbb{R}^+ := [0, \infty), \\
y(0) = y_0, \quad y'(0) = y_1,
\end{cases}
$$

where $\{A(t)\}_{0 \leq t < +\infty}$ is a family of linear closed operators from E into E that generates an evolution system of linear bounded operators $\{\mathcal{U}(t, s)\}_{(t,s) \in \mathbb{R}^+ \times \mathbb{R}^+}$ for $0 \leq s \leq t < +\infty$, $f : \mathbb{R}^+ \times E \to E$ is a Carathéodory function, and $(E, \|\cdot\|)$ is a real Banach space. The goal of Section 7.3 is to study the existence and attractivity of S-Asymptotic ω-Periodic mild solutions for integrodifferential system with nonlocal conditions via resolvent operators of the form:

$$
\begin{cases}
y'(t) = A_1 y(t) + f_1\left(t, y(t), x(t), \Psi_1(y(t), x(t))\right) \\
\qquad + \displaystyle\int_0^t B_1(t-s)y(s)ds, \text{ for } t \in \mathbb{R}^+, \\[4pt]
x'(t) = A_2 x(t) + f_2\left(t, y(t), x(t), \Psi_2(y(t), x(t))\right) \\
\qquad + \displaystyle\int_0^t B_2(t-s)x(s)ds, \text{ for } t \in \mathbb{R}^+, \\[8pt]
y(0) = y_0 + \Upsilon_1(y, x), \\
x(0) = x_0 + \Upsilon_2(y, x),
\end{cases}
$$

where $\mathbb{R}^+ = [0, +\infty)$, and for $i = 0, 1$, $A_i : D(A_i) \subset E \to E$ are the infinitesimal generators of strongly continuous semigroup $\{T_i(t)\}_{t \geq 0}$, $B_i(t)$ are a closed linear operators with domain $D(A_i) \subset D(B_i(t))$, the operators Ψ_i are defined by

$$
\Psi_i(y, x)(t) = \int_0^a g_i(t, s, y(s), x(s))ds, \quad a > 0.
$$

The nonlinear terms $f_i : \mathbb{R}^+ \times E \times E \times E \to E$, $\Upsilon_i \in BC(\mathbb{R}^+, E) \times BC(\mathbb{R}^+, E) \to E$, are given functions, and $(E, \|\cdot\|)$ a Banach space. In Section 7.4, we will establish a random version of the theory for the almost automorphic solution. More precisely we will verify the existence of mild random almost automorphic solutions for evolution equations with delay and random effects of the form:

$$
\begin{cases}
y'(t, \delta) = A(t)y(t, \delta) + f(t, y_t(\cdot, \delta), \delta), \quad \text{a.e. } t \in \mathbb{R}^+ := [0, \infty), \\
y(t, \delta) = \phi(t, \delta), \quad t \in (-\infty, 0],
\end{cases}
$$

where (Ω, F, P) is a complete probability space, $\delta \in \Omega$, $f : \mathbb{R}^+ \times \mathcal{B} \times \Omega \to E$, $\phi \in \mathcal{B} \times \Omega$ are given random functions, $\{A(t)\}_{0 \leq t < +\infty}$ is a family of linear closed operators from E into E that generates an evolution system of operators $\{\mathcal{U}(t, s)\}_{(t,s) \in \mathbb{R}^+ \times \mathbb{R}^+}$ for $0 \leq s \leq t < +\infty$, $\mathcal{B}$ is the phase space, and $(E, \|\cdot\|)$ is a Banach space. For y defined on $\mathbb{R} \times \Omega$ and any $t \in \mathbb{R}^+$ we denote by $y_t(\cdot, \delta)$ the element of $\mathcal{B} \times \Omega$ given by $y_t(\kappa, \delta) = y(t + \kappa, \delta)$, $\kappa \in (-\infty, 0]$.

Chapter 8 is dedicated to investigating the existence of mild solutions for classes of impulsive integro-differential equations on both bounded and unbounded domains in Banach spaces. We use resolvent operators to establish our results, drawing upon fixed point theory and the concept of measures of non-compactness. Additionally, we provide sufficient conditions that guarantee the controllability and attractivity of our problem. To validate the theoretical aspects, we offer examples that demonstrate the practical applicability of our findings. In Section 8.2, we first discuss the existence of mild solutions for the following nonlocal problem of impulsive integro-differential equations:

$$
\begin{cases}
y'(t) = Ay(t) + \displaystyle\int_0^t \Upsilon(t-s)y(s)ds + f(t,y(t)); \ t \in J_k, \ k = 0,\ldots,m, \\
y(t_k^+) = y(t_k^-) + L_k(y(t_k^-)); \ k = 1,\ldots,m, \\
y(0) + g(y) = y_0 \in E,
\end{cases}
$$

where $J_0 = [0,t_1]$, $J_k := (t_k, t_{k+1}]$; $k = 1,\ldots,m$, $0 = t_0 < t_1 < \cdots < t_m < t_{m+1} = T$, $f : J_k \times E \to E$; $k = 1,\ldots,m$, $L_k : E \to E$; $k = 1,\ldots,m$, $g : PC \to E$ are given functions, the set PC is given later, E is a real (or complex) Banach space with norm $\|\cdot\|$, $y'(t) := \frac{dy}{dt}$, $A : D(A) \subset E \to E$ generates a C_0-semigroup on the Banach space E, $\Upsilon(t)$ is a closed linear operator on E with $D(A) \subset D(\Upsilon)$. We next discuss the existence of mild solutions for the following nonlocal problem of not instantaneous impulsive integro-differential equations:

$$
\begin{cases}
y'(t) = Ay(t) + \displaystyle\int_0^t \Upsilon(t-s)y(s)ds + f(t,y(t)); \ t \in J_k, \ k = 0,\ldots,m, \\
y(t) = g_k(t,y(t_k^-)); \ t \in \tilde{J}_k, \ k = 1,\ldots,m, \\
y(s_k) + g(y) = y_k \in E; \ k = 0,\ldots,m,
\end{cases}
$$

where $J_0 := [0,t_1]$, $\tilde{J}_k := (t_k, s_k]$, $J_k := (s_k, t_{k+1}]$; $k = 1,\ldots,m$, $f : J_k \times E \to E$, $g_k : \tilde{J}_k \times E \to E$ are given functions such that $g_k(t,y(t_k^-))|_{t=s_k} = y_k \in E$; $k = 1,\ldots,m$, $g : \mathcal{PC} \to E$ is a given function, the set $\mathcal{PC}$ is given later, and $0 = s_0 < t_1 \leq s_1 < t_2 \leq s_2 < \cdots \leq s_{m-1} < t_m \leq s_m < t_{m+1} = T$. In Section 8.3, we investigate the existence and attractivity of mild solutions to the following impulsive integrodifferential equations using resolvent operators:

$$
\begin{cases}
y'(t) = Ay(t) + f(t,y(t),(Hy)(t)) \\
\qquad + \displaystyle\int_0^t B(t-s)y(s)ds; \quad \text{if } t \in J_k; k = 0,1,\ldots, \\
y(t) = g_k\left(t, y\left(t_k^-\right)\right); \quad\quad\quad \text{if } t \in \tilde{J}_k; k = 1,2,\ldots, \\
y(0) = y_0,
\end{cases}
$$

where $J_0 = [0, t_1]$, $J_k := (s_k, t_{k+1}]$ and $\tilde{J}_k = (t_k, s_k]$ with $0 = s_0 < t_1 \leq s_1 \leq t_2 < \cdots < s_{m-1} \leq t_m \leq s_m \leq t_{m+1} \leq \cdots \leq +\infty$, $A : D(A) \subset E \to E$ is the infinitesimal generator of a strongly continuous semigroup $\{T(t)\}_{t \geq 0}$, $B(t)$ is a closed linear operator with domain $D(A) \subset D(B(t))$, the operator H is defined by

$$(Hy)(t) = \int_0^a h(t, s, y(s))ds,$$

for $a > 0$, $D_h = \{(t, s) \in \mathbb{R}^2 \; ; \; 0 \leq s \leq t \leq a\}$, and $h : D_h \times E \to E$. The nonlinear term $f : J_k \times E \times E \to E$; $k = 0, 1, \ldots$, $g_k : \tilde{J}_k \times E \to E$; $k = 1, 2, \ldots$, are a given functions, and $(E, \| \cdot \|)$ is a Banach space. They base their arguments on the fixed point theory and the concept of measure of non-compactness with the help of the resolvent operator. Next, we will investigate the existence and controllability of our problem. The aim of Section 8.4 is to establish a result of the existence of mild solution for a class of the non-autonomous second order nonlinear differential equation with non-instantaneous impulses described in the form:

$$\begin{cases} y''(t) = A(t)y(t) + f\left(t, \int_0^t g(t, s, y(s))ds\right), \; t \in (s_i, t_{i+1}], i = 0, \ldots, m, \\ y(t) = \gamma_i(t, y(t^-)), \quad t \in (t_i, s_i], \quad i = 1, \ldots, m, \\ y'(t) = \zeta_i(t, y(t^-)), \quad t \in (t_i, s_i], \quad i = 1, \ldots, m, \\ y(0) = y_0, \; y'(0) = y_1, \end{cases}$$

where E is a Banach space endowed with a norm $\| \cdot \|$, $J = [0, a]$, $0 \leq s_0 \leq t_0 < t_1 < s_1 < t_2, \ldots, t_m < s_m < t_{m+1} \leq a < \infty$. We consider that $y \in C((s_i, t_{i+1}), E)$; $i = 0, 1, \ldots, m$. The functions $\gamma_i(t, y(t))$ and $\zeta_i(t, y(t))$ represent noninstantaneous impulses during the intervals $(t_i, s_i]$; $i = 1, \ldots, m$, so impulses at t_i have some duration, namely on intervals $(t_i, s_i]$. $\{A(t)\}_{0 \leq t \leq a}$ is a family of linear closed operators from E into E, that generate an evolution system of linear bounded operators $\{S(t, s)\}_{(t,s) \in \mathcal{U}}$, f is a given function $J \times E \times E \to E$, $g \in C(\mathcal{U}, E)$, $\mathcal{U} = \{(t, s) \in J \times J : s \leq t\}$.

    Chapter 9 deals with the existence of periodic mild solutions for a specific category of functional evolution equations featuring impulses and delay. The employed techniques involve several fixed point theorems in Banach spaces, namely the Darbo, Kuratowski, and Sadovskii fixed point theorems, as well as the utilization of the Poincaré operator and the measure of non-compactness. Additionally, illustrative examples are provided to showcase the validity of our findings. In Section 9.2, we discuss the existence of periodic mild solutions of the following class of functional differential equations with infinite delay and not instantaneous impulses:

$$\begin{cases} y'(t) + A(t)y(t) = f(t, y(t), y_t); & \text{if } t \in I_k; \ k = 0, 1, \ldots, \\ y(t) = g_k(t, y(t_k^-)); & \text{if } t \in J_k; \ k = 1, 2, \ldots, \\ y(t) = \phi(t); & \text{if } t \in \mathbb{R}_- := (-\infty, 0], \end{cases}$$

where $I_0 = [0, t_1]$, $I_k := (s_k, t_{k+1}]$, $J_k := (t_k, s_k]$, $0 = s_0 < t_1 \leq s_1 \leq t_2 < \cdots < s_{m-1} \leq t_m \leq s_m \leq t_{m+1} = T \leq s_{m+1} \leq t_{m+2} \leq \cdots < +\infty$, $f : I_k \times E \times \mathcal{B} \to E$; $k = 0, \ldots$, $g_k : J_k \times E \to E$; $k = 1, 2, \ldots$, are given functions T-periodic in t, $T > 0$, $\mathcal{B}$ is an abstract phase space to be specified later, $\phi : \mathbb{R}_- \to E$ is a given function, $\{A(t)\}_{t>0}$ is a T-periodic family of unbounded operators from E into E that generate an evolution system of operators $\{U(t, s)\}_{(t,s) \in \mathbb{R}_+ \times \mathbb{R}_+}$; for $(t, s) \in \Lambda := \{(t, s) \in \mathbb{R}_+ \times \mathbb{R}_+ : 0 \leq s \leq t < +\infty\}$, $\mathbb{R}_+ := [0, \infty)$, and $(E, \|\cdot\|_E)$ is a real Banach space. In Section 9.3, we discuss the existence of periodic mild solutions of the following class of second order evolution equations with infinite delay and not instantaneous impulses

$$\begin{cases} y''(t) + A(t)y(t) = f(t, y(t), y_t); & \text{if } t \in I_k; \ k = 0, 1, \ldots, \\ y(t) = g_k(t, y(t_k^-)); & \text{if } t \in J_k; \ k = 1, 2, \ldots, \\ y(t) = \phi(t); & \text{if } t \in \mathbb{R}_- := (-\infty, 0], \\ y'(s_k) = \psi_k \in E; & k = 0, \ldots, m, \ldots, \end{cases}$$

where $I_0 = [0, t_1]$, $I_k := (s_k, t_{k+1}]$, $J_k := (t_k, s_k]$, $0 = s_0 < t_1 \leq s_1 \leq t_2 < \cdots < s_{m-1} \leq t_m \leq s_m \leq t_{m+1} = T \leq s_{m+1} \leq t_{m+2} \leq \cdots < +\infty$, $f : I_k \times E \times \mathcal{B} \to E$; $k = 0, \ldots$, $g_k : J_k \times E \to E$; $k = 1, 2, \ldots$, are given functions T-periodic in t, $T > 0$, $\mathcal{B}$ is an abstract phase space to be specified later, $\phi : \mathbb{R}_- \to E$ is a given function, $\{A(t)\}_{t>0}$ is a T-periodic family of unbounded operators from E into E that generate an evolution system of operators $\{U(t, s)\}_{(t,s) \in \mathbb{R}_+ \times \mathbb{R}_+}$; for $(t, s) \in \Lambda := \{(t, s) \in \mathbb{R}_+ \times \mathbb{R}_+ : 0 \leq s \leq t < +\infty\}$, $\mathbb{R}_+ := [0, \infty)$, and $(E, \|\cdot\|_E)$ is a real Banach space.

Chapter 10 addresses the presence of periodic mild solutions in a specific category of functional impulsive evolution equations and inclusions. Our analysis relies on fixed point theory in conjunction with the measure of non-compactness approach, utilizing the resolvent operator. We establish that the Poincaré operator acts as a condensing operator in a defined phase space, as per Kuratowski's measure of non-compactness, and subsequently obtain periodic solutions from bounded solutions using Sadovskii's fixed point theorem. Additionally, we provide illustrative examples to showcase the feasibility of our findings. In Section 10.2, we discuss the existence of periodic mild solutions of the following class of first order differential inclusions with infinite delay and not instantaneous impulses:

$$\begin{cases} y'(t) + A(t)y(t) \in F(t, y(t), y_t); & \text{if } t \in I_k; \ k = 0, 1, \dots, \\ y(t) = g_k(t, y(t_k^-)); & \text{if } t \in J_k; \ k = 1, 2, \dots, \\ y(t) = \phi(t); & \text{if } t \in \mathbb{R}_- := (-\infty, 0], \end{cases}$$

where $I_0 = [0, t_1]$, $I_k := (s_k, t_{k+1}]$, $J_k := (t_k, s_k]$, $0 = s_0 < t_1 \leq s_1 \leq t_2 < \cdots < s_{m-1} \leq t_m \leq s_m \leq t_{m+1} = T \leq s_{m+1} \leq t_{m+2} \leq \cdots < +\infty$, $F : I_k \times E \times \mathcal{B} \to \mathcal{P}(E)$; $k = 0, \dots$, is compact valued multivalued map, and T-periodic in t, $T > 0$, $\mathcal{P}(E)$ is the family of all subsets of E, $\mathcal{B}$ is an abstract phase space to be specified later, $\phi : \mathbb{R}_- \to E$, $g_k : J_k \times E \to E$; $k = 1, 2, \dots$, are given functions, $\{A(t)\}_{t>0}$ is a T-periodic family of unbounded operators from E into E that generate an evolution system of operators $\{U(t, s)\}_{(t,s) \in \mathbb{R}_+ \times \mathbb{R}_+}$; for $(t, s) \in \Lambda := \{(t, s) \in \mathbb{R}_+ \times \mathbb{R}_+ : 0 \leq s \leq t < +\infty\}$, $\mathbb{R}_+ := [0, \infty)$, and $(E, \|\cdot\|_E)$ is a real separable Banach space. In Section 10.3, we consider the following problem:

$$\begin{cases} y'(t) - Ay(t) - \int_0^t \Psi(t-s)y(s)ds \in f(t, y(t), y_t); \ t \in J_j, \ j = 0, \dots, \\ y(t) = g_j(t, y(t_j^-)); \ t \in \tilde{J}_j, \ j = 1, \dots, \\ y(t) = \psi(t); \ \text{if } t \in \mathbb{R}_- := (-\infty, 0], \end{cases}$$

where $J_0 := [0, t_1]$, $\tilde{J}_j := (t_j, s_j]$, $J_j := (s_j, t_{j+1}]$; $j = 1, \dots, 0 = s_0 < t_1 \leq s_1 < t_2 \leq s_2 < \cdots \leq s_{m-1} < t_m \leq s_m < t_{m+1} = T \leq s_{m+1} < t_{m+2} \leq \cdots < +\infty$, $f : J_j \times E \times \mathcal{B} \to \mu(E)$; $j = 0, \dots$, is T-periodic compact multivalued map, t, $T > 0$, is the family of subsets of E is denoted by $\mu(E)$, $\mathcal{B}$ is a phase space given in the sequel, $g_j : \tilde{J}_j \times E \to E$ are given functions, and T-periodic in t, $T > 0$, $\psi : \mathbb{R}_- \to E$ is a given function, and $(E, \|\cdot\|)$ is a Banach space, $y'(t) := \frac{dy}{dt}$, $A : D(A) \subset E \to E$ generates a C_0-semigroup on the Banach space E, $\Psi(t)$ is a closed linear operator on E, and T-periodic in t, $T > 0$, with $D(A) \subset D(\Psi)$. For each continuous function y and any $t \in \mathbb{R}_+$, y_t is the element of $\mathcal{B}$ given by $y_t(\varepsilon) = y(t + \varepsilon)$

for $\varepsilon \in \mathbb{R}_-$. In Section 10.4, we consider the following problem:

$$\begin{cases} y''(t) + A(t)y(t) \in f(t, y(t), y_t); & \text{if } t \in J_k; \ k = 0, 1, \ldots, \\ y(t) = g_k(t, y(t_k^-)); & \text{if } t \in \tilde{J}_k; \ k = 1, 2, \ldots, \\ y(t) = \varkappa(t); & \text{if } t \in \mathbb{R}_-, \\ y'(s_k) = \wp_k \in E; & k = 0, \ldots, m, \ldots, \end{cases}$$

where $J_0 = [0, t_1]$, $J_k := (s_k, t_{k+1}]$, $\tilde{J}_k := (t_k, s_k]$, $0 = s_0 < t_1 \leq s_1 \leq t_2 < \cdots < s_{m-1} \leq t_m \leq s_m \leq t_{m+1} = T \leq s_{m+1} \leq t_{m+2} \leq \cdots < +\infty$, $f : J_k \times E \times \mathcal{B} \to \mu(E)$; $k = 0, \ldots,$ is compact multivalued map, and T-periodic in t, $T > 0$, $\mu(E)$ is the family of subsets of E, $\mathcal{B}$ is an abstract phase space, $\varkappa : \mathbb{R}_- \to E$, $g_k : \tilde{J}_k \times E \to E$; $k = 1, 2, \ldots,$ are given function, and T-periodic in t, $T > 0$, $\{A(t)\}_{t>0}$ is a T-periodic family of unbounded operators from E into E that generate an evolution system of operators $\{\mathcal{U}(t, s)\}_{(t,s) \in \mathbb{R}_+ \times \mathbb{R}_+}$; for $(t, s) \in \nabla := \{(t, s) \in \mathbb{R}_+ \times \mathbb{R}_+ : 0 \leq s \leq t < +\infty\}$, $\mathbb{R}_+ := [0, \infty)$, and $(E, \| \cdot \|_E)$ is a Banach space. y_t is the element of $\mathcal{B}$ given by $y_t(\iota) = y(t + \iota)$ for $\iota \in \mathbb{R}_-$.

In **Chapter 11**, we shift our attention to explore applications related to evolution problems. We delve into examples involving the population of mutated Lobesia botrana and the dynamics of fish density. Our exploration encompasses considerations of the existence, uniqueness, and various quantitative properties of the solutions, employing analytical tools like the fixed point theory. In Section 11.1, we consider the following model:

$$\begin{cases} \frac{\partial}{\partial t} u(t, a, \omega, x) + \frac{\partial}{\partial a}[v(t, a, \omega)u(t, a, \omega, x)] \\ \quad = -\mu(P(t, x), t, a)u(t, a, \omega, x) + dAu(t, a, \omega, x), \\ v(t, a = 0, \omega)u(t, 0, \omega, x) \\ \quad = \int_0^L \int_\Omega \beta(P(t, x), t, s)\gamma(\omega, \omega')\, u(t, s, \omega', x)\, d\omega' ds, \\ u(0, a, \omega, x) = u_0(a, \omega, x), (a, \omega, x) \in (0, L) \times \Omega \times D, \end{cases}$$

where $D \subset \mathbb{R}^m$ is a bounded domain with smooth boundary ∂D, $u(t, a, \omega, x)$ is the population density at time $t \in [0, T]$, size $a \in [0, L]$, position $x \in D$, an a continuous phenotype $\omega \in \Omega$, $T > 0$ is a given time, and L is the maximal size. We assume that the death μ and the fertility β rates depend on the total population P. The growth rate v depends on time t, size a and ω. In Section 11.2, we consider the following non-dimensionalized form of reaction-diffusion model:

$$
\begin{cases}
\dfrac{\partial u}{\partial t} - d_1 \Delta u = ru(1-u) - \chi(x)ue, & x \in \Omega, t > 0, \\[2mm]
\dfrac{\partial e}{\partial t} - d_2 \Delta e = -ce + p\chi(x)uedx, & x \in \Omega_*, t > 0, \\[2mm]
\dfrac{\partial u}{\partial \eta} = 0, & x \in \partial\Omega, t > 0, \\[2mm]
\dfrac{\partial e}{\partial \eta} = 0, & x \in \partial\Omega_*, t > 0, \\[2mm]
u(x,0) = u_0(x) \geq 0, & x \in \Omega, \\[2mm]
e(x,0) = e_0(x) \geq 0, & x \in \Omega_*,
\end{cases}
$$

where Ω is a bounded region in $\mathbb{R}^N$ for $N \geq 1$ with $C^{2,\alpha}$ smooth boundary $\partial\Omega$, the reserve zone (or no-harvesting zone) Ω_0 is a subdomain of Ω whose boundary $\partial\Omega_0$ is also smooth, thus the real space for the fishing effort is $\Omega_* = \Omega/\bar\Omega_0$, $\dfrac{\partial}{\partial\eta}$ is the outward normal derivative on the boundary, $u(x,t)$ is the density of fish and $e(x,t)$ is the fishing effort at position x and time t. The Laplace operator Δ represents the spatial diffusion with d_1 and d_2 are the diffusion rates of the fish population and fishing effort respectively. The boundary of the reserve zone does not affect the dispersal of fish, but it works as a barrier to block the fishing effort from entering Ω_0. Without the harvesting, we assume a logistic growth for the fish population with maximal intrinsic growth rate $r > 0$ and carrying capacity $K = 1$. c is the unit cost of fishing effort, and p is the per-unit price for fish. χ is the characteristic function of Ω_*, that is

$$
\chi(x) = \chi_{\Omega_*}(x) = \begin{cases} 0 & \text{if } x \in \bar\Omega_0, \\ q(x) \geq 0 & \text{if } x \in \Omega_*, \end{cases}
$$

where $q(x)$ is a catchability function. The fact of $\chi(x) = 0$ in $\bar\Omega_0$ implies that no fishing could take place there.

Keywords and Phrases: Asymptotically almost automorphic functions, second order nonautonomous differential equations, evolution system, Kuratowski measures of non-compactness, fixed point, solution operator, state-dependent nonlocal condition, nonlinear alternative, Fréchet spaces, abstract differential equations, state-dependent delay, approximate controllability, complete controllability, infinite delay, second-order integro-differential equation, resolvent operator, semilinear evolution equation, densely defined operator, nondensely defined operator, random solution, functional differential equation, cosine and sine families, controllability, semigroup theory, integral solution, finite delay, abstract random differential equations, ω-periodic, S-asymptotically ω-periodic, attractivity, S-asymptotic mild solution, generalized Banach space, non-instantaneous impulses, Darbo fixed point, neutral system, optimal control, M-set contractive, condensing operator, Poincaré operator, functional evolution inclusion, integro-differential inclusion.

Chapter 2

Preliminary Background

In this chapter, we will delve into various mathematical tools, notations, and concepts. These tools are essential for understanding the topics discussed later on. Specifically, we will explore significant properties of Banach spaces, phase spaces, evolution systems, and semigroups. Additionally, we will examine the fundamental aspects of the measure of non-compactness and fixed point theorems. These concepts play a crucial role in our study of fractional differential equations. To lay the groundwork, we will review some basic definitions and results concerning almost automorphic, asymptotically almost automorphic, and S-asymptotically ω-periodic functions. Following that, we will present necessary lemmas, theorems, and properties.

2.1   Notations and Definitions

In this section, we will present all of the notations and definitions of functional spaces that have been regarded as essential and constant throughout each of the preceding chapters. In fact, these are only referenced once in this section.

Let $J = [0, T]$. We denote by E a Banach space with the norm $\|\cdot\|$. Let $BC(\mathbb{R}^+, E)$ be the Banach space of all bounded and continuous functions y mapping $\mathbb{R}^+$ into E endowed with the usual supremum norm

$$\|y\|_\infty = \sup_{t \in \mathbb{R}^+} \|y(t)\|.$$

Let $C(J, E)$ be the Banach space of continuous functions y mapping J into E with the usual supremum norm

$$\|y\| = \|y\|_C = \sup_{t \in J} \|y(t)\|.$$

By BUC we denote the space of all bounded uniformly continuous functions defined from $(-\infty, 0]$ to E.

A measurable function $y : \mathbb{R}_+ \to E$ is Bochner integrable if and only if $\|y\|$ is Lebesgue integrable.

Denote $L^1(\mathbb{R}_+, E)$ the Banach space of measurable functions $y : \mathbb{R}_+ \to E$ which are Bochner integrable normed by

$$\|y\|_{L^1} = \int_0^{+\infty} \|y(t)\| \, dt.$$

Let $L^r(J, E)$ be the space of E-valued Bochner functions on $[0, a]$ with the norm

$$\|y\|_{L^r} = \left(\int_0^a \|y(t)\|^r dt \right)^{\frac{1}{r}}, \qquad r \geq 1.$$

Let $L^\infty(J)$ be the Banach space of measurable functions $v : J \to \mathbb{R}$ that are essentially bounded and equipped with the norm

$$\|v\|_{L^\infty} = \inf\{c > 0 : |v(t)| \leq c, \text{ a.e. } t \in J\}.$$

The Laplace transformation of a function $f \in L^1_{loc}(\mathbb{R}_+, E)$ is defined by

$$\mathcal{L}(f)(\gamma) :=: \widehat{f}(\gamma) = \int_0^\infty e^{-\gamma t} f(t) dt, \quad Re(\gamma) > \varpi,$$

if the integral is absolutely convergent for $Re(\gamma) > \varpi$.

2.2   Generalized Banach Space

Definition 2.1. Let X be a vector metric space on $\mathbb{K} = \mathbb{R}$ or $\mathbb{C}$. A map $\| \cdot \| : X \to \mathbb{R}^n_+$ is called a norm on X if it satisfies the following properties:

- (i) $\|x\| = 0$ then $x = (0, \ldots, 0)$;
- (ii) $\|\lambda x\| = |\lambda| \|x\|$ for $x \in X, \lambda \in \mathbb{K}$;
- (iii) $\|x + y\| \leq \|x\| + \|y\|$ for every $x, y \in X$.

Remark 2.1. The pair $(X, \| \cdot \|_X)$ is called a generalized normed space. If the generalized metric generated by $\| \cdot \|_X$ (*i.e* $d(x, y) = \|x - y\|_X$) is complete then the espace $(X, \| \cdot \|_X)$ is called a generalized Banach space, where

$$\|x - y\|_X = \begin{pmatrix} \|x - y\|_1 \\ \vdots \\ \|x - y\|_n \end{pmatrix}.$$

Let $X = BC(\mathbb{R}^+, E) \times BC(\mathbb{R}^+, E)$ be endowed with the vector norm $\| \cdot \|_{X \times X}$ defined by $\|v\|_{X \times X} = (\|u_1\|_{BC}, \|u_2\|_{BC})$ for $v = (u_1, u_2)$. It is clear that $(BC(\mathbb{R}^+, E) \times BC(\mathbb{R}^+, E), \| \cdot \|_{X \times X})$ is a generalized Banach space. In the case of generalized Banach spaces in the sense of Perov, the notations of convergent sequence, Cauchy sequence, completeness, open and closed subset are similar to those for usual metric spaces.

Definition 2.2. A square matrix M of real numbers is said to be convergent to zero if and only if, $M^n \to 0$ as $n \to +\infty$.

Theorem 2.1. *A square matrix M of real numbers is said to be convergent to zero if and only if its spectral radius $\rho(M)$ is strictly less than 1. In other words, this means that all the eigenvalues of M are in the open unit disc i.e. $|\lambda| < 1$, for every $\lambda \in \mathbb{C}$ with $\det(M - \lambda I) = 0$, where I denote the unit matrix of $\mathcal{M}_{n \times n}(\mathbb{R})$.*

Lemma 2.1 ([215]). *Let $M \in \mathcal{M}_{m \times m}(\mathbb{R}_+)$. Then, the following assertions are equivalent:*

- *M is convergent towards zero,*
- *$M^k \to 0$ as $k \to +\infty$,*
- *The matrix $(I - M)$ is non-singular and*
$$(I - M)^{-1} = I + M + M^2 + \ldots + M^k + \ldots$$
- *The matrix $(I - M)$ is non-singular and $(I - M)^{-1}$ has non-negative elements.*

Theorem 2.2 ([215]). *Let A and B be two $m \times m$ matrices with $O \leq |B| \leq A$. Then $\rho(B) \leq \rho(A)$.*

Lemma 2.2 ([216]). *If $A \in \mathcal{M}_{m \times m}(\mathbb{R}_+)$ is a matrix with $\rho(A) < 1$, then $\rho(A + B) < 1$ for every matrix $B \in \mathcal{M}_{m \times m}(\mathbb{R}_+)$ whose elements are small enough.*

Definition 2.3. Let $Q \in \mathcal{M}_{2 \times 2}(\mathbb{R})$ is said to be order preserving (or positive) if $p_1 \leq p_0, q_1 \leq q_0$ imply

$$Q \begin{pmatrix} p_0 \\ q_0 \end{pmatrix} \geq Q \begin{pmatrix} p_1 \\ q_1 \end{pmatrix}$$

in the sense of components.

Lemma 2.3 ([217]). *Let*

$$Q = \begin{pmatrix} a & -b \\ -c & d \end{pmatrix},$$

where $a, b, c, d \geq 0$ and $\det Q > 0$. Then, Q^{-1} is order preserving.

2.3   Phase Spaces

Let us adopt the definition of the phase space $\mathcal{B}$ presented in [218] and adhere to the terminology employed in [219]. Then, Let $(\mathcal{B}, \|\cdot\|_{\mathcal{B}})$ be a seminormed linear space of functions mapping $(-\infty, 0]$ into E, and verifying the following:

(Ax_1) If $y : (-\infty, T) \to E, T > 0$, is continuous on J and $y_0 \in \mathcal{B}$, then for every $t \in J$ the following hold:

  (a) $y_t \in \mathcal{B}$;

  (b) There exists $H > 0$ where $|y(t)| \leq H \|y_t\|_{\mathcal{B}}$ and y_t is a function defined by $y_t(\varkappa) = y(t + \varkappa)$ for $\varkappa \in \mathbb{R}_-$ for any $t \in \mathbb{R}_+$;

  (c) There exist two functions $L(\cdot), M(\cdot) : \mathbb{R}_+ \to \mathbb{R}_+$ independent of y with L continuous and bounded, and M locally bounded such that:

$$\|y_t\|_{\mathcal{B}} \leq L(t) \sup\{ |y(\tau)| : 0 \leq \tau \leq t\} + M(t)\|y_0\|_{\mathcal{B}}.$$

(Ax_2) For the function y in (Ax_1), y_t is a $\mathcal{B}$-valued continuous function on J.

(Ax_3) The space $\mathcal{B}$ is complete.

Denote

$$K_T = \sup\{L(t) : t \in J\},$$

and

$$M_T = \sup\{M(t) : t \in J\}.$$

Remark 2.2. We give the following observations:

1. (Ax_1) (b) is equivalent to $|\Psi(0)| \leq H\|\Psi\|_{\mathcal{B}}$ for every $\Psi \in \mathcal{B}$.
2. Since $\|\cdot\|_{\mathcal{B}}$ is a seminorm, two elements $\Psi, \varpi \in \mathcal{B}$ can verify $\|\Psi - \varpi\|_{\mathcal{B}} = 0$ without necessarily $\Psi(\varepsilon) = \varpi(\varepsilon)$ for all $\varepsilon \leq 0$.
3. We can see that for all $\Psi, \varpi \in \mathcal{B}$ where $\|\Psi - \varpi\|_{\mathcal{B}} = 0$: We necessarily have that $\Psi(0) = \varpi(0)$.

2.4   Measure of Non-Compactness

As motioned in the introduction part, the measure of non-compactness is one of the fundamental tools in the theory of nonlinear analysis. In this section, we recall some fundamental facts of the notion of measure of non-compactness. Particularly, we employ the Kuratowski measure of non-compactness in our studies throughout this book.

Let Ω_X be the class of all bounded subsets of a metric space X.

Definition 2.4 ([189]). A function $\alpha : \Omega_X \to [0, \infty)$ is said to be a measure of non-compactness on X if the following conditions are verified for all $B, B_1, B_2 \in \Omega_X$.

(a) Regularity, i.e., $\alpha(B) = 0$ if and only if B is precompact,
(b) invariance under closure, i.e., $\alpha(B) = \alpha(\overline{B})$,
(c) semi-additivity, i.e., $\alpha(B_1 \cup B_2) = \max\{\alpha(B_1), \alpha(B_2)\}$.

Definition 2.5 ([189]). Let X be a Banach space. The Kuratowski measure of non-compactness is the map $\alpha : \Omega_X \longrightarrow [0, \infty)$ defined by

$$\alpha(M) = inf\{\epsilon > 0 : M \subset \bigcup_{j=1}^{m} M_j, diam(M_j) \leq \epsilon\},$$

where $M \in \Omega_X$.

The map α satisfies the following properties:

- $\alpha(M) = 0 \Leftrightarrow \overline{M}$ is compact (M is relatively compact).
- $\alpha(M) = \alpha(\overline{M})$.
- $M_1 \subset M_2 \Rightarrow \alpha(M_1) \leq \alpha(M_2)$.
- $\alpha(M_1 + M_2) \leq \alpha(B_1) + \alpha(B_2)$.
- $\alpha(cM) = |c|\alpha(M), c \in \mathbb{R}$.
- $\alpha(convM) = \alpha(M)$.

Lemma 2.4 ([98]). *Let E be a Banach space, $D \subset E$ be bounded. Then there exists a countable set $D_0 \subset D$, such that*

$$\alpha(D) \leq 2\alpha(D_0).$$

We recall the following definition of the notion of a sequence of measures of non-compactness [220].

Definition 2.6. Let $\mathcal{M}_{\bar{E}}$ be the family of all non-empty and bounded subsets of a Fréchet space E. A family of functions $\{\alpha_n\}_{n \in \mathbb{N}}$ where $\alpha_n : \mathcal{M}_{\bar{E}} \to [0, \infty)$ is said to be a family of measures of non-compactness in the real Fréchet space $\bar{E}$ if it satisfies the following conditions for all $B, B_1, B_2 \in \mathcal{M}_{\bar{E}}$:

(a) $\{\alpha_n\}_{n \in \mathbb{N}}$ is full, i.e. $\alpha_n(B) = 0$ for $n \in \mathbb{N}$ if and only if B is precompact,
(b) $\alpha_n(B_1) \leq \alpha_n(B_2)$ for $B_1 \subset B_2$ and $n \in \mathbb{N}$,

(c) $\alpha_n(\operatorname{conv} B) = \alpha_n(B)$ for $n \in \mathbb{N}$,

(d) If $\{B_i\}_{j=1,\ldots}$ is a sequence of closed sets from $\mathcal{M}_{\bar{E}}$ where $B_{j+1} \subset B_j$; $j = 1,\ldots$ and if $\lim_{j \to \infty} \alpha_n(B_i) = 0$, for each $n \in \mathbb{N}$, then the intersection set $B_\infty := \cap_{j=1}^{\infty} B_i$ is non-empty.

Lemma 2.5 ([221]). *If Ω is bounded subset of Banach space E, then for $\nu > 0$ there is a sequence $\{\varkappa_n\}_{n=1}^{\infty} \subset \Omega$ such that*

$$\alpha(\Omega) \le 2\alpha(\{\varkappa_n\}_{n=1}^{\infty}) + \nu; \ for \ j \in \mathbb{N}.$$

Lemma 2.6. *[222] Let $D = \{y_n\}_{n=0}^{+\infty} \subset C(\mathbb{R}^+, E)$ be a bounded and countable set. Then $\alpha(D(t))$ is Lebesgue integrable on $\mathbb{R}^+$, and*

$$\alpha \left\{ \int_0^t y_n(s)) ds \right\}_{n=0}^{\infty} \le 2 \int_0^t \alpha(D(s)) ds, \quad t \in \mathbb{R}^+.$$

Definition 2.7. Let X be a Banach space. A continuous map $F : \Omega \subset X \to X$ is called condensing with respect to a measure of non-compactness α (or α-condensing) if for every bounded set $D \subset \Omega$ which is not relatively compact, we have

$$\alpha(F(D)) < \alpha(D).$$

For more information about measure of non-compactness, see [189, 193, 220, 223].

Lemma 2.7 ([189]). *If $\{D\}_{n=0}^{+\infty}$ is a sequence of non-empty, bounded and closed subsets of E such that $D_{n+1} \subset D_n$; $(n = 1, 2, 3, \ldots)$ and if $\lim_{n \to \infty} \alpha(D_n) = 0$, then the intersection*

$$D_\infty = \bigcap_{n=0}^{+\infty} D_n,$$

is non-empty and compact.

Definition 2.8. Let X be a generalized Banach space and $(\mathcal{A}, \le)$ be a partially ordered set. A map $\alpha : \mathcal{P}(X) \to \mathcal{A} \times \mathcal{A} \times \cdots \times \mathcal{A}$ is called a generalized measure of non-compactness (*M.N.C.*) on X if

$$\alpha(\overline{\operatorname{conv}} C) = \alpha(C) \ \text{for every} \ C \in \mathcal{P}(X),$$

where

$$\alpha(C) := \begin{pmatrix} \alpha_1(C) \\ \vdots \\ \alpha_n(C) \end{pmatrix}.$$

A typical example of a *M.N.C.* is the Hausdorff measure of non-compactness μ defined, for all $\Omega \subset X$, by

$$\mu(\Omega) := \inf \left\{ \epsilon \in \mathbb{R}^n_+ : \text{there exists } n \in \mathbb{N} \text{ such that } \Omega \text{ has finite } \epsilon\text{-net} \right\}.$$

Lemma 2.8 ([224]). *Let $\Omega \subset C(a,b)$ be bounded and equicontinuous. Then, $\overline{\mathrm{conv}}(\Omega) \subset C(a,b)$ is also bounded and equicontinuous.*

Lemma 2.9 ([225]). *Let $\Omega \subset C(a,b)$ be bounded and equicontinuous, and let α the Kuratowski's measure of non-compactness. Then, $y(t) = \alpha(\Omega(t))$ is continuous and*

$$\alpha \left(\int_a^b \Omega(s)ds \right) \leq \int_a^b \alpha(\Omega(\varsigma))ds.$$

Definition 2.9. Let X, Y be two generalized normed spaces and a map $N : X \to Y$. N is called an M-contraction (with respect to α) if there exists $M \in M_{n \times n}(\mathbb{R}_+)$ converging to zero such that, for every $\Omega \in \mathcal{P}(X)$, we have

$$\alpha(N(\Omega)) \leq M\alpha(\Omega).$$

2.5 Evolution System

Let $\{A(t), \; t \in \mathbb{R}^+\}$ be a family of closed linear operators on the Banach space E with domain $D(A(t))$ which is dense in E and independent of t.

Definition 2.10. A family $\mathcal{U}$ of bounded operators $\mathcal{U}(t,s) : E \to E$, $(t,s) \in \Delta := \{(t,s) \in \mathbb{R}^+ \times \mathbb{R}^+ : s \leq t\}$, is called an evolution operator of the equation

$$y''(t) = A(t)y(t), \qquad t \in \mathbb{R}^+, \tag{2.1}$$

if the following conditions hold:

(e_1) For any $x \in E$ the map $(t,s) \longmapsto \mathcal{U}(t,s)x$ is continuously differentiable and

   (a) for each $t \in \mathbb{R}$, $\mathcal{U}(t,t)x = 0, \forall x \in F$,

   (b) for all $(t,s) \in \Delta$ and for any $x \in E$, $\dfrac{\partial}{\partial t}\mathcal{U}(t,s)x\big|_{t=s} = x$ and $\dfrac{\partial}{\partial s}\mathcal{U}(t,s)x\big|_{t=s} = -x$.

(e_2) For all $(t,s) \in \Delta$, if $x \in D(A(t))$, then $\dfrac{\partial}{\partial s}\mathcal{U}(t,s)x \in D(A(t))$, the map $(t,s) \longmapsto \mathcal{U}(t,s)x$ is of class C^2 and

(a) $\dfrac{\partial^2}{\partial t^2}\mathcal{U}(t,s)x = A(t)\mathcal{U}(t,s)x,$

(b) $\dfrac{\partial^2}{\partial s^2}\mathcal{U}(t,s)x = \mathcal{U}(t,s)A(s)x,$

(c) $\dfrac{\partial^2}{\partial s\partial t}\mathcal{U}(t,s)x\big|_{t=s} = 0.$

(e_3) For all $(t,s)\ \in\ \Delta,$ then $\dfrac{\partial}{\partial s}\mathcal{U}(t,s)x\ \in\ D(A(t)),$ there exist

$\dfrac{\partial^3}{\partial t^2\partial s}\mathcal{U}(t,s)x,\ \dfrac{\partial^3}{\partial s^2\partial t}\mathcal{U}(t,s)x$ and

(a) $\dfrac{\partial^3}{\partial t^2\partial s}\mathcal{U}(t,s)x = A(t)\dfrac{\partial}{\partial s}(t)\mathcal{U}(t,s)x.$

    Moreover, the map $(t,s)\longmapsto A(t)\dfrac{\partial}{\partial s}(t)\mathcal{U}(t,s)x$ is continuous,

(b) $\dfrac{\partial^3}{\partial s^2\partial t}\mathcal{U}(t,s)x = \dfrac{\partial}{\partial t}\mathcal{U}(t,s)A(s)x.$

2.6   Semigroups

Let $B(E)$ be the space of all bounded linear operators from E into E with the norm

$$\|N\|_{B(E)} = \sup_{\|y\|=1}\|N(y)\|.$$

A semigroup of bounded linear operators $T(t)$, is uniformly continuous if

$$\lim_{t\to 0}\|T(t) - I\|_E = 0.$$

Here I denotes the identity operator in E.

    We note that if a semigroup $T(t)$ is of class (C_0) then it satisfies the growth condition $\|T(t)\|_{B(E)} \le Me^{\beta t}$, for $0 \le t < \infty$ with some constants $M > 0$ and $\beta \in \mathbb{R}$. If, in particular $M = 1$ and $\beta = 0$, i.e; $\|T(t)\|_{B(E)} \le 1$, for $t \ge 0$, then the semigroup $T(t)$ is called a *contraction semigroup*. For more details on strongly continuous operators, we refer the reader to the books [11, 226].

Definition 2.11. Let A be a closed and linear operator with a dense domain $D(A)$ defined on a Banach space E. A is the generator of a solution operator if there exists $\varpi > 0$ and a strongly continuous function $g : \mathbb{R}_+ \to B(E)$ where

$$\{\gamma^\varsigma : Re(\gamma) > \varpi\} \subset \varrho(A),$$

and

$$\gamma^{\varsigma-1}(\gamma^{\varsigma} - A)^{-1}z = \int_0^{\infty} e^{-\gamma t}g(t)z\,dt, \quad Re\gamma > \varpi, z \in E.$$

where $g(t)$ is the solution operator generated by A.

Proposition 2.1 ([227]). *Let $\{g(t)\}_{t\geq 0} \subset B(E)$ be the solution operator with generator A. Then the requirements listed below are met.*

(a) $g(t)$ is strongly continuous for $t \geq 0$ and $g(0) = I$;
(b) $g(t)D(A) \subset D(A)$ and $Ag(t)z = g(t)Az$ for all $z \in D(A)$, $t \geq 0$;
(c) for every $z \in D(A)$ and $t \geq 0$,

$$g(t)z = z + \int_0^t \frac{(t-s)^{\varsigma-1}}{\Gamma(\varsigma)} Ag(s)z\,ds.$$

(d) Let $z \in D(A)$. Then, $\int_0^t \frac{(t-s)^{\varsigma-1}}{\Gamma(\varsigma)}g(s)z\,ds \in D(A)$ and

$$g(t)z = z + A\int_0^t \frac{(t-s)^{\varsigma-1}}{\Gamma(\varsigma)}g(s)z\,ds.$$

Remark 2.3. The concept of a solution operator, as defined above, is closely related to the concept of a resolvent family (see Prüss [228]). Because of the uniqueness of the Laplace transform, in the border case $\varsigma = 1$, the family $g(t)$ corresponds to a C_0 semigroup (see [229]), whereas in the case $\varsigma = 2$ a solution operator corresponds to the concept of cosine family (see [230]).

In [228–231], one can find further information on C_0-semigroups and sine and cosine families.

Definition 2.12. A solution operator $\{g(t)\}_{t>0}$ is called uniformly continuous if

$$\lim_{t \to s} \|g(t) - g(s)\|_{B(E)} = 0.$$

Definition 2.13. A one parameter family $\{S_1(t) : t \in \mathbb{R}\}$ of bounded linear operators in the Banach space E is called a strongly continuous cosine family if and only if

- $S_1(0) = I$ (I is the identity operator);
- $S_1(t)w$ is strongly continuous in t on $\mathbb{R}$ for each fixed $w \in E$;
- $S_1(t+s) + S_1(t-s) = 2S_1(t)S_1(s)$ for all $t, s \in \mathbb{R}$.

Let $\{S_1(t) : t \in \mathbb{R}\}$ be a strongly continuous cosine family in E. Define the associated sine family $\{S_2(t) : t \in \mathbb{R}\}$ by

$$S_2(t)w = \int_0^t S_1(s)w \, ds, \quad w \in E, \ t \in \mathbb{R}.$$

The infinitesimal generator $A_1 : E \to E$ of the cosine family $\{S_1(t) : t \in \mathbb{R}\}$ is given by

$$A_1 w = \frac{d^2}{dt^2} S_1(t)w|_{t=0}, \quad w \in \Phi(A_1),$$

where

$$\Phi(A_1) = \{w \in E : S_1(\cdot)w \in C^2(\mathbb{R}, E)\}.$$

Definition 2.14 ([232]). Let E be a Banach space. An integrated semigroup is a family of operators $\zeta(t)_{t \geq 0}$ of bounded linear operators $\zeta(t)$ on E with:

(a) $\zeta(0) = 0$;
(b) $t \longrightarrow \zeta(t)$ is strongly continuous;
(c) $\zeta(\tau)\zeta(t) = \int_0^\tau (\zeta(t + \varrho) - \zeta(\varrho)) d\varrho$, for all $t, \tau \geq 0$.

Definition 2.15 ([232]). An operator A is called a generator of an integrated semigroup if there exists $\lambda \in \mathbb{R}$ such that $(\lambda, \infty) \subset \rho_0(A)$ $(\rho_0(A)$, is the resolvent set of $A)$ and there exists a strongly continuous exponentially bounded family $(\zeta(t))_{t \geq 0}$ of bounded operators such that $\zeta(0) = 0$ and $R(\eta, A) = (\eta I - A)^{-1} = \eta \int_0^\infty e^{-\eta t} \zeta(t) dt$ exists for all η with $\eta > \lambda$.

Proposition 2.2 ([232]). *Let A be the generator of an integrated semigroup $(\zeta(t))_{t \geq 0}$. Then for all $y \in E$ and $t \geq 0$,*

$$\int_0^t \zeta(\tau)y \, d\tau \in D(A) \quad \text{and} \quad \zeta(t)y = A \int_0^t \zeta(\tau)y \, d\tau + ty.$$

Definition 2.16 ([232]). A linear operator A verifies the Hille-Yosida requirement if there exists $M \geq 0$ and $\lambda \in \mathbb{R}$ such that $(\lambda, \infty) \subset \rho_0(A)$ and

$$\sup\{(\eta - \lambda)^n |(\eta I - A)^{-n}| : n \in \mathbb{N}, \eta > \lambda\} \leq M.$$

Definition 2.17 ([232]).

(a) An integrated semigroup $(\zeta(t))_{t \geq 0}$ is said to be locally Lipschitz continuous if, for all $\varrho > 0$, there exists a constant L where

$$|\zeta(t) - \zeta(\tau)| \leq L|t - \tau|, t, \tau \in [0, \varrho].$$

(b) $(\zeta(t))t \geq 0$ is called non degenerate if $\zeta(t)y = 0$, for all $t \geq 0$, implies that $y = 0$.

If A is the generator of an integrated semigroup $(\zeta(t))_{t\geq0}$ which is locally Lipschitz, then from [233], $\zeta(\cdot)y$ is continuously differentiable if and only if $y \in \overline{D(A)}$ and $(\zeta'(t))_{t\geq0}$ is a C_0-semigroup on $\overline{D(A)}$. Let $(\zeta(t))_{t\geq0}$ be the integrated semigroup generated by A. We note that, if A satisfies the Hille-Yosida condition, then $\|\zeta'(t)\|_{B(E)} \leq Me^{\lambda t}, t \geq 0$.

Proposition 2.3. *Let $f : J \longrightarrow E$ be a Bochner integrable function. Then, the function $B : J \longrightarrow E$ defined by $B(t) = \displaystyle\int_0^t \zeta(t - \tau)f(\tau)d\tau$, is continuously differentiable on J and satisfies*

$$|B'(t)| \leq M \int_0^t e^{\lambda(t-\tau)}|f(\tau)|d\tau, \ \text{for } t \in J.$$

For additional information on the differentiability of convolution product involving an integrated semigroup see, for example, [232, 234, 235].

2.7 Results on Almost Automorphic, Asymptotically Almost Automorphic and S-Asymptotically ω-Periodic Functions

In this section, we recall some basic definitions and results on almost automorphic, asymptotically almost automorphic and S-asymptotically ω-periodic functions (for more details, see [67, 73, 83, 236]).

Definition 2.18 ([73]). A continuous function $f : \mathbb{R} \to E$ is said to be almost automorphic if for every sequence of real numbers $\{\tau_n'\}$, there exists a subsequence $\{\tau_n\}$ such that

$$g(t) = \lim_{n\to\infty} f(t + \tau_n)$$

is well defined for each $t \in \mathbb{R}$ and

$$\lim_{n\to\infty} g(t - \tau_n) = f(t) \quad \text{for each } t \in \mathbb{R}.$$

Denote by $AA(\mathbb{R}, E)$ the set of all such functions.

Lemma 2.10 ([25]). *$AA(\mathbb{R}, E)$ is a Banach space with the supremum norm*

$$\|f\|_\infty = \sup_{t\in\mathbb{R}} \|f(t)\|.$$

Definition 2.19 ([73]). A continuous function $f : \mathbb{R} \times E \to E$ is said to be almost automorphic in $t \in \mathbb{R}$ for each $y \in E$ if for every sequence of real numbers $\{\tau'_n\}$, there exists a subsequence $\{\tau_n\}$ such that

$$\lim_{n \to \infty} f(t + \tau_n, y) = g(t, y)$$

is well defined for each $t \in \mathbb{R}$ and

$$\lim_{n \to \infty} g(t - \tau_n, y) = f(t, y)$$

for each $t \in \mathbb{R}$ and each $y \in E$. The collection of those functions is denoted by $AA(\mathbb{R} \times E, E)$.

Example 2.1 ([237]). The function $f : \mathbb{R} \times E \to E$ given by

$$f(t, y) = \sin\left(\frac{1}{2 + \cos t + \cos\sqrt{2}t}\right)\cos y$$

is almost automorphic in $t \in \mathbb{R}$ for each $y \in E$, where $E = L^2([0, 1])$.

The space of all continuous functions $h : \mathbb{R}^+ \to E$ such that $\lim_{t \to \infty} h(t) = 0$ is denoted by $C_0(\mathbb{R}^+, E)$. Moreover, we denote $C_0(\mathbb{R}^+ \times E, E)$; the space of all continuous functions from $\mathbb{R} \times E$ to E satisfying $\lim_{t \to \infty} h(t, y) = 0$ in t and uniformly in $y \in E$.

Remark 2.4. Note that if $\nu(t) \in C_0(\mathbb{R}^+, E)$, then

$$\int_0^t e^{-(t-s)}\nu(s)ds \in C_0(\mathbb{R}^+, E).$$

Definition 2.20 ([73]). A continuous function $f : \mathbb{R}^+ \to E$ is said to be asymptotically almost automorphic if it can be decomposed as

$$f(t) = g(t) + h(t),$$

where

$$g(t) \in AA(\mathbb{R}, E), \quad h(t) \in C_0(\mathbb{R}^+, E).$$

Denote by $AAA(\mathbb{R}^+, E)$ the set of all such functions.

Example 2.2. The function $f : \mathbb{R} \to \mathbb{R}$ defined by

$$f(t) = \sin\left(\frac{1}{2 + \cos t + \cos\sqrt{2}t}\right) + e^{-t}$$

is an asymptotically almost automorphic function with

$$g(t) = \sin\left(\frac{1}{2 + \cos t + \cos\sqrt{2}t}\right) \in AA(\mathbb{R}, \mathbb{R}), \quad h(t) = e^{-t} \in C_0(\mathbb{R}^+, \mathbb{R}).$$

Lemma 2.11 ([73, 238]). $AAA(\mathbb{R}^+, E)$ *is also a Banach space with the norm*

$$\|f\|_\infty = \sup_{t \in \mathbb{R}^+} \|f(t)\|.$$

Definition 2.21 ([73]). A continuous function $f : \mathbb{R}^+ \times E \to E$ is said to be asymptotically almost automorphic if it can be decomposed as

$$f(t, y) = g(t, y) + h(t, y),$$

where

$$g(t, y) \in AA(\mathbb{R} \times E, E), \quad h(t, y) \in C_0(\mathbb{R}^+ \times E, E).$$

Denote by $AAA(\mathbb{R}^+ \times E, E)$ the set of all such functions.

Example 2.3. The function $f : \mathbb{R}^+ \times E \to E$ given by

$$f(t, x) = \sin\left(\frac{1}{2 + \cos t + \cos \sqrt{2}t}\right) \cos y + e^{-t}\|y\|$$

is asymptotically almost automorphic in $t \in \mathbb{R}^+$ for each $y \in E$, where $E = L^2([0, 1])$.

$$g(t, y) = \sin\left(\frac{1}{2 + \cos t + \cos \sqrt{2}t}\right) \cos y \in AA(\mathbb{R} \times E, E),$$
$$h(t, y) = e^{-t}\|y\| \in C_0(\mathbb{R}^+ \times E, E).$$

Lemma 2.12 ([239]). $f : \mathbb{R} \times E \to E$ *is almost automorphic, and assume that $f(t, \cdot)$ is uniformly continuous on each bounded subset $K \subset E$ uniformly for $t \in \mathbb{R}$, that is for any $\varepsilon > 0$, there exists $\varrho > 0$ such that $y, z \in K$ and $\|y(t) - z(t)\| < \varrho$ imply that $\|f(t, y) - f(t, z)\| < \varepsilon$ for all $t \in \mathbb{R}$. Let $\varphi : \mathbb{R} \to E$ be almost automorphic. Then the function $F : \mathbb{R} \to E$ defined by $F(t) = f(t, \varphi(t))$ is almost automorphic.*

Definition 2.22 ([83]). A function $f \in BC(\mathbb{R}^+, X)$ is called S-asymptotically ω-periodic if there exists $\omega > 0$ such that $\lim_{t \to +\infty}(f(t + \omega) - f(t)) = 0$. In this case we say that ω is an asymptotic period of f and that f is S-asymptotically ω-periodic.

We will denote by $Y = SAP_\omega(X)$, the set of all S-asymptotically ω-periodic functions from $\mathbb{R}^+$ to X. Note that $SAP_\omega(X)$ is a Banach space with the sup-norm $\|\cdot\|_Y$.

The following result is due to Henriquez-Pierri-Táboas; Proposition 3.5 in [82].

Theorem 2.3. *Endowed with the norm $\|y\|_C$, $SAP_\omega(X)$ is a Banach space.*

Proof. We consider the translation operator τ_ω : $BC(\mathbb{R}^+, X) \rightarrow BC(\mathbb{R}^+, X)$ defined by $\tau_\omega f(t) = f(t + \omega)$. Clearly the function τ_ω is linear and continuous. We note that $SAP_\omega(X) = (\tau_\omega - I)^{-1}(C_0(\mathbb{R}^+, X))$. And then, since $(\tau_\omega - I)$ is linear continuous and since $C_0(\mathbb{R}^+, X)$ is a closed vector subspace of $BC(\mathbb{R}^+, X)$, $SAP_\omega(X)$ is a closed vector subspace of the Banach space $BC(\mathbb{R}^+, X)$. $\qquad\square$

Definition 2.23. A continuous function $f : [0, +\infty) \times X \rightarrow X$ is said to be uniformly S-asymptotically ω-periodic on bounded sets if for each bounded subset K of X, the set $\{f(t,x) : (t,x) \in [0,+\infty) \times K\}$ is bounded, and $\lim_{t \to +\infty}(f(t+\omega, x) - f(t,x)) = 0$ uniformly in $X \in K$.

Definition 2.24. A continuous function $f : [0, +\infty) \times X \rightarrow X$ is said to be asymptotically uniformly continuous on bounded sets if for every $\varepsilon > 0$ and any bounded set $K \subseteq X$, there exist constants $L_{\varepsilon,K} \geq 0$ and $\delta = \delta_{\varepsilon,K} > 0$ such that

$$\|f(t,x) - f(t,y)\| \leq \varepsilon,$$

for all $t \geq L_{\varepsilon,K}$ and $x, y \in K$ with $\|x - y\| \leq \delta_{\varepsilon,K}$.

Lemma 2.13 ([93]). *Let X, F be two Banach spaces, and $f : [0, +\infty) \times X \rightarrow F$ be a function uniformly S-asymptotically ω-periodic on bounded sets and asymptotically uniformly continuous on bounded sets. If $x \in SAP_\omega(X)$. Then*

$$\lim_{t \to +\infty} (f(t + \omega, x(t + \omega)) - f(t, x(t)) = 0.$$

2.8 Necessary Lemmas, Theorems and Properties

Lemma 2.14 ([240, p. 758]). *Let $u(t)$, $p(t)$ and $q(t)$ be real valued non-negative continuous functions defined on $\mathbb{R}^+$, for which the inequality*

$$u(t) \leq u_0 + \int_0^t p(s) \left[u(s) + \int_0^s q(\tau)u(\tau)d\tau \right] ds,$$

holds for all $t \in \mathbb{R}^+$, where u_0 is a non-negative constant, then

$$u(t) \leq u_0 \left[1 + \int_0^t p(s) \left[\exp \left(\int_0^s (p(\tau) + q(\tau)d\tau) \right) \right] ds \right],$$

for all $t \in \mathbb{R}^+$.

Lemma 2.15 ([175]). *Let $y(t)$ and $b(t)$ be non-negative continuous function for $t \geq \varepsilon$, and let*

$$y(t) \leq \sigma + \int_\varepsilon^t b(s)y(s)ds, \quad t \geq \varepsilon,$$

where $\sigma \geq 0$ is a constant. Then

$$y(t) \leq \sigma \exp\left(\int_\varepsilon^t b(s)ds\right), \quad t \geq \varepsilon.$$

Lemma 2.16 ([241]). *Let $y(t)$ be a non-negative piecewise continuous function that satisfies, for $t \geq t_0$, the inequality*

$$y(t) \leq C + \int_{t_0}^t V(s)y(s)ds + \sum_{t_0 < t_i < t} \beta_i y(t_i), \quad \text{for all } t \geq t_0,$$

where $C \geq 0, \beta_i \geq 0, V(\tau) > 0$, and τ_i are the first kind discontinuity points of the function $y(t)$. Then the following estimate holds for the function $y(t)$,

$$y(t) \leq C \prod_{t_0 < \tau_i < t} (1 + \beta_i) \exp\left[\int_{t_0}^t V(\tau)d\tau\right].$$

Definition 2.25. A function $f : \mathbb{R}_+ \times E \to E$ is L^1-Carathéodory if it verifies:

(i) for each $t \in \mathbb{R}_+$, $f(t, \cdot) : E \to E$ is continuous;
(ii) for each $y \in E$, $f(\cdot, y) : \mathbb{R}_+ \to E$ is measurable;
(iii) for every positive integer ι there exists $\varsigma_\iota \in L^1(\mathbb{R}_+, \mathbb{R}_+)$ where

$$\|f(t, y)\| \leq \varsigma_\iota(t)$$

for all $\|y\| \leq \iota$ and almost each $t \in \mathbb{R}_+$.

Definition 2.26 ([12]). A function $f : \Theta_1 \to \Theta_1$ is said to be *a contraction* if for each $\tau \in \mathbb{N}$ there exists $\iota_\tau \in (0, 1)$ where :

$$\|f(z) - f(y)\|_\tau \leq \iota_\tau \|z - y\|_\tau \quad \text{for all } z, y \in \Theta_1.$$

Definition 2.27. A non-empty subset $\Omega \subset E$ is bounded if

$$\sup_{y \in \Omega} \|y\|_j < \infty; \text{ for } j \in \mathbb{N}.$$

Lemma 2.17 ([242]). *Let $M \subset BC(\mathbb{R}^+, E)$ be a set verifying the requirements:*

(i) M *is bounded in* $BC(\mathbb{R}^+, E)$;
(ii) *the functions of* M *are equicontinuous on any compact interval of* $\mathbb{R}^+$;
(iii) *the set* $M(t) := \{y(t) : y \in M\}$ *is relatively compact on any compact interval of* $\mathbb{R}^+$;
(iv) *the functions from* M *are equiconvergent, i.e., given* $\upsilon > 0$, *there corresponds* $\varkappa(\upsilon) > 0$ *such that* $\|y(t) - y(+\infty)\| < \upsilon$ *for any* $t \geq \varkappa(\upsilon)$ *and* $y \in M$.

Then M *is relatively compact in* $BC(\mathbb{R}^+, E)$.

2.9   Results on Random Operators

Let (Ω, F, P) be a complete probability space and B_E be the σ-algebra of Borel subsets of E. $z : \Omega \to E$ is measurable if for all $B \in B_E$, we have

$$z^{-1}(B) = \{\delta \in \Omega : z(\delta) \in B\} \subset \mathcal{A}.$$

Definition 2.28. A mapping $N : \Omega \times E \to E$ is called jointly measurable if for any $B \in B_E$, we have

$$N^{-1}(B) = \{(\delta, z) \in \Omega \times E : N(\delta, z) \in B\} \subset \mathcal{A} \times B_E,$$

where $\mathcal{A} \times B_E$ is the product of the σ-algebras $\mathcal{A}$ and B_E.

Definition 2.29. A function $N : \Omega \times E \to E$ is jointly measurable if $N(\cdot, y)$ is measurable for all $y \in E$ and $N(\delta, \cdot)$ is continuous for all $\delta \in \Omega$.

Definition 2.30. A function $f : J \times E \times \Omega \to E$ is random Carathéodory if the assumptions that follows hold:

(i) The map $(t, \delta) \to f(t, z, y, \delta)$ is jointly measurable for all $y \in E$,
(ii) The map $y \to f(t, y, \delta)$ is continuous for all $t \in J$ and $\delta \in \Omega$.

Let $N : \Omega \times E \to E$ be a random operator (see [243] for more details). Then $N(\delta)$ is called continuous if $N(\delta, y)$ is continuous in y for all $\delta \in \Omega$. Let $\mathcal{P}(E)$ be the family of all non-empty subsets of E. In all the sequel, we apply the definition of the random operator with stochastic domain given in [244], as well as some definitions and properties of C_0-semigroups theory. More details can be found in [11].

Definition 2.31 ([244]). Let $\mathcal{P}(Y)$ be the family of all non-empty subsets of Y and C be a mapping from Ω into $\mathcal{P}(Y)$. A mapping $N : \{(\delta, z) : \delta \in \Omega, \ z \in C(\delta)\} \to Y$ is called random operator with stochastic domain C if C is measurable (i.e., for all closed $A \subset Y$, $\{\delta \in \Omega, C(\delta) \cap A \neq \emptyset\}$

is measurable) and for all open $D \subset Y$ and all $z \in Y$, $\{\delta \in \Omega : z \in C(\delta), N(\delta, z) \in D\}$ is measurable. N will be called continuous if every $N(\delta)$ is continuous. For a random operator N, a mapping $z : \Omega \to Y$ is called random (stochastic) fixed point of N if for P-almost all $\delta \in \Omega$, $z(\delta) \in C(\delta)$ and $N(\delta)z(\delta) = z(\delta)$ and for all open $D \subset Y$, $\{\delta \in \Omega : z(\delta) \in D\}$ is measurable.

Next, we will give a very useful random fixed point theorem with stochastic domain.

Definition 2.32 ([245]). Let $\tilde{N}$ be a mapping from Ω into 2^E. A mapping $N : \{(\delta, y) : \delta \in \Omega \wedge y \in \tilde{N}(\delta)\} \longrightarrow E$ is a random operator with stochastic domain $\tilde{N}$ if and only if for all closed $B_1 \subseteq E$, $\{\delta \in \Omega : \tilde{N}(\delta) \cap B_1 \neq \emptyset\} \in F)$ and for all open $B_2 \subseteq E$ and all $y \in E$, $\{\delta \in \Omega : y \in \tilde{N}(\delta) \wedge N(\delta, y) \in B_2\} \in F$. N is continuous if every $N(\delta)$ is continuous. A mapping $y : \Omega \longrightarrow E$ is a random fixed point of N if and only if for all $\delta \in \Omega$, $y(\delta) \in \tilde{N}(\delta)$ and $N(\delta)y(\delta) = y(\delta)$ and y is measurable if for all open $B_2 \subseteq E$, $\{\delta \in \Omega : y(\delta) \in B_2\} \in F$.

Remark 2.5. If $\tilde{N}(\delta) \equiv E$, then the definition of random operator with stochastic domain coincides with the definition of random operator.

Lemma 2.18 ([245]). *Let $\tilde{N} : \Omega \longrightarrow 2^E$ be measurable with $\tilde{N}(\delta)$ closed, convex and solid (i.e., int $\tilde{N}(\delta) \neq \emptyset$) for all $\delta \in \Omega$. We suppose that there exists measurable $y_0 : \Omega \longrightarrow E$ with $y_0 \in$ int $\tilde{N}(\delta)$ for all $\delta \in \Omega$. Let N be a continuous random operator with stochastic domain $\tilde{N}$ such that for every $\delta \in \Omega$, $\{y(\delta) \in \tilde{N}(\delta) : N(\delta)y = y\} \neq \emptyset$. Then N has a stochastic fixed point.*

The mapping y of $J \times \Omega$ into E is a stochastic process if for each $t \in J$, the function $y(t, \cdot)$ is measurable.

2.10 Fixed Point Theorems

In this part, we will go through all of the fixed point theorems that are employed in the various studies throughout the monograph. Is it commonly recognized that the results of fixed points have turned out to be the instruments for the solutions of differential equations. Fixed point theory has been one of the most intensely researched study subjects in recent decades. The fixed point notion dates back to the middle of the 18th century. While the fixed point theory seems to be a separate academic field nowadays, it first emerged in articles dealing with the solution of certain differential equations, see, e.g., Liouville (1837) [246], Picard (1890) [247], Poincaré (1886) [248]. One of the first independent fixed point results was obtained by Banach [249] by abstracting the successive approximation method of Picard.

Theorem 2.4 (Banach's fixed point theorem [216]). *Let D be a non-empty closed subset of a Banach space E, then any contraction mapping N of D into itself has a unique fixed point.*

Banach's fixed point theorem differs from previous fixed point theorems in that it ensures not only the existence but also the uniqueness of the fixed point. More crucially, it not only gives you the existence and uniqueness of a fixed point, but also how to obtain it.

In what follows, we list some other fixed point theorems that have turned out to be the instruments for the differential equations solutions.

Theorem 2.5 (Schaefer's fixed point theorem [216]). *Let E be a Banach space and $N : E \to E$ be a completely continuous operator. If the set*
$$D = \{u \in E : u = \lambda Nu, \text{ for some } \lambda \in (0,1)\}$$
is bounded, then N has a fixed point.

Theorem 2.6 ([250]). *Let Ω be a non-empty, bounded, closed and convex subset of a Banach space X and let $T : \Omega \to \Omega$ be a continuous mapping. Assume that there exists a constant $k \in [0,1)$, such that*
$$\alpha(TM) \le k\alpha(M),$$
for any non-empty subset M of Ω. Then, T has a fixed point in set Ω.

Theorem 2.7 ([251]). *Let X be a Banach space, Ω compact convex subset of X and $N : \Omega \to \Omega$ is a continuous map. Then N has at least one fixed point in Ω.*

Theorem 2.8 ([252]). *Let Ω be a non-empty, bounded, closed and convex subset of a Banach space E, and let $\Gamma : \Omega \to \Omega$ be a continuous operator satisfying the inequality*
$$\alpha(\Gamma(D)) \le \Psi(\alpha(D))$$
for any non-empty subset D of Ω, where $\Psi : \mathbb{R}^+ \to \mathbb{R}^+$ is a non-decreasing function such that
$$\lim_{n \to +\infty} \Psi^n(t) = 0 \text{ for each } t \ge 0.$$
Then Γ has at least one fixed point in the set Ω.

Theorem 2.9 (Darbo fixed point theorem [253]). *If $Q : W \subset E$ is closed and convex and $0 \in W$, the continuous map $Q : W \to W$ is a α-contraction, if the set $\{y \in W : y = \lambda \Gamma y\}$ is bounded for $0 < \lambda < 1$, then the map Q has at least one fixed point in W.*

Theorem 2.10 (Sadovskii's fixed point theorem [14]). *Let X be a Banach space, α be a mesure of non-compactness, and $P : X \to X$ be a condensing operator. If $P(H) \subset H$ for a convex, closed, and bounded set H of X then P has a fixed point in H.*

Theorem 2.11 (Darbo-Sadovskii [197]). *If $Q : W \subset Y$ is bounded, closed and convex, the continuous map $Q : W \to W$ is a α-contraction, then the map Q has at least one fixed point in W.*

Theorem 2.12 (Nonlinear alternative of Leray-Schauder type [254]). *Let X be a Banach space, Ω be a bounded open subset of X and $0 \in \Omega$. Suppose that $Q : \overline{\Omega} \to X$ is α-condensing and assume that $u \neq \lambda Q(u)$ for $u \in \partial\Omega$ and $\lambda \in (0, 1)$, hold. Then Q has a fixed point in $\overline{\Omega}$.*

Theorem 2.13 ([255]). *Let Θ_1 be a Fréchet space and $\Theta_2 \in \Theta_1$ a closed subset in Θ_2 and let $N : \Theta_2 \to \Theta_1$ be a contraction such that $N(\Theta_2)$ is bounded. Then one of the following statements holds:*

(C1) N has a unique fixed point;
(C2) There exists $\gamma \in [0, 1)$, $\tau \in \mathbb{N}$ and $z \in \partial_\tau \Theta_2{}^\tau$ such that $\|z - \gamma N(z)\| = 0$.

Theorem 2.14 ([220]). *Let Ω be a non-empty, bounded, closed, and convex subset of a Fréchet space F and let $V : \Omega \to \Omega$ be a continuous mapping. Suppose that V is a contraction with respect to a family of measures of non-compactness $\{\alpha_j\}_{j \in \mathbb{N}}$. Then V has at least one fixed point in the set Ω.*

Theorem 2.15 ([256]). *Let (X, d) be a non-empty complete metric space with a contraction mapping $T : X \to X$. Then T admits a unique fixed point x^* in X.*

Theorem 2.16 (Perov's fixed point theorem [124]). *Let (X, d) be a complete generalized metric space , with $d : X \times X \longrightarrow \mathbb{R}^n$ and let $N : X \longrightarrow X$, such that*

$$d(N(x), N(y)) \leq M d(x, y),$$

for all $x, y \in X$ and some square matrix M of nonnegative numbers. If the matrix M is convergent to zero, that is $M^k \longrightarrow 0$ as $k \longrightarrow +\infty$, then N

has a unique fixed point $x_* \in X$, *and we have*

$$d\left(N^k\left(x_0\right), x_*\right) \leq M^k(I-M)^{-1}d\left(N\left(x_0\right), x_0\right)$$

for every $x_0 \in X$ *and* $k \geq 1$.

Theorem 2.17 (Monch's fixed point theorem [7]). *Let* D *be a bounded, closed and convex subset of a Banach space such that* $0 \in D$, *and let* N *be a continuous mapping of* D *into itself. If the implication*

$$V = \overline{conv}N(V) \text{ or } V = N(V) \cup \{0\} \Rightarrow \overline{V} \text{ is compact,} \qquad (2.2)$$

holds for every subset V *of* D, *then* N *has a fixed point.*

Theorem 2.18 (Darbo's fixed point theorem [250]). *Let* Ω *be a non-empty, bounded, closed and convex subset of a Banach space* $\widetilde{E}$ *and let* $P : \Omega \to \Omega$ *be a continuous mapping. Assume that there exists a constant* $k \in [0, 1)$, *such that*

$$\alpha(P(M)) \leq k\alpha(M),$$

for any non-empty subset M *of* Ω. *Then,* P *has a fixed point in set* Ω.

The next result is concerned with α-condensing or M-contractivity.

Theorem 2.19 ([125]). *Let* $F \subset X$ *be a bounded closed convex subset and* $N : F \to F$ *be a generalized* α-*condensing continuous mapping, where* α *is a non-singular measure of non-compactness defined on the subsets of* X. *Then the set*

$$\text{Fix}(N) = \{x \in F : x = N(x)\}$$

is non-empty.

Theorem 2.20 ([257]). *Let* F *be a closed, bounded, and convex subset of* X, *and let* $N : F \to F$ *be a continuous operator. For any subset* Ω *of* F, *set*

$$N^1\Omega = N\Omega, \quad N^p\Omega = N\left(\overline{\text{conv}}\left(N^{p-1}\Omega\right)\right), \quad p = 2, 3, \ldots \qquad (2.3)$$

Suppose there exists a matrix M *that approaches zero and a positive integer* p_0 *such that for any subset* Ω *of* F, *we have*

$$\alpha\left(N^{p_0}\Omega\right) \leq M\alpha(\Omega),$$

where α *is an arbitrary generalized measure of non-compactness. Then,* N *has at least one fixed point in* F.

Now, we present versions of Schaefer's fixed point theorem and nonlinear alternative Leray-Schauder type theorem for α-condensing operators in a generalized Banach space.

Theorem 2.21 ([125]). *Let X be a generalized Banach space and $N : X \to X$ be a continuous and α-condensing operator. Assume that the set*

$$M_\lambda = \{y \in X : y = \lambda N(y) \text{ for some } \lambda \in (0, 1)\},$$

is bounded. Then N has a fixed point.

Theorem 2.22 ([125]). *Let X be a generalized Banach space, $U \subset X$ be a bounded, convex, open neighborhood of zero, and let $N : \overline{U} \to X$ be a continuous and α-condensing mapping. If N satisfies the boundary condition*

$$y \neq \lambda N(y), \quad \text{for all } y \in \partial U \text{ and } 0 < \lambda < 1,$$

then the set $Fiy(N) = \{y \in U : y = N(y)\}$ is non-empty.

Theorem 2.23 ([125]). *Let $F \subset X$ be a bounded closed convex subset and $N : F \to F$ be a generalized β-condensing continuous mapping, where β is a non-singular measure of non-compactness defined on the subsets of X. Then the set*

$$\text{Fix}(N) = \{x \in F : x = N(x)\},$$

is non-empty.

Theorem 2.24 ([258, 259]). *Let X be a non-empty, bounded, closed and convex subset of a Banach space E and $T : X \to \mathcal{P}_{cl,cv}(X)$ be a closed and nonlinear D-set contraction. Then, T has at least a fixed point.*

Theorem 2.25 ([14]). *Let $S : W \to W$ be a condensing operator where W a Banach space. If $S(V) \subset V$ for a bounded, closed and convex set V of W, then S admit a fixed point in V.*

Chapter 3

Semilinear Evolution Equations

3.1 Introduction and Motivations

The objective of this chapter is to examine the existence of mild solutions and asymptotically almost automorphic mild solutions for certain classes of second-order semilinear evolution equations. We will utilize the measure of non-compactness techniques to conduct our investigation. The analysis will be founded upon the Kuratowski measure of non-compactness and various fixed point techniques, including a new fixed point result that extends the well-known Darbo's fixed point theorem. To illustrate the practicality and relevance of our findings, examples will be provided at the conclusion of each section.

The results presented in this chapter are derived and explored as a direct consequence of the following:

- The publications of Aissani and Benchohra [194], Akhmerov *et al.* [195], Alváres [196], Banaś and Goebel [197], Olszowy and Wędrychowicz [198], Olszowy [199], and the references therein. In these papers, the authors shown the use of the technique of measures of non-compactness. It is well known that this method provides an excellent tool for obtaining existence of solutions of nonlinear differential equation. This technique works fruitfully for both integral and differential equations.
- The papers [434–437], where the authors investigated the same type of equations of these types and their special forms that commonly come across in almost all phases of physics and other areas of applied mathematics.

- The papers [260–265] and the references therein, where the existence of solutions to an evolution problem is related to the existence of an evolution operator $\mathcal{U}(t; s)$ for the homogeneous equation

$$y''(t) = A(t)y(t), \quad \text{for } t \geq 0.$$

For this purpose there are many techniques to show the existence of $\mathcal{U}(t, s)$ which has been developed by Kozak [266].

3.2 Second-Order Non-Autonomous Semilinear Evolution Equations

This section is mainly concerned with the existence of asymptotically almost automorphic mild solution for second differential equations. More precisely, we will consider the following problem:

$$y''(t) - A(t)y(t) = f(t, y(t)), \quad t \in \mathbb{R}^+ := [0, +\infty), \qquad (3.1)$$

$$y(0) = y_0, \ y'(0) = y_1, \qquad (3.2)$$

where $\{A(t)\}_{t \in \mathbb{R}^+}$ is a family of linear closed operators from E into E that generate an evolution system of linear bounded operators $\{\mathcal{U}(t, s)\}_{(t,s) \in \mathbb{R}^+ \times \mathbb{R}^+}$ for $0 \leq s \leq t < +\infty$, $f : \mathbb{R}^+ \times E \to E$ is a Carathéodory function, and $(E, |\cdot|)$ is a real Banach space.

3.2.1 *Existence Results*

In this work the existence of solution the problem (3.1)–(3.2) is related to the existence of an evolution operator $\mathcal{U}(t, s)$ for the following homogeneous problem

$$y''(t) = A(t)y(t), \quad t \in \mathbb{R}^+. \qquad (3.3)$$

This concept of evolution operator has been developed by Kozak [266] and recently used by Henríquez et al. [262].

Denote by $\omega^T(y, \varepsilon)$ the modulus of continuity of y on the interval $[0, T]$ i.e.

$$\omega^T(y, \varepsilon) = \sup \left\{ \|y(t) - y(s)\| ; t, s \in [0, T], \|t - s\| \leq \varepsilon \right\}.$$

Moreover, let us put

$$\omega^T(D, \varepsilon) = \sup \left\{ \omega^T(y, \varepsilon); y \in D \right\},$$

$$\omega_0^T(D) = \lim_{\varepsilon \to 0} \omega^T(D, \varepsilon).$$

Definition 3.1. A function $y \in BC(\mathbb{R}^+, E)$ is said to be a mild solution to the problem (3.1)–(3.2) if y satisfies the integral equation

$$y(t) = -\frac{\partial}{\partial s}\mathcal{U}(t,0)y_0 + \mathcal{U}(t,0)y_1 + \int_0^t \mathcal{U}(t,s)f(s,y(s))ds.$$

We need the following technical lemma.

Lemma 3.1. *Assume that the following hypothesis hold:*

(3.2.1) *There exist constants $M \geq 1$ and $\delta > 0$, such that*

$$\|\mathcal{U}(t,s)\|_{B(E)} \leq Me^{-\delta(t-s)} \quad \text{for any } (t,s) \in \Delta$$

and for any sequence of real numbers $\{\tau'_n\}$, we can extract a subsequence $\{\tau_n\}$ and for any $\varepsilon > 0$, there exists $N \in \mathbb{N}$ such that

$$\|U(t + \tau_n, s + \tau_n) - U(t,s)\|_{B(E)} \leq \varepsilon e^{-\delta(t-s)},$$

$$\|U(t - \tau_n, s - \tau_n) - U(t,s)\|_{B(E)} \leq \varepsilon e^{-\delta(t-s)}$$

for each $t, s \in \mathbb{R}$. for all $n > N$, for each $t, s \in \mathbb{R}$, $t \geq s$.

If $\varphi(t) \in AA(\mathbb{R}, E)$, then

$$\Lambda(t) := \int_{-\infty}^t U(t,s)\varphi(s)ds, \ t \in \mathbb{R},$$

belongs to $AA(\mathbb{R}, E)$.

Proof. From (3.2.1) it is clear that $\Lambda(t)$ is well-defined and continuous on $\mathbb{R}$. Since $\varphi(t) \in AA(\mathbb{R}, E)$, it follows that for every sequence of real numbers $\{\tau'_n\}$, we can extract a subsequence $\{\tau_n\}$ such that

(c_1) $\lim_{n\to\infty} \varphi(t + \tau_n) - \tilde{\varphi}(t) = 0$ for each $t \in \mathbb{R}$ and,
(c_2) $\lim_{n\to\infty} \tilde{\varphi}(t - \tau_n) - \varphi(t) = 0$ for each $t \in \mathbb{R}$.

Notes that $\tilde{\varphi}$ is also bounded on $\mathbb{R}$, and measurable. Define

$$\tilde{\Lambda}(t) = \int_{-\infty}^t \mathcal{U}(t,s)\tilde{\varphi}(s)ds, \quad t \in \mathbb{R}.$$

For $t \in \mathbb{R}$, Since $\tilde{\varphi}$ is measurable, $\tilde{\Lambda}$ is well-defined.

For $t \in \mathbb{R}$, we have

$$\left\| \Lambda y)(t + \tau_n) - (\tilde{\Lambda} y)(t) \right\|$$

$$= \left\| \int_{-\infty}^{t+\tau_n} \mathcal{U}(t + \tau_n, s)\varphi(s)ds - \int_{-\infty}^{t} \mathcal{U}(t, s)\tilde{\varphi}(s)ds \right\|$$

$$= \left\| \int_{-\infty}^{t} \mathcal{U}(t + \tau_n, s + \tau_n)\varphi(s + \tau_n)ds - \int_{-\infty}^{t} \mathcal{U}(t, s)\tilde{\varphi}(s)ds \right\|$$

$$\leq \int_{-\infty}^{t} \|\mathcal{U}(t + \tau_n, s + \tau_n)\|_{B(E)} \|\varphi(s + \tau_n) - \tilde{\varphi}(s))\| \, ds$$

$$+ \int_{-\infty}^{t} \|\mathcal{U}(t + \tau_n, s + \tau_n) - \mathcal{U}(t, s))\|_{B(E)} \tilde{\varphi}(s) ds$$

$$\leq \int_{-\infty}^{t} Me^{-\delta(t-s)} \|\varphi(s + \tau_n) - \tilde{\varphi}(s))\| \, ds$$

$$+ \int_{-\infty}^{t} \varepsilon e^{-\delta(t-s)} \|\tilde{\varphi}(s)\| \, ds$$

$$\leq M \int_{-\infty}^{t} e^{-\delta(t-s)} ds \sup_{s \in \mathbb{R}} \|\varphi(s + \tau_n) - \tilde{\varphi}(s))\|$$

$$+ \varepsilon \int_{-\infty}^{t} e^{-\delta(t-s)} ds \sup_{s \in \mathbb{R}} \|\tilde{\varphi}(s)\|$$

$$\leq \frac{M}{\delta} \sup_{s \in \mathbb{R}} \|\varphi(s + \tau_n) - \tilde{\varphi}(s))\| + \frac{\varepsilon}{\delta} \sup_{s \in \mathbb{R}} \|\tilde{\varphi}(s)\| \, .$$

Using (c_1), we obtain that for $n \to \infty$,

$$\Lambda(t + \tau_n) \to \tilde{\Lambda}(t).$$

Analogously, one can prove that,

$$\tilde{\Lambda}(t - \tau_n) \to \Lambda(t) \text{ for each } t \in \mathbb{R} \text{ as } n \to \infty.$$

This shows that

$$\Lambda \in AA(\mathbb{R}, E).$$

Theorem 3.1. *Assume that the hypotheses (3.2.1) and the following requirements are satisfied:*

(3.3.1) *There exist a constant $\tilde{M} \geq 0$ and $\delta > 0$, such that:*

$$\left\| \frac{\partial}{\partial s} \mathcal{U}(t, s) \right\|_{B(E)} \leq \tilde{M} e^{-\delta(t-s)}, (t, s) \in \Delta.$$

(3.3.2) *The function $f : \mathbb{R}^+ \times E \to E$ is Carathéodory and asymptotically almost automorphic i.e., $f(t, y) = g(t, y) + h(t, y)$ with*

$$g(t, y) \in AA(\mathbb{R} \times E, E), \quad h(t, y) \in C_0(\mathbb{R}^+ \times E, E),$$

and $g(t, y)$ *is uniformly continuous on any bounded subset* $K \subset E$ *uniformly for* $t \in \mathbb{R}$.

Moreover,

(a) *There exist* $p \in L^q(\mathbb{R}, \mathbb{R}^+)$, $q \in [1, \infty)$ *and a continuous non-decreasing function* $\psi : [0, \infty) \to (0, \infty)$ *such that for all* $t \in \mathbb{R}^+$ *and* $y \in E$,

$$\|g(t, y)\| \leq p(t)\psi(\|y\|) \quad and \quad \lim_{\|y\| \to +\infty} \inf \frac{\psi(\|y\|)}{\|y\|} = \rho_1.$$

(b) *There exist a function* $\beta(t) \in C_0(\mathbb{R}, \mathbb{R}^+)$ *and a non-decreasing function* $\Phi : \mathbb{R}^+ \to \mathbb{R}^+$ *such that for all* $t \in \mathbb{R}^+$ *and* $y \in E$ *with* $\|y\| \leq R$,

$$\|h(t, y)\| \leq \beta(t)\phi(\|y\|) \quad and \quad \lim_{R \to +\infty} \inf \frac{\phi(R)}{R} = \rho_2.$$

(3.3.3) *There exist a locally integrable function* $\eta : \mathbb{R} \to \mathbb{R}^+$ *and a continuous non-decreasing function* $\varphi : \mathbb{R}^+ \to \mathbb{R}^+$ *such that for any non-empty bounded set* $D \subset E$ *we have:*

$$\alpha(f(t, D)) \leq \eta(t)\varphi(\alpha(D)) \text{ for a.e } t \in \mathbb{R}^+.$$

Additionally we assume that $\lim_{n \to +\infty} (\psi + \phi)^n(t) = 0$ *for a.e* $t \in \mathbb{R}^+$. *Let* $\beta(t)$ *be the function involved in the assumption* (3.3.2), *then*

$$\int_0^t e^{-(t-s)}\beta(s)ds \in C_0(\mathbb{R}^+, \mathbb{R}^+).$$

Put

$$\rho = \sup_{t \in \mathbb{R}^+} \int_0^t e^{-(t-s)}\beta(s)ds.$$

If

$$M\rho_1\|p\|_{L^q} + M\delta^{-1}\rho\rho_2 < 1, \tag{3.4}$$

and

$$M \max(4\|\eta\|_{L^1}, \|p\|_{L^q}\delta^{-1+\frac{1}{q}}) < 1, \tag{3.5}$$

then the problem (3.1)–(3.2) *has a asymptotically almost automorphic mild solution.*

Proof. Consider the operator $N : AAA(\mathbb{R}^+, E) \to AAA(\mathbb{R}^+, E)$ defined by

$$(Ny)(t) = -\frac{\partial}{\partial s}\mathcal{U}(t,0)y_0 + \mathcal{U}(t,0)y_1 + \int_0^t \mathcal{U}(t,s)f(s,y(s))ds, \qquad (3.6)$$

where $y \in AAA(\mathbb{R}^+, E)$ with $y = \gamma + \zeta$, γ is the principal term and ζ the corrective term of y. We need to prove that N is well defined, that is $N(AAA(\mathbb{R}^+, E)) \subset AAA(\mathbb{R}^+, E)$. Let

$$\sigma(t) = -\frac{\partial}{\partial s}\mathcal{U}(t,0)y_0 + \mathcal{U}(t,0)y_1,$$

then

$$
\begin{aligned}
\|\sigma(t)\| &= \| -\tfrac{\partial}{\partial s}\mathcal{U}(t,0)y_0 + \mathcal{U}(t,0)y_1 \| \\
&\leq \|\tfrac{\partial}{\partial s}\mathcal{U}(t,0)y_0\| + \|\mathcal{U}(t,0)y_1\| \\
&\leq \tilde{M}e^{-\delta t}\|y_0\| + Me^{-\delta t}\|y_1\|.
\end{aligned}
$$

Since $\delta > 0$, we get $\lim_{t\to+\infty}\|(\sigma(t)\| = 0$. that is

$$\sigma \in C_0(\mathbb{R}^+, E). \qquad (3.7)$$

By assumption $f = g + h$ where g is the principal term and h the corrective term. So we can write

$$
\begin{aligned}
f(t,y(t)) &= g(t,\gamma(t)) + f(t,y(t)) - f(t,\gamma(t)) + h(t,\gamma(t)) \\
&= g(t,\gamma(t)) + H(t,y(t)), \qquad (3.8)
\end{aligned}
$$

In view of (3.8), we have

$$
\begin{aligned}
W(t) &= \int_0^t \mathcal{U}(t,s)f(s,y(s))ds \\
&= \int_0^t \mathcal{U}(t,s)g(s,\gamma(s))ds + \int_0^t \mathcal{U}(t,s)H(s,y(s))ds \\
&= \int_{-\infty}^t \mathcal{U}(t,s)g(s,\gamma(s))ds - \int_{-\infty}^0 \mathcal{U}(t,s)g(s,\gamma(s))ds \\
&\quad + \int_0^t \mathcal{U}(t,s)H(s,y(s))ds \\
&= (I_1 y)(t) + (I_2 y)(t),
\end{aligned}
$$

where

$$(I_1 y)(t) = \int_{-\infty}^t \mathcal{U}(t,s)g(s,\gamma(s))ds,$$

$$(I_2 y)(t) = \int_0^t \mathcal{U}(t,s)H(s,y(s))ds$$

$$-\int_{-\infty}^{0} \mathcal{U}(t,s)g(s,y(s))ds$$

$$= (J_1 y)(t) + (J_2 y)(t),$$

where

$$(J_1 y)(t) = \int_0^t \mathcal{U}(t,s)H(s,y(s))ds,$$

$$(J_2 y)(t) = \int_{-\infty}^t \mathcal{U}(t,s)g(s,\gamma(s))ds.$$

Using (3.3.2) and Lemma 2.12, we deduce that $s \to g(s,\gamma(s))$ is in $AA(\mathbb{R}, E)$. Thus, by Lemma 3.1 we obtain

$$(I_1 y)(t) \in AA(\mathbb{R}, E). \tag{3.9}$$

Let us prove that $J_1 \in C_0(\mathbb{R}^+, E), J_2 \in C_0(\mathbb{R}^+, E)$.

Indeed, by definition $H \in C_0(\mathbb{R}^+, E)$, that means given $\varepsilon > 0$, there exists $T > 0$ such that if $t \geq T$, we have $\|H(t,y)\| \leq \varepsilon$. Therefore if $t \geq T$, we get

$$\int_T^t \|\mathcal{U}(t,s)\|_{B(E)} \|H(s,y(s))\| ds \leq M\varepsilon \int_T^t e^{-\delta(t-s)} ds$$

$$\leq \frac{M}{\delta}\varepsilon,$$

then

$$\|(J_1 y)(t)\| \leq \frac{M}{\delta}\varepsilon \quad \text{if } t \geq T.$$

So,

$$J_1 \in C_0(\mathbb{R}^+, E). \tag{3.10}$$

Next, let us show that $J_2 \in C_0(\mathbb{R}^+, E)$.

$$\|(J_2 y)(t)\| \leq \int_{-\infty}^0 \|\mathcal{U}(t,s)\|_{B(E)} \|g(s,y(s))\| ds$$

$$\leq M \sup_{t\in\mathbb{R}} \|g(t,y(t))\| \int_0^T e^{-\delta(t-s)} ds$$

$$+ M\|g\|_\infty \frac{e^{-\delta(t}}{\delta} \quad \to 0 \text{ as } \to \infty.$$

So,

$$J_2 \in C_0(\mathbb{R}^+, E). \tag{3.11}$$

Finally combining (3.7),(3.9), (3.10) and (3.11) proves our claim that $N \in AAA(\mathbb{R}^+, E)$. Next, we will prove that the operator N satisfies all the assumptions of Theorem 2.8. We will break the proof into several steps.

Let

$$B_R = \left\{ y \in AAA(\mathbb{R}^+, E) : \|y\|_\infty \leq R \right\},$$

where R be any positive constant. Then B_R is a bounded, closed and convex subset of $AAA(\mathbb{R}^+, E)$.

Step 1: $N(y) \in B_R$ for any $y \in B_R$.

In fact, if we assume that the assertion is false, then $R < \|(Ny)(t)\|$. This yields that

$$
\begin{aligned}
R < \|(Ny)(t)\| &\leq \int_0^t \|\mathcal{U}(t,s)\|_{B(E)} \|g(s, y(s))\| ds \\
&\quad + \int_0^t \|\mathcal{U}(t,s)\|_{B(E)} \|h(s, y(s))\| ds \\
&\leq \int_0^t \|\mathcal{U}(t,s)\|_{B(E)}\, p(s)\psi(\|y(s)\|) ds \\
&\quad + \int_0^t \|\mathcal{U}(t,s)\|_{B(E)}\beta(s)\phi(\|y(s)\|) ds \\
&\leq M\psi(R) \int_0^t e^{-\delta(t-s)} p(s) ds \\
&\quad + M\,\phi(R) \int_0^t e^{-\delta(t-s)} \beta(s) ds.
\end{aligned}
$$

For $t \geq 0$, it follows from the Hölder inequality that

$$R < \|(Ny)(t)\| \leq M\psi(R)\|p\|_{L^q} + M\rho_2\phi(R).$$

Dividing both sides by R and taking the lim inf as $R \to +\infty$, we have

$$M\rho_1\|p\|_{L^q} + M\delta^{-1}\rho\rho_2 > 1,$$

which contradicts (3.4). Hence, the operator N transforms the set B_R into itself.

Step 2. N is continuous.

Let $(y_n)_{n \in N}$ be a sequence in B_R such that $y_n \to y$ in B_R.

    Case 1. If $t \in [0, T]$; $T > 0$, then, we have

$$\|(Ny_n)(t) - (Ny)(t)\| \leq M \int_0^t \|f(s, y_n(s)) - f(s, y(s))\|\, ds.$$

Since the functions f is Carathéodory, the Lebesgue dominated convergence theorem implies that

$$\|Ny_n - Ny\|_\infty \to 0 \quad \text{as } n \to +\infty.$$

Case 2. Since the functions f is Carathéodory, we can see that

$$\|f(s, y_n(s)) - f(s, y(s))\| \le \frac{\delta\varepsilon}{M} \quad \text{for } t \ge T. \tag{3.12}$$

If $t \in (T, \infty)$, $T > 0$, then (3.12) and the hypotheses give us that

$$\begin{aligned}
\|Ny_n(t) - Ny(t)\| &\le \int_0^t \|\mathcal{U}(t,s)\|_{B(E)} \Big\| f(s, y_n(s)) - f(s, y(s)) \Big\| ds \\
&\le M \frac{\delta\varepsilon}{M} \int_0^t e^{-\delta(t-s)} ds \\
&\le \frac{M}{\delta} \frac{\delta\varepsilon}{M} \\
&\le \varepsilon.
\end{aligned} \tag{3.13}$$

Then the inequality (3.13) reduces to

$$\|N(y_n) - N(y)\|_\infty \to 0 \quad \text{as } n \to \infty.$$

Now, we conclude that N is continuous from B_R to B_R.

Step 3: $N(B_R)$ is equicontinuous.

Let $t_1, t_2 \in [0, T]$ with $t_2 > t_1$ and $y \in B_R$. Then, we have

$$\begin{aligned}
&\|(N_1 y)(t_2) - (N_1 y(t_1)\| \\
&= \left\| \int_0^{t_1} (\mathcal{U}(t_2, s) - \mathcal{U}(t_1, s))g(s, y(s)) \right. \\
&\quad \left. + \int_{t_1}^{t_2} \mathcal{U}(t_2, s)g(s, y(s)) ds \right\| \\
&\quad + \left\| \int_0^{t_1} (\mathcal{U}(t_2, s) - \mathcal{U}(t_1, s))h(s, y(s)) \right. \\
&\quad \left. + \int_{t_1}^{t_2} \mathcal{U}(t_2, s)h(s, y(s)) ds \right\| \\
&\le \int_0^{t_1} \|\mathcal{U}(t_2, s) - \mathcal{U}(t_1, s)\|_{B(E)}\, p(s)\psi(\|y(s)\|) ds \\
&\quad + M \int_{t_1}^{t_2} e^{-\delta(t-s)} p(s)\psi(\|y(s)\|) ds. \\
&\quad + \int_0^{t_1} \|\mathcal{U}(t_2, s) - \mathcal{U}(t_1, s)\|_{B(E)}\, \beta(s)\phi(\|y(s)\|) ds \\
&\quad + M \int_{t_1}^{t_2} e^{-\delta(t-s)} \beta(s)\phi(\|y(s)\|) ds.
\end{aligned}$$

It follows from the Hölder inequality that

$$
\begin{aligned}
\|(N_1 y)&(t_2) - (N_1 y(t_1))\| \\
&\leq \int_0^{t_1} \|\mathcal{U}(t_2, s) - \mathcal{U}(t_1, s)\|_{B(E)} \, p(s)\psi(\|y(s)\|)ds \\
&+ \frac{M\|p\|_{L^q}\psi(R)}{\delta^{1-\frac{1}{q}}} \left(e^{-\frac{q\delta}{q-1}(t-t_2)} - e^{-\frac{q\delta}{q-1}(t-t_2)} \right)^{1-\frac{1}{q}} \\
&+ \int_0^{t_1} \|\mathcal{U}(t_2, s) - \mathcal{U}(t_1, s)\|_{B(E)} \, \beta(s)\phi(\|y(s)\|)ds \\
&+ \frac{M\phi(R) \sup_{t\in\mathbb{R}} \beta(t)}{\delta} (e^{-\delta(t-t_2)} - e^{-\delta(t-t_1)}).
\end{aligned}
$$

The right-hand side of the above inequality tends to zero as $t_2 - t_1 \to 0$, which implies that $N(B_R)$ is equicontinuous.

Consider the measure of non-compactness $\mu(B)$ defined on the family of bounded subsets of the space $AAA(\mathbb{R}^+, E)$ (see [198]) by

$$
\mu(B) = \omega_0^T(B) + \sup_{t\in J} \alpha(B(t)) + \lim_{T\to+\infty} \sup\{\|y(t)\| : t \geq T, y \in E\}.
$$

Step 4: $\mu(N(B)) \leq M \max(4\|\eta\|_{L^1}, \|p\|_{L^q}\delta^{-1+\frac{1}{q}})(\varphi + \psi)(\mu(B))$ for all $B \subset B_R$. For all $B \subset B_R$, $N(B)$ is bounded. Hence, by Lemma 2.4, there exists a countable set $B_1 = \{y\}_{n=1}^\infty \subset B$, such that

$$
(N(B)) \leq 2\alpha(N(B_1)). \tag{3.14}
$$

Using the properties of α, Lemma 2.4, Lemma 2.6 and assumptions (3.2.1) and (3.3.3), we get

$$
\begin{aligned}
\alpha(NB_1(t)) &\leq \alpha\left(\left\{ \int_0^t \mathcal{U}(t, s)f(s, y_n(s))ds \right\}_{n=0}^\infty \right) \\
&\leq 2M \int_0^t \{\alpha\left(f(s, y_n(s))ds\right)\}_{n=0}^\infty \, ds \\
&\leq 2M \int_0^t \eta(s)\varphi\left(\{(\alpha(y_n(s))\}_{n=0}^\infty)\right) ds \\
&\leq 2M \int_0^t \eta(s)\varphi(\alpha(B(s)))ds.
\end{aligned}
$$

Form inequality (3.14), it follows that

$$
\alpha(NB(t)) \leq 4M \int_0^t \eta(s)\varphi(\alpha(B(s)))ds,
$$

then

$$\alpha\big(N(B(t))\big) \leq 4M\|\eta\|_{L^1}\varphi\left(\sup_{t\in\mathbb{R}^+}\alpha(B(t))\right).$$

Since

$$\sup_{t\in\mathbb{R}^+}\alpha(B(t)) \leq \sup_{t\in\mathbb{R}^+}\alpha(B(t)) + \lim_{t\to+\infty}\sup\{\|y(t)\| : t \geq T, y \in E\}),$$

then

$$\alpha(N(B(t)) \leq 4M\|\eta\|_{L^1}\varphi(\sup_{t\in\mathbb{R}^+}\alpha(B(t)) + \lim_{t\to+\infty}\sup\{\|y(t)\| : t \geq T, y \in E\}).$$

$$(3.15)$$

On the other hand, we have

$$\|(Ny)(t)\| \leq \tilde{M}e^{-\delta t}\|y_1\| + Me^{-\delta t}\|y_0\|$$

$$+ M\int_{-\infty}^{t} e^{-\delta(t-s)}p(s)\psi(\|\gamma(s)\|)ds + \|(I_2y)(t)\|$$

$$+ M\int_{-\infty}^{T} e^{-\delta(t-s)}p(s)\psi(\|\gamma(s)\|)ds.$$

$$+ M\int_{T}^{t} e^{-\delta(t-s)}p(s)\psi(\|\gamma(s)\|)ds + \|I_2(t)\|.$$

$$\leq \tilde{M}e^{-\delta t}\|y_1\| + Me^{-\delta t}\|y_0\|$$

$$+ M\int_{-\infty}^{T} e^{-\delta(t-s)}p(s)ds\psi(\sup_{s\in\mathbb{R}}\|\gamma(s)\|)$$

$$+ M\int_{T}^{t} e^{-\delta(t-s)}p(s)ds\psi(\sup\{\|\gamma(t)\| : t \geq T, y \in E\})$$

$$+ \sup\{\|(I_2y)(t)\| : t \geq T, y \in E\}).$$

Next, applying the Hölder inequality we derive

$$\|(Ny)(t)\| \leq \tilde{M}e^{-\delta t}\|y_1\| + Me^{-\delta t}\|y_0\|$$

$$+ \frac{M\|p\|_{L^q}}{\delta^{1-\frac{1}{q}}}e^{-\delta(t-T)}\psi(\|y\|_\infty).$$

$$+ \frac{M\|p\|_{L^q}}{\delta^{1-\frac{1}{q}}}(1 - e^{-\frac{q\delta}{q-1}t})^{1-\frac{1}{q}}\psi(\sup\{\|y(t)\| : t \geq T, y \in E\})$$

$$+ \sup\{\|(I_2y)(t)\| : t \geq T, y \in E\}).$$

Then

$$\|(Ny)(t)\| \leq \tilde{M}e^{-\delta t}\|y_1\| + Me^{-\delta t}\|y_0\|$$
$$+ \frac{M\|p\|_{L^q}}{\delta^{1-\frac{1}{q}}}e^{-\delta T}\psi(\|y\|_\infty).$$
$$+ \frac{M\|p\|_{L^q}}{\delta^{1-\frac{1}{q}}}\psi(\sup\{\|y(t)\| : t \geq T, y \in E\})$$
$$+ \sup\{\|(I_2 y)(t)\| : t \geq T, y \in E\}).$$

Since $\delta \geq 0$, $I_2 \in C_0(\mathbb{R}^+, E)$ and

$$\lim_{T\to+\infty} \sup\{\|y(t)\| : t \geq T, y \in E\}$$
$$\leq \sup_{t\in\mathbb{R}} \alpha(B(t)) + \lim_{T\to+\infty} \sup\{\|y(t)\| : t \geq T, y \in E\},$$

then

$$\lim_{T\to+\infty} \sup \|(Ny)(t)\| : t \geq T, y \in E\})$$
$$\leq \frac{M\|p\|_{L^q}}{\delta^{1-\frac{1}{q}}}\psi(\sup_{t\in J} \alpha(B(t)) + \lim_{T\to+\infty} \sup\{\|y(t)\| : t \geq T, y \in E\}). \quad (3.16)$$

Further, combining (3.15) and (3.16), we get

$$\sup_{t\in J} \alpha((NB)(t)) + \lim_{T\to+\infty} \sup \|(Ny)(t)\| : t \geq T, y \in E\})$$
$$\leq 4M\|\eta\|_{L^1}\varphi(\sup_{t\in J} \alpha(B(t)) + \lim_{T\to+\infty} \sup\{\|y(t)\| : t \geq T, y \in E\})$$
$$+ \frac{M\|p\|_{L^q}}{\delta^{1-\frac{1}{q}}}\psi(\sup_{t\in J} \alpha(B(t)) + \lim_{T\to+\infty} \sup\{\|y(t)\| : t \geq T, y \in E\}$$
$$\leq M \max(4\|\eta\|_{L^1}, \frac{\|p\|_{L^q}}{\delta^{1-\frac{1}{q}}})(\varphi + \psi)$$
$$\times (\sup_{t\in J} \alpha(B(t)) + \lim_{T\to+\infty} \sup\{\|y(t)\| : t \geq T, y \in E\}). \quad (3.17)$$

From **Step 3** and inequality (3.17), we conclude that

$$\mu(N(B)) \leq M \max\left(4\|\eta\|_{L^1}, \frac{\|p\|_{L^q}}{\delta^{1-\frac{1}{q}}}\right)(\varphi + \psi)(\mu(B)).$$

It follows from Lemma 2.8 that N has at least one fixed point $y \in B_R$, which is just a asymptotically almost automorphic mild solution of problem (3.1)–(3.2) on $\mathbb{R}^+$.

3.2.2 An Example

Consider the second order differential equation of the form:

$$\begin{cases} \dfrac{\partial^2}{\partial t^2}z(t,\tau) = \dfrac{\partial^2}{\partial \tau^2}z(t,\tau) + 2\sin\left(\dfrac{1}{2+\cos t+\cos\sqrt{2}t}\right)\dfrac{\partial}{\partial t}z(t,\tau) \\ \qquad + \dfrac{\sin^2 t}{12\sqrt{1+t^2}}\sin\left(\dfrac{1}{2+\cos t+\cos\sqrt{2}t}\right) \\ \qquad \times (\|z(t,\tau)\| + \ln\left(1+\|z(t,\tau)\|\right)) \\ \qquad + \dfrac{\sin^2 t}{15\sqrt{1+t^2}}\dfrac{\sin\pi z(t,\tau)}{(1+\|z(t,\tau)\|)}, \quad t\in\mathbb{R}^+,\ \tau\in[0,\pi], \\ z(t,0) = z(t,\pi) = 0, \quad t\in\mathbb{R}^+, \\ \dfrac{\partial}{\partial t}z(0,\tau) = \psi(\tau), \quad \tau\in[0,\pi]. \end{cases} \qquad (3.18)$$

Let $E = L^2([0,\pi],\mathbb{R}^+)$ be the space of 2-integrable functions from $[0,\pi]$ into $\mathbb{R}^+$, and let $H^2([0,\pi],\mathbb{R}^+)$ be the Sobolev space of functions $x:[0,\pi]\to\mathbb{R}^+$, such that $x''\in L^2([0,\pi],\mathbb{R}^+)$. We consider the operator $A_1 z(\tau) = z''(\tau)$ with domain $D(A_1) = H^2(\mathbb{R}^+,C)$, which is the infinitesimal generator of strongly continuous cosine function $C(t)$ on E. Moreover, A_1 has discrete spectrum, the spectrum of A_1 consists of eigenvalues $-n^2$ for $n\in\mathbb{Z}$, with associated eigenvector

$$w_n(\xi) = \frac{1}{\sqrt{2\pi}}e^{in\xi}, n\in\mathbb{Z},$$

the set $\{w_n\in\mathbb{Z}\}$ is an orthonormal basis of E. In particular,

$$A_1 x = -\sum_{n=1}^{\infty} n^2\langle x,w_n\rangle w_n \text{ for } x\in D(A).$$

The cosine function $C(t)$ is given by

$$C(t)x = \sum_{n=1}^{\infty}\cos(nt)\langle x,w_n\rangle w_n \text{ for } x\in D(A), t\in\mathbb{R}^+,$$

form a cosine function on H, with associated sine function

$$S(t)x = \sum_{n=1}^{\infty}\frac{\sin(nt)}{n}\langle x,w_n\rangle w_n \text{ for } x\in D(A), t\in\mathbb{R}^+.$$

From [8], for all $x\in H^2([0,\pi],\mathbb{R}^+), t\in\mathbb{R}^+$, $\|C(t)\|_{B(E)}\le e^{-t}$ and $\|S(t)\|_{B(E)}\le e^{-t}$.

Now, we define an operator $A(t) : D(A) \subset H \to H$ by

$$\begin{cases} D(A(t)) = D(A) \\ A(t) = A_1 + b(t, \tau). \end{cases}$$

where $b(t, \tau) = 2 \sin \left(\dfrac{1}{2 + \cos t + \cos \sqrt{2t}} \right)$.

Note that $A(t)$ generates an evolutionary process $\mathcal{U}(t, s)$ of the form

$$\mathcal{U}(t, s) = S(t - s) e^{\int_s^t b(t, s) ds}$$

Since $b(t, \tau) = 2 \sin \left(\dfrac{1}{2 + \cos t + \cos \sqrt{2t}} \right) \leq -2$, we have

$$\mathcal{U}(t, s) = S(t - s) e^{-2(t-s)} \tag{3.19}$$

and

$$\| \mathcal{U} \|_{B(E)} \leq \| S \|_{B(E)} e^{-2(t-s)} \leq e^{-3(t-s)}$$

We conclude that $\mathcal{U}(t, s)$ is a evolutionary process exponentially stable with $M = 1$ and $\delta = 3$.

It follows from the estimate (3.19) that $\mathcal{U}(t, s) : E \to E$ is well defined and satisfies the conditions of Definition 2.10.

Hence conditions (3.2.1) and (3.3.1) are satisfied.

Now, let

$$z(t)(\tau) = w(t)(\tau), \ t \geq 0, \ \tau \in [0, \pi],$$

$$g(t, z)(\tau) = \frac{\sin^2 t}{12\sqrt{1 + t^2}} \sin \left(\frac{1}{2 + \cos t + \cos \sqrt{2t}} \right)$$
$$\times \left(\| z(t, \tau) \| + \ln \left(1 + \| z(t, \tau) \| \right) \right),$$

$$h(t, z)(\tau) = \frac{\sin^2 t \sin \pi z(t, \tau)}{15\sqrt{1 + t^2}(1 + \| z(t, \tau) \|)}.$$

Then it is easy to verify that $g : \mathbb{R} \times E \times E$ is continuous and

$$g \in AA(\mathbb{R} \times E; E).$$

We can estimate for the functions g:

$$g(t, z)(\tau) \leq \frac{\sin^2 t}{12\sqrt{1 + t^2}} \left(\| z(t, \tau) \| + \ln \left(1 + \| z(t, \tau) \| \right) \right).$$

Hence conditions (3.3.2)(a) is satisfied with

$$p(t) = \frac{\sin^2 t}{3\sqrt{1 + t^2}}, \quad \psi(t) = \frac{1}{4}(t + \ln(1 + t)).$$

Then it is easy to verify that $p \in L^2(\mathbb{R})$ and $\rho_1 = \dfrac{1}{4}$.

On the other hand, it is clear that $h : \mathbb{R}^+ \times E \times E$ is continuous and

$$h \in C_0(\mathbb{R}^+ \times E; E).$$

We can also estimate for the functions h:

$$h(t, z)(\tau) \leq \frac{\pi}{15\sqrt{1 + t^2}} \|z(t, \tau)\|.$$

Hence conditions $(3.3.2)(b)$ is satisfied with

$$\beta(t) = \frac{\pi}{15\sqrt{1 + t^2}}, \quad \phi(R) = R.$$

Then it is easy to verify that $\beta \in C_0(\mathbb{R}^+, \mathbb{R})$, $\rho_2 = 1$ and $\rho \leq \dfrac{\pi}{15}$.

Furthermore,

$$f(t; z) = g(t; z) + h(t; z) \in AA(\mathbb{R}^+ \times E; E).$$

We can also estimate for the functions f:

$$f(t, z)(\tau) \leq \frac{2\sin^2 t}{\sqrt{1 + t^2}} \|z(t, \tau)\|. \tag{3.20}$$

By (3.20), for every $t \in J$, and $B \in D \subset E$, we have

$$\alpha(f(t, D) \leq \frac{\sin^2 t}{12\sqrt{1 + t^2}} \alpha(D),$$

Hence conditions $(3.3.3)$ is satisfied with

$$\eta(t) = \frac{1}{6\sqrt{1 + t^2}}, \quad \varphi(t) = \frac{\sin^2 t}{2}.$$

Moreover, we have

$$(\psi + \varphi)(t) = \frac{\sin^2 t}{2} + \frac{1}{4}(t + \ln(1 + t)) \leq t.$$

We conclude that (see Lemma 2.1 [252])

$$\lim_{n \to +\infty} (\psi + \phi)^n(t) = 0 \text{ for a.e } t \in \mathbb{R}^+.$$

Consequently, can be written in the abstract form (3.1)–(3.2) with $A(t)$ and f as defined above. Thus, Theorem 3.1 yields that equation (3.18) has a asymptotically almost automorphic mild solution.

3.3   Semilinear Integro-Differential Evolution Equations

In this section, we discuss the non-autonomous second order semilinear evolution equation of the type:

$$y''(t) - A(t)y(t) = f\left(t, y(t), \int_0^t K(t, s, y(s))ds\right), \quad t \in J := [0, T], \quad (3.21)$$

$$y(0) = y_0, \ y'(0) = y_1, \qquad (3.22)$$

where $\{A(t)\}_{0 \leq t < +\infty}$ is a family of linear closed operators from E into E, that generate an evolution system of linear bounded operators $\{\mathcal{U}(t, s)\}_{(t,s) \in \Delta}$, $f : J \times E \times E \to E$ is a Carathéodory function, $K : \Delta \times E \to E$ is a continuous function, $\Delta := \{(t, s) \in J \times J : s \leq t\}$, and $(E, |\cdot|)$ is a real Banach space.

3.3.1   *Existence Results*

Definition 3.2. A function $y \in C(J, E)$ is called a mild solution to the problem (3.21)–(3.22) if y satisfies the integral equation

$$y(t) = -\frac{\partial}{\partial s}\mathcal{U}(t, 0)y_0 + \mathcal{U}(t, 0)y_1 + \int_0^t \mathcal{U}(t, s)f\left(s, y(s), \int_0^s K(s, \tau, y(\tau))d\tau\right)ds. \qquad (3.23)$$

To prove our results we introduce the following conditions:

(3.4.1) There exists a constant $M \geq 1$, such that

$$\|\mathcal{U}(t, s)\|_{B(E)} \leq M, (t, s) \in \Delta.$$

(3.4.2) There exists a constant $\tilde{M} \geq 0$, such that

$$\left\|\frac{\partial}{\partial s}\mathcal{U}(t, s)\right\|_{B(E)} \leq \tilde{M}, (t, s) \in \Delta.$$

(3.4.3) There exists a continuous function $p : J \to \mathbb{R}^+$, such that:

$$\|f(t, u, v)\| \leq p(t)(\|u\| + \|v\|) \text{ for a.e } t \in J \text{ and each } u, v \in E.$$

(3.4.4) There exist integrable functions $\sigma_i (i = 1, 2) : J \to \mathbb{R}^+$, such that:

$$\alpha(f(t, D_1, D_2)) \leq \sigma_1(t)\alpha(D_1) + \sigma_2(t)\alpha(D_2) \text{ for a.e } t \in J \text{ and } D_1, D_2 \subset E.$$

(3.4.5) There exists a continuous function $q : J \to \mathbb{R}_+$, such that:
$$\|K(t, s, u)\| \le q(t) \|u\| \text{ for a.e } (t, s) \in \Delta \text{ and each } u \in E.$$
(3.4.6) There exists a constant $K^* > 0$, such that
$$\alpha(K(t, s, D)) \le K^* \alpha(D) \text{ for a.e } (t, s) \in \Delta \text{ and } D \subset E.$$

Theorem 3.2. *Assume that the hypotheses (3.4.1)–(3.4.6) are satisfied. Then the problem (3.21)–(3.22) admits at least one mild solution.*

Proof. Consider the operator $N : C(J, E) \to C(J, E)$, defined by
$$(Ny)(t) = -\frac{\partial}{\partial s}\mathcal{U}(t, 0)y_0 + \mathcal{U}(t, 0)y_1$$
$$+ \int_0^t \mathcal{U}(t, s)f\left(s, y(s), \int_0^s K(s, \tau, y(\tau))d\tau\right) ds.$$
We observe that the mild solution of problem (3.21)–(3.22) is a fixed point of the operator N. Set
$$B_R = \{y \in C(J, E) : \|y\| < R\},$$
where R be any positive constant.
Step 1. $N(B_R)$ is bounded for any bounded set B_R.

Now, for $t \in J$, we have
$$\|Ny(t)\| \le \left\|\frac{\partial}{\partial s}\mathcal{U}(t, 0)\right\|_{B(E)} \|y_0\| + \|\mathcal{U}(t, s)\|_{B(E)} \|y_1\|$$
$$+ \int_0^t \|\mathcal{U}(t, s)\|_{B(E)} \, p(s) \left(\|y(s)\| + \int_0^s q(\tau)\|y(\tau)\|d\tau\right) ds$$
$$\le \tilde{M} \|y_0\| + M \|y_1\| + MRT\|p\| \left(1 + \frac{T}{2}\|q\|\right)$$
$$< +\infty.$$

Step 2. N is continuous.
Let $(y_n)_{n \in N}$ be a sequence in B_R such that $y_n \to y$. For $t \in J$, we have
$$\|(Ny_n)(t) - (Ny)(t)\| \le M \int_0^t \left\|f\left(s, y_n(s), \int_0^s K(s, \tau, y_n(\tau))d\tau\right)\right.$$
$$\left. - f\left(s, y(s), \int_0^s K(s, \tau, y(\tau))d\tau\right)\right\| ds.$$
Since the functions f is Carathéodory and K is continuous function, the Lebesgue dominated convergence theorem implies that
$$\|Ny_n - Ny\| \to 0 \quad \text{as } n \to +\infty.$$
So N is continuous.

Step 3. $N(B_R)$ is equicontinuous.

Let $t_1, t_2 \in J$ with $t_2 > t_1$ and $y \in B_R$, then, we have

$$\|(Ny)(t_2) - (Ny)(t_1)\|$$

$$= \left\| \int_0^{t_1} (\mathcal{U}(t_2, s) - \mathcal{U}(t_1, s)) f\left(s, y(s), \int_0^s K(s, \tau, y(\tau)) d\tau\right) ds \right.$$

$$\left. + \int_{t_1}^{t_2} \mathcal{U}(t_2, \tau) f\left(s, y(s), \int_0^s K(s, \tau, y(\tau)) d\tau\right) ds \right\|$$

$$\leq \int_0^{t_1} \|\mathcal{U}(t_2, \tau) - \mathcal{U}(t_1, \tau)\|_{B(E)} \, p(\tau) \left(\|y(s)\| + \int_0^s q(\tau) \|y(\tau)\| d\tau \right) ds$$

$$+ M \int_{t_1}^{t_2} p(s) \left(\|y(s)\| + \int_0^s q(\tau) \|y(\tau)\| d\tau \right) ds.$$

We get

$$\|(Ny)(t_2) - (Ny)(t_1)\|$$

$$\leq R \int_0^{t_1} \|\mathcal{U}(t_2, \tau) - \mathcal{U}(t_1, \tau)\|_{B(E)} \, p(\tau) \left(1 + \int_0^s q(\tau) d\tau \right) ds$$

$$+ MR \int_{t_1}^{t_2} p(s) \left(1 + \int_0^s q(\tau) d\tau \right) ds$$

$$\leq R \int_0^{t_1} \|\mathcal{U}(t_2, \tau) - \mathcal{U}(t_1, \tau)\|_{B(E)} \, p(\tau) \left(1 + \int_0^s q(\tau) d\tau \right) ds$$

$$+ MR \int_0^{t_1} \sup_{t \in J} \|p(t)\| \left(1 + \int_0^s \sup_{t \in J} \|q(t)\| d\tau \right) ds$$

$$\leq R \int_0^{t_1} \|\mathcal{U}(t_2, \tau) - \mathcal{U}(t_1, \tau)\|_{B(E)} \, p(\tau) \left(1 + \int_0^s q(\tau) d\tau \right) ds$$

$$+ MR\|p\| \int_{t_1}^{t_2} (1 + s\|q\|) ds$$

$$\leq R \int_0^{t_1} \|\mathcal{U}(t_2, \tau) - \mathcal{U}(t_1, \tau)\|_{B(E)} \, p(\tau) \left(1 + \int_0^s q(\tau) d\tau \right) ds$$

$$+ MR\|p\| \left((t_2 - t_1) + \frac{1}{2} \|q\| (t_2 - t_1)^2 \right)$$

The right-hand side of the above inequality tends to zero as $t_2 - t_1 \to 0$, which implies that $N(B_R)$ is equicontinuous.

Step 4. N is a condensing operator.

Consider the measure of non-compactness $\mu^*(B)$, defined on the family of bounded subsets of the space $C(J, E)$ by

$$\mu^*(B) = \omega_0^T(B) + \sup_{t \in J} e^{-\tau \sigma(t)} \overline{\alpha}(B(t)),$$

where

$$\sigma(t) = 4M \int_0^t (\sigma_1(s) + 2K^* s \sigma_2(s)) ds, \tau \geq 1, \quad \overline{\alpha}(B(t)) = \sup_{s \in [0,t]} \alpha(B(s)).$$

For all $B \subset B_R$, $N(B)$ is bounded. Hence, by Lemma 2.4, there exists a countable set $B_1 = \{y\}_{n=1}^\infty \subset B$, such that for each $t \in J$, we have

$$\alpha(N(B)(t)) \leq 2\alpha(N(B_1)(t)). \tag{3.24}$$

Using the properties of α, Lemma 2.4, Lemma 2.6 and assumptions (3.4.1), (3.4.3) and (3.4.5), we get

$$\alpha(NB_1(t)) \leq \alpha \left(\left\{ \int_0^t \mathcal{U}(t,s) f(s, y_n(s), \int_0^s K(s, \tau, y_n(s)) d\tau) ds \right\}_{n=0}^\infty \right)$$

$$\leq 2M \int_0^t \left\{ \alpha \left(f(s, y_n(s), \int_0^s K(s, \tau, y_n(\tau)) d\tau) ds \right) \right\}_{n=0}^\infty ds$$

$$\leq 2M \int_0^t \sigma_1(s) \left\{ \alpha(y_n(s)) \right\}_{n=0}^\infty ds$$

$$+ \sigma_2(s) \left\{ \alpha \left(\int_0^s K(s, \tau, y_n(\tau)) d\tau \right) \right\}_{n=0}^\infty ds$$

$$\leq 2M \int_0^t \sigma_1(s) \left\{ \alpha(y_n(s)) \right\}_{n=0}^\infty)$$

$$+ 2K^* \sigma_2(s) \left\{ \int_0^s \alpha(y_n(\tau)) d\tau \right\}_{n=0}^\infty ds$$

$$\leq 2M \int_0^t \left(\sigma_1(s) \alpha(B_1(s)) + 2K^* \sigma_2(s) \int_0^s \alpha(B_1(\tau)) d\tau \right) ds.$$

$$\leq 2M \int_0^t \left(\sigma_1(s) \alpha(B_1(s)) + 2K^* \sigma_2(s) s \sup_{\tau \in [0,s]} \alpha(B_1(\tau)) \right) ds.$$

$$\leq 2M \int_0^t \left(\sigma_1(s) \sup_{s \in [0,t]} \alpha(B_1(s)) + 2K^* \sigma_2(s) s \sup_{\tau \in [0,t]} \alpha(B_1(\tau)) \right) ds.$$

$$\leq 2M \int_0^t (\sigma_1(s) + 2K^* s \sigma_2(s)) \sup_{s \in [0,t]} \alpha(B_1(s)) ds.$$

Form inequality (3.24), it follows that

$$\alpha(NB(t)) \leq 4M \int_0^t (\sigma_1(s) + 2K^* \sigma_2(s) s) \overline{\alpha}(B(s)) ds.$$

Therefore, we know

$$\alpha(NB(t)) \leq 4M \int_0^t (\sigma_1(s) + 2K^* s \sigma_2(s)) e^{\tau \sigma(s)} e^{-\tau \sigma(s)} \overline{\alpha}(B(s)) ds,$$

then

$$e^{-\tau\sigma(t)}\alpha(N(B(t)) \le \frac{1}{\tau}\sup_{t\in J} e^{-\tau\sigma(t)}\overline{\alpha}(B(t)),$$

hence

$$e^{-\tau\sigma(t)}\sup_{t\in J}\alpha(N(B(t)) \le \frac{1}{\tau}\sup_{t\in J} e^{-\tau\sigma(t)}\overline{\alpha}(B(t)).$$

Since

$$e^{-\tau\sigma(t)}\sup_{s\in[0,t]}\alpha(N(B(s)) \le e^{-\tau\sigma(t)}\sup_{t\in J}\alpha(N(B(t))),$$

we get

$$e^{-\tau\sigma(t)}\sup_{s\in[0,t]}\alpha(N(B(s)) \le \frac{1}{\tau}\sup_{t\in J} e^{-\tau\sigma(t)}\overline{\alpha}(B(t)).$$

Then

$$e^{-\tau\sigma(t)}\overline{\alpha}(N(B(t)) \le \frac{1}{\tau}\sup_{t\in J} e^{-\tau\sigma(t)}\overline{\alpha}(B(t)). \tag{3.25}$$

From **Step 3** and inequality (3.25), we obtain

$$\mu^*(ND) \le \mu^*(D).$$

Thus, we find that N is a condensing operator.

Step 5. A priori bounds.

We now show there exists an open set $Y \subseteq C(J, E)$ with $y \ne \lambda N(y)$, for $\lambda \in (0,1)$ and $y \in \partial Y$. Let $y \in C(J, E)$ be such that $y = \lambda N(y)$ for some $0 < \lambda < 1$. Then,

$$y(t) = -\lambda\frac{\partial}{\partial s}\mathcal{U}(t,0)y_0 + \lambda\mathcal{U}(t,0)y_1$$
$$+ \lambda\int_0^t \mathcal{U}(t,s)f\left(s, y(s), \int_0^s K(s,\tau,y(\tau))d\tau\right)ds.$$

This implies by (3.4.1)–(3.4.3) and (3.4.5) that, for each $t \in J$, we have

$$\|y(t)\| \le \left\|\frac{\partial}{\partial s}\mathcal{U}(t,0)\right\|_{B(E)}\|y_0\| + \|\mathcal{U}(t,s)\|_{B(E)}\|y_1\|$$
$$+ \int_0^t \|\mathcal{U}(t,s)\|_{B(E)}\,p(s)\left(\|y(s)\| + \int_0^s q(\tau)\|y(\tau)\|)d\tau\right)ds$$
$$\le \tilde{M}\|y_0\| + M\|y_1\| + \int_0^t Mp(s)\left(\|y(s)\| + \int_0^s q(\tau)\|y(\tau)\|d\tau\right)ds.$$

By Lemma 2.14, we have

$$\|y(t)\| \le \left[\tilde{M}\,\|y_0\| + M\,\|y_1\|\right]\left[1 + \int_0^t Mp(s)\exp\left(\int_0^s Mp(\tau) + q(\tau)d\tau\right)ds\right]$$

$$\le \left[\tilde{M}\,\|y_0\| + M\,\|y_1\|\right]$$

$$\times \left[1 + M\int_0^t \sup_{t\in J}\|p(t)\|\exp\left(\int_0^s M\sup_{t\in J}\|p(t)\| + \sup_{t\in J}\|q(t)\|d\tau\right)ds\right]$$

$$\le \left[\tilde{M}\,\|y_0\| + M\,\|y_1\|\right]\left[1 + TM\|p\|\exp T\left(M\|p\| + \|q\|\right)\right]$$

$$= \eta.$$

Thus, we have $\|y\| \le \eta$. Set

$$Y = \{y \in C(J, E) : \|y\| < \eta + 1\}.$$

By the choice of Y, there is no $y \in \partial Y$ such that $u = \lambda N(u)$, for $\lambda \in (0,1)$. As a consequence of the nonlinear alternative of Leray-Schauder type [254], we deduce that N has a fixed point $y \in C(J, E)$ which is a solution to problem (3.21)–(3.22).

3.3.2 *An Example*

Consider the second-order differential equation of the form:

$$
\begin{cases}
\dfrac{\partial^2}{\partial t^2}z(t,\tau) = \dfrac{\partial^2}{\partial \tau^2}z(t,\tau) + a(t)\dfrac{\partial}{\partial t}z(t,\tau) \\
\qquad + \dfrac{z(t,\tau)}{(\sqrt{t}+1)(1+\|z(t,\tau)\|)} + \dfrac{e^{-t}}{(\sqrt{t}+1)(t+1)} \\
\qquad \times \displaystyle\int_0^t \dfrac{\sqrt{t}z(s,\tau)}{(1+s^2+t)(1+z^2(s,\tau))}ds,\ t \in J,\ \tau \in [0,\pi], \\[2mm]
z(t,0) = z(t,\pi) = 0, \quad t \in J, \\[2mm]
\dfrac{\partial}{\partial t}z(0,\tau) = \psi(\tau), \quad \tau \in [0,\pi],
\end{cases}
$$

$$(3.26)$$

where $a : J \to \mathbb{R}$ is a Hölder continuous function.

Let $E = L^2([0,\pi], \mathbb{R})$ be the space of 2-integrable functions from $[0,\pi]$ into $\mathbb{R}$, and let $H^2([0,\pi], \mathbb{R})$ be the Sobolev space of functions $x : [0,\pi] \to \mathbb{R}$, such that $x'' \in L^2([0,\pi], \mathbb{R})$. We consider the operator $A_1 z(\tau) = z''(\tau)$ with domain $D(A_1) = H^2(\mathbb{R}, C)$, which is the infinitesimal generator of strongly continuous cosine function $C(t)$ on E. Moreover, A_1 has discrete spectrum, the spectrum of A_1 consists of eigenvalues $-n^2$ for $n \in \mathbb{Z}$, with associated eigenvector

$$w_n(\xi) = \frac{1}{\sqrt{2\pi}} e^{in\xi}, n \in \mathbb{Z},$$

the set $\{w_n \in \mathbb{Z}\}$ is an orthonormal basis of E. In particular,

$$A_1 x = -\sum_{n=1}^{\infty} n^2 \langle x, w_n \rangle w_n \text{ for } x \in D(A).$$

The cosine function $C(t)$ is given by

$$C(t)x = \sum_{n=1}^{\infty} \cos(nt) \langle x, w_n \rangle w_n \text{ for } x \in D(A), t \in \mathbb{R},$$

and the associated sine function is defined by

$$S(t)x = \sum_{n=1}^{\infty} \frac{\sin(nt)}{n} \langle x, w_n \rangle w_n \text{ for } x \in D(A), t \in \mathbb{R}.$$

From [8], for all $x \in H^2([0, \pi], \mathbb{R}), t \in \mathbb{R}, \|C(t)\|_{B(E)} \leq 1$ and $\|S(t)\|_{B(E)} \leq 1$. Thus, $C(\cdot)$ is uniformly bounded on J.

We take $A_2(t)z(s) = a(t)z'(s)$, defined on $H^1([0, \pi], \mathbb{R})$, and consider the closed linear operator $A(t) = A_1 + A_2(t)$. Initially we will show that $A_1 + A_2(t)$ generates an evolution operator $\mathcal{U}$.

It is well known that the solution of the scalar initial value problem

$$\begin{cases} v''(t) = -n^2 v(t) + ina(t)v(t) \\ v(s) = 0, \quad v'(s) = v_1. \end{cases} \tag{3.27}$$

satisfies the integral equation

$$v(t) = \frac{v_1}{n} \sin n(t - s) + i \int_s^t \sin n(t - \tau) a(\tau) v(\tau) ds.$$

Applying the Gronwall-Bellman lemma we can affirm that

$$\|v(t)\| \leq \frac{|v_1|}{n} e^{\beta(t-s)}, \tag{3.28}$$

where

$$\beta = \sup_{t \in J} |a(t)|.$$

We denote by $v_n(t, s)$ the solution of (3.27). We define

$$\mathcal{U}(t, s) = \sum_{n \in \mathbb{Z}} v_n(t, s) \langle x, w_n \rangle w_n,$$

It follows from the estimate (3.28) that $\mathcal{U}(t,s) : E \to E$ is well defined and satisfies the conditions of Definition 2.10, and there exist constants $M \geq 1$ and $\tilde{M} \geq 0$, such that:

$$\|\mathcal{U}(t,s)\|_{B(E)} \leq M \text{ and } \left\|\frac{\partial}{\partial s}\mathcal{U}(t,s)\right\|_{B(E)} \leq \tilde{M} \text{ for every } (t,s) \in \Delta.$$

Hence conditions (3.4.1) and (3.4.2) are satisfied.

Set

$$z(t)(\tau) = w(t)(\tau), \ t \geq 0, \ \tau \in [0, \pi],$$

$$f(t,u,v)(\tau) = \frac{1}{(\sqrt{t}+1)(1+\|u(t,\tau)\|)}u(t,\tau) + \frac{e^{-t}}{(\sqrt{t}+1)(t+1)}v(t,\tau),$$

$$K(t,s,u)(\tau) = \frac{\sqrt{t}u(t,\tau)}{(1+s^2+t)(1+u^2(t,\tau))},$$

and

$$\frac{\partial}{\partial t}z(0)(\tau) = \frac{d}{dt}w(0)(\tau), \ \tau \in [0, \pi].$$

We have

$$\|f(t,u,v)(\tau)\| \leq \frac{1}{1+\sqrt{t}}(\|u(t,\tau)\| + \|v(t,\tau)\|), \tag{3.29}$$

and

$$|K(t,s,u)(\tau)\| \leq \frac{\sqrt{t}}{1+t}\|u(t,\tau)\|. \tag{3.30}$$

Hence conditions (3.4.3) and (3.4.5) are satisfied with

$$p(t) = \frac{1}{1+\sqrt{t}}, \quad q(t) = \frac{\sqrt{t}}{1+t}.$$

From the definition of f, for every $t \in J$, and $D_1, D_2 \in D \subset E$, we have

$$\alpha(f(t,D_1,D_2)) \leq \frac{1}{(\sqrt{t}+1)}\alpha(D_1) + \frac{e^{-t}}{(\sqrt{t}+1)(t+1)}\alpha(D_2),$$

Hence condition (3.4.4) is satisfied with

$$\sigma_1(t) = \frac{1}{(\sqrt{t}+1)}, \quad \sigma_2(t) = \frac{e^{-t}}{(\sqrt{t}+1)(t+1)}.$$

By (3.30), for every $t \in J$ and $D \subset E$, we have

$$\alpha(K(t,s,D)) \leq \sup_{t \in J} \frac{\sqrt{t}}{1+t} \alpha(D),$$

then

$$\alpha(K(t,s,D)) \leq \frac{\sqrt{2}}{3} \alpha(D).$$

Hence (3.4.6) is satisfied with $K^* = \dfrac{\sqrt{2}}{3}$.

Consequently, (3.26) can be written in the abstract form (3.21)–(3.22) with $A(t)$ and f as defined above. The existence of mild solutions can be deduced from an application of Theorem 3.2.

3.4   Notes and Remarks

Using the articles [267, 268], we presented and proved the results of Chapter 3. For more relevant results and studies, one can see the monographs [269] and the papers [260–265, 270].

Chapter 4

Semilinear Evolution Equations with State-Dependent Delay

4.1 Introduction and Motivations

The objective of this chapter is to investigate the existence of mild solutions and approximate controllability for a broad class of abstract nonlinear differential and integro-differential equations with state-dependent delay. In order to establish our criteria, we utilize fixed point theorems such as the nonlinear alternative of Granas-Frigon for contraction and a generalized version of the classical Darbo fixed point theorem in Fréchet spaces, along with measures of non-compactness. Each section includes examples that demonstrate the practical application of our results.

The investigations conducted in this chapter were motivated by the following works, which we will mention as references for interested readers:

- For the importance of nonlocal conditions in different fields, we refer to [271, 272] and the reference therein. Hernandez and O'Regan were the first to establish a new sort of nonlocal condition in [273], which we now refer to as state dependent nonlocal conditions. More information can be found in the paper [274]. For basic results and recent development on differential equations, one can refer to [38, 39, 44, 275–281].
- The problem of the existence of solutions for the Cauchy problem of fractional integro-differential equations was investigated in numerous works; we refer the reader to books by Kilbas *et al.* [282], Lakshmikantham *et al.* [283], and the papers by Agarwal *et al.* [284], Anguraj *et al.* [285], Balachandran *et al.* [286]. Cuevas *et al.* [287], studied S-asymptotically ω-periodic solutions. Many authors have employed this approach to explore different types of differential problems; for additional details, see the works [288–290] and the references therein.

- The literature on the topic of this chapter is mostly concerned with first order ordinary differential equations on finite dimensional spaces; we mention the works [291–293]. While there are several papers in which authors discuss different problems with various forms of delays (see [38, 39, 44, 275]), there are few studies on abstract second order ordinary differential equations with state-dependent delay applied to partial differential equations; we cite [227, 273, 274, 294].

- The study of integro-differential systems, on the other hand, has piqued the curiosity of numerous researchers. Grimmer was an early pioneer in this subject, making substantial advances to the understanding of complex systems by applying resolvent operators. Grimmer's study established the existence of solutions of integro-differential systems and gave critical insights into their behavior and dynamics. He wrote numerous papers on the subject [295, 296]. Resolvent operators are mathematical tools used to investigate the properties of integro-differential systems. They allow these systems' solutions to be expressed in terms of their initial conditions and input function. These operators are very relevant for investigating the stability and controllability of integro-differential systems. We propose numerous publications that cover resolvent operators and integro-differential systems in depth for those who want to learn more about these topics. These publications include [107, 297–300], as well as the references mentioned therein.

- In [301], Hernandez studied the global existence and uniqueness of solutions and well posedness of the following general class of abstract second order differential equations with state dependent delay:

$$\begin{cases} y''(t) = Ay(t) + f\left(t, y\left(t - \sigma_1\left(t, y_t\right)\right), y'\left(t - \sigma_2\left(t, y_t\right)\right)\right), & t \in [0, a], \\ y_0 = \Psi \in C([-r, 0]; E), \ (y')_0 = \Psi' \in C([-r, 0]; E), \end{cases}$$

where $A : D(A) \subset E \to E$ is the generator of a strongly continuous cosine family of bounded linear operators $(C(t))_{t \in \mathbb{R}}$ defined on a Banach space $(E, \| \cdot \|), \sigma_i \in C([0, a] \times C([-r, 0]; E); [0, r])$, for $i = 1, 2$, and $f \in C([0, a] \times E \times E; E)$.

- The authors of [302] presented the existence of mild solutions for the following partial integro-differential equations with state-dependent nonlocal conditions:

$$\begin{cases} y'(t) = Ay(t) + \int_0^t Y(t - s)y(s)ds + f\left(t, y_{\rho(t, y_t)}\right), & t \in \mathbb{R}^+, \\ y_0 = \mathrm{K}(\zeta(y), y) \in C([-r, 0], E), \end{cases}$$

where $A : D(A) \subset E \to E$ is the infinitesimal generator of C_0-semigroup $(T(t))_{t \geq 0}$ on a Banach space E.

4.2 Integro-Differential Equations with Nonlocal Conditions in Fréchet Spaces

In this section, we consider the existence and uniqueness of mild solutions defined on unbounded interval for semilinear integro-differential equations of fractional order of the form

$$y'(t) - \int_0^t \frac{(t-s)^{\zeta-2}}{\Gamma(\zeta-1)} Ay(s)ds = f(t, y_{\varrho(t,y_t)}), \quad \text{a.e.} \ \ t \in \mathbb{R}_+ := [0, +\infty),$$
$$(4.1)$$

$$y_0 = \Psi(\sigma(y), y) \in C([-r, 0], E),$$
$$(4.2)$$

where $1 < \zeta < 2$, $A : D(A) \subset E \to E$ is a closed linear operator, and $(E, \| \cdot \|)$ is a Banach space. $f : \mathbb{R}_+ \times C([-r, 0], E) \to E$, $\sigma : C([-r, +\infty), E) \to \mathbb{R}_+$, $\Psi : \mathbb{R}_+ \times C([-r, +\infty), E) \to E$ and $\varrho : \mathbb{R}_+ \times C([-r, 0], E) \to \mathbb{R}$, are suitable functions. If $y \in C([-r, +\infty), E)$, then for any $t \in \mathbb{R}_+$, define y_t by $y_t(\varkappa) = y(t + \varkappa)$ for $\varkappa \in [-r, 0]$. Here we establish sufficient conditions to get the existence of the unique mild solution for fractional integro-differential equations with state dependent nonlocal conditions in Fréchet spaces. A nonlinear alternative of Leray-Schauder type for contraction maps in Fréchet spaces due to Frigon-Granas is employed in our study.

4.2.1 *Existence Results*

Let Θ_1 be a Fréchet space with a family of semi-norms $\{\| \cdot \|_\tau\}_{\tau \in \mathbb{N}}$ and let $\Theta_2 \subset \Theta_1$. For every $\tau \in \mathbb{N}$, let the equivalence relation $\sim_\tau$ given by: $z \sim_\tau y$ if and only if $\|z - y\|_\tau = 0$ for all $z, y \in \Theta_1$. We denote $\Theta_1^\tau = (\Theta_1|_{\sim_\tau}, \| \cdot \|_\tau)$ the quotient space, the completion of Θ_1^τ with respect to $\| \cdot \|_\tau$. To every $\Theta_2 \subset \Theta_1$, we associate a sequence $\{\Theta_2^\tau\}$ of subsets $\Theta_2^\tau \subset \Theta_1^\tau$ as follows: for every $z \in \Theta_1$, we denote $[z]_\tau$ the equivalence class of z of subset Θ_1^τ and we defined $\Theta_2^\tau = \{[z]_\tau : z \in \Theta_2\}$. We denote $\overline{\Theta_2^\tau}$, $int_\tau(\Theta_2^\tau)$ and $\partial_\tau \Theta_2^\tau$, respectively, the closure, the interior and the boundary of Θ_2^τ with respect to $\| \cdot \|$ in Θ_1^τ. We suppose that the family of semi-norms $\{\| \cdot \|_\tau\}$ satisfies:

$$\|z\|_1 \leq \|z\|_2 \leq \|z\|_3 \leq \cdots \quad for \ every \ z \in \Theta_1.$$

Before presenting the proof of the main result, we give first the definition of mild solution of the problem (4.1)–(4.2).

Definition 4.1. A function $y \in C([-r, +\infty], E)$ is said to be a mild solution of (4.1)–(4.2) if $y_0 = \Psi(\sigma(y), y)$ for all $t \in [-r, 0]$ and y satisfies

$$y(t) = g(t)\Psi(\sigma(y), y)(0) + \int_0^t g(t-s)\, f(s, y_{\varrho(s,y_s)})\, ds \quad \text{for each } t \in \mathbb{R}_+.$$
(4.3)

Theorem 4.1. *Assume that the following requirements are verified:*

(4.2.1) *There exists a constant $\lambda_1 > 1$ such that*
$$\|g(t)\|_{B(E)} \le \lambda_1 \text{ for every } t \in \mathbb{R}_+;$$

(4.2.2) *There exist a continuous non-decreasing function $\chi : \mathbb{R}_+ \to (0, \infty)$ and $\delta \in L^1_{Loc}(\mathbb{R}_+; \mathbb{R}_+)$ where*
$$\|f(t, y)\| \le \delta(t)\, \chi(\|y\|) \text{ for a.e. } t \in \mathbb{R}_+ \text{ and each } y \in C([-r, 0], E).$$

(4.2.3) *There exists $\lambda_2 > 0$ such that*
$$\|\Psi(\sigma(y), y)\| \le \lambda_2(1 + \|y\|) \quad \text{for each } y \in C([-r, 0], E)$$

(4.2.4) *For all $\Lambda > 0$, there exists $\beta_\Lambda \in L^1_{Loc}([-r, +\infty); \mathbb{R}_+)$ such that*
$$\|f(t, y) - f(t, z)\| \le \beta_\Lambda(t)\|y - z\| \quad \text{for all } y, z \in C([-r, 0], E);$$

(4.2.5) *There exists $\lambda_3 > 0$ such that*
$$|\Psi(\sigma(y), y) - \Psi(\sigma(z), z)| \le \lambda_3\|y - z\| \quad \text{for all } y, z \in C([-r, 0], E),$$
   where
$$\lambda_\tau^* = \int_0^\tau \lambda_1 \beta_\tau(s) ds.$$

If
$$\lambda_1 \lambda_3 + \lambda_\tau^* < 1,$$
then the problem (4.1)–(4.2) has a unique mild solution.

Proof. For every $\tau \in \mathbb{N}$, we define in $C([-r, +\infty), E)$ the semi-norms by:
$$\|y\|_\tau := \sup\{\|y(t)\| : t \in [0, \tau]\}.$$
Now we turn the problem (4.1)–(4.2) into fixed-point problem. Consider the operator $N : C([-r, +\infty), E) \to C([-r, +\infty), E)$ defined by:

$$(Ny)(t) = \begin{cases} \Psi(\sigma(y), y), & \text{if } t \in [-r, 0], \\ g(t)\Psi(\sigma(y), y)(0) + \displaystyle\int_0^t g(t-s)\, f(s, y_{\varrho(s,y_s)})\, ds & \text{if } t \in \mathbb{R}_+. \end{cases}$$
(4.4)

It is clear that the fixed points of N are mild solutions of the problem (4.1)–(4.2). By (4.2.1)–(4.2.3), we have

$$|y(t)| \leq |g(t)||\Psi(\sigma(y), y)(0)| + \int_0^t \|g(t-s)\|_{B(E)}|f(s, y_{\varrho(s,y_s)})|\ ds$$

$$\leq \lambda_1\lambda_2(1 + \|y\|) + \lambda_1 \int_0^t \delta(s)\chi(\|y\|_\tau)ds.$$

Then

$$\|y\| \leq \frac{\lambda_1\lambda_2}{1 - \lambda_1\lambda_2} + \frac{\lambda_1}{1 - \lambda_1\lambda_2} \int_0^t \delta(s)\chi(\|y\|_\tau)ds.$$

Consider the function ϕ given by

$$\phi(t) := \sup\{\|y(s)\| : 0 \leq s \leq t\},\ 0 \leq t \leq +\infty.$$

Let $t^* \in [-r, t]$ be such that $\phi(t) = \|y(t^*)\|$.
If $t^* \in [0, \tau]$, we have

$$\phi(t) \leq \frac{\lambda_1\lambda_2}{1 - \lambda_1\lambda_2} + \frac{\lambda_1}{1 - \lambda_1\lambda_2} \int_0^t \delta(s)\chi(\|\phi\|_\tau)ds,\quad t \in [0, \tau].$$

If $t^* \in [-r, 0]$, then $\phi(t) = \|y_0\|$ and the previous inequality holds.

Let us take the right-hand side of the above inequality as $z(t)$. Then we have

$$\phi(t) \leq z(t)\quad for\ all\ t \in [0, \tau].$$

From the definition of z, we have

$$c := z(0) = \frac{\lambda_1\lambda_2}{1 - \lambda_1\lambda_2}$$

and

$$z'(t) = \frac{\lambda_1}{1 - \lambda_1\lambda_2}\delta(s)\chi(\phi(t))\quad a.e\ t \in [0, \tau].$$

Using the non-decreasing character of χ, we obtain

$$z'(t) \leq \frac{\lambda_1}{1 - \lambda_1\lambda_2}\delta(t)\chi(z(t))\quad a.e\ t \in [0, \tau].$$

By integration, we get

$$z(t) - z(0) \leq \frac{\lambda_1}{1 - \lambda_1\lambda_2} \int_0^t \delta(s)\chi(z(s))ds.$$

Thus

$$z(t) \leq \frac{\lambda_1\lambda_2}{1 - \lambda_1\lambda_2} + \frac{\lambda_1}{1 - \lambda_1\lambda_2} \int_0^t \delta(s)\chi(z(s))ds.$$

Bihari's inequality implies that

$$z(t) \leq \Gamma^{-1}\left(\frac{\lambda_1}{1 - \lambda_1\lambda_2}\int_0^\tau \delta(s)ds\right),$$

where

$$\Gamma(t) = \int_{z(0)}^t \frac{dx}{\chi(x)}.$$

Then there exists a constant Δ_τ such that $z(t) \leq \Delta_\tau, t \in [0, \tau]$ and thus $\phi(t) \leq \Delta_\tau, t \in [0, \tau]$. Since for every $t \in [0, \tau]$, $\|y_t\| \leq \phi(t)$, we have

$$\|y\|_\tau \leq max\{\|y_0\|, \Delta_\tau\} := \lambda_{1\tau}.$$

Set

$$\Theta = \{y \in C([-r, +\infty); E) : sup\{|y(t)| : 0 \leq t \leq \tau\} \leq \lambda_{1\tau}+1 \; for \; all \; \tau \in \mathbb{N}\}.$$

Clearly, Θ is a closed subset of $C([-r, +\infty); E)$.

We will demonstrate that $N : \Theta \to C([-r, +\infty); E)$ is a contraction operator.

Consider $y, \overline{y} \in C([-r, +\infty); E)$; then using $(4.2.1), (4.2.4)$ *and* $(4.2.5)$ for each $t \in [0, \tau]$ and $\tau \in \mathbb{N}$, we get

$$|N(y)(t) - N(\overline{y})(t)| = \Big|g(t)[\Psi(\sigma(y), y)(0) - \Psi(\sigma(\overline{y}), \overline{y})(0)]$$
$$+ \int_0^t g(t - s)[f(s, y_{\varrho(s,y_s)}) - f(s, \overline{y}_{\varrho(s,\overline{y}_s)})]ds\Big|$$
$$\leq \lambda_1\|\Psi(\sigma(y), y)(0) - \Psi(\sigma(\overline{y}), \overline{y})(0)\|$$
$$+ \int_0^t \lambda_1|f(s, y_{\varrho(s,y_s)}) - f(s, \overline{y}_{\varrho(s,\overline{y}_s)})|ds$$
$$\leq \lambda_1\lambda_3\|y - \overline{y}\|_\tau + \int_0^t \lambda_1\beta_\tau\|y(s) - \overline{y}(s)\|ds$$
$$\leq \lambda_1\lambda_3\|y - \overline{y}\|_\tau + \lambda_\tau^*\|y - \overline{y}\|_\tau$$
$$\leq (\lambda_1\lambda_3 + \lambda_\tau^*)\|y - \overline{y}\|_\tau.$$

Therefore,

$$\|N(y) - N(\overline{y})\|_\tau \leq (\lambda_1\lambda_3 + \lambda_\tau^*)\|y - \overline{y}\|_\tau,$$

Consequently, N is a contraction for all $\tau \in \mathbb{N}$. By the choice of Θ there is no $y \in \partial\Theta^\tau$ where $y = \gamma N(y)$ for some $\gamma \in (0, 1)$. Thus the statement $(C2)$ in Theorem 2.13 is not met. As a result of the nonlinear alternative of Frigon and Granas that $(C1)$ is met. We conclude that N has unique fixed-point which is the unique mild solution of the problem (4.1)–(4.2).

4.2.2 An Example

To illustrate our result, we consider the following problem:

$$
\begin{cases}
\dfrac{\partial u}{\partial t}(t,v) - \dfrac{1}{\Gamma(\zeta-1)} \displaystyle\int_t^0 (t-s)^{\zeta-2} L_v u(s,v)\, ds \\
\qquad = Q(t)|u(t-\eta u(t,v),v)|^\delta, \qquad t \in \mathbb{R}_+,\ v \in [0,\pi], \\
u_0(\varkappa,v) = \zeta(u_{\sigma(u)}(\varkappa,v)),\ \varkappa \in [-r,0],\ v \in [0,\pi],
\end{cases}
\tag{4.5}
$$

where $1 < \zeta < 2$, $\eta \in C(\mathbb{R},[0,\infty))$, $\zeta \in C(\mathbb{R},\mathbb{R})$, $\sigma \in C((C[-r,+\infty),E),\mathbb{R}_+)$, $Q : \mathbb{R}_+ \to \mathbb{R}$ is a continuous function and λ_v is the operator with respect to the spatial variable v which is given by:

$$
\lambda_v = \frac{\partial^2}{\partial v^2} - r, \quad \text{with } r > 0.
$$

Consider $E = L^2([0,\pi],\mathbb{R})$ and the operator $A := \lambda_v : D(A) \subset E \to E$ with domain

$$
D(A) := \{u \in E : u, u' \text{ are absolutely continuous}, u'' \in E, u(0) = u(\pi) = 0\,\}.
$$

The operator A is densely defined in E and is sectorial. Thus A is a generator of a solution operator on E.

Set

$$
y(t)(v) = u(t,v), \quad t \in \mathbb{R}_+,\ v \in [0,\pi].
$$

$$
\Psi(t,v) = \zeta(v_t(\varkappa)), \quad t \in \mathbb{R}_+,\ \varkappa \in [-r,0].
$$

$$
f(t,\mu)(v) = Q(t)|u(v)|^\delta, \quad \text{for } t \in \mathbb{R}_+,\ v \in [0,\pi],\ \mu \in E.
$$

Thus, under the above definitions of f and $A(\cdot)$, the system (4.5) can be represented by the problem (4.1)–(4.2). Furthermore, more appropriate conditions on Q ensure the existence of a unique mild solution for (4.5) by Theorem 4.1.

4.3 Abstract Differential Equations with State-Dependent Delay on the Half Line

Motivated by the above-mentioned papers and the works of Büger and Martin [16] and Si and Wang [303], we discuss the existence of mild solutions defined on unbounded interval for a general class of abstract second-order differential equations with state dependent delay of the form:

$$y''(t) = Ay(t) + f(t, y_{\sigma_1(t,y_t)}, y'_{\sigma_2(t,y_t)}), \quad \text{a.e. } t \in \mathbb{R}_+ := [0, +\infty), \quad (4.6)$$

$$y_0 = \Psi \in C([-r, 0], E), \tag{4.7}$$

$$y'(0) = \Psi'(0) \in E, \tag{4.8}$$

where A is the generator of a strongly continuous cosine family of bounded linear operators $(\Phi(t))_{t \in \mathbb{R}}$ on a Banach space $(E, \| \cdot \|)$. $f(\cdot), \sigma_i(\cdot)$ are given functions. In this work we set outs to examine some existence results for abstract second order differential equations with state dependent delay, our method relies on the technique of generalization of the classical fixed point theorem for Fréchet spaces associated with the concept of measures of non-compactness.

4.3.1   *Existence Results*

Let $(S(t))_{t \in \mathbb{R}}$ be the associated sine family given by

$$S(t) := \int_0^t \Phi(s) ds.$$

Let $\Upsilon = \{y \in E : \Phi(\cdot)y \in C^1(\mathbb{R} ; E)\}$. From Kisynski [304], we have that E is a Banach space with the norm

$$\|y\|_\Upsilon = \sup_{t \in [0,+\infty)} \|AS(t)y\| + \|y\|,$$

$AS(t)$ is a bounded linear operator from Υ into E for all $t \in \mathbb{R}_+$. $AS(s)y \to 0$ as $s \to 0$, for $y \in \Upsilon$.

Now, we offer some observations on the following second-order abstract Cauchy problem (see [6] for more details):

$$y''(t) = Ay(t) + \varphi(t), \quad t \in [0, +\infty), \tag{4.9}$$

$$y_0 = \varkappa_1, \quad y'(0) = \varkappa_2, \tag{4.10}$$

where $\varphi : [0, \infty) \to E$ is an integrable function and $\varkappa_1, \varkappa_2 \in E$. The function

$$y(t) = \Phi(t)\varkappa_1 + S(t)\varkappa_2 + \int_0^t S(t - s)\varphi(s) ds, \quad t \in [0, +\infty), \tag{4.11}$$

is called a mild solution of (4.9)–(4.10), and that when $\varkappa_1 \in \Upsilon$, $y(\cdot)$ is a C^1 function on $[0, +\infty)$ and

$$y'(t) = AS(t)\varkappa_1 + \Phi(t)\varkappa_2 + \int_0^t \Phi(t-s)\varphi(s)ds, \quad t \in [0, +\infty).$$

For additional details on second order abstract Cauchy problems, we refer the reader to Fottorini [6].

Let $C(\mathbb{R}_+)$ be the Fréchet space of continuous functions y from $\mathbb{R}_+$ into E, with the family of seminorms

$$\|y\|_j = \sup_{t\in[0,j]} \|y(t)\|; j \in \mathbb{N},$$

and the distance

$$d(y, \varkappa) = \sum_{j=1}^{\infty} 2^{-j} \frac{\|y - \varkappa\|_j}{1 + \|y - \varkappa\|_j}; y, \varkappa \in C(\mathbb{R}_+).$$

The definitions of the notions of family of measures of non-compactness $\mathcal{M}_E$ in the real Fréchet space E and a sequence of measures of non-compactness included in the publications [189, 220, 305, 306] will be used throughout the rest of this paper.

Example 4.1. For $\Omega \in \mathcal{M}_E$, $y \in \Omega, j \in \mathbb{N}$ and $\nu > 0$, let us denote by $\omega^j(y, \nu)$ for $j \in \mathbb{N}$ the modulus of continuity of y on $[0, j]$; that is

$$\omega^j(y, \nu) = \sup\{|y(t) - y(s)| \, t, s \in [0, j], |t - s| \le \nu\}.$$

Further, let us put

$$\omega^j(\Omega, \nu) = \sup\{\omega^j(y, \nu) : y \in \Omega\},$$

$$\omega_0^j(\Omega) = \lim_{\nu \to 0^+} \omega^j(\Omega, \nu),$$

$$\bar{\mu}^j(\Omega) = \sup_{t\in[0,j]} \mu(\Omega(t)) := \sup_{t\in[0,j]} \mu(\{y(t) : y \in \Omega\}),$$

and

$$b_j(\Omega) = \omega_0^j(\Omega) + \bar{\mu}^j(\Omega).$$

Thus, $\{b_j\}_{j\in\mathbb{N}}$ where $b_j : \mathcal{M}_E \to [0, \infty)$ is a family of measures of non-compactness.

For brevity, in the following, we consider $\sigma_1(\cdot) = \sigma_2(\cdot)$ and we note that the case $\sigma_1 \neq \sigma_2$ can be studied.

In this section, we present the main results of the existence of solutions for our problem.

Let us introduce the followings concepts of mild solution of the problem (4.6)–(4.8).

Definition 4.2. A function $y \in C^1([-r, +\infty], E)$ is said to be a mild solution of (4.6)–(4.7) if $y_0 = \Psi$ for all $t \in [-r, 0]$ and y satisfies the following integral equation

$$y(t) = \Phi(t)\Psi(0) + S(t)\Psi'(0) + \int_0^t S(t-s)\, f(s, y_{\sigma(s,y_s)}, y'_{\sigma(s,y_s)})\, ds,\ t \in \mathbb{R}_+.$$

$$(4.12)$$

We can now establish our first result by introducing the following hypotheses:

(4.4.1) There exists $\eta_1 > 1$ where

$$\|S(t)\|_{B(E)} \leq \eta_1 \text{ for each } t \in \mathbb{R}_+.$$

(4.4.2) There exists $\eta_2 > 0$ where

$$\|\Phi(t)\| \leq \eta_2 \text{ for each } t \in \mathbb{R}_+.$$

(4.4.3) The function $t \longmapsto f(t, y, v)$ is measurable on $\mathbb{R}_+$ for each $y, v \in C([-r, 0], E)$, and $(y, v) \longmapsto f(t, y, v)$ is continuous on $C([-r, 0], E) \times C([-r, 0], E)$ for a.e $t \in \mathbb{R}_+$.

(4.4.4) For each $j \in \mathbb{N}$, there exists a function $\Im_j \in L^1([0, j], \mathbb{R}_+)$ and a continuous non-decreasing function $\psi : \mathbb{R}_+ \to (0, \infty)$ where

$$\|f(t, y, v)\| \leq \Im_j(t)\, \psi(\|y\| + \|v\|),$$

for a.e. $t \in [0, j]$ and each $y,\ v \in C([-r, 0], E)$.

(4.4.5) For each bounded sets $B, D \subset C([-r, 0], E)$ and for each $t \in \mathbb{R}_+$, we have

$$\alpha(f(t, B, D)) \leq \Im_j(t) \sup_{s \in [-r, 0]} \alpha(B(s) + D(s)),$$

where α is a Kuratowski measure of non-compactness on the Banach space E.

(4.4.6) For each $j \in \mathbb{N}$, there exists $R_j > 0$ where

$$\max(\eta_2, \eta_1)\Big[\|y_0\| + \|y'_0\| + \psi(R_j)\Im_j{}^*\Big] \leq R_j.$$

For $j \in \mathbb{N}$, let

$$p_j^* = \int_0^j \Im_j(s)ds;$$

and define on $C(([-r, \infty), E)$ the family of measure of non-compactness by

$$\alpha_j(D) = \sup_{t \in [-r,j]} \alpha(D(t)) + w_0^j(D);$$

and

$$D(t) = \{v(t) \in E; v \in D\}; t \in [-r, j].$$

Theorem 4.2. *Assume* (4.4.1)–(4.4.6) *are satisfied, and*

$$\max(\eta_1, \eta_2)\left[4\Im_j^*\right] < 1$$

for each $j \in \mathbb{N}$. Then the problem (4.6)–(4.8) *has at least one mild solution.*

Proof. Consider the operators $N, N' : C^1([-r, +\infty), E) \to C^1([-r, +\infty), E)$ defined by

$$(Ny)(t) = \Phi(t)\Psi(0) + S(t)\Psi'(0) + \int_0^t S(t-s)\, f(s, y_{\sigma(s,y_s)}, y'_{\sigma(s,y_s)})\, ds,\ t \in \mathbb{R}_+, \tag{4.13}$$

and

$$(Ny)'(t) = S(t)\Psi(0) + \Phi(t)\Psi'(0) + \int_0^t \Phi(t-s)\, f(s, y_{\sigma(s,y_s)}, y'_{\sigma(s,y_s)})\, ds,\ t \in \mathbb{R}_+. \tag{4.14}$$

Clearly, the fixed points of the operator N are solutions of the problem (4.6)–(4.8). We define the ball

$$\Omega_{R_j} = \Omega(0, R_j) = \{y \in C^1([-r, +\infty), E) : \|y\|_j + \|y'\|_j \leq R_j\},$$

For any $j \in \mathbb{N}$, and $y \in \Omega_{R_j}$ and $t \in [0, j]$, By (4.4.1)–(4.4.4), we have

$$\|(Ny)(t)\| \leq \|\Phi(t)\Psi(0) + S(t)\Psi'(0) + \int_0^t S(t-s)\, f(s, y_{\sigma(s,y_s)}, y'_{\sigma(s,y_s)})\| ds$$

$$\leq \eta_2\|y_0\| + \eta_1\|y_0'\| + \eta_1 \int_0^t \Im_j(s)\psi(\|y\|_j + \|y'\|_j)ds$$

$$\leq \eta_2\|y_0\| + \eta_1\|y_0'\| + \eta_1\psi(R_j) \int_0^t \Im_j(s)ds$$

$$\leq \eta_2\|y_0\| + \eta_1\|y_0'\| + \eta_1\psi(R_j)\Im_j^*$$

$$\leq R_j.$$

Thus

$$\|(Ny)\|_j \le R_j$$

and

$$\|(Ny)'(t)\| \le \|S(t)\Psi(0) + \Phi(t)\Psi'(0) + \int_0^t \Phi(t-s) \, f(s, y_{\sigma(s,y_s)}, y'_{\sigma(s,y_s)})\|ds$$

$$\le \eta_1\|y_0\| + \eta_2\|y_0'\| + \eta_2 \int_0^t \Im_j(s)\psi(\|y\|_j + \|y'\|_j)ds$$

$$\le \eta_1\|y_0\| + \eta_2\|y_0'\| + \eta_2\psi(R_j) \int_0^t \Im_j(s)ds$$

$$\le \eta_1\|y_0\| + \eta_2\|y_0'\| + \eta_2\psi(R_j)\Im_j{}^*$$

$$\le R_j.$$

Thus

$$\|(Ny)'\|_j \le R_j.$$

Consequently, we deduce that N and N' transform the ball Ω_{R_j} into itself.

We will now demonstrate that $N, N' : \Omega_{R_j} \to \Omega_{R_j}$ satisfy all the assumptions of Theorem 2.14.

Claim 1. $N, N' : \Omega_{R_j} \to \Omega_{R_j}$ are continuous.
Let $\{y^\beta\}_{\beta \in \mathbb{N}}$ be a sequence where $y^\beta \to y$ in Ω_{R_j}. Then for $t \in [0, j]$, we get

$$\|N(y^\beta)(t) - N(y)(t)\|$$

$$= \left\| \int_0^t S(t-s)\Big[f(s, y^\beta_{\sigma(s,y_s)}, y'^\beta_{\sigma(s,y_s)}) - f(s, y_{\sigma(s,y_s)}, y'_{\sigma(s,y_s)})\Big]ds \right\|$$

$$\le \|S(t-s)\|_{B(E)} \int_0^t \left\| f(s, y^\beta_{\sigma(s,y_s)}, y'^\beta_{\sigma(s,y_s)}) - f(s, y_{\sigma(s,y_s)}, y'_{\sigma(s,y_s)}) \right\| ds$$

$$\le \eta_1 \int_0^t \| f(s, y^\beta_{\sigma(s,y_s)}, y'^\beta_{\sigma(s,y_s)}) - f(s, y_{\sigma(s,y_s)}, y'_{\sigma(s,y_s)}) \| \, ds.$$

Since $y^\beta \longrightarrow y$ as $\beta \longrightarrow \infty$, then

$$\|N(y^\beta) - N(y)\|_j \longrightarrow 0 \text{ as } \beta \longrightarrow +\infty.$$

and

$$\|N'(y^\beta)(t) - N'(y)(t)\|$$

$$= \left\| \int_0^t \Phi(t-s)\Big[f(s, y^\beta_{\sigma(s,y_s)}, y'^\beta_{\sigma(s,y_s)}) - f(s, y_{\sigma(s,y_s)}, y'_{\sigma(s,y_s)})\Big]ds \right\|$$

$$\leq \|\Phi(t-s)\|_{B(E)} \int_0^t \left\| f(s, y^\beta_{\sigma(s,y_s)}, y'^\beta_{\sigma(s,y_s)}) - f(s, y_{\sigma(s,y_s)}, y'_{\sigma(s,y_s)}) \right\| ds$$

$$\leq \eta_2 \int_0^t \left\| f(s, y^\beta_{\sigma(s,y_s)}, y'^\beta_{\sigma(s,y_s)}) - f(s, y_{\sigma(s,y_s)}, y'_{\sigma(s,y_s)}) \right\| ds.$$

Since $y^\beta \longrightarrow y$ as $\beta \longrightarrow \infty$, then

$$\|N'(y^\beta) - N'(y)\|_j \longrightarrow 0 \ \text{ as } \ \beta \longrightarrow +\infty.$$

Thus N and N' are continuous.

Claim 2. $N(\Omega_{R_j})$ and $N'(\Omega_{R_j})$ are bounded.
Since $N(\Omega_{R_j}) \subset \Omega_{R_j}, N'(\Omega_{R_j}) \subset \Omega_{R_j}$ and Ω_{R_j} is bounded, then $N(\Omega_{R_j})$ and $N'(\Omega_{R_j})$ are bounded.

Claim 3. For each equicontinuous subset Θ of $\Omega_{R_j}, \alpha_j(N(\Theta)) \leq l_j \alpha_j(\Theta)$
From Lemma 2.5 and 2.6, for any equicontinuous set $\Theta \subset \Omega_{R_j}$ and any $\kappa > 0$, there exist $\{\varkappa_\beta\}_{\beta=1}^\infty \subset \Theta$, where for $t \in [0,j]$, we get

$$\alpha((N\Theta)(t)) = \alpha\left(\left\{ \Phi(t)\Psi(0) + S(t)\Psi'(0) \right.\right.$$

$$\left.\left. + \int_0^t S(t-s) \ f(s, y_{\sigma(s,y_s)}, y'_{\sigma(s,y_s)}) ds; \ y, y' \in \Theta \right\} \right)$$

$$\leq \alpha\left(\left\{ \int_0^t S(t-s) \ f(s, y_{\sigma(s,y_s)}, y'_{\sigma(s,y_s)}) ds; \ y, y' \in \Theta \right\} \right)$$

$$\leq 2\alpha\left(\left\{ \int_0^t S(t-s) f(s, y^\beta_{\sigma(s,y^\beta_s)}, y'^\beta_{\sigma(s,y^\beta_s)}) ds \right\}_{\beta=1}^\infty \right) + \kappa$$

$$\leq 4 \int_0^t \alpha\left(\|S(t-s)\|_{B(E)} \left\{ f(s, y^\beta_{\sigma(s,y^\beta_s)}, y'^\beta_{\sigma(s,y^\beta_s)}) \right\}_{\beta=1}^\infty \right) ds + \kappa$$

$$\leq 4\eta_1 \int_0^t \alpha\left\{ f(s, y^\beta_{\sigma(s,y^\beta_s)}, y'^\beta_{\sigma(s,y^\beta_s)}) \right\}_{\beta=1}^\infty \right) ds + \kappa$$

$$\leq 4\eta_1 \int_0^t \Im_j(s)\alpha\left(\left\{ y^\beta_{\sigma(s,y^\beta_s)} \right\}_{\beta=1}^\infty + \left\{ y'^\beta_{\sigma(s,y^\beta_s)} \right\}_{\beta=1}^\infty \right) ds + \kappa$$

$$\leq 4\eta_1\alpha(\Theta) \int_0^t \Im_j(s) ds + \kappa$$

$$\leq 4\eta_1 \Im_j{}^* \alpha_j(\Theta) + \kappa.$$

Since $\kappa > 0$ is arbitrary, then

$$\alpha((N\Theta)(t)) \leq 4\eta_1 \Im_j^* \alpha_j(\Theta),$$

Thus

$$\alpha_j(N(\Theta)) \leq 4\eta_1 \Im_j^* \alpha_j(\Theta)$$

and

$$\alpha((N'\Theta)(t)) = \alpha\left(\left\{ S(t)\Psi(0) + \Phi(t)\Psi'(0) \right.\right.$$

$$\left.\left. + \int_0^t \Phi(t-s) \, f(s, y_{\sigma(s,y_s)}, y'_{\sigma(s,y_s)})ds; \ y, y' \in \Theta \right\} \right)$$

$$\leq \alpha\left(\left\{ \int_0^t \Phi(t-s) \, f(s, y_{\sigma(s,y_s)}, y'_{\sigma(s,y_s)})ds; \ y, y' \in \Theta \right\} \right)$$

$$\leq 2\alpha\left(\left\{ \int_0^t \Phi(t-s) f(s, y^\beta_{\sigma(s,y_s^\beta)}, y'^\beta_{\sigma(s,y_s^\beta)})ds \right\}_{\beta=1}^\infty \right) + \kappa$$

$$\leq 4\int_0^t \alpha\left(\|\Phi(t-s)\|_{B(E)} \left\{ f(s, y^\beta_{\sigma(s,y_s^\beta)}, y'^\beta_{\sigma(s,y_s^\beta)}) \right\}_{\beta=1}^\infty \right)ds + \kappa$$

$$\leq 4\eta_2 \int_0^t \alpha\left\{ f(s, y^\beta_{\sigma(s,y_s^\beta)}, y'^\beta_{\sigma(s,y_s^\beta)}) \right\}_{\beta=1}^\infty \right)ds + \kappa$$

$$\leq 4\eta_2 \int_0^t \Im_j(s)\alpha\left(\left\{ y^\beta_{\sigma(s,y_s^\beta)} \right\}_{\beta=1}^\infty + \left\{ y'^\beta_{\sigma(s,y_s^\beta)} \right\}_{\beta=1}^\infty \right)ds + \kappa$$

$$\leq 4\eta_2\alpha(\Theta) \int_0^t \Im_j(s)ds + \kappa$$

$$\leq 4\eta_2 \Im_j^* \alpha_j(\Theta) + \kappa.$$

Since $\kappa > 0$ is arbitrary, then

$$\alpha((N'\Theta)(t)) \leq 4\eta_2 \Im_j^* \alpha_j(\Theta),$$

Thus

$$\alpha_j(N'(\Theta)) \leq 4\eta_2 \Im_j^* \alpha_j(\Theta).$$

Consequently, Theorem 2.14 implies that N and N' have at least one fixed point in Ω_{R_j} which is a mild solution of (4.6)–(4.7). □

4.3.2   An Example

We consider the following abstract differential equation with state dependent delay:

$$\begin{cases} \dfrac{\partial^2 y}{\partial t^2}(t,\gamma) = \dfrac{\partial^2 y}{\partial \gamma^2}(t,\gamma) + f\left(t, y(t - \zeta(t,y(t))), y'(t - \zeta(t,y(t)),\gamma)\right), \\ \qquad t \in [0,+\infty), \quad \gamma \in [0,\pi], \\ y(t,0) = y(t,\pi) = 0, \ t \in [0,+\infty), \\ y(s,\gamma) = \Psi(s,\gamma), \quad s \in [-r,0], \\ y'(0) = \Psi'(0), \end{cases}$$

$$(4.15)$$

where $\Psi \in C^1([-r,0];E)$, $\zeta \in C([0,+\infty) \times \mathbb{R}; \mathbb{R}_+)$, $\zeta(0,\Psi) = 0$ and $f \in C([0,+\infty) \times \mathbb{R} \times \mathbb{R}, \mathbb{R})$. To make system (4.15) as problem (4.6)–(4.8), we need to define $f : [0,+\infty) \times \mathbb{R} \times \mathbb{R} \to E$ and $\sigma : [0,+\infty) \times C([-r,0];\mathbb{R}) \to \mathbb{R}$ by

$$f(t,\lambda_1,\lambda_2)(\gamma) = f(t,\lambda_1(0,\gamma),\lambda_2(0,\gamma))$$

and

$$\sigma(s,\lambda_1) = s - \zeta(s,\lambda_1(0))$$

It is easy to show that $f(\cdot)$ and $\sigma(\cdot)$ are Lipschitz.

Consider $E = L^2([0,\pi],\mathbb{R})$ and the domain

$$D(A) := \{\, y \in E \ : y'' \in E, \ y(0) = y(\pi) = 0 \,\}.$$

Let $A : D(A) \subset E \to E$ be the operator given by $Ay = y''$ Clearly A is the infinitesimal generator of a strongly continuous cosine family $(\Phi(t))_{t \in \mathbb{R}}$ on E. The spectrum of A is discrete with eigenvalues $-j^2, j \in \mathbb{N}$, and eigenvectors

$$\varkappa_j(\zeta) := \left(\frac{1}{\pi}\right)^{\frac{1}{2}} \sin(j\zeta).$$

The set $\{\varkappa_j : j \in \mathbb{N}\}$ is an orthonormal basis of E. We have

$$\Phi(t)y = \sum_{j-1}^{\infty} \cos(jt)\langle y, \varkappa_j\rangle \varkappa_j,$$

$$S(t)y = \sum_{j=1}^{\infty} \frac{\sin(jt)}{j}\langle y, \varkappa_j\rangle \varkappa_j,$$

the sine family $S(\cdot)$ is compact, $\|\Phi(t)\| = \|S(t)\| = 1$ for all $t \in \mathbb{R}$.

Consequently, Theorem 4.2 implies that the system (4.15) has at least one mild solution on $[-r,+\infty)$.

4.4 Controllability Results for Second-Order Integro-Differential Equations

In this section, inspired by the above works, we shall discuss the approximate controllability and complete controllability for second-order integro-differential equations with state-dependent delay described in the form:

$$
\begin{cases}
y''(t) = A(t)y(t) + f\left(t, y_{\rho(t,y_t)}, (\sigma y)(t)\right) \\
\qquad + \displaystyle\int_0^t \Upsilon(t,s)y(s)ds + \mathcal{P}u(t), \text{ if } t \in J, \\
y'(0) = \zeta_0 \in E, \quad y(t) = \Psi(t), \text{ if } t \in \mathbb{R}_-,
\end{cases}
\tag{4.16}
$$

where $J = [0,T]$, $A(t) : D(A(t)) \subset E \to E$, $\Upsilon(t,s)$ are closed linear operators on E, with dense domain $D(A(t))$, which is independent of t, and $D(A(s)) \subset D(\Upsilon(t,s))$, the operator σ is defined by

$$
(\sigma y)(t) = \int_0^T \Xi(t,s,y(s))ds, \quad a > 0,
$$

the nonlinear terms $\Xi : J \times J \times E \to E$, $f : J \times \mathbb{B} \times E \to E$, $\Psi : \mathbb{R}_- \to E$, $\rho : J \times \mathbb{B} \to (-\infty, \infty)$, are a given functions, the control function u is given in $L^2(J, U)$ Banach space of admissible control with U as a Banach space. $\mathcal{P}$ is a bounded linear operator from U into E, and $(E, \|\cdot\|)$ is a Banach space.

4.4.1 *Existence Results*

Next, we consider the second-order integro-differential systems

$$
\begin{aligned}
z''(t) &= A(t)z(t) + \int_s^t \Upsilon(t,\tau)z(\tau)d\tau, \quad s \le t \le T, \\
z(s) &= 0, \quad z'(s) = x \in E,
\end{aligned}
\tag{4.17}
$$

for $0 \le s \le T$. This problem was discussed in [307]. We denote $\Delta = D_\Xi = \{(t,s) : 0 \le s \le t \le T\}$. We now introduce some conditions fulfilling the operator Υ:

(B1) For each $0 \le s \le t \le T$, $\Upsilon(t,s) : D(A) \to E$ is a bounded linear operator, for every $z \in D(A)$, $\Upsilon(\cdot, \cdot)z$ is continuous and

$$
\|\Upsilon(t,s)z\| \le b\|z\|_{[D(A)]},
$$

for $b > 0$ which is a constant independent of $(s,t) \in \Delta$.

(B2) There exists $L_\Upsilon > 0$ such that
$$\|\Upsilon(t_2, s) z - \Upsilon(t_1, s) z\| \le L_\Upsilon |t_2 - t_1| \|z\|_{[D(A)]},$$
for all $z \in D(A), 0 \le s \le t_1 \le t_2 \le T$.

(B3) There exists $b_1 > 0$ such that
$$\left\| \int_\sigma^t S(t, s) \Upsilon(s, \sigma) z ds \right\| \le b_1 \|z\|, \text{ for all } z \in D(A).$$

Under these conditions, it has been established that there exists a resolvent operator $(\mathcal{Q}(t, s))_{t \ge s}$ associated with the systems (4.17). From now on, we are going to consider that such a resolvent operator exists, and we adopt its properties as a definition.

Definition 4.3 ([307]). A family of bounded linear operators $(\mathcal{Q}(t, s))_{t \ge s}$ on E is said to be a resolvent operator for the systems (4.17) if it satisfies:

(a) The map $\mathcal{Q} : \Delta \to \mathcal{L}(E)$ is strongly continuous, $\mathcal{Q}(t, \cdot)z$ is continuously differentiable for all $z \in E, \mathcal{Q}(s, s) = 0, \frac{\partial}{\partial t} \mathcal{Q}(t, s)\big|_{t=s} = I$ and $\frac{\partial}{\partial s} \mathcal{Q}(t, s)\big|_{s=t} = -I$.

(b) Assume $x \in D(A)$. The function $\mathcal{Q}(\cdot, s)x$ is a solution for the systems (6) and (7). This means that
$$\frac{\partial^2}{\partial t^2} \mathcal{Q}(t, s)x = A(t)\mathcal{Q}(t, s)x + \int_s^t \Upsilon(t, \tau)\mathcal{Q}(\tau, s)x d\tau,$$
for all $0 \le s \le t \le T$.

It follows from condition (a) that there are constants $M_\mathcal{Q} > 0$ and $\widetilde{M_\mathcal{Q}} > 0$ such that
$$\|\mathcal{Q}(t, s)\| \le M_\mathcal{Q}, \quad \left\| \frac{\partial}{\partial s} \mathcal{Q}(t, s) \right\| \le \widetilde{M_\mathcal{Q}}, \quad (t, s) \in \Delta.$$

Moreover, the linear operator
$$G(t, \tau)x = \int_\tau^t \Upsilon(t, s)\mathcal{Q}(s, \tau)x ds, x \in D(A), 0 \le \tau \le t \le T,$$
can be extended to E. Portraying this expansion by the similar notation $G(t, \tau), G : \Delta \to \mathcal{L}(E)$ is strongly continuous, and it is verified that
$$\mathcal{Q}(t, \tau)x = S(t, \tau)x + \int_\tau^t S(t, s)G(s, \tau)x ds, \text{ for all } x \in E.$$

It follows from this property that $\mathcal{Q}(\cdot)$ is uniformly Lipschitz continuous, that is, there exists a constant $L_\mathcal{Q} > 0$ such that
$$\|\mathcal{Q}(t + h, \tau) - \mathcal{Q}(t, \tau)\| \le L_\mathcal{Q}|h|, \text{ for all } t, t + h, \tau \in [0, T].$$

In this paper, we assume that the state space $(\mathcal{B}, \|\cdot\|_\mathcal{B})$ is a seminormed linear space of functions mapping $(-\infty, 0]$ into $\mathbb{R}$, and satisfying the following fundamental axioms which were introduced by Hale and Kato in [308].

(A_1) If $y \in C$ and $y_0 \in \mathcal{B}$, then for every $t \in J$, the following conditions hold:

   (i) $y_t \in \mathcal{B}$,
   (ii) There exists a positive constant H such that $|y(t)| \leq H \|y_t\|_{\mathcal{B}}$,
   (iii) There exist two functions $L(\cdot)$ and $M(\cdot) : \mathbb{R}_+ \to \mathbb{R}_+$ independent of y with L continuous and bounded and M locally bounded such that:

$$\|y_t\|_{\mathcal{B}} \leq L(t) \sup\{|y(s)| : 0 \leq s \leq t\} + M(t) \|y_0\|_{\mathcal{B}}.$$

(A_2) For the function y in (A_1), y_t is a $\mathcal{B}$ - valued continuous function on $\mathbb{R}^+$.

(A_3) The space $\mathcal{B}$ is complete.

Denote

$$L_* = \sup\{L(t) : t \in J\},$$

$$M_* = \sup\{M(t) : t \in J\},$$

and

$$N = \max\{L_*, M_*\}.$$

We define the space

$$C_\theta := \{\phi \in C(\mathbb{R}^-, E) : \lim_{\tau \to -\infty} \phi(\tau) \ \text{exist in} \ E\},$$

endowed with the norm

$$\|\phi\|_\theta = \sup\{|\phi(\tau)| : \tau \leq 0\}.$$

Then, the axioms (A_1) $-$ (A_3) are satisfied in the space C_θ. So in all what follows, we consider the phase space $\mathcal{B} = C_\theta$, and let

$$\mathcal{X} = C(\widetilde{J}, E) = \{y : \ \widetilde{J} \ \to E \ : \ y|_{\mathbb{R}^-} \in \mathcal{B}, \ y|_J \in C(J, E)\},$$

such that

$$\|y\|_{\mathcal{X}} = \sup_{t \in \widetilde{J}} \{\|y(t)\|\}.$$

In this part, we prove the existence of mild solutions of the problem:

$$\begin{cases} y''(t) = Ay(t) + f\left(t, y_{\rho(t, y_t)}, (\sigma y)(t)\right) + \displaystyle\int_0^t \Upsilon(t, s)y(s)ds, \ \text{if } t \in J, \\ y'(0) = \zeta_0 \in E, \ \ y(t) = \Psi(t), \ \text{if } t \in \mathbb{R}_-. \end{cases}$$

$$(4.18)$$

In [300], the authors have investigated the existence of mild solution of system (4.18) and they used the Leray-Schauder's alternative theorem and Krasnoselskii's theorem. So we will weaken the conditions (in particular the compactness property) by using Darbo fixed point theorem.

Definition 4.4. A function $y \in \mathcal{X}$ is called a mild solution of problem (4.18), if it satisfies

$$
y(t) = \begin{cases}
-\dfrac{\partial \mathcal{Q}(t,s)\Psi(0)}{\partial s}\bigg|_{s=0} + \mathcal{Q}(t,0)\zeta_0 \\[2mm]
+ \displaystyle\int_0^t \mathcal{Q}(t,s)f(s, y_{\rho(s,y_s)}, (\sigma y)(s))ds; & \text{if } t \in J, \\[4mm]
\Psi(t); & \text{if } t \in \mathbb{R}_-.
\end{cases}
$$

The following assumption will be needed throughout the paper:

(4.7.1) $f : J \times \mathcal{B} \times E \to E$ is a Carathéodory function and there exist positive constants ξ_1, ξ_2 and continuous non-decreasing functions $\psi_f^1, \psi_f^2 : J \to (0, +\infty)$ such that:

$$
\|f(t, y_1, y_2)\| \leq \xi_1 \psi_f^1(\|y_1\|_{\mathcal{B}}) + \xi_2 \psi_f^2(\|y_2\|), \quad \text{for } y_1 \in \mathcal{B}, \ y_2 \in E.
$$

And there exists a positive constant l_f, such that for any bounded set $B \subset E$, and $B_t \in \mathcal{B}$ and each $t \in \mathbb{R}$, we have

$$
\mu(f(t, B_t, \sigma(B(t)))) \leq l_f \mu(B).
$$

(4.7.2) The function $\Xi : D_\Xi \times E \to E$ is continuous and there exists $\Xi_{c_1} > 0$, such that

$$
\|\Xi(t, s, y_1) - \Xi(t, s, y_2)\| \leq \Xi_{c_1} \|y_1 - y_2\|,
$$

for each $(t, s) \in D_\Xi$ and $y_1, y_2 \in E$, where

$$
\sup_{D_\Xi}\{\|\Xi(t, s, 0)\|\} = \Xi^* < \infty.
$$

(4.7.3) Assume that $(B1) - (B3)$ hold, and there exist $M_\mathcal{Q}, \widetilde{M_\mathcal{Q}} \geq 1$ and $\mu \geq 0$, such that

$$
\|\mathcal{Q}(t,s)\|_{\Upsilon(E)} \leq M_\mathcal{Q} e^{-\mu t},
$$

and

$$
\left\|\frac{\partial \mathcal{Q}(t,s)}{\partial s}\right\|_{\Upsilon(E)} \leq \widetilde{M_\mathcal{Q}} e^{-\mu t}.
$$

(4.7.4) Set $\mathcal{R}\left(\rho^-\right) = \{\rho(s,\varphi) : (s,\varphi) \in J \times \mathcal{B}, \rho(s,\varphi) \leq 0\}$. We assume that $\rho : J \times \mathcal{B} \to \mathbb{R}$ is continuous. Moreover we assume the following assumption:

• (4.7.4.1) The function $t \to \Psi_t$ is continuous from $\mathcal{R}\left(\rho^-\right)$ into $\mathcal{B}$ and there exists a continuous and bounded function $L^\Psi : \mathcal{R}\left(\rho^-\right) \to (0, \infty)$ such that

$$\|\Psi_t\|_{\mathcal{B}} \leq L^\Psi(t)\|\Psi\|_{\mathcal{B}}, \quad \text{for every } t \in \mathcal{R}\left(\rho^-\right).$$

Remark 4.1. The condition (4.7.4.1) is frequently verified by continuous and bounded functions. For more details, see for instance [309].

Lemma 4.1 ([310]). *If $y : (-\infty, +\infty) \to E$ is a function such that $y_0 = \Psi$, then*

$$\|y_s\|_{\mathcal{B}} \leq \left(M + \mathcal{L}^\Psi\right)\|\Psi\|_{\mathcal{B}} + l \sup\{|y(\theta)| \; ; \; \theta \in [0, \max\{0, s\}]\}, \quad s \in \mathcal{R}\left(\rho^-\right) \cup J,$$

where $\mathcal{L}^\Psi = \sup_{t \in \mathcal{R}(\rho^-)} L^\Psi(t)$.

Theorem 4.3. *Assume that the conditions (4.7.1)–(4.7.3) and (4.7.4) are satisfied. Then, the system (4.18) has at least one mild solution.*

Proof. Firstly we define on $\mathcal{X}$ measures of non compactness by

$$\mu_C(S) = \omega_0(S) + \sup\left\{e^{-\tau \Sigma(t)}\mu(S(t))\right\},$$

with $\tau > 1$, $\Sigma(t) = 4M_\varrho l_f t$, $S(t) = \{v(t) \in E \; ; \; v \in S\}$, and $\omega^T(v, \epsilon)$ denote the modulus of continuity of the function v on the interval $[-T, T]$, namely,

$$\omega^T(v, \epsilon) = \sup\{\|e^{-\kappa_1}v(\kappa_1) - e^{-\kappa_2}v(\kappa_2)\|; \kappa_1, \kappa_2 \in [-T, T], |\kappa_1 - \kappa_2| \leq \epsilon\},$$
$$\omega^T(S, \epsilon) = \sup\{\omega^T(v, \epsilon) \; ; \; v \in S\},$$
$$\omega_0(S) = \lim_{\epsilon \to 0}\{\omega^T(S, \epsilon)\}.$$

Notice that if the set S is equicontinuous, then $\omega_0(S) = 0$.

Now, transform the problem (4.18) into a fixed point problem and define the operator $\Theta_1 : \mathcal{X} \to \mathcal{X}$ by:

$$\Theta_1 y(t) = \begin{cases} \left. -\dfrac{\partial \mathcal{Q}(t,s)\Psi(0)}{\partial s}\right|_{s=0} + \mathcal{Q}(t,0)\zeta_0 \\[2ex] +\displaystyle\int_0^t \mathcal{Q}(t,s)f(s, y_{\rho(s,y_s)}, (\sigma y)(s))ds; \text{ if } t \in J, \\[2ex] \Psi(t), \text{ if } t \in \mathbb{R}_-. \end{cases} \qquad (4.19)$$

Let $x(\cdot) : (-\infty, T] \to E$ be the function defined by:

$$x(t) = \begin{cases} -\dfrac{\partial \mathcal{Q}(t,s)\Psi(0)}{\partial s}\bigg|_{s=0} + \mathcal{Q}(t,0)\zeta_0, & \text{if } t \in J, \\ \Psi(t), & \text{if } t \in \mathbb{R}_-. \end{cases}$$

Then, $x_0 = \Psi$, and for each $w \in \mathcal{X}$, with $w(0) = 0$, we denote by $\overline{w}$ the function

$$\overline{w}(t) = \begin{cases} w(t), & \text{if } t \in \mathbb{R}^+, \\ 0, & \text{if } t \in \mathbb{R}_-. \end{cases}$$

If y satisfies (4.19), we can decompose it as $y(t) = w(t) + x(t)$, which implies $y_t = w_t + x_t$, and the function $w(\cdot)$ satisfies

$$w(t) = \int_0^t \mathcal{Q}(t,s) f(s, w_{\rho(s,w_s + x_s)} + x_{\rho(s,w_s + x_s)}, \sigma(w+x)(s)) ds; \text{ if } t \in J.$$

Set

$$\Omega = \{ w \in \mathcal{X} \ : \ w(0) = 0 \}.$$

Let the operator $\widetilde{\Theta}_1 : \Omega \to \Omega$ defined by

$$\widetilde{\Theta}_1 w(t) = \int_0^t \mathcal{Q}(t,s) f(s, w_{\rho(s,w_s + x_s)} + x_{\rho(s,w_s + x_s)}, \sigma(w+x)(s)) ds, \text{ if } t \in J.$$

The operator Θ_1 has a fixed point is equivalent to say that $\widetilde{\Theta}_1$ has one, so it turns to prove that $\widetilde{\Theta}_1$ has a fixed point. We shall check that operator $\widetilde{\Theta}_1$ satisfies all conditions of Darbo's theorem.

Let $\Pi_{\theta'} = \{ w \in \Omega : \|w\|_\Omega \leq \theta' \}$, with

$$M_{\mathcal{Q}}\big(\xi_1 \psi_f^1(\eta_{\theta'}^*) + \xi_2 \psi_f^2(\overline{\eta}^*)\big)T \leq \theta',$$

such that $\eta_{\theta'}^*$, $\overline{\eta}^*$ are constants, they will be specific later.

The set $\Delta_{\theta'}$ is bounded, closed and convex. We have divided the proof into four steps.

Step 1: $\widetilde{\Theta}_1(\Pi_{\theta'}) \subset \Pi_{\theta'}$.

For $w \in \Pi_{\theta'}$, $t \in J$ and by (4.7.1)–(4.7.3), we have

$$\big\| w_{\rho(s,w_s + x_s)} + x_{\rho(s,w_s + x_s)} \big\|_{\mathcal{B}}$$

$$\leq \big\| w_{\rho(s,w_s + x_s)} \big\|_{\mathcal{B}} + \big\| x_{\rho(s,w_s + x_s)} \big\|_{\mathcal{B}}$$

$$\leq L(t) \sup_{[0,s]} |w(t)| + \big(M(t) + \mathcal{L}^\Psi\big)\|\Psi\|_{\mathcal{B}} + L(t) \sup_{[0,s]} \|x(\theta)\|$$

$$\leq L_*\theta' + (M_* + \mathcal{L}^\Psi)\|\Psi\|_\mathcal{B}$$
$$+ L_*\big(\widetilde{M_\mathcal{Q}}\|\Psi_0\| + M_\mathcal{Q}\|\zeta_0\|\big)H\|\Psi\|_\mathcal{B}$$
$$\leq L_*\theta' + \Big[M_* + \mathcal{L}^\Psi + L_*\big(\widetilde{M_\mathcal{Q}}\|\Psi_0\| + M_\mathcal{Q}\|\zeta_0\|\big)H\Big]\|\Psi\|_\mathcal{B}$$
$$= \eta^*_{\theta'},$$

and

$$\|\sigma(w+x)(s)\| \leq a\Xi_{c_1}\big(\theta' + \widetilde{M_\mathcal{Q}}\|\Psi_0\| + M_\mathcal{Q}\|\zeta_0\|\big) + a\Xi^* = \overline{\eta}^*.$$

Then,

$$\|\widetilde{\Theta}_1 w(t)\| \leq M_\mathcal{Q}\Big[\psi_f^1(\eta^*_{\theta'})\xi_1 + \psi_f^2(\overline{\eta}^*)\xi_2\Big]T.$$

Thus,

$$\|\widetilde{\Theta}_1 w\|_\Omega \leq \theta'.$$

Therefore $\widetilde{\Theta}_1(\Pi_{\theta'}) \subset \Pi_{\theta'}$, implies that $\widetilde{\Theta}_1(\Pi_{\theta'})$ is bounded.

Step 2: $\widetilde{\Theta}_1$ is continuous.

Let $\{w_m\}_{m\in\mathbb{N}}$ be a sequence such that $w_m \to w^*$ in $\Pi_{\theta'}$. At the first, we study the convergence of the sequences $\big(w^m_{\rho(s,w_s^m)}\big)_{m\in\mathbb{N}}$, $s \in J$. If $s \in J$ is such that $\rho(s, w_s) > 0$, then we have

$$\left\|w^m_{\rho(s,w_s^m)} - w^*_{\rho(s,w_s^*)}\right\|_\mathcal{B} \leq \left\|w^m_{\rho(s,w_s^m)} - w^*_{\rho(s,w_s^m)}\right\|_\mathcal{B} + \left\|w^*_{\rho(s,w_s^m)} - w^*_{\rho(s,w_s^*)}\right\|_\mathcal{B}$$
$$\leq L\|w_m - w^*\| + \left\|w^*_{\rho(s,w_s^m)} - w^*_{\rho(s,w_s^*)}\right\|_\mathcal{B},$$

which proves that $w^m_{\rho(s,w_s^m)} \to w^*_{\rho(s,w_s)}$ in $\mathcal{B}$, as $m \to \infty$, for every $s \in J$ such that $\rho(s, w_s) > 0$. Similarly, if $\rho(s, w_s) < 0$, we get

$$\left\|w^m_{\rho(s,w_s^m)} - w^*_{\rho(s,w_s)}\right\|_\mathcal{B} = \left\|\Psi^m_{\rho(s,w_s^m)} - \Psi_{\rho(s,w_s^*)}\right\|_\mathcal{B} = 0,$$

which also shows that $w^m_{\rho(s,w_s^m)} \to w^*_{\rho(s,w_s)}$ in $\mathcal{B}$, as $m \to \infty$, for every $s \in J$ such that $\rho(s, w_s) < 0$. Then for $t \in J$, we have

$$\|(\widetilde{\Theta}_1 w^m)(t) - (\widetilde{\Theta}_1 w^*)(t)\|$$
$$\leq M_\mathcal{Q} \int_0^t \|f(s, w^m_{\rho(s,w_s^m)} + x_{\rho(s,w_s^m+x_s)}, H(w^m + x)(s))$$
$$- f(s, (w^*_{\rho(s,w_s^*)} + x_{\rho(s,w_s^*+x_s)}), H(w^* + x)(s))\|ds.$$

Since Ξ and f are continuous, we obtain that

$$\Xi(t, s, (w^m + x)(s)) \to \Xi(t, s, (w^* + x)(s)), \quad as \ m \to +\infty,$$

and

$$\|\Xi(t, s, (w^m + x)(s)) - \Xi(t, s, (w^* + x)(s))\| \leq \Xi_{c_1}^* \|w^m(s) - w^*(s)\|.$$

By the Lebesgue dominated convergence theorem, we have

$$\int_0^t \Xi(t, s, (w^m + x)(s))ds \xrightarrow[m \to +\infty]{} \int_0^t \Xi(t, s, (w^* + x)(s))ds.$$

Then, by (4.7.1), we get

$$f(s, w^m_{\rho(s,w^m_s)} + x_{\rho(s,w^m_s + x_s)}), \sigma(w^m + x)(s))$$

$$\xrightarrow[m \to +\infty]{} f(s, (w^*_{\rho(s,w^*_s)} + x_{\rho(s,w^*_s + x_s)}), \sigma(w^* + x)(s)).$$

By Lebesgue dominated convergence theorem, we obtain

$$\|(\widetilde{\Theta}_1 w^m)(t) - (\widetilde{\Theta}_1 w^*)(t)\| \to 0, \quad as \ m \to +\infty.$$

Thus, $\widetilde{\Theta}_1$ is continuous.

Step 3: $\widetilde{\Theta}_1$ is μ_C-contraction.
Let Π be a bounded equicontinuous subset of $\Pi_{\theta'}$, $w \in \Pi$, and $\kappa_1, \kappa_2 \in J$, with $\kappa_2 > \kappa_1$, we have

$$\left\| \widetilde{\Theta}_1 w(\kappa_1) - \widetilde{\Theta}_1 w(\kappa_2) \right\|$$

$$\leq \int_{\kappa_1}^{\kappa_2} \|\mathcal{Q}(\kappa_2, s)\| \left\| f(s, w_{\rho(s,w_s + x_s)} + x_{\rho(s,w_s + x_s)}), \sigma(w + x)(s)) \right\| ds$$

$$+ \int_0^{\kappa_1} \|\mathcal{Q}(\kappa_2, s) - \mathcal{Q}(\kappa_1, s)\|$$

$$\times \left\| f(s, w_{\rho(s,w_s + x_s)} + x_{\rho(s,w_s + x_s)}), \sigma(w + x)(s)) \right\| ds$$

$$\leq \left[\psi_f^1(\eta_{\theta'}^*)\xi_1 + \psi_f^2(\overline{\eta}^*)\xi_2 \right]$$

$$\times \left(M_{\mathcal{Q}}|\kappa_2 - \kappa_1| + \int_0^{\kappa_1} \|\mathcal{Q}(\kappa_2, s) - \mathcal{Q}(\kappa_1, s)\| ds \right).$$

By the strong continuity of $\mathcal{Q}(\cdot)$, we get

$$\left\| \widetilde{\Theta}_1 w(\kappa_1) - \widetilde{\Theta}_1 w(\kappa_2) \right\| \to 0, \ as \ \kappa_1 \to \kappa_2.$$

Thus $\widetilde{\Theta}_1 (\Pi)$ is equicontinuous, then $\omega_0 \left(\widetilde{\Theta}_1(\Pi) \right) = 0$.

Now, for $w \in \Pi$, and for any $\varrho > 0$, there exist a sequence $\{w^k\}_{k=0}^\infty \subset \Pi$ such that for $t \in J$. We have

$$\mu(\widetilde{\Theta}_1(\Pi)(t))$$

$$\leq \mu\left(\left\{\int_0^t \mathcal{Q}(t,s)f(s,w_{\rho(s,w_s)} + x_{\rho(s,w_s+x_s)}), \sigma(w+x)(s))ds \ ; \ w \in \Pi\right\}\right)$$

$$\leq 2\mu\left(\left\{\int_0^t \mathcal{Q}(t,s)f(s,w^k_{\rho(s,w^k_s)} + x_{\rho(s,w^k_s+x_s)}), \sigma(w^k+x)(s))ds \ ; \ w \in \Pi\right\}\right)$$

$$+ \varrho$$

$$\leq \int_0^t 4M_\mathcal{Q}l_f\mu(\{\Pi(s)\})ds + \varrho$$

$$\leq \int_0^t e^{4\tau M_\mathcal{Q}l_f s}e^{-4\tau M_\mathcal{Q}l_f s}4M_\mathcal{Q}l_f\mu(\Pi(s))ds + \varrho$$

$$\leq \int_0^t 4M_\mathcal{Q}l_f e^{4\tau M_\mathcal{Q}l_f s} \sup_{s\in[0,t]} e^{-4\tau M_\mathcal{Q}l_f s}\mu(\Pi(s))ds + \varrho$$

$$\leq \mu_C(\Pi)\int_0^t \left(\frac{e^{4\tau M_\mathcal{Q}l_f s}}{\tau}\right)' ds + \varrho$$

$$\leq \frac{e^{4\tau M_\mathcal{Q}l_f t}}{\tau}\mu_C(\Pi) + \varrho.$$

Since ϱ is arbitrary, we get

$$\mu(\widetilde{\Theta}_1(\Pi)(t)) \leq \frac{e^{4\tau M_\mathcal{Q}l_f t}}{\tau}\mu_C(\Pi).$$

Thus,

$$\mu_C(\widetilde{\Theta}_1(\Pi)) \leq \frac{1}{\tau}\mu_C(\Pi).$$

As a consequence of Theorem 2.6, we deduce that $\widetilde{\Theta}_1$ has at least one fixed point w^*. Then $y^* = w^* + x$ is a fixed point of the operator Θ_1, which is a mild solution of problem (4.18). □

4.4.2   *Controllability Results*

4.4.2.1   *Complete controllability*

Definition 4.5. The system (4.16) is said to be exactly controllable on the interval J, if for every function $\Psi \in \mathcal{B}$ and ζ_0, $\hat{v} \in E$, there is some control $u \in L^2(J,E)$ such that the mild solution v of this problem satisfies the terminal condition $v(T) = \hat{v}$.

We will need to introduce the following hypotheses:

(4.11.1) (*i*) The linear operator $W : L^2(J,U) \to X$, defined by

$$Wu = \int_0^T \mathcal{Q}(T,s)\mathcal{P}u(s)ds,$$

has a pseudo-inverse operator W^{-1}, which takes values in

$$L^2(J,U) \backslash Ker(W),$$

(ii) There exist positive constants m_1, m_2, such that

$$\|\mathcal{P}\| \leq m_1 \quad and \quad \|W^{-1}\| \leq m_2.$$

(iii) There exist $q_w > 0$, $m_\mathcal{P} > 0$, such that for any bounded sets $\widetilde{M_1} \subset E$, $\widetilde{M_2} \subset U$,

$$\mu((W^{-1}\widetilde{M_1})(t)) \leq q_w \mu(\widetilde{M_1}), \quad \mu((\mathcal{P}\widetilde{M_2})(t)) \leq m_\mathcal{P}\mu(\widetilde{M_2}(t)).$$

(4.11.2) There exists a positive constant ρ, such that $\varphi_1^\rho \leq \rho$, with

$$\varphi_1^\rho = M_\mathcal{Q} Bigg[\psi_f^1(\eta_\rho^*)\xi_1 + \psi_f^2(\widetilde{\eta}^*)\xi_2 + m_1 m_2 \left(\rho + \widetilde{M_\mathcal{Q}}\|\Psi_0\| + M_\mathcal{Q}\|\zeta_0\| \right.$$

$$\left. + M_\mathcal{Q}\psi_f^1(\eta_\rho^*)\xi_1 + M_\mathcal{Q}\psi_f^2(\widetilde{\eta}^*)\xi_2 \right) Bigg],$$

$$\eta_\rho^* = L_*\rho + \left[M_* + \mathcal{L}^\Psi + L_*(\widetilde{M_\mathcal{Q}}\|\Psi_0\| + M_\mathcal{Q}\|\zeta_0\|)H\right]\|\Psi\|_\mathcal{B},$$

and

$$\widetilde{\eta}^* = a\Xi_{c_1}^*\left(\rho + \widetilde{M_\mathcal{Q}}\|\Psi_0\| + M_\mathcal{Q}\|\zeta_0\|\right) + a\Xi^*.$$

Theorem 4.4. *Suppose that the hypotheses (4.7.1)–(4.7.4), (4.11.1) and (4.11.2) are valid. Then the problem (4.16) is exactly controllable.*

Proof. Since the calculating techniques were covered in-depth in the previous proofs, the steps of the proof won't be described in detail. We define in $\mathcal{X}$ measures of non-compactness as in Section 4, but we change Σ by $\varkappa$, such that

$$\varkappa(t) = 4M_\mathcal{Q}\left(l_f + m_\mathcal{P}(M_\mathcal{Q}l_f T)q_w\right)t.$$

Now, using (4.11.1) we define the control:

$$u_y(t) - W^{-1}\left(y(T) + \left.\frac{\partial \mathcal{Q}(T,s)\Psi(0)}{\partial s}\right|_{s=0} - \mathcal{Q}(T,0)\zeta_0 \right.$$

$$\left. - \int_0^T \mathcal{Q}(T,s)f(s, y_{\rho(s,y_s)}, (\sigma y)(s))ds\right).$$

We shall show that when using the control $u(\cdot)$, the operator $\Upsilon_3' : \mathcal{X} \to \mathcal{X}$ defined by:

$$\Upsilon_3'y(t) = -\frac{\partial \mathcal{Q}(t,s)\Psi(0)}{\partial s}\bigg|_{s=0} + \mathcal{Q}(t,0)\zeta_0 + \int_0^t \mathcal{Q}(t,s)f(s,y_{\rho(s,y_s)},(\sigma y)(s))ds$$

$$+ \int_0^t \mathcal{Q}(t,s)\mathcal{P}u_y(s)ds; \text{ if } t \in J,$$

has fixed point, this fixed point is a mild solution of system (4.16), and this implies that the system is controllable.

If y is a fixed point of Υ_3', then similar transformation to that in the Proof of Theorem 4.3, give the following decomposition $y(t) = y(t) + x(t)$, which implies $y_t = y_t + x_t$.

Let the operator $\Upsilon_3 : \Omega \to \Omega$ defined by

$$\Upsilon_3y(t) = \int_0^t \mathcal{Q}(t,s)f(s,y_{\rho(s,y_s)},(\sigma y)(s))ds + \int_0^t \mathcal{Q}(t,s)\mathcal{P}u_y(s)ds; \text{ if } t \in J.$$

It thus becomes necessary to demonstrate that Υ_3 has a fixed point since the operator Υ_3' having a fixed point is similar to saying that Υ_3 has one. We will make sure operator Υ_3' satisfies all of the conditions of Darbo's theorem.

Let $B_\rho = B(0,\rho) = \{y \in \Omega \; : \; \|y\|_\Omega \leq \rho\}$, then the set B_ρ is closed, bounded and convex.

Step 1: $\Upsilon_3(B_\rho) \subset B_\rho$.
For $t \in J$ and $y \in B_\rho$, we have

$$\|\Upsilon_3y(t)\| \leq \int_0^t \|\mathcal{Q}(t,s)\| \, \|f(s,y_{\rho(s,y_s)},(\sigma y)(s))\| \, ds$$

$$+ \int_0^t \|\mathcal{Q}(t,s)\| \, \|\mathcal{P}u_y(s)\| \, ds$$

$$\leq M_{\mathcal{Q}}\Big(\psi_f^1(\eta_\rho^*)\xi_1 + \psi_f^2(\tilde{\eta}^*)\xi_2 + m_1m_2$$

$$\times \Big(\rho + \widetilde{M_{\mathcal{Q}}}\|\Psi_0\| + M_{\mathcal{Q}}\|\zeta_0\| + M_{\mathcal{Q}}\psi_f^1(\eta_\rho^*)\xi_1 + M_{\mathcal{Q}}\psi_f^2(\tilde{\eta}^*)\xi_2\Big)\Big).$$

Thus, we deduce from (4.11.2) that $\Upsilon_3(B_\rho) \subset B_\rho$ and $\Upsilon_3(B_\rho)$ is bounded.

Step 2: Υ_3 is continuous.
Let $\{y_n\}_{n\in\mathbb{N}}$ be a sequence such that $y_n \to y_*$ in B_ρ.

Since f, Ξ, $\mathcal{P}$ are continuous, and by the Lebegue dominated convergence theorem, we have

$$\int_0^t \mathcal{Q}(t,s)\mathcal{P}u_{y_n+x}(s)ds \xrightarrow[n\to+\infty]{} \int_0^t \mathcal{Q}(t,s)\mathcal{P}u_{y_*+x}(s)ds.$$

Then, similar to Step 2 in Proof of Theorem 4.3, we get

$$\|(\Upsilon_3 y_n)(t) - (\Upsilon_3 y_*)(t)\| \to 0, \quad as \quad n \to +\infty.$$

Consequently, Υ_3 is continuous.

Step 3: Υ_3 is μ_C-contraction operator.
Let Π be a bounded equicontinuous subset of B_ρ, $y \in \Pi$, and $\kappa_1, \kappa_2 \in J$, with $\kappa_2 > \kappa_1$, we have

$$\left\| \int_0^{\kappa_2} \mathcal{Q}(\kappa_2, s) \mathcal{P} u_{y_n + x}(s) ds - \int_0^{\kappa_1} \mathcal{Q}(\kappa_1, s) \mathcal{P} u_{y_n + x}(s) ds \right\|$$

$$\leq \int_{\kappa_1}^{\kappa_2} \|\mathcal{Q}(\kappa_2, s)\| \|\mathcal{P} u_{y_n + x}(s)\| ds$$

$$+ \int_0^{\kappa_1} \|\mathcal{Q}(\kappa_2, s) - \mathcal{Q}(\kappa_1, s)\| \|\mathcal{P} u_{y_n + x}(s)\| ds$$

$$\leq m_1 m_2 \left(\rho + \widetilde{M_\mathcal{Q}} \|\Psi_0\| + M_\mathcal{Q} \|\zeta_0\| + M_\mathcal{Q} \psi_f^1(\eta_\rho^*) \xi_1 + M_\mathcal{Q} \psi_f^2(\widetilde{\eta}^*) \xi_2 \right)$$

$$\times \left(M_\mathcal{Q} |\kappa_2 - \kappa_1| + \int_0^{\kappa_1} \|\mathcal{Q}(\kappa_2, s) - \mathcal{Q}(\kappa_1, s)\| ds \right) \xrightarrow[\kappa_1 \to \kappa_2]{} 0.$$

Thus $\{\Upsilon_3(\Pi)\}$ is equicontinuous, then $\omega_0\left(\Upsilon_3(\Pi)\right) = 0$. Now for any $\varrho > 0$ there exist a sequence $\{y_k\}_{k=0}^\infty \subset \Pi$, such that for $t \in J$, we get

$$\mu(\Upsilon_3(\Pi)(t)) \leq 4 \int_0^t M_\mathcal{Q}(l_f + m_\mathcal{P}(M_\mathcal{Q} l_f T) q_y) \mu(\{\Pi(s)\}) ds + \varrho$$

$$\leq \frac{e^{\tau \varkappa(t)}}{\tau} \mu_C(\Pi) + \varrho.$$

Therefore,

$$\mu_C(\Upsilon_3(\Pi)) \leq \frac{1}{\tau} \mu_C(\Pi).$$

We come to the conclusion that Υ_3 has at least one fixed point y^* according to Darbo's fixed point theorem. Consequently, $y^* = y^* + x$ is a fixed point of the operator Υ_3', which implies that the system is exactly controllable. $\square$

4.4.2.2 *Approximate controllability*

Definition 4.6. For $(\Psi, \zeta_0) \in \mathcal{B} \times E$, system (4.16) is said to be approximately controllable on the interval $J = [0, T]$ if $\mathcal{R}(T, \Psi, \zeta_0)$ is dense in E, i.e. $\overline{\mathcal{R}(T, \Psi, \zeta_0)} = E$, where $\mathcal{R}(T, \Psi, \zeta_0) = \left\{ x(T, \Psi, \zeta_0, u), u(\cdot) \in L^2(J; U) \right\}$.

As mentioned in Section 1, we shall study the approximate controllability by using a so-called resolvent operator condition. For this purpose, we introduce the following controllability operator $\Gamma_0^T : E \to E$ and resolvent operator $\mathcal{W}\left(\lambda, \Gamma_0^T\right) : E \to E$ defined by

$$\Gamma_0^T = \int_0^T \mathcal{Q}(T,s)\mathcal{P}\mathcal{P}^*\mathcal{Q}^*(T,s)ds, \quad \mathcal{W}\left(\lambda,\Gamma_0^T\right) = \left(\lambda I + \Gamma_0^T\right)^{-1},$$

where $\mathcal{P}^*$ and $\mathcal{Q}^*$ denote the adjoints of the operators $\mathcal{P}$ and $\mathcal{Q}$ respectively, It is straightforward to see that the operator Γ_0^T is a linear bounded operator. So we assume that the operator $\mathcal{W}\left(\lambda, \Gamma_0^T\right)$ satisfies:

(4.13.1) $\lambda \mathcal{W}\left(\lambda, \Gamma_0^T\right) \longrightarrow 0$ as $\lambda \longrightarrow 0^+$ in the strong operator topology.

From [64], hypothesis (4.13.1) is equivalent to the fact that the linear control system corresponding to system (4.16) is approximately controllable on $[0,T]$.

Theorem 4.5. *The following statements are equivalent:*

(i) *The linear control system corresponding to system (4.16) is approximately controllable on $[0,T]$.*

(ii) *If $\mathcal{W}^*\mathcal{Q}^*(t,s)z = 0$ for all s, $t \in [0,T]$, with $s \le t$, then $z = 0$.*

(iii) *The condition (C_0) holds.*

The proof of this theorem is similar with the one of ([311], Theorem 2) and ([64], Theorem 4.4.17), so we omit it here. Right now, we can demonstrate that the system (4.16) is approximately controllable.

For any given $\delta^T \in E$, $\lambda \in (0,1]$, we take the control function $u^\lambda(t)$ as follows:

$$u^\lambda(t) = \mathcal{P}^*\mathcal{Q}^*(T,s)\mathcal{W}\left(\lambda, \Gamma_0^T\right)\Delta(\delta^T, t),$$

where

$$\Delta(\delta^T, t) = \delta^T + \left.\frac{\partial \mathcal{Q}(t,s)\Psi(0)}{\partial s}\right|_{s=0} - \mathcal{Q}(t,0)\zeta_0$$

$$- \int_0^t \mathcal{Q}(t,s)f(s, y_{\rho(s,y_s)}, (\sigma y)(s))ds.$$

Theorem 4.6. *Assume that the hypotheses (4.7.1)–(4.7.4) and (4.13.1) are satisfied, in addition, the function f is uniformly bounded. Then, equation (4.16) is approximately controllable on $[0,T]$.*

Proof. We can observe that system (4.16) has at least one mild solution ρ^λ, based on Theorem 4.3. Then, we have

$$\rho^\lambda(T) = -\frac{\partial \mathcal{Q}(T,s)\Psi(0)}{\partial s}\bigg|_{s=0} + \mathcal{Q}(T,0)\zeta_0$$

$$+ \int_0^T \Big(\mathcal{Q}(T,s)f(s,y^\lambda_{\rho(s,y_s)},(\sigma y^\lambda)(s)) + \mathcal{P}u(s) \Big)\, ds$$

$$= -\frac{\partial \mathcal{Q}(T,s)\Psi(0)}{\partial s}\bigg|_{s=0} + \mathcal{Q}(T,0)\zeta_0$$

$$+ \int_0^T \Big(\mathcal{Q}(T,s)f(s,y^\lambda_{\rho(s,y_s)},(\sigma y^\lambda)(s)) \Big)\, ds$$

$$+ \int_0^T \mathcal{Q}(T,s)\left(\mathcal{P}^*\mathcal{Q}^*(T,s)\mathcal{W}\left(\lambda,\Gamma_0^T\right)\Delta(\delta^T,T)\right) ds$$

$$= \delta^T + (\Gamma_0^T\mathcal{W}\left(\lambda,\Gamma_0^T\right) - I)\Delta(\delta^T,T)$$

$$= \delta^T + \lambda\mathcal{W}\left(\lambda,\Gamma_0^T\right)\Delta(\delta^T,T).$$

Furthermore, we infer from the uniform boundedness of $f(\cdot,\cdot,\cdot)$ that there exists $M_f > 0$, such that

$$\int_0^T \|f(s,y^\lambda_{\rho(s,y_s)},(\sigma y^\lambda)(s))\|^2 ds \leq T(M_f)^2.$$

Therefore, the sequence $\{f(s,y^\lambda_{\rho(s,y_s)},(\sigma y^\lambda)(s))\}_\lambda$ is bounded in $L^2(J,E)$, then there exists a subsequence still indicated by $\{f(s,y^\lambda_{\rho(s,y_s)},(\sigma y^\lambda)(s))\}_\lambda$ that weakly converge to the limit $\tilde{f}(s)$ in $L^2(J,E)$. Then, we have

$$\int_0^T \|f(s,y^\lambda_{\rho(s,y_s)},(\sigma y^\lambda)(s)) - \tilde{f}(s)\|ds \xrightarrow[\lambda\to 0]{} 0.$$

Thus,

$$\|\rho^\lambda(T) - \delta^T\|$$

$$\leq \left\|\mathcal{W}\left(\lambda,\Gamma_0^T\right)\left[\delta^T + \frac{\partial \mathcal{Q}(T,s)\Psi(0)}{\partial s}\bigg|_{s=0} - \mathcal{Q}(T,0)\zeta_0\right]\right\|$$

$$+ \left\|\mathcal{W}\left(\lambda,\Gamma_0^T\right)\left[+\int_0^T \left(\mathcal{Q}(T,s)f(s,y_{\rho(s,y_s)},(\sigma y)(s))\right) ds\right]\right\|$$

$$\leq \left\|\mathcal{W}\left(\lambda,\Gamma_0^T\right)\left[\delta^T + \frac{\partial \mathcal{Q}(T,s)\Psi(0)}{\partial s}\bigg|_{s=0} - \mathcal{Q}(T,0)\zeta_0\right]\right\|$$

$$+ \left\|\mathcal{W}\left(\lambda,\Gamma_0^T\right)\left[\int_0^T \mathcal{Q}(T,s)\left(f(s,y_{\rho(s,y_s)},(\sigma y)(s)) - \tilde{f}(s)\right) ds\right]\right\|$$

$$+ \left\|\mathcal{W}\left(\lambda,\Gamma_0^T\right)\left[\int_0^T \mathcal{Q}(T,s)\tilde{f}(s)ds\right]\right\| \xrightarrow[\lambda\to 0]{} 0.$$

Thus, $\rho^\lambda(t) \to \delta^T$ holds, and consequently system (4.16) is approximately controllable on J. $\qquad\square$

4.4.3   An Example

Consider the following class of partial integro-differential system:

$$
\begin{cases}
\dfrac{\partial^2 \zeta(t,x)}{\partial^2 t} = \dfrac{\partial^2 \zeta(t,x)}{\partial^2 x} - \displaystyle\int_0^t \Gamma(t-s)\dfrac{\partial^2 \zeta(s,x)}{\partial^2 x}\,ds \\
\quad + \displaystyle\int_{-\infty}^{-t} \dfrac{e^{-8\tau}\|\zeta(t+\sigma(t,\zeta(t+\tau,x)),x)\|_{L^2}}{83\left((t+\tau)^2 + 2t + 1\right)}\,d\tau \\
\quad - \dfrac{1 - e^{-16\pi}}{332\,(t+1)^2} + \displaystyle\int_0^a \dfrac{\cos(t)\ln(1+e^{-t^2})(1+\zeta(s,x))}{177(1+2t^2+s^2)e^{4t}}\,ds \\
\quad + \widetilde{\sigma}(t)\zeta(t,x) + \mathcal{L}(t,x), \quad \text{if } t \in I \ \text{ and } \ x \in (0,\pi), \\[4pt]
\zeta(t,0) = \zeta(t,1) = 0, \qquad \text{for } t \in I, \\[4pt]
\left.\dfrac{\partial \zeta(t,x)}{\partial t}\right|_{t=0} = \zeta_1(x), \ \zeta(t,x) = \Psi(t,x), \quad \text{if } t \in \mathbb{R}_- \ \text{ and } \ x \in (0,\pi),
\end{cases}
$$

$$(4.20)$$

where $I = [0,1], \sigma : J \times \mathbb{R} \to \mathbb{R}, \mathcal{L} : [0,1] \times [0,\pi] \to [0,\pi]$.

Let

$$
\mathcal{H} := L^2(0,\pi) = \left\{ u : (0,\pi) \longrightarrow \mathbb{R} : \int_0^\pi |u(x)|^2 dx < \infty \right\},
$$

be the Hilbert space with the scalar product $\langle u, v \rangle = \displaystyle\int_0^\pi u(x)v(x)dx$, and the norm

$$
\|u\|_2 = \left(\int_0^\pi |u(x)|^2 dx \right)^{1/2}.
$$

Let the phase space $\mathcal{B}$ be $BUC\,(\mathbb{R}^-, \mathcal{H})$, the space of bounded uniformly continuous functions endowed with the following norm:

$$
\|\psi\|_{\mathcal{B}} = \sup_{-\infty < \tau \le 0} \|\psi(\tau)\|_{L^2}, \psi \in \mathcal{B}.
$$

It is well known that $\mathcal{B}$ satisfies the axioms (A_1) and (A_2) with $K = 1$ and $L(t) = M(t) = 1$, (see [309]). We define the operator $\widehat{A}$ induced on $\mathcal{H}$ as follows:

$$
\widehat{A}z = z'', \text{ and } D(A) = \{z \in H^2(0,\pi) \ : \ z(0) = z(\pi) = 0\}.
$$

Then $\widehat{A}$ is the infinitesimal generator of a cosine function of operators $(C_0(t))_{t\in\mathbb{R}}$ on H associated with sine function $(S_0(t))_{t\in\mathbb{R}}$. Additionally,

$\widehat{A}$ has discrete spectrum which consists of eigenvalues $-n^2$ for $n \in \mathbb{N}$, with corresponding eigenvectors

$$w_n(x) = \frac{1}{\sqrt{2\pi}} e^{inx}, \quad n \in \mathbb{N}.$$

The set $\{w_n : n \in \mathbb{N}\}$ is an orthonormal basis of H. Applying this idea, we can write

$$\widehat{A}z = \sum_{n=1}^{\infty} -n^2 \langle z, w_n \rangle w_n,$$

for $z \in D(A)$, $(C_0(t))_{t \in \mathbb{R}}$ is given by

$$C_0(t)z = \sum_{n=1}^{\infty} \cos(nt) \langle z, w_n \rangle w_n, \quad t \in \mathbb{R},$$

and the sine function is given by

$$S_0(t)z = \sum_{n=1}^{\infty} \frac{\sin(nt)}{n} \langle z, w_n \rangle w_n, \quad t \in \mathbb{R}.$$

It is immediate from these representations that $\|C_0(t)\| \leq 1$ and that $S_0(t)$ is compact for all $t \in \mathbb{R}$. We define $A(t)z = \widehat{A}z + \widetilde{\sigma}(t)z$ on $D(A)$. Clearly, $A(t)$ is a closed linear operator. Therefore, $A(t)$ generates $(S(t,s))_{(t,s) \in \Delta}$ such that $S(t,s)$ is compact and self-adjoint for all $(t,s) \in \Delta = \{(t,s) : 0 \leq s \leq t \leq 1\}$, (see [307]).

We define the operators $\Lambda(t,s) : D(A) \subset \mathcal{H} \mapsto \mathcal{H}$ as follows:

$$\Lambda(t,s)z = \Gamma(t-s)\widehat{A}z, \ for \ 0 \leq s \leq t \leq 1, \ z \in D(A).$$

The assumption (4.11.1) holds under more suitable conditions on the operator B. Furthermore, it is not difficult to see that conditions $(B1) - (B3)$ are fulfilled, which in turn implies that there exists a resolvent operator and it's a compact operator. More details about these facts can be seen from the monograph [229, 231, 307, 312].

Now let $\mathcal{P} : U \to \mathcal{H}$ be defined by $\mathcal{P}u(t)(x) = \mathcal{L}(t,x)$, $x \in [0, \pi], u \in U$, where $\mathcal{L} : [0,1] \times [0, \pi] \to \mathcal{H}$ is linear continuous and for $\Psi \in BUC(\mathbb{R}^-, H)$, we put $\rho(t, \Psi)(\zeta) = \sigma(t, \zeta(t+\tau, x))$, such that (4.7.4.1) hold, and let $t \to \Psi_t$ be continuous on $\mathcal{R}(\rho^-)$.

We put $\zeta(t)(x) = \zeta(t,x)$, for $t \in [0,1]$, and define

$$f(t, y_1, y_2)(x) = \int_{-\infty}^{-t} \frac{e^{-8\tau} \|y_1(t + \sigma(t, \zeta(t+\tau, x)), x)\|_{L^2}}{83\left((t+\tau)^2 + 2t + 1\right)} d\tau$$
$$- \frac{1 - e^{-16\pi}}{332(t+1)^2} + \frac{\cos(t)y_2(t)(x)}{e^{-4t}},$$

and

$$y_2(t)(x) = \sigma(y_1)(x) = \int_0^a \frac{\ln(1 + e^{-t^2})(1 + y_1(s,x))}{177(1 + 2t^2 + s^2)} ds.$$

These definitions allow us to depict the system (4.20) in the abstract form (4.16).

Now, for $t \in [0,1]$, we have

$$\|f(t, \varkappa_{1(t)}, \varkappa_2(t))\| \leq \frac{1 - e^{-16\pi}}{332(t+1)^2} (1 + \|\varkappa_1\|_\mathcal{B}) + \cos(t)e^{-4t} (\|\varkappa_2(t)\|).$$

So, $\psi_{i+1}(t) = t + i;\ i = 0, 1$ are continuous non-decreasing functions, and we have

$$\xi_1 = \frac{(1 - e^{-16\pi})(1 - (1+\pi)^{-3})}{332\sqrt{3}}, \quad \text{and } \xi_2 = \frac{1}{4}\sqrt{\frac{33}{17}(1 - e^{-8\pi})}.$$

And for any bounded set $\Pi \subset \mathcal{H}$, and $\Pi_t \in \mathcal{B}$, we get

$$\chi(f(t, \Pi_t, \sigma(\Pi(t)))) \leq (\xi_1 + \xi_2)\, \chi(\Pi).$$

Now, about Ξ, we obtain

$$\|\Xi(t, s, \varkappa_1) - \Xi(t, s, \varkappa_2)\|_2 \leq \frac{\ln(2)}{177}\|\varkappa_1 - \varkappa_2\|_2.$$

Now, similar reasoning as in [313], if the corresponding linear system is approximately controllable, then from Theorem 4.5 we obtain

$$\lambda \left(\lambda I + \int_0^1 \mathcal{Q}(1,s)\mathcal{L}(s,x)\mathcal{L}(t,x)^* \mathcal{Q}^*(1,s)ds\right)^{-1} \xrightarrow[\lambda \to 0^+]{} 0.$$

And for $p_3 = \|\varkappa_1\|_\mathcal{B}$, $p_4 = \|\varkappa_2\|_2$, for all $\varkappa_1 \in \mathcal{B}$, $\varkappa_2 \in H$, we get

$$\|f(\cdot, \varkappa_{1(\cdot)}, \varkappa_2(\cdot))\|_2 \leq \frac{1}{332\sqrt{3}}(1 - e^{-16\pi})(1 - (1+\pi)^{-3})(1 + p_3 + p_4).$$

Thus, all the assumptions of Theorem 4.6 are fulfilled. Consequently, the problem (4.20) is approximately controllable on $[0, 1]$.

Remark 4.2. We can take the same example but we change the operator $A(t)$ by another operator such that $(S(t,s))_{(t,s)\in\Delta}$ will be not compact. On the other hand, from [314] the operator W given by

$$Wu = \int_0^1 \mathcal{Q}(1,s)\mathcal{P}u(s)ds,$$

is a bounded linear operator but not necessarily one-to-one. Let

$$\text{Ker } W = \{u \in L^2([0,1], U), Wu = 0\}$$

be the null space of W and $[\mathrm{Ker}\,W]^{\perp}$ be its orthogonal complement in $L^2([0,1],U)$. Let $\widetilde{W} : [\mathrm{Ker}\,W]^{\perp} \longrightarrow \mathrm{Range}(W)$ be the restriction of W to $[\mathrm{Ker}\,W]^{\perp}$, $\widetilde{W}$ is necessarily one-to-one operator. The inverse mapping theorem says that $\widetilde{W}^{-1}$ is bounded since $[\mathrm{Ker}W]^{\perp}$ and Range (W) are Banach spaces. So that W^{-1} is bounded and takes values in $L^2([0,1],U)\backslash \mathrm{Ker}\,W$, hypothesis (4.11.1) is satisfied. Then, all the assumptions given in Theorem (4.4) are verified. Therefore, the problem (4.20) is exactly controllable on $[0,1]$.

4.5 Notes and Remarks

This chapter's contents are based on the articles [315–317]. The monographs [160, 192, 350], and the papers [44, 275–279] provide more important conclusions and analyses about the subject.

Chapter 5

Semilinear Differential Evolution Equations with Random Effects

5.1 Introduction and Motivations

The present chapter deals with the study of some existence and controllability results for a class of first and second order semilinear differential evolution equations with random effects in Fréchet spaces. The technique used to show the existence of random mild solutions is a generalization of the original Darbo fixed point theorem for Fréchet spaces combined with the notion of measure of non-compactness. We provide illustrations to demonstrate the applicability of our results for each section.

The inquiries carried out in this chapter were driven by the following studies, which we will cite as sources for readers who are curious about further information:

- Considerable progress has been made in semilinear functional evolution equations, along with significant findings on the global existence for functional evolution equations and inclusions on Fréchet spaces. This progress has been documented in various sources, including the papers cited as [10–13]. Notably, recent advancements have been achieved through different approaches. It is worth mentioning that earlier papers often imposed specific constraints, such as Lipschitz conditions and the compactness of the semigroup.
- In certain cases, it is advantageous to handle second-order abstract differential equations directly, without the need to always transform them into first-order systems. The theory of strongly continuous cosine families provides a valuable framework for investigating such second-order problems. To facilitate our analysis, we will employ several fundamental principles from cosine family theory, as outlined in the reference [318].

- The authors in the papers [319, 320] presented sufficient conditions for establishing the controllability of second-order systems in Banach spaces. These conditions apply to both deterministic and stochastic systems and make use of alternative fixed point theorems, as well as the concept of strongly continuous cosine families. Additionally, the nonlinearity involved in these systems satisfies the Lipschitz condition.
- The concept of controllability has been extended to infinite-dimensional systems in Banach spaces with bounded operators by numerous researchers (refer to [47, 48]). In order to address the controllability problem, the authors in [49–51] approached it by transforming it into a fixed point problem. In a similar way, Balachandran and Dauer explored various classes of semilinear ordinary differential equations in [52]. Additionally, Benchohra et al. examined numerous classes of functional differential equations and inclusions, employing fixed point arguments to establish certain controllability results in [53–57].

5.2   Semilinear Random Differential Evolution Equations in Fréchet Spaces

Motivated by the above mentioned works, we study the following equation

$$y'(t,\delta) = Ay(t,\delta) + f(t,y(t,\delta),\delta); \quad \text{if } t \in \mathbb{R}_+ := [0,\infty),\ \delta \in \Omega, \qquad (5.1)$$

with the initial condition

$$y(0,\delta) = y_0(\delta) \in E;\ \delta \in \Omega, \qquad (5.2)$$

where (Ω, F, P) is a complete probability space, $y_0 : \Omega \to E$ is a given function, $f : \mathbb{R}_+ \times E \times \Omega \to E$ is a given function, $(E, \|\cdot\|)$ is a Banach space, and $A : D(A) \subset E \to E$ is the infinitesimal generator of a C_0-semigroup $S(t),\ t \geq 0$.

Next, we shall be concerned with existence of integral solutions for problem (5.1)–(5.2), in the case where $A : D(A) \subset E \to E$ is a nondensely defined closed linear operator on the Banach space E generating an integrated semigroup.

5.2.1   *Existence Results*

Consider the Fréchet space $\bar{E} := C(\mathbb{R}_+, E)$ of continuous functions z from $\mathbb{R}_+$ into E, with the seminorm

$$\|z\|_\beta = \sup_{t \in [0,\beta]} \|z(t)\|;\ \beta \in \mathbb{N},$$

and the distance

$$d(y, z) = \sum_{\beta=1}^{\infty} 2^{-\beta} \frac{\|y - z\|_\beta}{1 + \|y - z\|_\beta}; \ y, z \in \bar{E}.$$

5.2.2   Global Existence of Random Integral Solutions

Definition 5.1. The continuous function $y(\cdot, \delta) : \mathbb{R}_+ \times \Omega \to E$ is a mild random solution of the problem (5.1)–(5.2), if y verifies:

$$y(t, \delta) = S(t)y_0(\delta) + \int_0^t S(t-s) \, f(s, y(s, \delta), \delta) ds; \ for \ each \ t \in \mathbb{R}_+, \ and \ \delta \in \Omega.$$

We assume the following needed conditions:

(5.1.1) $A : D(A) \subset E \to E$ is the infinitesimal generator of a C_0-semigroup $\{S(t)\}_{t \geq 0}$.

(5.1.2) The function f is random Carathéeodory on $\mathbb{R}_+ \times E \times \Omega$.

(5.1.3) There exists a measurable and essentially bounded function $\lambda : \mathbb{R}_+ \to \mathbb{R}_+$ such that for any $\delta \in \Omega$, we have

$$\|f(t, y, \delta)\| \leq \lambda(t)(1 + \|y\|); \ for \ a.e. \ t \in \mathbb{R}_+, \ and \ each \ y \in E.$$

(5.1.4) For each bounded set $B \subset E$ and for any $\delta \in \Omega$, we have

$$\alpha(f(t, B, \delta)) \leq \lambda(t)\alpha(B); \ for \ a.e. \ t \in \mathbb{R}_+.$$

Set

$$\gamma = \sup_{t \in \mathbb{R}_+} \|S(t)\|_{B(E)} < \infty, \quad \lambda_\beta^* = esssup_{t \in [0, \beta]} \lambda(t); \ for \ \beta \in \mathbb{N}.$$

Consider

$$\alpha_\beta(D) = \sup_{t \in [0, \beta]} e^{-4\gamma\lambda_\beta^* \zeta t} \alpha(D(t)) + \omega_0(D),$$

where $\zeta > 1$, and $D(t) = \{z(t) \in E : z \in D\}; \ t \in [0, \beta]$, and

$$\omega_0(D) = \lim_{\epsilon \to 0} \sup_{z \in D} \max\{\|z(t_1) - z(t_2)\| : |t_1 - t_2| \leq \epsilon\},$$

which represents the modulus of equicontinuity of the set D.

Remark 5.1. Notice that $\omega_0(D) = 0$, when D is equicontinuous.

Theorem 5.1. *Assume that the hypotheses* (5.1.1)–(5.1.4) *are satisfied, and* $\beta\gamma\lambda_\beta^* < 1$ *for each* $\beta \in \mathbb{N}$. *Then the problem* (5.1)–(5.2) *has at least one random mild solution in* $\bar{E}$.

Proof. Consider the operator $T_1 : \Omega \times \bar{E} \to \bar{E}$ given by:

$$(T_1(\delta)y)(t) = S(t)y_0(\delta) + \int_0^t S(t-s) \; f(s, y(s, \delta), \delta)ds. \qquad (5.3)$$

We will demonstrate that T_1 verifies all assumptions of Lemma 2.14.

Claim 1. $T_1(\delta)$ *is a random operator with stochastic domain on* $\bar{E}$.
Using Definition 2.29 and since $f(t, y, \delta)$ is random Carathéodory, we deduce that $\delta \to f(t, y, \delta)$ is measurable. Thus,

$$\delta \mapsto S(t)y_0(\delta) + \int_0^t S(t-s)f(s, y(s, \delta), \delta)ds,$$

is measurable. As a result, T_1 is a random operator on $\Omega \times \bar{E}$ into $\bar{E}$.

Let the ball $\Upsilon : \Omega \to \mathcal{P}(\bar{E})$ given by

$$\Upsilon(\delta) := B(0, R_\beta(\delta)) = \{z \in \bar{E} : \|z\|_\beta \le R_\beta(\delta)\}; \; \delta \in \Omega, \; \beta \in \mathbb{N},$$

where $R_\beta : \Omega \to (0, \infty)$ is a function such that

$$R_\beta(\delta) \ge \frac{\gamma\|y_0(\delta)\| + \beta\gamma\lambda_\beta^*}{1 - \beta\gamma\lambda_\beta^*}.$$

Using Lemma 17 form [244] and since $\Upsilon(\delta)$ is closed, convex, bounded and solid for all $\delta \in \Omega$, then Υ is measurable. Let $\delta \in \Omega$, then from (5.1.3), for any $y \in \delta(\delta)$, and each $t \in [0, \beta]$ we get

$$\|(T_1(\delta)y)(t)\| \le \|S(t)y_0(\delta) + \int_0^t S(t-s) \; f(s, y(s, \delta), \delta)ds\|$$

$$\le \gamma\|y_0(\delta)\| + \gamma \left(\int_0^t \lambda(s)(1 + \|y(s, \delta)\|)ds \right)$$

$$\le \gamma\|y_0(\delta)\| + \beta\gamma\lambda_\beta^* + \beta\gamma\lambda_\beta^* R_\beta$$

$$\le R_\beta(\delta).$$

Thus, T_1 is a random operator with stochastic domain Υ and $T_1(\delta) : \Upsilon(\delta) \to \Upsilon(\delta)$. Moreover, $T_1(\delta)$ maps bounded sets into bounded sets in $\bar{E}$.

Claim 2. $T_1(\delta) : \Upsilon(\delta) \to \Upsilon(\delta)$ *is continuous.*
Let $\{y_n\}_{n \in \mathbb{N}}$ be a sequence where $y_n \to y$ in $B_{R_\beta}(\delta)$. Then, for each $t \in [0, \beta]$ and $\delta \in \Omega$, we obtain

$$\|(T_1(\delta)y_n)(t) - (T_1(\delta)y)(t)\|$$

$$\le \int_0^t \|S(t-s)\|_{B(E)} \|f(s, y_n(s, \delta), \delta) - f(s, y(s, \delta), \delta)\|ds$$

$$\leq \gamma \int_0^t \|f(s, y_n(s, \delta), \delta) - f(s, y(s, \delta), \delta)\| ds.$$

Since $y_n \to y$ as $n \to \infty$, we conclude that

$$\|T_1(\delta)(y_n) - T_1(\delta)(y)\|_\beta \to 0 \quad \text{as } n \to \infty.$$

Consequently, we deduce that $T_1(\delta) : \Upsilon(\delta) \to T_1(\delta)$ is a continuous random operator with stochastic domain Υ, and $T_1(\delta)(\Upsilon(\delta))$ is bounded.

Claim 3. *For each equicontinuous subset Φ of $\Upsilon(\delta)$, $\alpha_\beta(T_1(\delta)(\Phi)) \leq \ell_\beta \alpha_\beta(\Phi)$.*

Let $\Phi \subset \Upsilon(\delta)$, $\epsilon > 0$, then by Lemmas 2.5 and 2.6, there exists a sequence $\{y_n\}_{n=0}^\infty \subset \Phi$, where for $t \in [0, \beta]$ and $\delta \in \Omega$, we get

$$\alpha((T_1(\delta)\Phi)(t)) = \alpha\left(\left\{S(t)y_0 + \int_0^t S(t-s) \; f(s, y(s, \delta), \delta) ds; \; y \in \Phi\right\}\right)$$

$$\leq 2\alpha\left(\left\{\int_0^t S(t-s) f(s, y_n(s, \delta), \delta) ds\right\}_{n=1}^\infty\right) + \epsilon$$

$$\leq 4 \int_0^t \alpha\left(\|S(t-s)\|_{B(E)}\{f(s, y_n(s, \delta), \delta)\}_{n=1}^\infty\right) ds + \epsilon$$

$$\leq 4\gamma \int_0^t \alpha\left(\{f(s, y_n(s, \delta), \delta)\}_{n=1}^\infty\right) ds + \epsilon$$

$$\leq 4\gamma \int_0^t \lambda_\beta(s)\alpha\left(\{y_n(s, \delta)\}_{n=1}^\infty\right) ds + \epsilon$$

$$\leq 4\gamma\lambda_\beta^* \int_0^t e^{4\gamma\lambda_\beta^*\zeta s} e^{-4\gamma\lambda_\beta^*\zeta s}\alpha\left(\{y_n(s, \delta)\}_{n=1}^\infty\right) ds + \epsilon$$

$$\leq \frac{\left(e^{4\gamma\lambda_\beta^*\zeta t} - 1\right)}{\zeta}\alpha_\beta(\Phi) + \epsilon$$

$$\leq \frac{e^{4\gamma\lambda_\beta^*\zeta t}}{\zeta}\alpha_\beta(\Phi) + \epsilon.$$

Since $\epsilon > 0$ is arbitrary, then

$$\alpha((T_1(\delta)\Phi)(t)) \leq \frac{e^{4\gamma\lambda_\beta^*\zeta t}}{\zeta}\alpha_\beta(\Phi).$$

Thus

$$\alpha_\beta(T_1(\delta)(\Phi)) \leq \frac{1}{\zeta}\alpha_\beta(\Phi).$$

Consequently, Theorem 2.14 implies that T_1 has at least one fixed point in $\Upsilon(\delta)$ which is a random mild solution of problem (5.1)–(5.2).

5.2.3 *Integral Solutions for Random Evolution Equations*

Now, we study the case where $A : D(A) \subset E \to E$ is a nondensely defined closed linear operator on the Banach space E.

Definition 5.2 ([321]). An integrated semigroup is a family of operators $(S(t))_{t \geq 0}$ of bounded linear operators $S(t)$ on E with the following properties:

(i) $S(0) = 0$;

(ii) $t \to S(t)$ is strongly continuous;

(iii) $S(s)S(t) = \displaystyle\int_0^s (S(t+r) - S(r))dr$; for all $t, s \geq 0$.

Definition 5.3 ([322]). An operator A is called a generator of an integrated semigroup if there exists $\omega \in \mathbb{R}$ such that $(\omega, \infty) \subset \rho(A)$ ($\rho(A)$, is the resolvent set of A) and there exists a strongly continuous exponentially bounded family $(S(t))_{t \geq 0}$ of bounded operators such that $S(0) = 0$ and
$$R(\chi, A) := (\chi I - A)^{-1} = \chi \int_0^\infty e^{-\chi t} S(t)dt \text{ exists for all } \chi \text{ with } \chi > \omega.$$

Definition 5.4. We say that $y(\cdot, \delta) : \mathbb{R}_+ \times \Omega \to E$ is an integral random solution of problem (5.1)–(5.2) if

(i) $y(t, \delta) = y_0(\delta) + A \displaystyle\int_0^t y(s, \delta)ds + \int_0^t f(s, y(s, \delta), \delta)ds$; for any $\delta \in \Omega$, and each $t \in \mathbb{R}_+$;

(ii) For any $\delta \in \Omega$, $\displaystyle\int_0^t y(s, \delta)ds \in D(A)$; for each $t \in \mathbb{R}_+$,

From the above definition it follows that for any $\delta \in \Omega$, $y(t, \delta) \in \overline{D(A)}$, for each $t \in \mathbb{R}_+$, in particular $y_0(\delta) \in \overline{D(A)}$. Moreover, $y(\cdot, \delta)$ satisfies the following variation of constants formula:

$$y(t, \delta) = S'(t)y_0(\delta) + \frac{d}{dt}\int_0^t S(t-s)f(s, y(s, \delta), \delta)ds; \quad t \in \mathbb{R}_+. \tag{5.4}$$

Notice that, if $y(\cdot, \delta)$ satisfies (5.4), then

$$y(t, \delta) = S'(t)y_0(\delta) + \lim_{\chi \to \infty} \int_0^t S'(t-s)B_\chi f(s, y(s, \delta), \delta)ds; \quad t \in \mathbb{R}_+.$$

Consider the Fréchet space $\tilde{E} := C(\mathbb{R}_+, \overline{D(A)})$ of continuous functions z from $\mathbb{R}_+$ into $\overline{D(A)}$, with the seminorm

$$\|z\|_\beta = \sup_{t \in [0,\beta]} \|z(t)\|; \quad \beta \in \mathbb{N}.$$

Consider

$$\alpha_\beta(B) = \sup_{t \in [0,\beta]} e^{-(\omega + 4\gamma\lambda_\beta^*\zeta)t} \alpha(B(t)) + \omega_0(B),$$

where $\zeta > 1$, and $B(t) = \{z(t) \in \overline{D(A)} : z \in B\}; \ t \in [0, \beta]$.

Theorem 5.2. *If the hypotheses* (5.1.2)–(5.1.4) *and the following hypothesis hold:*

(5.7.1) *A satisfies Hille-Yosida condition.*

If

$$\frac{\gamma}{\omega} e^{\omega\beta} \lambda_\beta^* (1 - e^{-\omega\beta}) < 1;$$

for each $\beta \in \mathbb{N}$, then the problem (5.1)–(5.2) *has at least one random integral solution.*

Proof. Consider the operator $G : \Omega \times \tilde{E} \to \tilde{E}$ given by:

$$(G(\delta)y)(t) = y(t, \delta) = S'(t)y_0(\delta) + \frac{d}{dt} \int_0^t S(t - s)f(s, y(s, \delta), \delta)ds. \quad (5.5)$$

Let us demonstrate that G verifies all the assumptions of Lemma 2.14.

Claim 1. $G(\delta)$ *is a random operator with stochastic domain on* $\tilde{E}$.
Using Definition 2.29 and since $f(t, y, \delta)$ is random Carathéodory, we obtain that $\delta \to f(t, y, \delta)$ is measurable. Thus,

$$\delta \mapsto S'(t)y_0(\delta) + \frac{d}{dt} \int_0^t S(t - s)f(s, y(s, \delta), \delta)ds,$$

is measurable. Further, G is a random operator on $\Omega \times \tilde{E}$ into $\tilde{E}$.

Consider the ball $\Upsilon : \Omega \to \mathcal{P}(\tilde{E})$ given by

$$\Upsilon(\delta) := B(0, \rho_\beta(\delta)) = \{z \in \tilde{E} : \|z\|_\beta \le \rho_\beta(\delta)\}; \ \delta \in \Omega, \ \beta \in \mathbb{N},$$

where $\rho_\beta : \Omega \to (0, \infty)$ is a function such that

$$\rho_\beta(\delta) \ge \frac{\gamma e^{\omega\beta} \|y_0(\delta)\| + \frac{\gamma}{\omega} e^{\omega\beta} \lambda_\beta^* (1 - e^{-\omega\beta})}{1 - \frac{\gamma}{\omega} e^{\omega\beta} \lambda_\beta^* (1 - e^{-\omega\beta})}.$$

By Lemma 17 from [244] and since $\Upsilon(\delta)$ is closed, convex, bounded and solid for all $\delta \in \Omega$, we deduce that Υ is measurable. Let $\delta \in \Omega$, then from (5.1.3), for any $y \in \Upsilon(\delta)$, and each $t \in [0, \beta]$ we get

$$\|(G(\delta)y)(t)\| \leq \|S'(t)y_0(\delta) + \frac{d}{dt}\int_0^t S(t-s)f(s,y(s,\delta),\delta)ds\|$$

$$\leq \gamma e^{\omega\beta}\|y_0(\delta)\| + \gamma e^{\omega\beta}\left(\int_0^t e^{-\omega s}\lambda(s)(1+\|y(s,\delta)\|)ds\right)$$

$$\leq \gamma e^{\omega\beta}\|y_0(\delta)\| + \gamma e^{\omega\beta}\lambda_\beta^*(1+\rho_\beta)\left(\int_0^\beta e^{-\omega s}ds\right)$$

$$\leq \gamma e^{\omega\beta}\|y_0(\delta)\| + \frac{\gamma}{\omega}e^{\omega\beta}\lambda_\beta^*(1+\rho_\beta)(1-e^{-\omega\beta})$$

$$\leq \rho_\beta(\delta).$$

Therefore, G is a random operator with stochastic domain Υ and $G(\delta)$: $\Upsilon(\delta) \to \Upsilon(\delta)$. Moreover, $G(\delta)$ maps bounded sets into bounded sets in $\tilde{E}$.

Claim 2. $G(\delta) : \Upsilon(\delta) \to \Upsilon(\delta)$ *is continuous.*
Let $\{y_n\}_{n\in\mathbb{N}}$ be a sequence where $y_n \to y$ in $\Upsilon(\delta)$. Then, for each $t \in [0,\beta]$ and $\delta \in \Omega$, we get

$$\|(G(\delta)y_n)(t) - (G(\delta)y)(t)\|$$

$$\leq \frac{d}{dt}\int_0^t \|S(t-s)\|_{B(E)}\|f(s,y_n(s.\delta),\delta) - f(s,y(s,\delta),\delta)\|ds$$

$$\leq \gamma e^{\omega\beta}\int_0^t e^{-\omega s}\|f(s,y_n(s,\delta),\delta) - f(s,y(s,\delta),\delta)\|ds.$$

Since $y_n \to y$ as $n \to \infty$, we deduce that $\|G(\delta)(y_n) - G(\delta)(y)\|_\beta \to 0$ as $n \to \infty$. Consequently, we have that $G(\delta) : \Upsilon(\delta) \to \Upsilon(\delta)$ is a continuous random operator with stochastic domain Υ, and $G(\delta)(\Upsilon(\delta))$ is bounded.

Claim 3. *For all bounded subset B of $\Upsilon(\delta)$, $\alpha_\beta(G(\delta)(B)) \leq \ell_\beta'\alpha_\beta(B)$.*
From Lemmas 2.5 and 2.6, for any $B \subset \Upsilon(\delta)$ and any $\epsilon > 0$, there exists a sequence $\{y_n\}_{n=1}^\infty \subset B$, such that for all $t \in [0,\beta]$ and $\delta \in \Omega$, we have

$$\alpha((G(\delta)B)(t)) = \alpha\left(\left\{S'(t)y_0(\delta) + \frac{d}{dt}\int_0^t S(t-s)f(s,y(s,\delta),\delta)ds; \ y \in B\right\}\right)$$

$$\leq 2\alpha\left(\left\{\frac{d}{dt}\int_0^t S(t-s)f(s,y(s,\delta),\delta)ds\right\}_{n=1}^\infty\right) + \epsilon$$

$$\leq 4\gamma e^{\omega\beta}\int_0^t \alpha\left(\{e^{-\omega s}f(s,y_n(s,\delta),\delta)\}_{n=1}^\infty\right)ds + \epsilon$$

$$\leq 4\gamma e^{\omega\beta}\int_0^t \alpha\left(\{e^{-\omega s}f(s,y_n(s,\delta),\delta)\}_{n=1}^\infty\right)ds + \epsilon$$

$$\leq 4\gamma e^{\omega\beta}\int_0^t e^{-\omega s}\lambda_\beta(s)\alpha_\beta\left(\{y_n(s,\delta)\}_{n=1}^\infty\right)ds + \epsilon$$

$$\leq 4\gamma e^{\omega\beta}\lambda_\beta^* \int_0^t e^{4\gamma\lambda_\beta^*\zeta s}e^{-(\omega+4\gamma\lambda_\beta^*\zeta)s}\alpha\left(\{y_n(s,\delta)\}_{n=1}^\infty\right)ds + \epsilon$$

$$\leq \frac{e^{\omega t}\left(e^{4\gamma\lambda_\beta^*\zeta t}-1\right)}{\zeta}\alpha_\beta(B) + \epsilon$$

$$\leq \frac{e^{(\omega+4\gamma\lambda_\beta^*\zeta)t}}{\zeta}\alpha_\beta(B) + \epsilon.$$

Since $\epsilon > 0$ is arbitrary, then

$$\alpha((G(\delta)B)(t)) \leq \frac{e^{(\omega+4\gamma\lambda_\beta^*\zeta)t}}{\zeta}\alpha_\beta(B).$$

Thus

$$\alpha_\beta(G(\delta)(B)) \leq \frac{1}{\zeta}\alpha_\beta(B).$$

Thus, Theorem 2.14 implies that G has at least one fixed point in $\Upsilon(\delta)$ which is a random integral solution of problem (5.1)–(5.2).

5.2.4 *Examples*

Let $\Omega = (-\infty, 0)$ be equipped with the usual σ-algebra consisting of Lebesgue measurable subsets of $(-\infty, 0)$.

Example 1. Consider a measurable function $y : \Omega \to L^2([0,\pi], \mathbb{R})$, and the following problem:

$$\frac{\partial y}{\partial t}(t,w,\delta) = \frac{\partial^2 y}{\partial w^2}(t,w,\delta)$$

$$+ G(t, y(t,w,\delta), \delta); \ w \in [0,\pi], \ t \in \mathbb{R}_+, \ \delta \in \Omega, \quad (5.6)$$

$$y(t,0,\delta) = y(t,\pi,\delta) = 0; \ t \in \mathbb{R}_+, \ \delta \in \Omega, \quad (5.7)$$

$$y(0,w,\delta) = V(w,\delta); \ w \in [0,\pi], \ \delta \in \Omega, \quad (5.8)$$

where $V : [0,\pi] \times \Omega \to \mathbb{R}$, and $G : \mathbb{R}_+ \times \mathbb{R} \times \Omega \to \mathbb{R}$ are given functions.
Let

$$y(t,\delta)(w) = y(t,w,\delta), \ t \in \mathbb{R}_+ \times \Omega \to \mathbb{R}, \ w \in [0,\pi],$$

$$f(t,y,\delta)(w) = G(t, y(t,w,\delta), \delta); \ w \in [0,\pi],$$

$$y(0,\delta)(w) = V(w,\delta); \ w \in [0,\pi].$$

Take $E = L^2[0,\pi]$ and define $A : D(A) \subset E \to E$ by $A\delta = \delta''$ with domain $D(A) = \{\delta \in E, \delta, \delta' \text{ are absolutely continuous}, \delta'' \in E, \delta(0) = \delta(\pi) = 0\}$.

Then

$$A\delta = \sum_{\beta=1}^{\infty} \beta^2 (\delta, \delta_\beta) \delta_\beta, \ \delta \in D(A)$$

where $(\cdot, \cdot)$ is the inner product in L^2 and $\delta_\beta(s) = \sqrt{\frac{2}{\pi}} \sin ns$, $\beta = 1, 2, \ldots$ is the orthogonal set of eigenvectors in A. It is well known (see [11]) that A is the infinitesimal generator of an analytic semigroup $S(t)$, $t \geq 0$ in E and is given by

$$S(t)\delta = \sum_{\beta=1}^{\infty} exp(-\beta^2 t)(\delta, \delta_\beta)\delta_\beta, \ \delta \in E.$$

Since the analytic semigroup $S(t)$ is compact, there exists a constant $\gamma \geq 1$ such that

$$\|S(t)\|_{B(E)} \leq \gamma.$$

Also assume that there exists a continuous function $\lambda : \mathbb{R}_+ \to \mathbb{R}_+$ such that

$$|G(t, y(t, w, \delta), \delta)| \leq \lambda(t)(1 + |y(t, w, \delta)|).$$

Assume that there exists a function $L \in L^1(\mathbb{R}_+, \mathbb{R}_+)$ such that

$$|G(t, y, \delta) - G(t, \bar{y}, \delta)| \leq L(t)|y - \bar{y}|, \ t \geq 0, \ y, \bar{y} \in \mathbb{R}.$$

We can show that problem (5.1)–(5.2) is an abstract formulation of problem (5.6)–(5.8). Since all the conditions of Theorem 2.14 are satisfied, the problem (5.6)–(5.8) has a random mild solution.

Example 2. Consider now the following partial random evolution equation

$$\frac{\partial y}{\partial t}(t, w, \delta) = \frac{\partial^2 y}{\partial w^2}(t, w, \delta)$$

$$+ G(t, y(t, w, \delta), \delta); \ w \in [0, \pi], \ t \in \mathbb{R}_+, \ \delta \in \Omega, \quad (5.9)$$

$$y(t, 0, \delta) = y(t, \pi, \delta) = 0; \ t \in \mathbb{R}_+, \ \delta \in \Omega, \quad\quad (5.10)$$

$$y(0, w, \delta) = V(w, \delta); \ w \in [0, \pi], \ \delta \in \Omega, \quad\quad\quad (5.11)$$

where $V : [0, \pi] \times \Omega \to \mathbb{R}$, *and* $G : \mathbb{R}_+ \times \mathbb{R} \times \Omega \to \mathbb{R}$ are given functions.
     Let

$$y(t, \delta)(w) = y(t, w, \delta), \ t \in \mathbb{R}_+ \times \Omega \to \mathbb{R}, \ w \in [0, \pi],$$

$$f(t, y, \delta)(w) = G(t, y(t, w, \delta), \delta); \ w \in [0, \pi],$$

$$y(0, \delta)(w) = V(w, \delta); \ w \in [0, \pi].$$

Take $E = C(\overline{\Omega})$, the Banach space of continuous function on $\overline{\Omega}$ with values in $\mathbb{R}$. Define the linear operator A on E by

$$Ay = \frac{\partial^2 y}{\partial w^2}, \quad \text{in} \quad D(A) = \{y \in C(\overline{\Omega}) : y = 0 \text{ on } \partial\Omega, \ \frac{\partial^2 y}{\partial w^2} \in C(\overline{\Omega}\}.$$

Now, we have

$$\overline{D(A)} = C_0(\overline{\Omega}) = \{z \in C(\overline{\Omega}) : z = 0 \text{ on } \partial\Omega\} \neq C(\overline{\Omega}).$$

It is well known from [323] that $(0, +\infty) \subseteq \rho(A)$ and for $\chi > 0$

$$\|R(\chi, A)\|_{B(E)} \leq \frac{1}{\chi}.$$

It follows that A generates an integrated semigroup $(S(t))_{t \geq 0}$ and that

$$\|S'(t)\|_{B(E)} \leq e^{-\alpha t};$$

for $t \geq 0$ and some constant $\alpha > 0$, and A satisfied the Hille-Yosida condition.

Assume that there exists a continuous function $\lambda : \mathbb{R}_+ \to \mathbb{R}_+$ such that

$$|G(t, y(t, w, \delta), \delta)| \leq \lambda(t)(1 + |y(t, w, \delta)|).$$

Assume that there exists a function $\tilde{l} \in L^1(\mathbb{R}_+, \mathbb{R}_+)$ such that

$$|G(t, y, \delta) - G(t, \overline{y}, \delta)| \leq \tilde{l}(t)|y - \overline{y}|, \ t \in \mathbb{R}_+, \ y, \overline{y} \in \mathbb{R}.$$

We can show that problem (5.1)–(5.2) is an abstract formulation of problem (5.9)–(5.11). Since all the conditions of Theorem 5.2 are satisfied, the problem (5.9)–(5.11) has a random integral solution solution.

5.3 Controllability of Second Order Semilinear Random Differential Equations in Fréchet Spaces

This paper deals with the controllability of the functional differential equation with random effect:

$$\begin{cases} y''(t, \delta) = A_1 y(t, \delta) + f(t, y(t, \delta), \delta) + A_2 g(t, \delta), \ \text{a.e.} \ t \in \mathbb{R}_+ := [0, \infty), \\ y(0, \delta) = \varpi_1(\delta), \ y'(0, \delta) = \varpi_2(\delta), \ \delta \in \Omega, \end{cases}$$

$$(5.12)$$

where (Ω, F, P) is a complete probability space, $f : \mathbb{R}_+ \times E \times \Omega \to E$ is a given function, $\varpi_1, \varpi_2 : \Omega \to E$ are given measurable functions, $A_1 : \Phi(A_1) \subset E \to E$ is the infinitesimal generator of a strongly continuous cosine family of bounded linear operators $(S_1(t))_{t\in\mathbb{R}}$ on E, and $(E, \|\cdot\|)$ is a real Banach space. $g(\cdot, \delta)$ is the control function defined in $L^2(\mathbb{R}_+, U)$, a Banach space of admissible control functions with U as a Banach space, and A_2 is a bounded linear operator from U into E. The main result is based upon a generalization of the classical Darbo fixed point theorem and the concept of measure of non-compactness combined with the family of cosine operators.

5.3.1 Existence Results

Definition 5.5. The problem (5.12) is controllable on $\mathbb{R}_+$, if for every initial states $\varpi_1(\delta)$, $\varpi_2(\delta)$, and final state $y^1(\delta)$, and for some $m \in \mathbb{N}$, there is a control $g(\cdot, \delta)$ in $L^2(\mathbb{R}_+, U)$, where the solution $y(t, \delta)$ of (5.12) verifies $y(m, \delta) = y^1(\delta)$.

Let $\tilde{E} = C(\mathbb{R}_+, E)$ be a Fréchet space with the family of seminorms

$$\|z_1\|_m = \sup_{t\in[0,m]} |z_1(t)|; \ m \in \mathbb{N},$$

and the distance

$$d(z_2, z_1) = \sum_{m=1}^{\infty} 2^{-m} \frac{\|z_2 - z_1\|_m}{1 + \|z_2 - z_1\|_m}; \ z_2, z_1 \in \tilde{E}.$$

Definition 5.6. A stochastic process $y : \mathbb{R}_+ \times \Omega \to E$ is a random mild solution of (5.12) if $y(0, \delta) = \varpi_1(\delta)$, $y'(0, \delta) = \varpi_2(\delta)$, and $y(\cdot, \delta)$ is continuous and verifies:

$$y(t, \delta) = S_1(t)\varpi_1(\delta) + S_2(t)\varpi_2(\delta) + \int_0^t S_2(t - s)f(s, y(s, \delta), \delta)ds$$

$$+ \int_0^t S_2(t - s)A_2 g(t, \delta)ds.$$

Let

$$\gamma = \sup\left\{ \|S_1(t)\|_{B(E)} : t \geq 0 \right\} \text{ and } \gamma' = \sup\left\{ \|S_2(t)\|_{B(E)} : t \geq 0 \right\}.$$

We will need the following conditions in our proof of the main result:

(5.9.1) The function f is random Carathéodory on $\mathbb{R}_+ \times E \times \Omega$.

(5.9.2) There exists a function $\lambda : \mathbb{R}_+ \times \Omega \to \mathbb{R}_+$ which is continuous in t and measurable and essentially bounded in δ such that

$$\|f(t, g, \delta)\| \leq \lambda(t, \delta)(1 + \|g\|); \text{ for a.e. } t \in \mathbb{R}_+, \text{ and each } g \in E.$$

(5.9.3) For each bounded set $B \subset E$ and for any $\delta \in \Omega$, we have

$$\alpha(f(t, B, \delta)) \leq \lambda(t, \delta)\alpha(B); \text{ for a.e. } t \in \mathbb{R}_+.$$

(5.9.4) For any $m \in \mathbb{N}^*$, the linear operator $\Upsilon : L^2([0, m], U) \to E$ defined by:

$$\Upsilon(g) = \int_0^m S_2(m - s)A_2 g(s, \delta)ds$$

has a pseudo inverse operator Υ^{-1} which takes values in the space $L^2([0, m], U)/ker(\Upsilon)$ and there exists a positive constant η such that $\|A_2\Upsilon^{-1}\| \leq \eta$.

Set

$$\lambda_m^*(\delta) = \sup_{t \in [0,m]} \lambda(t, \delta); \text{ for } m \in \mathbb{N}.$$

and let $\tilde{E}$ be the family of measure of non-compactness given by

$$\alpha_m(B) = \sup_{t \in [0,m]} e^{-4\gamma\lambda_m^*(\delta)(1+m\gamma\eta)\varsigma t}\alpha(B(t)),$$

where $\varsigma > 1$, B is bounded and equicontinuous, and

$$B(t) = \{z(t) \in E : z \in B\}; \ t \in [0, m].$$

Theorem 5.3. *Suppose that hypotheses* (5.9.1)–(5.9.4) *are met, and*

$$m\gamma'\lambda_m^*(\delta)(1 + m\gamma'\eta) < 1,$$

for each $m \in \mathbb{N}$, *and* $\delta \in \Omega$. *Then the problem* (5.12) *is controllable.*

Proof. Using (5.9.4) we define the control

$$g(t, \delta) = \Upsilon^{-1}\bigg(y^1(\delta) - S_1(m)\varpi_1(0, \delta) - S_2(m)\varpi_2(\delta)$$
$$- \int_0^m S_2(m - s)f(s, y(s, \delta), \delta)ds\bigg),$$

for $m \in \mathbb{N}$. We will demonstrate that the operator $T : \Omega \times \tilde{E} \longrightarrow \tilde{E}$ given by:

$$(T(\delta)y)(t) = S_1(t)\varpi_1(\delta) + S_2(t)\varpi_2(\delta) + \int_0^t S_2(t - s) \ f(s, y(s, \delta), \delta)ds$$
$$+ \int_0^t S_2(t - s)A_2\Upsilon^{-1}\bigg(y^1(\delta) - S_1(m)\varpi_1(\delta) - S_2(m)\varpi_2(\delta)$$
$$- \int_0^m S_2(m - \varsigma)f(\varsigma, y(\varsigma, \delta), \delta)d\varsigma\bigg)ds,$$

$$(5.13)$$

has a fixed point $y(t, \delta)$ which is a random mild solution of the problem (5.12). This implies that the problem (5.12) is controllable. Now, proving that T defined in (5.13) verifies all the requirements of Lemma 2.14.

Step 1. $T(\delta)$ *is a random operator with stochastic domain on* $\tilde{E}$.
To demonstrate that for any $y \in \tilde{E}$, $T(\cdot)(y) : \Omega \longrightarrow \tilde{E}$ is a random operator, we need to prove that $T(\cdot)(y) : \Omega \longrightarrow \tilde{E}$ is measurable. As the mapping $f(t, y, \cdot)$, $t \in \mathbb{R}_+$, $y \in \tilde{E}$ is measurable by (5.9.1) and (5.9.4). Let $\Phi : \Omega \longrightarrow 2^{\tilde{E}}$ be given by:

$$\Phi(\delta) = \{y \in \tilde{E} : \|y\|_m \leq R_m(\delta), \ m \in \mathbb{N}\},$$

where $R_m : \Omega \longrightarrow (0, \infty)$ is a random function such that

$$R_m(\delta) \geq \frac{(\gamma|\varpi_1(\delta)| + \gamma'|\varpi_2(\delta)| + m\gamma'\lambda_m^*(\delta))(1 + m\gamma'\eta) + m\gamma'\eta \left|y^1(\delta)\right|}{1 - m\gamma'\lambda_m^*(\delta)(1 + m\gamma'\eta)}.$$

Since $\Phi(\delta)$ is bounded, closed, convex and solid for all $\delta \in \Omega$, then Φ is measurable. (See [243] for more details.) For any $y \in \Phi(\delta)$ and from (5.9.2) and (5.9.3), we have

$$
\begin{aligned}
|(T(\delta)y)(t)| &\leq \gamma|\varpi_1(\delta)| + \gamma'|\varpi_2(\delta)| + \gamma' \int_0^t |f(s, y(s, \delta), \delta)|ds \\
&\quad + \gamma'\eta \int_0^t \left|y^1(\delta)\right| + \gamma|\varpi_1(\delta)| + \gamma'|\varpi_2(\delta)|ds \\
&\quad + \gamma'\eta \int_0^t \int_0^m \|S_2(\varsigma - s)\|_{B(E)} \, |f(\varsigma, y(\varsigma, \delta), \delta)| \, d\varsigma ds \\
&\leq \gamma|\varpi_1(\delta)| + \gamma'|\varpi_2(\delta)| + \gamma' \int_0^m \lambda(s, \delta)(1 + |y(s, \delta)|) \, ds \\
&\quad + m\gamma'\eta \left|y^1(\delta)\right| + m\gamma\gamma'\eta|\varpi_1(\delta)| + m{\gamma'}^2\eta|\varpi_2(\delta)| \\
&\quad + m{\gamma'}^2\eta \int_0^m \lambda(\varsigma, \delta)(1 + |y(\varsigma, \delta)|) \, d\varsigma \\
&\leq \gamma(1 + m\gamma'\eta)|\varpi_1(\delta)| + m\gamma'\eta \left|y^1(\delta)\right| + \gamma'(1 + m\gamma'\eta)|\varpi_2(\delta)| \\
&\quad + \gamma'\left(1 + m\gamma'\eta\right) \int_0^m \lambda(s, \delta)(1 + |y(s, \delta)|)ds \\
&\leq (\gamma|\varpi_1(\delta)| + \gamma'|\varpi_2(\delta)| + m\gamma'\lambda_m^*(\delta))(1 + m\gamma'\eta) + m\gamma'\eta \left|y^1(\delta)\right| \\
&\quad + m\gamma'\lambda_m^*(\delta)(1 + m\gamma'\eta)R_m(\delta) \\
&\leq R_m(\delta).
\end{aligned}
$$

Thus,

$$\|(T(\delta)y)\|_m \leq R_m(\delta).$$

Consequently, T is a random operator with stochastic domain Φ and $T(\delta) : \Phi(\delta) \longrightarrow \Phi(\delta)$.

Step 2. $T(\delta) : \Phi(\delta) \to \Phi(\delta)$ *is continuous.*
Let $\{y_n\}_{n \in \mathbb{N}}$ be a sequence where $y_n \to y$ in $\Phi(\delta)$. Then, for each $t \in [0, m]$ and $\delta \in \Omega$, we get

$$|(T(\delta)y_n)(t) - (T(\delta)y)(t)|$$

$$\leq \gamma' \int_0^t |f(s, y_n(s, \delta)) - f(s, y(s, \delta))| \, ds$$

$$+ \eta \gamma' \int_0^t \int_0^m \|S_2(m - \varsigma)\|_{B(E)} |f(\varsigma, y_n(\varsigma, \delta), \delta) - f(\varsigma, y(\varsigma, \delta), \delta)| d\varsigma ds$$

$$\leq \gamma' \int_0^t |f(s, y_n(s, \delta), \delta) - f(s, y(s, \delta), \delta)| \, ds$$

$$+ m\gamma'^2 \eta \int_0^m |f(\varsigma, y_n(\varsigma, \delta), \delta) - f(\varsigma, y(\varsigma, \delta), \delta)| d\varsigma$$

$$\leq \gamma' \left(1 + m\gamma' \eta \right) \int_0^m |f(\varsigma, y_n(\varsigma, \delta), \delta) - f(\varsigma, y(\varsigma, \delta), \delta)| d\varsigma.$$

Since $f(s, \cdot, \delta)$ is continuous, and $y_n \to y$ as $n \to \infty$, then

$$\|T(\delta)(y_n) - T(\delta)(y)\|_m \to 0 \quad \text{as } n \to \infty.$$

Consequently, we deduce that $T(\delta) : \Phi(\delta) \to \Phi(\delta)$ is a continuous random operator with stochastic domain Φ, and $T(\delta)(\Phi(\delta))$ is bounded.

Step 3. *For each bounded and equicontinuous subset B of $\Phi(\delta)$,*

$$\alpha_m(T(\delta)(B)) \leq \ell_m \alpha_m(B).$$

From Lemmas 2.5 and 2.6, for any $B \subset \Phi(\delta)$ and any $\varkappa > 0$, there exists a sequence $\{y_n\}_{n=0}^\infty \subset B$. Then, for all $t \in [0, m]$ and $\delta \in \Omega$, we have

$$\alpha((T(\delta)B)(t))$$

$$= \alpha(\{\gamma|\varpi_1(\delta)| + \gamma'|\varpi_2(\delta)| + \gamma' \int_0^t |f(s, y(s, \delta), \delta)| ds$$

$$+ \gamma' \eta \int_0^t |y^1(\delta)| + \gamma|\varpi_1(\delta)| + \gamma'|\varpi_2(\delta)| ds$$

$$+ \gamma' \eta \int_0^t \int_0^m \|S_2(\varsigma - s)\|_{B(E)} |f(\varsigma, y(\varsigma, \delta), \delta)| \, d\varsigma ds; \; y \in \Phi\})$$

$$\leq 2\alpha \left(\left\{ \gamma' \int_0^t |f(s, y_n(s, \delta), \delta)| ds \right. \right.$$

$$\left. \left. + \gamma' \eta \int_0^t \int_0^m \|S_2(\varsigma - s)\|_{B(E)} |f(\varsigma, y_n(\varsigma, \delta), \delta)| \, d\varsigma ds \right\}_{n=1}^{\infty} \right) + \varkappa$$

$$\leq 4\gamma' \int_0^t \alpha \left(\{ f(s, y_n(s, \delta), \delta) \}_{n=1}^{\infty} \right) ds$$

$$+ 4\gamma' \eta \int_0^t \int_0^m \alpha \left(\{ \|S_2(\varsigma - s)\|_{B(E)} f(\varsigma, y_n(\varsigma, \delta), \delta) \}_{n=1}^{\infty} \right) d\varsigma ds + \varkappa$$

$$\leq 4\gamma' \int_0^t \lambda_m(s) \alpha \left(\{ y_n(s, \delta) \}_{n=1}^{\infty} \right) ds$$

$$+ 4m\gamma'^2 \eta \int_0^m \lambda_m(\varsigma) \alpha \left(\{ y_n(\varsigma, \delta) \}_{n=1}^{\infty} \right) ds + \varkappa$$

$$\leq 4\gamma' \lambda_m^*(\delta)(1 + m\gamma' \eta) \int_0^t e^{4\gamma' \lambda_m^*(\delta)(1 + m\gamma' \eta)\varsigma s} e^{-4\gamma' \lambda_m^*(\delta)(1 + m\gamma' \eta)\varsigma s}$$

$$\times \alpha \left(\{ y_n(s, \delta) \}_{n=1}^{\infty} \right) ds + \varkappa$$

$$\leq \frac{1}{\varsigma} \left(e^{4\gamma' \lambda_m^*(1 + m\gamma' \eta)\varsigma t} - 1 \right) \alpha_m(\Phi) + \varkappa$$

$$\leq \frac{1}{\varsigma} e^{4\gamma' \lambda_m^*(\delta)(1 + m\gamma' \eta)\varsigma t} \alpha_m(\Phi) + \varkappa.$$

Since $\varkappa > 0$ is arbitrary, then

$$\alpha((T(\delta)B)(t)) \leq \frac{e^{4\gamma' \lambda_m^*(\delta)(1 + m\gamma' \eta)\varsigma t}}{\varsigma} \alpha_m(B).$$

Thus

$$\alpha_m(T(\delta)(B)) \leq \frac{1}{\varsigma} \alpha_m(B).$$

Thus, Theorem 2.14 implies that T has at least one fixed point in $\Phi(\delta)$ which is a random mild solution of problem (5.12). $\qquad \square$

5.3.2   *An Example*

Consider the following problem:

$$\frac{\partial^2 y}{\partial t^2}(t, w, \delta) = \frac{\partial^2 y}{\partial w^2}(t, w, \delta) + f(t, y(t, w, \delta), \delta) + A_2 g(t, \delta) \quad w \in [0, \pi]; \; t \in \mathbb{R}_+,$$
$$(5.14)$$

$$y(t, 0, \delta) = y(t, \pi, \delta) = 0; \; t \in \mathbb{R}_+, \; \delta \in \Omega, \tag{5.15}$$

$$y(0, w, \delta) = \varpi_1(w, \delta), \frac{\partial y}{\partial t}(0, w, \delta) = \varpi_2(w, \delta); \; t \in \mathbb{R}_+, \; \delta \in \Omega, \tag{5.16}$$

where $f : \mathbb{R}_+ \times \mathbb{R} \times \Omega \longrightarrow \mathbb{R}$ is a given function. Let $E = L^2[0, \pi]$ and consider the operator $A_1 : E \longrightarrow E$ by $A_1 z = z''$ with domain

$$\Phi(A_1) = \{z \in E; z, z' \text{ are absolutely continuous, } z'' \in E, \ z(0) = z(\pi) = 0\}.$$

It is commonly established that A_1 is the infinitesimal generator of a strongly continuous cosine function $(S_1(t))_{t \in \mathbb{R}}$ on E, respectively. Moreover, A_1 has discrete spectrum, the eigenvalues are $-m^2, m \in \mathbb{N}$ with corresponding normalized eigenvectors

$$y_m(\varsigma) := \left(\frac{2}{\pi}\right)^{\frac{1}{2}} \sin(m\varsigma),$$

and the following properties are met:

(a) $\{y_m : m \in \mathbb{N}\}$ is an orthonormal basis of E,

(b) If $y \in E$, then $A_1 y = - \sum\limits_{m=1}^{\infty} m^2 \langle y, y_m \rangle y_m$,

(c) For $y \in E$,

$$S_1(t)y = \sum_{m=1}^{\infty} \cos(nt) \langle y, y_m \rangle y_m,$$

and

$$S_2(t)y = \sum_{m=1}^{\infty} \frac{\sin(nt)}{m} \langle y, y_m \rangle y_m,$$

and that

$$\|S_1(t)\| = \|S_2(t)\| \le 1, \text{ for all } t \ge 0.$$

(d) If we denote the group of translations on E given by

$$\chi(t)y(\varsigma, \delta) = \tilde{y}(\varsigma + t, \delta),$$

where $\tilde{y}$ is the extension of y with period 2π, then

$$S_1(t) = \frac{1}{2}(\chi(t) + \chi(-t)); A_1 = \Phi,$$

where Φ is the infinitesimal generator of the group on

$$\tilde{E} = \{y(\cdot, \delta) \in H^1(0, \pi) : y(0, \delta) = y(\pi, \delta) = 0\}.$$

Suppose that A_2 is a bounded linear from U into E and $\Upsilon : L^2(\mathbb{R}_+, U) \to E$ defined by:

$$\Upsilon(g) = \int_0^m S_2(m - s)A_2 g(s, \delta)ds; \ m \in \mathbb{N},$$

has a pseudo inverse operator Υ^{-1} in $L^2(\mathbb{R}_+, U)/ker\,\Upsilon$. Then the problem (5.12) is an abstract formulation of the problem (5.14)–(5.16). If (5.9.1)–(5.9.4) hold, then Theorem 5.3 implies that the problem (5.14)–(5.16) is controllable.

5.4 Notes and Remarks

The findings in Chapter 5 are based on the papers [324, 325]. We suggest to the reader the monographs [119–123], and the papers [44, 291, 326–333], for more information on the concepts studied in this chapter.

Chapter 6

Semilinear Differential Evolution Equations with Random Effects and Delay

6.1 Introduction and Motivations

This chapter is devoted to studying some existence and controllability results for some classes of functional differential equations with delay and random effects. We base our arguments on some suitable random fixed point theorem with stochastic domain and the integrated semigroup and a generalization of the classical Darbo fixed point theorem for Fréchet spaces associated with the concept of measures of non-compactness. Finally, examples are given to validate the theoretical part.

The following works served as our motivation, and we refer interested readers to these references:

- Functional differential equation theory has evolved as a significant field of nonlinear analysis. For several years, functional differential delay equations have been employed to represent scientific phenomena. It is sometimes believed that the delay may well be a fixed constant or supplied as an integral, in which instance it is referred to as a distributed delay [218, 219, 334].

- Several papers on random differential equations have been published in recent years; see [38, 44, 119, 123, 327] and references therein.

- The controllability of linear and nonlinear systems described by ordinary differential equations in finite-dimensional space has received a great deal of attention. Numerous researchers have expanded the notion to infinite-dimensional systems with bounded operators in Banach spaces, see [47, 48, 335]. The authors of [336] demonstrated how to transform the controllability problem into a fixed point problem. We suggest the papers [50, 51] for more details. In [10, 32, 52, 53], the authors explored a wide range of functional differential equations and inclusions and suggested various controllability results. Dilao *et al.* [337] considered the controllability of a class of integrodifferential evolution equations.

- Many researchers have expressed interest in the study of differential equations with state-dependent delay since they are fundamental in applications and qualitative theory, and they describe many physical, chemical, and biological problems, for more details, see the papers of Büger and Martin [16], Si and Wang [303]. The literature on this topic is mostly concerned with first order ordinary differential equations on finite dimensional spaces, we mention the works [291–293]. While there are several papers in which authors discuss differential problems with various forms of delays (see [38, 44, 275]), there are few studies on abstract second order ordinary differential equations with state-dependent delay applied to partial differential equations, we cite [273, 274, 294]. The authors of [39, 338–341] investigated multiple differential problems using various tools and approaches, one of which was the fixed point theory.

- In [328], Diop *et al.* considered the following random partial integro-differential equations with unbounded delay:

$$\begin{cases} y'(t,\delta) = Ay(t,\delta) + \displaystyle\int_0^t \beta(t-r)y(r,\delta)dr + \mathrm{F}\left(t, y_t(\cdot,\delta),\delta\right), & t \in [0,\kappa], \\ y(t,\delta) = \phi(t,\delta), & t \in (-\infty, 0], \end{cases}$$

where A is a generator of a C_0 semigroup $(\mathrm{S}(t)_{t\geq 0})$ from a Banach space J into J, $\beta(t)$ is a closed linear operator with domain $\mathcal{D}(\mathrm{A}) \subset \mathcal{D}(\beta(t))$ and δ is a random variable. $\mathrm{F} : I \times \mathcal{B} \times J \to \mathrm{X}$ is a Carathéodory function, and $\phi \in \mathcal{B} \times J$, where $\mathcal{B}$ denotes an abstract phase space described axiomatically. For any $t \in I, y_t(\cdot,\delta)$ denotes the function in $\mathcal{B}$ defined by $y_t(t,\delta) = y(t+t,\delta), t \in (-\infty; 0]$. The authors used a random fixed point theorem with a stochastic domain combined with Schauder's fixed point theorem and Grimmer's resolvent operator theory in their proofs.

- The authors of [38] studied the existence and controllability results for the second order functional differential equation with delay and random effects that follows:

$$\begin{cases} y''(t,\delta) = A_1 y(t,\delta) + \psi\left(t, y_t(\cdot,\delta),\delta\right) + A_2 f(t,\delta), & \text{a.e. } t \in [0,\kappa], \\ y(t,\delta) = \varpi_1(t,\delta); t \in (-\infty, 0], \\ y'(0,\delta) = \varpi_2(\delta), \end{cases}$$

with results based on a random fixed point theorem with a stochastic domain.

- The authors of [302] presented the existence of mild solutions for the following partial integro-differential equations with state-dependent nonlocal conditions:

$$
\begin{cases}
y'(t) = Ay(t) + \displaystyle\int_0^t Y(t-s)y(s)ds + f\left(t, y_{\rho(t,y_t)}\right), \; t \in \mathbb{R}^+, \\
y_0 = \mathrm{K}(\zeta(y), y) \in C([-\varpi, 0], E),
\end{cases}
$$

where $A : D(A) \subset E \to E$ is the infinitesimal generator of C_0-semigroup $(\mathrm{T}(t))_{t \geq 0}$ on a Banach space E.

- In [301], Hernandez studied the global existence and uniqueness of solutions and well posedness of the following general class of abstract second order differential equations with state dependent delay:

$$
\begin{cases}
y''(t) = Ay(t) + f\left(t, y\left(t - \sigma_1\left(t, y_t\right)\right), y'\left(t - \sigma_2\left(t, y_t\right)\right)\right), \quad t \in [0, a], \\
y_0 = \chi \in C([-\varpi, 0]; E), \; (y')_0 = \chi' \in C([-\varpi, 0]; E),
\end{cases}
$$

where $A : D(A) \subset E \to E$ is the generator of a strongly continuous cosine family of bounded linear operators $(S(t))_{t \in \mathbb{R}}$ defined on a Banach space $(E, \| \cdot \|), \sigma_i \in C([0, a] \times C([-\varpi, 0]; E); [0, \varpi])$, for $i = 1, 2$, and $f \in C([0, a] \times E \times E; E)$.

- In [342], Balachandran and Sakthivel considered the following integrodifferential system:

$$
y'(t) = Ay(t) + (Bu)(t) + f\left(t, y(t), \int_0^t g(t, s, y(s))ds\right),
$$

$$
y(0) = y_0, \quad t \in J = [0, b],
$$

where the state $y(\cdot)$ takes values in a Banach space X with the norm $\| \cdot \|$ and the control function $u(\cdot)$ is given in $L^2(J, U)$, a Banach space of admissible control functions, with U as a Banach space. Here, A is the infinitesimal generator of a strongly continuous semigroup $T(t), t \geq 0$ in the Banach space X and $g : \Delta \times X \to X, f : J \times X \times X \to X$ are given functions and B is a bounded linear operator from U into X. Here $\Delta = \{(t, s) : 0 \leq s \leq t \leq b\}$. The authors employed a fixed-point theorem due to Schaefer.

- In [343], Yan investigated the controllability of the following fractionalorder partial neutral functional integrodifferential inclusions with infinite delay in Banach spaces:

$$\begin{cases} {}^cD^q\left[y(t) - g\left(t, y_t\right)\right] \in Ay(t) + (Bu)(t) \\ \qquad\qquad + F\left(t, y_t, \displaystyle\int_0^t h\left(t, s, y_s\right) ds\right), \quad t \in J = [0, b], \\ y(t) = \phi(t), \quad t \in (-\infty, 0], \end{cases}$$

where the unknown $y(\cdot)$ takes values in Banach space X with norm $\|\cdot\|$, ${}^cD^q$ is the Caputo fractional derivative of order $0 < q < 1$, A is the infinitesimal generator of a compact analytic semigroup of uniformly bounded linear operators $\{T(t), t \geq 0\}$ in X. The authors established sufficient conditions for the controllability for the problem in Banach spaces by relying on analytic semigroups and fractional powers of closed operators and nonlinear alternative of Leray-Schauder type for multi-valued maps due to D. O'Regan.

6.2 Functional Delay Random Semilinear Differential Equations

6.2.1 *Introduction*

In this section, we show that there are integral solutions to the following problem:

$$y'(t, \delta) = Ay(t, \delta) + f(t, y_t(\cdot, \delta), \delta), \quad \text{a.e. } t \in J := [0, T], \qquad (6.1)$$

$$y(t, \delta) = \Psi(t, \delta), \quad t \in (-\infty, 0], \qquad (6.2)$$

where (Ω, G, P) is a complete probability space, δ is a random variable in Ω, $f : J \times \mathcal{B} \times \Omega \to E$, $\Psi \in \mathcal{B} \times \Omega$ are given random functions, $A : D(A) \subset E \to E$ is nondensely defined closed linear operator on E, $\mathcal{B}$ is the phase space and $(E, |\cdot|)$ is a real Banach space. For any function y defined on $(-\infty, T] \times \Omega$ and any $t \in J$ we denote by $y_t(\cdot, \delta)$ the element of $\mathcal{B} \times \Omega$ defined by $y_t(\varepsilon, \delta) = y(t + \varepsilon, \delta), \varepsilon \in (-\infty, 0]$. Some notations will be given later.

6.2.2 *Existence Results*

By $\mathcal{B}_{UC}((-\infty, 0], E)$, we denote the space of bounded uniformly continuous functions defined from $(-\infty, 0]$ to E.

Definition 6.1. A stochastic process $y : J \times \Omega \to E$ is called a random integral solution of (6.1)–(6.2) if:

(i) $y(\cdot, \delta)$ is continuous on J, and $\Psi(0, \delta) \in \overline{D(A)}$

(ii) $\int_0^t y(\tau, \delta) d\tau \in D(A)$ for $t \in J$,

(iii) $y(t, \delta) = \Psi(t, \delta), t \in (-\infty, 0]$, and

(vi) $y(t, \delta) = \Psi(0, \delta) + A \int_0^t y(\tau, \delta) d\tau + \int_0^t f(\tau, y_\tau(\cdot, \delta), \delta) d\tau, \ t \in J$,

where we have the following hypothesis:

(6.1.1) *A satisfies the Hille-Yosida condition.*

It follows that $y(t, \delta) \in \overline{D(A)}$, for $t \geq 0$. Furthermore, y verifies the following:

$$y(t, \delta) = \zeta'(t)\Psi(0, \delta) + \frac{d}{dt} \int_0^t \zeta(t - \tau) f(\tau, y_\tau(\cdot, \delta), \delta) d\tau, \quad t \geq 0 \qquad (6.3)$$

Let $B_\eta = \eta R(\eta, A) = \eta(\eta I - A)^{-1}$, then for all $x \in \overline{D(A)}, B_\eta x \longrightarrow x$ as $\eta \longrightarrow \infty$. Also from the Hille-Yosida condition (with n=1) it is easy to see that $\lim_{\eta \to \infty} |B_\eta x| \leq M|x|$, since

$$|B_\eta| = |\eta(\eta I - A)^{-1}| \leq \frac{M\eta}{\eta - \lambda}.$$

Thus $\lim_{\eta \to \infty} |B_\eta| \leq M$.

We notice also that if y satisfies (6.3), then

$$y(t, \delta) = \zeta'(t)\Psi(0, \delta) + \lim_{\eta \to \infty} \int_0^t \zeta'(t - \tau) B_\eta f(\tau, y_\tau(\cdot, \delta), \delta) d\tau, \quad t \geq 0.$$

Suppose that $\lambda > 0$.

Theorem 6.1. *Suppose that hypothesis (6.1.1) and the following requirements are valid:*

(6.2.1) *The operator $\zeta'(t)$ is compact in $\overline{D(A)}$, for $t > 0$.*

(6.2.2) *The function $f : J \times \mathcal{B} \times \Omega \to E$ is Carathéodory.*

(6.2.3) *There exist a functions $\varpi : J \times \Omega \to \mathbb{R}^+$ and $p : J \times \Omega \to \mathbb{R}^+$ such that for each $\delta \in \Omega$, $\varpi(\cdot, \delta)$ is a continuous non-decreasing function and $p(\cdot, \delta)$ integrable with:*

$$|f(t, u, \delta)| \leq p(t, \delta) \, \varpi(\|u\|_{\mathcal{B}}, \delta) \text{ for a.e. } t \in J \text{ and each } u \in \mathcal{B}.$$

(6.2.4) *There exists a random function $R : \Omega \to \mathbb{R}^+ \backslash \{0\}$ such that:*

$$Me^{\lambda T} \|\Psi\|_{\mathcal{B}} + Me^{\lambda T} \varpi(D_T, \delta) \|p\|_{L^1} \leq R(\delta),$$

where

$$D_T := K_T R(\delta) + M_T \|\Psi\|_{\mathcal{B}}.$$

(6.2.5) *For each $\delta \in \Omega, \Psi(\cdot, \delta)$ is continuous and for each $t, \Psi(t, \cdot)$ is measurable.*

Then the problem (6.1)–(6.2) *has at least one random integral solution on* $(-\infty, T]$.

Proof. Let

$$\Delta = \{u \in C((-\infty, T], \overline{D(A)}) : u|_{(-\infty, 0]} \in \mathcal{B} \text{ and } u \in C\}$$

and $\mathcal{T} : \Omega \times \Delta \to \Delta$ be the random operator given by

$$(\mathcal{T}(\delta)y)(t) = \begin{cases} \Psi(t, \delta), & t \in (-\infty, 0], \\ \zeta'(0) \, \Psi(0, \delta) + \dfrac{d}{dt} \displaystyle\int_0^t \zeta(t - \tau) \, f(\tau, y_\tau(\cdot, \delta), \delta) \, d\tau, & t \in J. \end{cases}$$
$$(6.4)$$

The fixed points of $\mathcal{T}$ are solutions to (6.1)–(6.2). We demonstrate that $\mathcal{T}$ is a random operator, i.e. for any $y \in \Delta$, $\mathcal{T}(\cdot)(y) : \Omega \longrightarrow \Delta$ is a random variable. Thus, we need to demonstrate that $\mathcal{T}(\cdot)(y) : \Omega \longrightarrow \Delta$ is measurable since the mapping $f(t, y, \cdot)$, $t \in J$, $y \in \Delta$ is measurable by conditions (6.2.2) and (6.2.5).

Let $D : \Omega \longrightarrow 2^\Delta$ be defined by:

$$D(\delta) = \{y \in \Delta : \|y\| \leq R(\delta)\}.$$

$D(\delta)$ is closed, convex, bounded and solid for all $\delta \in \Omega$. Then D is measurable (see [243]).

Let $\delta \in \Omega$, then for $y \in D(\delta)$, and by (6.1.1), we have

$$\|y_\tau\|_{\mathcal{B}} \leq L(\tau)|y(\tau)| + M(\tau)\|y_0\|_{\mathcal{B}}$$
$$\leq K_T |y(\tau)| + M_T \|\Psi\|_{\mathcal{B}}.$$

By (6.2.3) and (6.2.2), we have

$$|(\mathcal{T}(\delta)y)(t)| \leq |\zeta'(t)\Psi(0, \delta)| + \left| \frac{d}{dt} \int_0^t \zeta(t - \tau) f(\tau, y_\tau, \delta) \right| d\tau$$

$$\leq Me^{\lambda t}|\Psi(0,\delta)| + M_T e^{\lambda t} \int_0^t e^{-\lambda \tau} p(\tau,\delta)\, \varpi\, (\|y_\tau\|_{\mathcal{B}},\delta)\, d\tau$$

$$\leq Me^{\lambda T}\|\Psi\|_{\mathcal{B}} + M_T e^{\lambda T} \int_0^t p(\tau,\delta)\, \varpi\, (\|y_\tau\|_{\mathcal{B}},\delta)\, d\tau.$$

Set

$$D_T := K_T R(\delta) + M_T\|\Psi\|_{\mathcal{B}}.$$

Thus, we have

$$|(\mathcal{T}(\delta)y)(t)| \leq Me^{\lambda T}\|\Psi\|_{\mathcal{B}} + M_T e^{\lambda T} \varpi(D_T,\delta) \int_0^T p(\tau,\delta)\, d\tau.$$

Thus

$$\|(\mathcal{T}(\delta)y)\| \leq Me^{\lambda T}\|\Psi\|_{\mathcal{B}} + Me^{\lambda T} \varpi(D_T,\delta)\|p\|_{L^1} \leq R(\delta).$$

This implies that $\mathcal{T}$ is a random operator with stochastic domain D and $F(\delta) : D(\delta) \longrightarrow D(\delta)$ for each $\delta \in \Omega$.

Step 1: $\mathcal{T}$ is continuous.
Let y^n be a sequence such that $y^n \longrightarrow y$ in Δ. Then

$$|(\mathcal{T}(\delta)y^n)(t) - (\mathcal{T}(\delta)y)(t)| = \left| \frac{d}{dt} \int_0^t \zeta(t-\tau)[f(\tau,y_\tau^n,\delta) - f(\tau,y_\tau,\delta)]\, d\tau \right|$$

$$\leq Me^{\lambda T} \int_0^t e^{-\lambda \tau}\, |f(\tau,y_\tau^n,\delta) - f(\tau,y_\tau,\delta)|\, d\tau.$$

$$\leq Me^{\lambda T} \int_0^t |f(\tau,y_\tau^n,\delta) - f(\tau,y_\tau,\delta)|\, d\tau.$$

Since $f(\tau,\cdot,\delta)$ is continuous, we have by the Lebesgue dominated convergence theorem

$$\|f(\cdot,y^n,\delta) - f(\cdot,y.,\delta)\|_{L^1} \to 0 \text{ as } n \to +\infty,$$

which implies that $\mathcal{T}$ is continuous.

Step 2: We demonstrate that for $\delta \in \Omega, \{y \in D(\delta) : \mathcal{T}(\delta)y = y\} \neq \emptyset$. To prove this we apply Schauder's theorem [214].

(a) $\mathcal{T}$ maps bounded sets into equicontinuous sets in $D(\delta)$.
    Let $\varrho_1, \varrho_2 \in J$ with $\varrho_2 > \varrho_1$, $D(\delta)$ be a bounded set, and $y \in D(\delta)$. Then

$$|(\mathcal{T}(\delta)y)(\varrho_2) - (\mathcal{T}(\delta)y)(\varrho_1)|$$

$$= \left| \zeta'(\varrho_2)\Psi(0,\delta) - \zeta'(\varrho_1)\Psi(0,\delta) \right.$$

$$+ \frac{d}{dt}\int_0^{\varrho_2} \zeta(\varrho_2 - \tau)f(\tau, y_\tau, \delta)d\tau$$

$$\left. - \frac{d}{dt}\int_0^{\varrho_1} \zeta(\varrho_1 - \tau)f(\tau, y_\tau, \delta)\,d\tau \right|$$

$$\leq |\zeta'(\varrho_2) - \zeta'(\varrho_1)|\|\Psi\|_{\mathcal{B}}$$

$$+ \left| \lim_{\eta\to\infty}\int_0^{\varrho_1} [\zeta'(\varrho_2 - \tau) - \zeta'(\varrho_1 - \tau)]B_\eta f(\tau, y_\tau, \delta)\,d\tau \right|$$

$$+ \left| \lim_{\eta\to\infty}\int_{\varrho_1}^{\varrho_2} \zeta'(\varrho_2 - \tau)B_\eta f(\tau, y_\tau, \delta)\,d\tau \right|$$

$$\leq |\zeta'(\varrho_2) - \zeta'(\varrho_1)|\|\Psi\|_{\mathcal{B}}$$

$$+ \varpi(D_T, \delta)\left| \lim_{\eta\to\infty}\int_0^{\varrho_1} [\zeta'(\varrho_2 - \tau) - \zeta'(\varrho_1 - \tau)]B_\eta p(\tau, \delta)\,d\tau \right|$$

$$+ M_T e^{\lambda T}\varpi(D_T, \delta)\left| \lim_{\eta\to\infty}\int_{\varrho_1}^{\varrho_2} B_\eta p(\tau, \delta)\,d\tau \right|.$$

The right-hand of the above inequality tends to zero as $\varrho_2 - \varrho_1 \to 0$, since $T(t)$ is uniformly continuous. As $\mathcal{T}$ is bounded and equicontinuous together with the Arzelá-Ascoli theorem it suffices to show that the operator $\mathcal{T}$ maps $D(\delta)$ into a precompact set in E.

(b) Let $t \in J$ be fixed and let ϵ be a real number satisfying $0 < \epsilon < t$. For $y \in D(\delta)$ we define

$$(\mathcal{T}_\epsilon(\delta)y)(t) = \zeta'(t)\Psi(0,\delta) + \zeta'(\epsilon)\lim_{\eta\to\infty}\int_0^{t-\epsilon} \zeta'(t-\tau-\epsilon)B_\eta f(\tau, y_\tau, \delta)\,d\tau.$$

Since $T(t)$ is a compact operator, the set $Z_\epsilon(t,\delta) = \{(\mathcal{T}_\epsilon(\delta)y)(t) : y \in D(\delta)\}$ is pre-compact in E for every ϵ, $0 < \epsilon < t$. Moreover

$$|(\mathcal{T}(\delta)y)(t) - (\mathcal{T}_\epsilon(\delta)y)(t)| \leq \int_{t-\epsilon}^t \|\zeta'(t-\tau)\|\|f(\tau, y_\tau, \delta)|d\tau$$

$$\leq M e^{\lambda T}\varpi(D_T, \delta)\lim_{\eta\to\infty}\left| \int_{t-\epsilon}^t B_\eta p(\tau, \delta)d\tau \right|.$$

Therefore the set $Z(t,\delta) = \{(\mathcal{T}(\delta)y)(t) : y \in D(\delta)\}$ is precompact in E.

Consequently, we can deduce that $\mathcal{T}(\delta) : D(\delta) \to D(\delta)$ is continuous and compact. We may conclude by Schauder's theorem [214] that $\mathcal{T}(\delta)$ has a fixed point $y(\delta)$ in $D(\delta)$. Since $\bigcap_{\delta\in\Omega} D(\delta) \neq \emptyset$, there exists a measurable selector of int$D$. By Lemma 2.18, the random operator $\mathcal{T}$ has a stochastic fixed point $y^*(\delta)$, which is a random integral solution of the random problem (6.1)–(6.2). $\qquad\square$

6.2.3   An Example

We consider the following problem:

$$\frac{\partial\chi(\varsigma,\beta,\delta)}{\partial\varsigma} = \frac{\partial^2\chi(\varsigma,\beta,\delta)}{\partial\beta^2} + C_0(\delta)K(\delta)e^{-\varsigma}\int_{-\infty}^{0}\frac{\exp(\chi(\varsigma+\tau,\beta,\delta))}{1+\tau^2}d\tau,$$

(6.5)

$$\chi(\varsigma,0,\delta) = \chi(\varsigma,\pi,\delta) = 0,$$

(6.6)

$$\chi(\tau,\beta,\delta) = \chi_0(\tau,\beta,\delta), \quad \tau \in (-\infty,0],$$

(6.7)

where K and C_0 are a real-valued random variable, $\beta \in [0,\pi]$, $\varsigma \in [0,1]$, $\delta \in \Omega$ and $\chi_0 : (-\infty,0] \times [0,\pi] \times \Omega \to \mathbb{R}$.

We choose $E = C([0,\pi];\mathbb{R})$ equipped with the uniform topology and let $A : D(A) \subset E \to E$ given by:

$$D(A) = \{y \in C^2([0,\pi],\mathbb{R}) : y(0) = y(\pi) = 0\}, \quad Ay = y''.$$

We know that A verifies the Hille-Yosida condition (see [344]), with

$$(0,+\infty) \subset \rho(A),$$

$$\|(\eta I - A)^{-1}\| \le \frac{1}{\eta}, \quad \eta > 0,$$

and

$$\overline{D(A)} = \{y \in E; y(0) = y(\pi) = 0\} \neq E.$$

which implies that A generates an integrated semigroup $\zeta(\varsigma))_{\varsigma \ge 0}$ and

$$\|\zeta'(\varsigma)\| \le e^{-\epsilon\varsigma}$$

for $\varsigma \ge 0$ and for some constant $\epsilon > 0$.

The function $f(\varsigma,\varphi(\beta),\delta)$ is of Carathéodory, and verifies (6.2.3) with

$$p(\varsigma,\delta) = K(\delta)\frac{\pi}{2}e^{-\varsigma} \text{ and } \varpi(\beta,\delta) = |C_0(\delta)|e^{\beta}.$$

Let $\mathcal{B} = \mathcal{B}_{UC}((-\infty, 0]; E)$ with

$$\|\Psi\| = \sup_{\tau \leq 0} |\Psi(\tau)|, \ \ \Psi \in \mathcal{B}.$$

Set

$$y(\varsigma, \beta, \delta) = \chi(\varsigma, \beta, \delta), \varsigma \in [0, 1],$$

$$\Psi(\tau, \beta, \delta) = \chi_0(\tau, \beta, \delta), \ \ \tau \in (-\infty, 0], \ \ \beta \in [0, \pi], \delta \in \Omega,$$

$$f(\varsigma, \varphi(\beta), \delta) = \int_{-\infty}^{0} e^{-\varsigma} \varphi(\tau, \beta, \delta) d\tau,$$

$$\varphi(\tau, \beta, \delta) = \exp(\chi(\varsigma + \tau, \beta, \delta)).$$

It is clear that problem (6.1)–(6.2) is an abstract formulation of problem (6.5)–(6.7). Also, we can show that conditions (6.1.1)–(6.2.5) are satisfied. Hence Theorem 6.1 implies that the random problem (6.5)–(6.7) has at least one random integral solution.

6.3   Abstract Random Differential Equations with State-Dependent Delay Using Measures of Non-Compactness

6.3.1   *Introduction*

Motivated by the above-mentioned papers and the works [329, 330], where the authors presented some existence results for a random fractional equation, we discuss in this section the existence of mild solutions defined on unbounded interval for general class of abstract second-order differential equations with state dependent delay of the form:

$$y''(t, \delta) = Ay(t, \delta) + f(t, y_{\sigma_1(t, y_t, \delta)}, y'_{\sigma_2(t, y_t, \delta)}, \delta), \ t \in \mathbb{R}_+ := [0, +\infty), \ \delta \in \Omega, \tag{6.8}$$

$$y(t, \delta) = \Psi(t, \delta), \ t \in [-r, 0], \tag{6.9}$$

$$y'(0, \delta) = \Psi'(0, \delta), \tag{6.10}$$

where $A : D(A) \subset E \rightarrow E$ is the generator of a strongly continuous cosine family of bounded linear operators $(\Phi(t))_{t \in \mathbb{R}}$ on a Banach space $(E, \| \cdot \|)$, and $f : \mathbb{R}_+ \times C([-r, 0], E) \times C([-r, 0], E) \times \Omega \rightarrow E, \ \sigma_i : \mathbb{R}_+ \times C([-r, 0], E) \times \Omega \rightarrow E$ and $\Psi \in C([-r, 0], E) \times \Omega$ are suitable functions.

6.3.2 Existence Results

We denote by $(S(t))_{t\in\mathbb{R}}$ the associated sine family by

$$S(t) := \int_0^t \Phi(s)ds.$$

Let $\Upsilon = \{y \in E : \Phi(\cdot)y \in C^1(\mathbb{R}, E)\}$ be a Banach space (see [304]), with the norm

$$\|y\|_\Upsilon = \sup_{t\in[0,+\infty)} \|AS(t)y\| + \|y\|,$$

$AS(t)$ is a bounded linear operator from Υ into E for $t \in \mathbb{R}_+$ and $AS(s)y \to 0$ as $s \to 0$, for $y \in \Upsilon$.

Consider now the second-order abstract Cauchy problem:

$$y''(t) = Ay(t) + \psi(t), \quad t \in [0, +\infty), \tag{6.11}$$

$$y(0) = y, \quad y'(0) = z, \tag{6.12}$$

where $\psi : [0, \infty) \to E$ is an integrable function and $y, z \in E$. The function

$$y(t) = \Phi(t)y + S(t)z + \int_0^t S(t-s)\psi(s)ds, \quad t \in [0, +\infty), \tag{6.13}$$

is a mild solution of (6.11)–(6.12), and when $y \in \Upsilon$, $y(\cdot)$ is a C^1 function on $[0, +\infty)$ and

$$y'(t) = AS(t)y + \Phi(t)z + \int_0^t \Phi(t-s)\psi(s)ds, \quad t \in [0, +\infty).$$

We refer the reader to the papers of Fottorini [6] and Vasil'ev Piskarev [345] for additional details.

For sake of simplicity, in the following, we always assume that $\sigma_1(\cdot) = \sigma_2(\cdot)$ and we note that the case $\sigma_1 \neq \sigma_2$ can be studied.

In this section, we present the main results of the existence of solutions for our problem. Let us introduce the followings concepts of mild solution of the problem (6.8)–(6.10).

Definition 6.2. A function $y \in C^1([-r, +\infty], E) \times \Omega$ is said to be a mild solution of (6.8)–(6.10) if y satisfies condition (6.9) for all $t \in [-r, 0], \delta \in \Omega$ and y is solution of the following integral equation

$$y(t, \delta) = \Phi(t)\Psi(0, \delta) + S(t)\Psi'(0, \delta)$$

$$+ \int_0^t S(t-s) \, f(s, y_{\sigma(s,y_s,\delta)}, y'_{\sigma(s,y_s,\delta)}, \delta) \, ds, \quad t \in \mathbb{R}_+, \delta \in \Omega. \tag{6.14}$$

Theorem 6.2. *Assume that the following hypotheses are satisfied:*

(6.4.1) *There exists a constant $\Im_1 > 1$ where*

$$\|S(t)\|_{B(E)} \leq \Im_1 \text{ for } t \in \mathbb{R}_+.$$

(6.4.2) *There exists $\Im_2 > 0$ such that*

$$\|\Phi(t)\| \leq \Im_2 \quad \text{for each } t \in \mathbb{R}_+.$$

(6.4.3) *The function f is random Carathéeodory on $\mathbb{R}_+ \times C([-r,0], E) \times C([-r,0], E) \times \Omega$.*

(6.4.4) *There exist a continuous function $p : \mathbb{R}_+ \times \Omega \to \mathbb{R}_+$ with $p(\cdot, \delta) \in L^1_{Loc}(\mathbb{R}_+; \mathbb{R}_+)$ and a continuous non-decreasing function $\lambda : \mathbb{R}_+ \to (0, \infty)$ such that for any $\delta \in \Omega$*

$$\|f(t, y, \varkappa, \delta)\| \leq p(t, \delta)\, \lambda(\|y\| + \|\varkappa\|)$$

for a.e. $t \in \mathbb{R}_+$ and y, $\varkappa \in C([-r, 0], E)$.

(6.4.5) *For each bounded and measurable sets $\mathcal{K}, D \subset C([-r,0], E)$ and for any $\delta \in \Omega$, we have*

$$\alpha(f(t, \mathcal{K}, D, \delta)) \leq p(t, \delta) \sup_{\tau \in [-r, 0]} \alpha(\mathcal{K}(\tau) + D(\tau)); \text{ for a.e. } t \in \mathbb{R}_+.$$

where α is a Kuratowski measure of non-compactness on the Banach space E,

(6.4.6) *There exists a random function $R : \Omega \to (0, \infty)$ such that*

$$\max(\Im_2, \Im_1)[\|\Psi(0, \delta)\| + \|\Psi'(0, \delta)\| + \lambda(R_j(\delta))p_j^*(\delta)] \leq R_j(\delta).$$

For $j \in \mathbb{N}$, let

$$p_j^*(\delta) = \int_0^j p(s, \delta)ds$$

and define on $C(([-r, \infty), E)$ the family of measure of non-compactness by

$$\alpha_j(D) = \sup_{t \in [-r, j]} \alpha(D(t)) + \delta_0^j(D);$$

and $D(t) = \{\varkappa(t) \in E; \varkappa \in D\}; \ t \in [-r, j]$.

If

$$\max\{\Im_1, \Im_2\}[4p_j^*(\delta)] < 1,$$

for each $j \in \mathbb{N}$, then, (6.8)–(6.10) has at least one random mild solution.

Proof. Consider the operators $N, N' : \Omega \times C^1 \to C^1$ defined by:

$$(N(\delta)y)(t) = \Phi(t)\Psi(0,\delta) + S(t)\Psi'(0,\delta)$$

$$+ \int_0^t S(t-s) \ f(s, y_{\sigma(s,y_s,\delta)}, y'_{\sigma(s,y_s,\delta)}, \delta) \ ds \quad t \in \mathbb{R}_+,$$

and

$$(N'(\delta)y)(t) = S(t)\Psi(0,\delta) + \Phi(t)\Psi'(0,\delta)$$

$$+ \int_0^t \Phi(t-s) \ f(s, y_{\sigma(s,y_s,\delta)}, y'_{\sigma(s,y_s,\delta)}, \delta) \ ds \quad t \in \mathbb{R}_+.$$

Since the function f is continuous on $\mathbb{R}_+$, then $N(\delta)$ and $N'(\delta)$ define the mappings $N, N' : \Omega \times C^1 \to C^1$. Thus, y is a random solution for (6.8)–(6.10) if and only if $y = (N(\delta))y$. We will demonstrate that N and N' verify all requirements of Darbo's fixed point theorem for Fréchet spaces [255].

Step 1: $N(\delta)$ and $N'(\delta)$ are a random operators with stochastic domain on C^1.

Since $f(t, y, \varkappa, \delta)$ is random Carathéodory, $\delta \to f(t, y, \varkappa, \delta)$ is measurable in view of Definition 2.29. Therefore, the map

$$\delta \mapsto \Phi(t)\Psi(0,\delta) + S(t)\Psi'(0,\delta) + \int_0^t S(t-s) \ f(s, y_{\sigma(s,y_s,\delta)}, y'_{\sigma(s,y_s,\delta)}, \delta) \ ds$$

is measurable. Thus, N and N' are a random operators on $\Omega \times C^1$ into C^1.

Let $\Lambda : \Omega \to \mathcal{P}(C^1)$ be the ball given by

$$\Lambda(\delta) = \mathcal{K}_{R_j}(\delta) = \mathcal{K}(0, R_j(\delta))$$

$$= \{y \in C^1 : \|y\|_j + \|y'\|_j \le R_j(\delta)\}, \ \delta \in \Omega, \ j \in \mathbb{N},$$

Then, $\Lambda(\delta)$ is a bounded, closed, convex and solid for all $\delta \in \Omega$. Consequently, Λ is measurable by [244]. Let $\delta \in \Omega$ be fixed, then from (6.4.1)–(6.4.4), for any $j \in \mathbb{N}$, and each $y, y' \in \delta(\delta)$ and $t \in [0, j]$, we get

$$\|(N(\delta)y)(t)\| \le \|\Phi(t)\Psi(0,\delta) + S(t)\Psi'(0,\delta)$$

$$+ \int_0^t S(t-s) \ f(s, y_{\sigma(s,y_s,\delta)}, y'_{\sigma(s,y_s,\delta)}, \delta) \ ds$$

$$\le \Im_2\|\Psi(0,\delta)\| + \Im_1\|\Psi'(0,\delta)\| + \Im_1 \int_0^t p(s,\delta)\lambda(\|y\|_j + \|y'\|_j) ds$$

$$\le \Im_2\|\Psi(0,\delta)\| + \Im_1\|\Psi'(0,\delta)\| + \Im_1\lambda(R_j(\delta)) \int_0^t p(s,\delta) ds$$

$$\le \Im_2\|\Psi(0,\delta)\| + \Im_1\|\Psi'(0,\delta)\| + \Im_1\lambda(R_j(\delta)) p_j^*(\delta)$$

$$\le R_j(\delta),$$

and

$$\|(N'(\delta)y)(t)\| \le \|S(t)\Psi(0,\delta) + \Phi(t)\Psi'(0,\delta)$$

$$+ \int_0^t \Phi(t-s)\, f(s, y_{\sigma(s,y_s,\delta)}, y'_{\sigma(s,y_s,\delta)}, \delta)\, ds$$

$$\le \Im_1\|\Psi(0,\delta)\| + \Im_2\|\Psi'(0,\delta)\| + \Im_2 \int_0^t p(s,\delta)\lambda(\|y\|_j + \|y'\|_j)ds$$

$$\le \Im_1\|\Psi(0,\delta)\| + \Im_2\|\Psi'(0,\delta)\| + \Im_2\lambda(R_j(\delta)) \int_0^t p(s,\delta)ds$$

$$\le \Im_1\|\Psi(0,\delta)\| + \Im_2\|\Psi'(0,\delta)\| + \Im_2\lambda(R_j(\delta))p_j^*(\delta)$$

$$\le R_j(\delta).$$

Therefore, N and N' are random operators with stochastic domain Λ and $N(\delta): \Lambda(\delta) \to N(\delta)$, $N'(\delta): \Lambda(\delta) \to N'(\delta)$. Furthermore, $N(\delta)$ and $N'(\delta)$ map bounded sets into bounded sets in C^1.

Step 2. $N(\delta), N'(\delta): \mathcal{K}_{R_j}(\delta) \to \mathcal{K}_{R_j}(\delta)$ are continuous.
Let $\{y^k\}_{k\in\mathbb{N}}$ be a sequence such that $y^k \to y$ in $\mathcal{K}_{R_j}(\delta)$. Then for each $t \in [0,j]$ and $\delta \in \Omega$, we have

$$\|(N(\delta)y^k(t)) - (N(\delta)y)(t)\|$$

$$= \left\| \int_0^t S(t-s)\Big[f(s, y^k_{\sigma(s,y_s,\delta)}, y'^k_{\sigma(s,y_s,\delta)}, \delta) - f(s, y_{\sigma(s,y_s,\delta)}, y'_{\sigma(s,y_s,\delta)})\Big]ds \right\|$$

$$\le \int_0^t \left\| f(s, y^k_{\sigma(s,y_s,\delta)}, y'^k_{\sigma(s,y_s,\delta)}, \delta) - f(s, y_{\sigma(s,y_s,\delta)}, y'_{\sigma(s,y_s,\delta)}, \delta) \right\| ds$$

$$\times \|S(t-s)\|_{B(E)}$$

$$\le \Im_1 \int_0^t \left\| f(s, y^k_{\sigma(s,y_s,\delta)}, y'^k_{\sigma(s,y_s,\delta)}, \delta) - f(s, y_{\sigma(s,y_s,\delta)}, y'_{\sigma(s,y_s,\delta)}, \delta) \right\| ds.$$

Since $y^k \longrightarrow y$ as $k \longrightarrow \infty$, the Lebesgue dominated convergence theorem implies that

$$\|N(y^k) - N(y)\|_j \longrightarrow 0 \text{ as } k \longrightarrow +\infty.$$

and

$$\|(N'(\delta)y^k(t)) - (N'(\delta)y)(t)\|$$

$$= \left\| \int_0^t \Phi(t-s)\Big[f(s, y^k_{\sigma(s,y_s,\delta)}, y'^k_{\sigma(s,y_s,\delta)}, \delta) - f(s, y_{\sigma(s,y_s,\delta)}, y'_{\sigma(s,y_s,\delta)})\Big]ds \right\|$$

$$\leq \int_0^t \left\| f(s, y^k_{\sigma(s,y_s,\delta)}, y'^k_{\sigma(s,y_s,\delta)}, \delta) - f(s, y_{\sigma(s,y_s,\delta)}, y'_{\sigma(s,y_s,\delta)}, \delta) \right\| ds$$

$$\times \, \|S(t-s)\|_{B(E)}$$

$$\leq \Im_2 \int_0^t \| f(s, y^k_{\sigma(s,y_s,\delta)}, y'^k_{\sigma(s,y_s,\delta)}, \delta) - f(s, y_{\sigma(s,y_s,\delta)}, y'_{\sigma(s,y_s,\delta)}, \delta) \| \, ds.$$

Since $y^k \longrightarrow y$ as $k \longrightarrow \infty$, the Lebesgue dominated convergence theorem implies that

$$\|N'(y^k) - N'(y)\|_j \longrightarrow 0 \text{ as } k \longrightarrow +\infty.$$

As a consequence of Steps 1 and 2, we can conclude that $N(\delta) : \Lambda(\delta) \to N(\delta)$ and $N'(\delta) : \Lambda(\delta) \to N'(\delta)$ are continuous random operators with stochastic domain Λ, and $N(\delta)(\Lambda(\delta))$ and $N'(\delta)(\Lambda(\delta))$ are bounded.

Step 3: For each bounded subset D of $\Lambda(\delta), \alpha_j(N(\delta)(D)) \leq l_j \alpha_j(D)$ and $\alpha_j(N'(\delta)(D)) \leq l_j \alpha_j(D)$.

From Lemma 2.5 and 2.6, for any $D \subset \mathcal{K}_{R_j}(\delta)$ and any $\epsilon > 0$, there exist a sequence $\{y_k\}_{k=1}^\infty \subset D$, such that for all $t \in [0, j]$ and $\delta \in \Omega$, we have

$$\alpha((N(\delta)D)(t))$$

$$= \alpha\Big(\Big\{ \Phi(t)\Psi(0,\delta) + S(t)\Psi'(0,\delta)$$

$$+ \int_0^t S(t-s) \, f(s, y_{\sigma(s,y_s,\delta)}, y'_{\sigma(s,y_s,\delta)}, \delta) ds; \; y, y' \in D \Big\}\Big)$$

$$\leq \alpha\Big(\Big\{ \int_0^t S(t-s) \, f(s, y_{\sigma(s,y_s,\delta)}, y'_{\sigma(s,y_s)}, \delta, \delta) ds; \; y, y' \in D \Big\}\Big)$$

$$\leq 2\alpha\Big(\Big\{ \int_0^t S(t-s) f(s, y^k_{\sigma(s,y^k_s,\delta)}, y'^k_{\sigma(s,y^k_s,\delta)}, \delta) ds \Big\}_{k=1}^\infty\Big) + \epsilon$$

$$\leq 4 \int_0^t \alpha\Big(\|S(t-s)\|_{B(E)} \Big\{ f(s, y^k_{\sigma(s,y^k_s,\delta)}, y'^k_{\sigma(s,y^k_s,\delta)}, \delta) \Big\}_{k=1}^\infty \Big) ds + \epsilon$$

$$\leq 4\Im_1 \int_0^t \alpha\Big\{ f(s, y^k_{\sigma(s,y^k_s,\delta)}, y'^k_{\sigma(s,y^k_s,\delta)}, \delta) \Big\}_{k=1}^\infty \Big) ds + \epsilon$$

$$\leq 4\Im_1 \int_0^t p(s,\delta)\alpha\Big(\Big\{ y^k_{\sigma(s,y^k_s,\delta)} \Big\}_{k=1}^\infty + \Big\{ y'^k_{\sigma(s,y^k_s,\delta)} \Big\}_{k=1}^\infty \Big) ds + \epsilon$$

$$\leq 4\Im_1 \alpha(D) \int_0^t p(s,\delta) ds + \epsilon$$

$$\leq 4\Im_1 p_j^*(\delta)\alpha_j(D) + \epsilon.$$

Since $\epsilon > 0$ is arbitrary, then

$$\alpha((N(\delta)D)(t)) \le 4\Im_1 p_j^*(\delta)\alpha_j(D).$$

Thus

$$\alpha_j(N(\delta)D) \le 4\Im_1 p_j^*\alpha_j(D).$$

On the other hand, we have

$$\alpha((N'(\delta)D)(t))$$
$$= \alpha\Big(\Big\{S(t)\Psi(0,\delta) + \Phi(t)\Psi'(0,\delta)$$
$$+ \int_0^t \Phi(t-s)\,f(s,y_{\sigma(s,y_s,\delta)},y'_{\sigma(s,y_s,\delta)},\delta)ds;\ y,y' \in D\Big\}\Big)$$
$$\le \alpha\Big(\Big\{\int_0^t \Phi(t-s)\,f(s,y_{\sigma(s,y_s,\delta)},y'_{\sigma(s,y_s)},\delta,\delta)ds;\ y,y' \in D\Big\}\Big)$$
$$\le 2\alpha\Big(\Big\{\int_0^t \Phi(t-s)f(s,y^k_{\sigma(s,y_s^k,\delta)},y'^k_{\sigma(s,y_s^k,\delta)},\delta)ds\Big\}_{k=1}^\infty\Big) + \epsilon$$
$$\le 4\int_0^t \alpha\Big(\|S(t-s)\|_{B(E)}\Big\{f(s,y^k_{\sigma(s,y_s^k,\delta)},y'^k_{\sigma(s,y_s^k,\delta)},\delta)\Big\}_{k=1}^\infty\Big)ds + \epsilon$$
$$\le 4\Im_2 \int_0^t \alpha\Big(\Big\{f(s,y^k_{\sigma(s,y_s^k,\delta)},y'^k_{\sigma(s,y_s^k,\delta)},\delta)\Big\}_{k=1}^\infty\Big)ds + \epsilon$$
$$\le 4\Im_2 \int_0^t p(s,\delta)\alpha\Big(\Big\{y^k_{\sigma(s,y_s^k,\delta)}\Big\}_{k=1}^\infty + \Big\{y'^k_{\sigma(s,y_s^k,\delta)}\Big\}_{k=1}^\infty\Big)ds + \epsilon$$
$$\le 4\Im_2\alpha(D)\int_0^t p(s,\delta)ds + \epsilon$$
$$\le 4\Im_2 p_j^*(\delta)\alpha_j(D) + \epsilon.$$

Since $\epsilon > 0$ is arbitrary, then

$$\alpha((N'(\delta)D)(t)) \le 4\Im_2 p_j^*(\delta)\alpha_j(D),$$

Thus

$$\alpha_j(N'(\delta)D) \le 4\Im_2 p_j^*\alpha_j(D).$$

Consequently, by Darbo's fixed point theorem for Fréchet spaces [255], we can conclude that N and N' have at least one fixed point in $\Lambda(\delta)$ which is a random mild solution of problem (6.8)–(6.10). $\qquad\square$

6.3.3 *An Example*

We consider the following abstract differential equation with state dependent delay:

$$
\begin{cases}
\dfrac{\partial^2 y}{\partial t^2}(t,\gamma,\delta) = \dfrac{\partial^2 y}{\partial \gamma^2}(t,\gamma,\delta) \\
\qquad\qquad + f(t, y(t - \zeta(t, y(t,\delta),\delta)), y'(t - \zeta(t, y(t,\delta),\delta),\gamma,\delta)), \\
\qquad\qquad t \in [0,+\infty),\ \gamma \in [0,\pi], \delta \in \Omega, \\[4pt]
y(t,0,\delta) = y(t,\pi,\delta) = 0,\ t \in [0,+\infty), \delta \in \Omega, \\[4pt]
y(\tau,\gamma,\delta) = \Psi(\tau,\gamma,\delta), \tau \in [-r,0], \delta \in \Omega, \\[4pt]
y'(0,\delta) = \Psi'(0,\delta),
\end{cases}
$$

$$(6.15)$$

where $\Psi \in C([-r,0]; E) \times \Omega$, $\zeta \in C([0,+\infty) \times \mathbb{R}; \mathbb{R}_+)$, $\zeta(0, \Psi, \delta) = 0$ and f is Carathéodory on $[0,+\infty) \times \mathbb{R} \times \mathbb{R} \times \Omega$. To make system (6.15) as problem (6.8)–(6.10), we need to define $f : [0,+\infty) \times \mathbb{R} \times \mathbb{R} \times \Omega \to E$ and $\sigma : [0,+\infty) \times C([-r,0]; \mathbb{R}) \times \Omega \to \mathbb{R}$ by

$$
f(t, \lambda, \phi, \delta)(\gamma) = f(t, \lambda(0,\gamma,\delta), \phi(0,\gamma,\delta), \delta),
$$

and

$$
\sigma(s, \lambda, \delta) = s - \zeta(s, \lambda(0,\delta), \delta).
$$

It is easy to show that $f(\cdot)$ and $\sigma(\cdot)$ are Lipschitz.

Consider $E = L^2([0,\pi], \mathbb{R})$ and the domain

$$
D(A) := \{\, y \in E\ : y'' \in E,\ y(0) = y(\pi) = 0\,\}.
$$

Let the operator $A : D(A) \subset E \to E$ be the operator given by $Ay = y''$. Clearly, A is the infinitesimal generator of a strongly continuous cosine family $(\Phi(t))_{t \in \mathbb{R}}$ on E. The spectrum of A is discrete with eigenvalues $-j^2, j \in \mathbb{N}$, and eigenvectors

$$
z_j(\zeta) := \left(\frac{1}{\pi}\right)^{\frac{1}{2}} \sin(j\zeta).
$$

The set of functions $\{z_j : j \in \mathbb{N}\}$ is an orthonormal basis of E. We note that

$$
\Phi(t)y = \sum_{j=1}^{\infty} \cos(jt)\langle y, z_j\rangle z_j,
$$

$$S(t)y = \sum_{j=1}^{\infty} \frac{\sin(jt)}{j} \langle y, z_j \rangle z_j,$$

the sine family $S(\cdot)$ is compact, $\|\Phi(t)\| = \|S(t)\| = 1$ for all $t \in \mathbb{R}$.

Furthermore, we can verify that the requirements of Theorem 6.2 hold. Thus, Theorem 6.2 implies that (6.15) has at least one mild solution on $[-r, +\infty)$.

6.4 Controllability of Second-Order Functional Random Differential Equations with Delay

6.4.1 *Introduction*

As a continuation of the studies in the preceding publications and in order to expand the controllability results to more problems, in this section, we consider the following functional differential equation with delay and random effect:

$$\begin{cases} y''(t,\delta) = A_1 y(t,\delta) + f(t, y_t(\cdot,\delta), \delta) + A_2 g(t,\delta), & \text{a.e. } t \in J := [0,T], \\ y(t,\delta) = \varpi_1(t,\delta); \ t \in (-\infty, 0], \\ y'(0,\delta) = \varpi_2(\delta), \end{cases}$$

(6.16)

where (Ω, F, P) is a complete probability space, $f : J \times \mathcal{B} \times \Omega \to E$, $\varpi_1 \in \mathcal{B} \times \Omega$ are given functions, $A_1 : D(A_1) \subset E \to E$ is the infinitesimal generator of a strongly continuous cosine family of bounded linear operators $(S_1(t))_{t \in \mathbb{R}}$ on E, $\mathcal{B}$ is the phase space, and $(E, |\cdot|)$ is a real Banach space. The control function $g(\cdot, \delta)$ is given in $L^2(J, U)$, a Banach space of admissible control functions with U as a Banach space, and A_2 is a bounded linear operator from U into E.

For a function y defined on $(-\infty, T] \times \Omega$ and each $t \in J$, we denote by $y_t(\cdot, \delta)$ the element of $\mathcal{B} \times \Omega$ given by $y_t(\iota, \delta) = y(t + \iota, \delta), \iota \in (-\infty, 0]$. Here $y_t(\cdot, \delta)$ represents the history of the state from time $-\infty$, up to the present time t. We assume that the histories $y_t(\cdot, \delta)$ belong to some abstract phases $\mathcal{B}$.

Next, we consider the following random problem:

$$\begin{cases} y''(t,\delta) = A_1 y(t,\delta) + f(t, y_{\sigma(t, y_t)}(\cdot, \delta), \delta) + A_2 g(t,\delta); & \text{a.e. } t \in J, \\ y(t,\delta) = \varpi_1(t,\delta); \quad t \in (-\infty, 0], \\ y'(0,\delta) = \varpi_2(\delta), \end{cases}$$

(6.17)

where $f : J \times \mathcal{B} \times \Omega \to E$, $\varpi_1 \in \mathcal{B} \times \Omega$ are given random functions, $A_1 : D(A_1) \subset E \to E$ is as in problem (6.16), $\mathcal{B}$ is the phase space, $\sigma : J \times \mathcal{B} \to (-\infty, T]$, and $(E, |\cdot|)$ is a real Banach space. We based our arguments for the main results on Schauder's fixed theorem [214] and random fixed point theorem combined with the family of cosine operators.

6.4.2 *Existence Results*

Consider the space

$$\Lambda := \{y : (-\infty, T] : y|_{(-\infty,0]} \in \mathcal{B} \text{ and } y|_J \in \mathcal{C}\}.$$

Let $\|y\|_\Lambda$ be the seminorm in Λ given by

$$\|y\|_\Lambda = \|\varpi_1\|_\mathcal{B} + \|y\|_\mathcal{C}.$$

We define the infinitesimal generator $A_1 : E \to E$ of the cosine family $\{S_1(t) : t \in \mathbb{R}\}$ by

$$A_1 y = \frac{d^2}{dt^2} S_1(t) y|_{t=0}, \; y \in D(A_1),$$

where

$$D(A_1) = \{y \in E : S_1(\cdot)y \in C^2(\mathbb{R}, E)\}.$$

6.4.3 *Controllability Results for the Constant Delay Case*

Definition 6.3. The problem (6.16) is controllable on the interval $(-\infty, T]$, if for every final state $y^1(\delta)$, there exists a control $g(\cdot, \delta)$ in $L^2(J, U)$, such that the solution $y(t, \delta)$ of (6.16) verifies $y(T, \delta) = y^1(\delta)$.

Definition 6.4. A stochastic process $y : (-\infty, T] \times \Omega \to E$ is a random mild solution of problem (6.16) if $y(t, \delta) = \varpi_1(t, \delta)$; $t \in (-\infty, 0]$, $y'(0, \delta) = \varpi_2(\delta)$ and the restriction of $y(\cdot, \delta)$ to the interval J is continuous and verifies:

$$y(t, \delta) = S_1(t)\varpi_1(0, \delta) + S_2(t)\varpi_2(\delta) + \int_0^t S_1(t - s)f(s, y_s(\cdot, \delta), \delta)ds$$

$$+ \int_0^t S_1(t - s)A_2 g(t, \delta)ds$$

Let

$$M = \sup\{\|S_1(t)\|_{B(E)} : t \geq 0\} \text{ and } M' = \sup\{\|S_2(t)\|_{B(E)} : t \geq 0\}.$$

We will need to introduce the following hypotheses:

(6.6.1) $S_1(t)$ is compact for $t > 0$,

(6.6.2) The function $f : J \times \mathcal{B} \times \Omega \to E$ is random Carathéodory,

(6.6.3) There exist functions $\varkappa : J \times \Omega \to \mathbb{R}^+$ and $p : J \times \Omega \to \mathbb{R}^+$ such that for each $\delta \in \Omega$, $\varkappa(\cdot, \delta)$ is continuous non-decreasing and $p(\cdot, \delta)$ integrable with:

$$|f(t, g, \delta)| \leq p(t, \delta) \, \varkappa(\|g\|_{\mathcal{B}}, \delta) \text{ for a.e. } t \in J \text{ and each } g \in \mathcal{B},$$

(6.6.4) There exists a random function $Q : \Omega \longrightarrow \mathbb{R}^+ \backslash \{0\}$ where:

$$M\left(1 + TM\zeta\right)\left(\|\varpi_1\|_{\mathcal{B}} + \varkappa(D, \delta)\|p\|_{L^1}\right) + TM\zeta \left\|y^1\right\|$$
$$+ M'\left(1 + TM\zeta\right)|\varpi_2| \leq Q(\delta),$$

where

$$D := \zeta Q(\delta) + M\|\varpi_1\|_{\mathcal{B}},$$

(6.6.5) The linear operator $\mathcal{K} : L^2(J, U) \to E$ given by:

$$\mathcal{K}g = \int_0^T S_1(T - s)A_2 g(s, \delta) ds$$

has an inverse operator $\mathcal{K}^{-1}$ in $L^2(J, U)/ker\mathcal{K}$ and there exists a positive constant ζ such that $\left\|A_2 \mathcal{K}^{-1}\right\| \leq \zeta$,

(6.6.6) For each $\delta \in \Omega$, $\varpi_1(\cdot, \delta)$ is continuous and for each t, $\varpi_1(t, \cdot)$ is measurable, and for each $\delta \in \Omega$, $\varpi_2(\delta)$ is measurable.

Theorem 6.3. *If (6.6.1)–(6.6.6) are satisfied, then the problem (6.16) is controllable on J.*

Proof. Define the control:

$$g(t, \delta) = \mathcal{K}^{-1}\left(y^1(\delta) - S_1(T)\varpi_1(0, \delta) - S_2(T)\varpi_2(\delta)\right.$$
$$\left. - \int_0^T S_1(T - s)f(s, y_s(\cdot, \delta), \delta) ds\right).$$

We define the operator $T : \Omega \times \Lambda \longrightarrow \Lambda$ by: $(T(\delta)y)(t) = \varpi_1(t, \delta)$, if $t \in (-\infty, 0]$, and for $t \in J$:

$$(T(\delta)y)(t) = S_1(t)\varpi_1(0, \delta) + S_2(t)\varpi_2(\delta) + \int_0^t S_1(t - s) \, f(s, y_s(\cdot, \delta), \delta) ds$$

$$+ \int_0^t S_1(t - s)A_2\mathcal{K}^{-1}\left(y^1(\delta) - S_1(T)\varpi_1(0, \delta) - S_2(T)\varpi_2(\delta)\right.$$

$$\left. - \int_0^T S_1(T - \varepsilon)f(\varepsilon, y_\varepsilon(\cdot, \delta), \delta) d\varepsilon\right) ds. \tag{6.18}$$

Using (6.6.5), we will demonstrate that T has a fixed point $y(t, \delta)$ which is a mild solution of (6.16). This implies that the problem (6.16) is controllable on J. Further, we prove that $T(\cdot)$ is a random operator. For that, we demonstrate that for any $y \in \Lambda$, $T(\cdot)(y) : \Omega \longrightarrow \Lambda$ is a random variable. Then we demonstrate that $T(\cdot)(y) : \Omega \longrightarrow \Lambda$ is measurable. As the mapping $f(t, y, \cdot)$, $t \in J, y \in \Lambda$ is measurable by assumption (6.6.2) and (6.6.6). Let $D : \Omega \longrightarrow 2^\Lambda$ be given by:

$$D(\delta) = \{ y \in \Lambda : \|y\|_\Lambda \le Q(\delta) \}.$$

$D(\delta)$ is bounded, closed, convex and solid for all $\delta \in \Omega$. Then D is measurable by Lemma 17 (see [243]). Let $\delta \in \Omega$ be fixed, then for any $y \in D(\delta)$ and by (A_1), we obtain:

$$\|y_s\|_{\mathcal{B}} \le L(s)|y(s)| + M(s)\|y_0\|_D$$
$$\le \zeta_T |y(s)| + M_T \|\varpi_1\|_{\mathcal{B}},$$

and by (6.6.3) and (6.6.4), we have

$$|(T(\delta)y)(t)| \le M \|\varpi_1\|_{\mathcal{B}} + M' |\varpi_2| + M \int_0^t |f(s, y_s, \delta)| ds$$

$$+ M\zeta \int_0^t |y^1(\delta)| + M \|\varpi_1\|_{\mathcal{B}} + M' |\varpi_2| \, ds$$

$$+ M\zeta \int_0^t \int_0^T \|S_1(\varepsilon - s)\| |f(\varepsilon, y_\varepsilon, \delta)| \, d\varepsilon ds$$

$$\le M\|\varpi_1\|_{\mathcal{B}} + M' |\varpi_2| + M \int_0^T p(s, \delta) \, \varkappa(\|y_s\|_{\mathcal{B}}, \delta) \, ds$$

$$+ TM\zeta |y^1(\delta)| + TM^2\zeta\|\varpi_1\|_{\mathcal{B}} + TMM'\zeta |\varpi_2|$$

$$+ TM^2\zeta \int_0^T p(\varepsilon, \delta) \, \varkappa(\|y_\varepsilon\|_{\mathcal{B}}, \delta) \, d\varepsilon$$

$$\le M(1 + TM\zeta) \|\varpi_1\|_{\mathcal{B}} + TM\zeta |y^1(\delta)| + M'(1 + TM\zeta) |\varpi_2|$$

$$+ M(1 + TM\zeta) \int_0^T p(s, \delta) \, \varkappa(\|y_s\|_{\mathcal{B}}, \delta) \, ds$$

$$\le M(1 + TM\zeta) \left(\|\varpi_1\|_{\mathcal{B}} + \varkappa(D_T, \delta) \int_0^T p(s, \delta) \, ds \right)$$

$$+ TM\zeta \|y^1(\delta)\| + M'(1 + TM\zeta) |\varpi_2|.$$

Set

$$D_T := \zeta_T Q(\delta) + M_T \|\varpi_1\|_{\mathcal{B}}.$$

Then, we have

$$|(T(\delta)y)(t)| \leq M\left(1 + TM\zeta\right)\left(\|\varpi_1\|_{\mathcal{B}} + \varkappa(D_T,\delta)\int_0^T p(s,\delta)\ ds\right)$$
$$+ TM\zeta\left|y^1\left(\delta\right)\right| + M'\left|\varpi_2\right|\left(1 + TM\zeta\right).$$

Thus

$$\|(T(\delta)y)\|_\Lambda \leq M\left(1 + TM\zeta\right)\left(\|\varpi_1\|_{\mathcal{B}} + \varkappa(D_T,\delta)\|p\|_{L^1}\right)$$
$$+ TM\zeta\left|y^1\left(\delta\right)\right| + M'\left(1 + TM\zeta\right)\left|\varpi_2\right|$$
$$\leq Q(\delta).$$

Thus, we deduce that T is a random operator with stochastic domain D and $T(\delta) : D(\delta) \longrightarrow D(\delta)$ for each $\delta \in \Omega$.

Claim 1: T is continuous.
Let y^n be a sequence where $y^n \longrightarrow y$. Then

$$|(T(\delta)y^n)(t) - (T(\delta)y)(t)|$$
$$\leq M\int_0^t |f(s,y_s^n,\delta) - f(s,y_s,\delta)|\ ds$$
$$+ \zeta M\int_0^t \int_0^T \|S_1(T-\varepsilon)\|\ |f(\varepsilon,y_\varepsilon^n,\delta) - f(\varepsilon,y_\varepsilon,\delta)|d\varepsilon ds$$
$$\leq M\int_0^t |f(s,y_s^n,\delta) - f(s,y_s,\delta)|\ ds$$
$$+ TM^2\zeta\int_0^T |f(\varepsilon,y_\varepsilon^n,\delta) - f(\varepsilon,y_\varepsilon,\delta)|d\varepsilon$$
$$\leq M\left(1 + TM\zeta\right)\int_0^T |f(\varepsilon,y_\varepsilon^n,\delta) - f(\varepsilon,y_\varepsilon,\delta)|d\varepsilon.$$

As $f(s,\cdot,\delta)$ is continuous, we obtain

$$\|f(\cdot,y^n_\cdot,\delta) - f(\cdot,y_\cdot,\delta)\|_{L^1} \to 0 \text{ as } n \to +\infty.$$

Thus T is continuous.

Claim 2: We demonstrate that for every $\delta \in \Omega, \{y \in D(\delta) : T(\delta)y = y\} \neq \emptyset$.
We apply Schauder's theorem.

(a) T maps bounded sets into equicontinuous sets in $D(\delta)$.
Let $\varepsilon_1, \varepsilon_2 \in [0,T]$ with $\varepsilon_2 > \varepsilon_1$, $D(\delta)$ be a bounded set as in Claim 2, and $y \in D(\delta)$. Then

$$|(T(\delta)y)(\varepsilon_2) - (T(\delta)y)(\varepsilon_1)|$$

$$\leq \|S_1(\varepsilon_2) - S_1(\varepsilon_1)\|_{B(E)} \|\varpi_1\|_{\mathcal{B}} + \|S_2(\varepsilon_2) - S_2(\varepsilon_1)\|_{B(E)} |\varpi_2|$$

$$+ \int_0^{\varepsilon_1} \|S_1(\varepsilon_2 - s) - S_1(\varepsilon_1 - s)\|_{B(E)} |f(s, y_s, \delta)| \, ds$$

$$+ \int_{\varepsilon_1}^{\varepsilon_2} \|C(\varepsilon_2 - s)\|_{B(E)} |f(s, y_s, \delta)| \, ds$$

$$+ \zeta \int_0^{\varepsilon_1} \|S_1(\varepsilon_2 - s) - S_1(\varepsilon_1 - s)\|_{B(E)}$$

$$\times \left[|y^1(\delta)| + \|S_1(T)\|_{B(E)} \|\varpi_1\|_{\mathcal{B}} + \|S_2(T)\|_{B(E)} |\varpi_2| \right] ds$$

$$+ \zeta \int_0^{\varepsilon_1} \|S_1(\varepsilon_2 - s) - S_1(\varepsilon_1 - s)\|_{B(E)}$$

$$\times \int_0^T \|S_1(T - \varepsilon)\|_{B(E)} |f(\varepsilon, y_\varepsilon(\cdot, \delta), \delta)| \, d\varepsilon ds$$

$$+ \zeta \int_{\varepsilon_1}^{\varepsilon_2} \|C(\varepsilon_2 - s)\|_{B(E)}$$

$$\times \left[|y^1(\delta)| + \|S_1(T)\|_{B(E)} \|\varpi_1\|_{\mathcal{B}} + \|S_2(T)\|_{B(E)} |\varpi_2| \right] ds$$

$$+ \zeta \int_{\varepsilon_1}^{\varepsilon_2} \|C(\varepsilon_2 - s)\|_{B(E)} \int_0^T \|S_1(T - \varepsilon)\|_{B(E)} |f(\varepsilon, y_\varepsilon(\cdot, \delta), \delta)| \, d\varepsilon ds$$

$$\leq \|S_1(\varepsilon_2 - s) - S_1(\varepsilon_1 - s)\|_{B(E)} \|\varpi_1\|_{\mathcal{B}} + \|S_2(\varepsilon_2) - S_2(\varepsilon_1)\|_{B(E)} |\varpi_2|$$

$$+ \varkappa(D_T, \delta) \int_0^{\varepsilon_1} \|S_1(\varepsilon_2 - s) - S_1(\varepsilon_1 - s)\|_{B(E)} p(s, \delta) ds$$

$$+ M\varkappa(D_T, \delta) \int_{\varepsilon_1}^{\varepsilon_2} p(s, \delta) ds$$

$$+ \zeta \int_0^{\varepsilon_1} \|S_1(\varepsilon_2 - s) - S_1(\varepsilon_1 - s)\|_{B(E)}$$

$$\times \left[|y^1(\delta)| + \|S_1(T)\|_{B(E)} \|\varpi_1\|_{\mathcal{B}} + \|S_2(T)\|_{B(E)} |\varpi_2| \right] ds$$

$$+ \zeta M \varkappa(D_T, \delta) \int_0^{\varepsilon_1} \|S_1(\varepsilon_2 - s) - S_1(\varepsilon_1 - s)\|_{B(E)} \int_0^T p(\varepsilon, \delta) d\varepsilon ds$$

$$+ \zeta M \int_{\varepsilon_1}^{\varepsilon_2} (|y^1(\delta)| + \|S_1(T)\|_{B(E)} \|\varpi_1\|_{\mathcal{B}} + \|S_2(T)\|_{B(E)} |\varpi_2|$$

$$+ M\varkappa(D_T, \delta) \int_0^T p(\varepsilon, \delta) d\varepsilon) ds.$$

The right-hand of the above inequality tends to zero as $\varepsilon_2 - \varepsilon_1 \to 0$, since $S_1(t), S_2(t)$ are compact for $t > 0$, and strongly continuous, then we obtain the continuity in the uniform operator topology (see [318, 346]).

(b) Let $t \in [0,T]$ be fixed and let $y \in D(\delta)$. From assumptions (6.6.3), (6.6.5) and since $S_1(t)$ is compact, the set

$$\left\{ \int_0^t S_1(t-s)f(s,y_s(\cdot,\delta),\delta)ds + \int_0^t S_1(t-s)A_2g(t,\delta)ds \right\}$$

is precompact in E, and the set

$$\left\{ S_1(t)\varpi_1(0,\delta) + S_2(t)\varpi_2(\delta) + \int_0^t S_1(t-s)f(s,y_s(\cdot,\delta),\delta)ds \right.$$
$$\left. + \int_0^t S_1(t-s)A_2g(t,\delta)ds \right\}$$

is precompact in E. Thus, $T(\delta) : D(\delta) \to D(\delta)$ is continuous and compact. Schauder's theorem implies that $T(\delta)$ has a fixed point $y(\delta)$ in $D(\delta)$. Since $\bigcap_{\delta \in \Omega} D(\delta) \neq \emptyset$, and a measurable selector of int$D$ exists, then by Lemma 2.18, we conclude that T has a stochastic fixed point $y^*(\delta)$, which is a random mild solution of (6.16).

6.4.4　*Controllability Results for the State-Dependent Delay Case*

Definition 6.5. A stochastic process $y : (-\infty, T] \times \Omega \to E$ is said to be a random mild solution of problem (6.17) if $y(t,\delta) = \varpi_1(t,\delta)$; $t \in (-\infty, 0]$, $y'(0,\delta) = \varpi_2(\delta)$ and the restriction of $y(\cdot,\delta)$ to the interval J is continuous and verifies equation:

$$y(t,\delta) = S_1(t)\varpi_1(0,\delta) + S_2(t)\varpi_2(\delta) + \int_0^t S_1(t-s)f(s,y_{\sigma(s,y_s)}(\cdot,\delta),\delta)ds$$
$$+ \int_0^t S_1(t-s)A_2g(t,\delta)ds.$$

Set

$$\mathcal{Q}(\sigma^-) = \{\sigma(s,\varpi_2) : (s,\varpi_2) \in J \times \mathcal{B}, \sigma(s,\varpi_2) \leq 0\}.$$

Suppose that $\sigma : J \times \mathcal{B} \to (-\infty, T]$ is continuous. And,

(6.8.1) The function $t \to \varpi_{1t}$ is continuous from $\mathcal{Q}(\sigma^-)$ into $\mathcal{B}$ and there exists a continuous and bounded function $L^{\varpi_1} : \mathcal{Q}(\sigma^-) \to (0,\infty)$ where

$$\|\varpi_{1t}\|_{\mathcal{B}} \leq L^{\varpi_1}(t)\|\varpi_1\|_{\mathcal{B}} \quad \text{for every } t \in \mathcal{Q}(\sigma^-).$$

Remark 6.1 ([219]). *The hypothesis* (6.8.1) *is satisfied by continuous and bounded functions.*

Lemma 6.1 ([347]). *If* $y : (-\infty, T] \to E$ *is a function such that* $y_0 = \varpi_1$, *then*

$$\|y_s\|_\mathcal{B} \leq (M_T + L^{\varpi_1})\|\varpi_1\|_\mathcal{B} + \zeta_T \sup\{|y(\iota)|; \iota \in [0, max\{0, s\}]\}, \ s \in \mathcal{Q}(\sigma^-) \cup J,$$

where $L^{\varpi_1} = \sup_{t \in \mathcal{Q}(\sigma^-)} L^{\varpi_1}(t).$

We consider now the hypotheses:

(6.10.1) $S_1(t)$ is compact for $t > 0$ in E.

(6.10.2) The function $f : J \times \mathcal{B} \times \Omega \to E$ is random Carathéodory.

(6.10.3) There exist a function $\varkappa : J \times \Omega \to \mathbb{R}^+$ and $p : J \times \Omega \to \mathbb{R}^+$ such that for each $\delta \in \Omega$, $\varkappa(\cdot, \delta)$ is a continuous non-decreasing function and $p(\cdot, \delta)$ integrable with:

$$|f(t, g, \delta)| \leq p(t, \delta) \ \varkappa(\|g\|_\mathcal{B}, \delta) \text{ for a.e. } t \in J \text{ and each } g \in \mathcal{B}.$$

(6.10.4) There exists a function $\beta : J \times \Omega \longrightarrow \mathbb{R}^+$ with $\beta(\cdot, \delta) \in L^1(J, \mathbb{R}^+)$ for each $\delta \in \Omega$ such that for any bounded $B \subseteq E$.

$$\alpha(f(t, B, \delta)) \leq \beta(t, \delta)\alpha(B).$$

(6.10.5) There exists a random function $Q : \Omega \longrightarrow \mathbb{R}^+ \backslash \{0\}$ where:

$$\left(\|\varpi_1\|_\mathcal{B} + \varkappa((M_T + L^{\varpi_1})\|\varpi_1\|_\mathcal{B} + \zeta_T Q(\delta), \delta) \int_0^T p(s, \delta) \ ds\right)$$

$$\times M (1 + TM\lambda) + TM\lambda \left\|y^1 (\delta)\right\| + M' (1 + TM\lambda) |\varpi_2| \leq Q(\delta).$$

(6.10.6) The linear operator $\mathcal{K} : L^2(J, U) \to E$ defined by:

$$\mathcal{K}g = \int_0^T S_1(T - s)A_2 g(s, \delta)ds$$

has an inverse operator $\mathcal{K}^{-1}$ which takes values in $L^2(J, U)/ker\mathcal{K}$ and there exists a positive constant λ such that $\|A_2\mathcal{K}^{-1}\| \leq \lambda$.

(6.10.7) For each $\delta \in \Omega, \varpi_1(\cdot, \delta)$ is continuous and for each $t, \varpi_1(t, \cdot)$ is measurable and for each $\delta \in \Omega, \varpi_2(\delta)$ is measurable.

Theorem 6.4. *Assume that* (6.10.1)–(6.10.7) *and* (6.8.1) *hold. If*

$$M (1 + M\lambda T) \int_0^T \beta(s)\zeta(s)ds < 1, \tag{6.19}$$

then the random problem (6.17) *is controllable on* J.

Proof. Using (6.10.6), we define the control:

$$g(t, \delta) = \mathcal{K}^{-1} \left(y^1(\delta) - S_1(T)\varpi_1(0, \delta) - S_2(T)\varpi_2(\delta) \right.$$

$$\left. - \int_0^T S_1(T - s)f(s, y_{\sigma(s, y_s)}, \delta)ds \right).$$

We define the operator $T : \Omega \times \Lambda \longrightarrow \Lambda$ by: $(T(\delta)y)(t) = \varpi_1(t, \delta)$, if $t \in (-\infty, 0]$, and for $t \in J$ by:

$$(T(\delta)y)(t) = S_1(t)\varpi_1(0, \delta) + S_2(t)\varpi_2(\delta)$$

$$+ \int_0^t S_1(t - s) \, f(s, y_{\sigma(s, y_s)}, \delta), \delta)ds + \int_0^t S_1(t - s)A_2\mathcal{K}^{-1}$$

$$\times \left(y^1(\delta) - S_1(T)\varpi_1(0, \delta) - S_2(T)\varpi_2(\delta) \right.$$

$$\left. - \int_0^T S_1(T - \varepsilon)f(\varepsilon, y_{\sigma(\varepsilon, y_\varepsilon)}, \delta)d\varepsilon \right)ds.$$

$$(6.20)$$

Proving that $T(\cdot)$ has a fixed point $y(t, \delta)$ and that (6.17) is controllable. Further, we demonstrate that $T(\cdot)$ is a random operator by showing that for any $y \in \Lambda$, $T(\cdot)(y)$: $\Omega \longrightarrow \Lambda$ is a random variable. Also, we show that $T(\cdot)(y) : \Omega \longrightarrow \Lambda$ is measurable, as a mapping $f(t, y, \cdot)$, $t \in J, y \in \Lambda$ is measurable by (6.10.2) and (6.10.6). Let $D : \Omega \longrightarrow 2^\Lambda$ be given by:

$$D(\delta) = \{y \in \Lambda : \|y\|_\Lambda \le Q(\delta)\}.$$

$D(\delta)$ is bounded, closed, convex and solid for all $\delta \in \Omega$. Then D is measurable. Let $\delta \in \Omega$ be fixed. If $y \in D(\delta)$, then

$$\left\| y_{\sigma(t, y_t)} \right\|_\mathcal{B} = (M_T + L^{\varpi_1})\|\varpi_1\|_\mathcal{B} + \zeta_T Q(\delta),$$

and for each $y \in D(\delta)$, from (6.10.3) and (6.10.5), for each $t \in J$, we have

$$|(T(\delta)y)(t)|$$

$$\le M \|\varpi_1\|_\mathcal{B} + M' |\varpi_2| + M \int_0^t |f(s, y_{\sigma(s, y_s)}, \delta)|ds$$

$$+ M\lambda \int_0^t |y^1(\delta)| + M \|\varpi_1\|_\mathcal{B} + M' |\varpi_2| \, ds$$

$$+ M\lambda \int_0^t \int_0^T \|S_1(\varepsilon - s)\| |f(\varepsilon, y_{\sigma(\varepsilon, y_\varepsilon)}, \delta)| \, d\varepsilon ds$$

$$\leq M\|\varpi_1\|_{\mathcal{B}} + M'\,|\varpi_2| + M\int_0^T p(s,\delta)\,\varkappa\left(\|y_{\sigma(s,y_s)}\|_{\mathcal{B}},\delta\right)\,ds$$

$$+ TM\lambda\,|y^1(\delta)| + TM^2\zeta\|\varpi_1\|_{\mathcal{B}} + TMM'\zeta\,|\varpi_2|$$

$$+ TM^2\lambda\int_0^T p(\varepsilon,\delta)\,\varkappa\left(\|y_{\sigma(\varepsilon,y_\varepsilon)}\|_{\mathcal{B}},\delta\right)\,d\varepsilon$$

$$\leq M\,(1+TM\lambda)\,\|\varpi_1\|_{\mathcal{B}} + TM\lambda\,|y^1(\delta)| + M'\,(1+TM\lambda)\,|\varpi_2|$$

$$+ M\,(1+TM\lambda)\int_0^T p(s,\delta)\,\varkappa\left(\|y_{\sigma(s,y_s)}\|_{\mathcal{B}},\delta\right)\,ds$$

$$\leq M\,(1+TM\lambda)$$

$$\times\left(\|\varpi_1\|_{\mathcal{B}} + \varkappa((M_T+L^{\varpi_1})\|\varpi_1\|_{\mathcal{B}} + \zeta_T Q(\delta),\delta)\int_0^T p(s,\delta)\,ds\right)$$

$$+ TM\lambda\,|y^1(\delta)| + M'\,(1+TM\lambda)\,|\varpi_2|.$$

Thus, T is a random operator with stochastic domain D and $T(\delta): D(\delta) \to D(\delta)$ for each $\delta \in \Omega$.

Claim 1: T is continuous.
Let y^n be a sequence such that $y^n \longrightarrow y$ in Λ. Then

$$|(T(\delta)y^n)(t) - (T(\delta)y)(t)|$$

$$\leq M\int_0^t |f(s,y^n_{\sigma(s,y^n_s)},\delta) - f(s,y_{\sigma(s,y_s)},\delta)|\,ds$$

$$+ \lambda M\int_0^t\int_0^T \|S_1(T-\varepsilon)\|\,|f(\varepsilon,y^n_{\sigma(s,y^n_s)},\delta) - f(\varepsilon,y_{\sigma(s,y_s)},\delta)|d\varepsilon ds$$

$$\leq M\int_0^t |f(s,y^n_{\sigma(s,y^n_s)},\delta) - f(s,y_{\sigma(s,y_s)},\delta)|\,ds$$

$$+ TM^2\lambda\int_0^T |f(s,y^n_{\sigma(s,y^n_s)},\delta) - f(s,y_{\sigma(s,y_s)},\delta)|\,ds$$

$$\leq M\,(1+TM\lambda)\int_0^T |f(s,y^n_{\sigma(s,y^n_s)},\delta) - f(s,y_{\sigma(s,y_s)},\delta)|\,ds.$$

As $f(s,\cdot,\delta)$ is continuous, then $\|(T(\delta)y^n)(t) - (T(\delta)y)(t)\|_\Lambda \to 0$ as $n \to +\infty$. Thus T is continuous.

Claim 2: We demonstrate that for every $\delta \in \Omega, \{y \in D(\delta) : T(\delta)y = y\} \neq \emptyset$ by employing Mönch fixed point theorem [7,348].

(a) T maps bounded sets into equicontinuous sets in $D(\delta)$.
 Let $\varepsilon_1,\varepsilon_2 \in [0,T]$ with $\varepsilon_2 > \varepsilon_1$, $D(\delta)$ be a bounded set, and $y \in D(\delta)$. Then

$$|(T(\delta)y)(\varepsilon_2) - (T(\delta)y)(\varepsilon_1)|$$
$$\leq \|S_1(\varepsilon_2) - S_1(\varepsilon_1)\|_{B(E)} \|\varpi_1\|_{\mathcal{B}} + \|S_2(\varepsilon_2) - S_2(\varepsilon_1)\|_{B(E)} |\varpi_2|$$
$$+ \int_0^{\varepsilon_1} \|S_1(\varepsilon_2 - s) - S_1(\varepsilon_1 - s)\|_{B(E)} f(s, y_{\sigma(s,y_s)}, \delta) ds$$
$$+ \int_{\varepsilon_1}^{\varepsilon_2} \|S_1(\varepsilon_2 - s)\|_{B(E)} f(s, y_{\sigma(s,y_s)}, \delta) ds$$
$$+ \lambda \int_0^{\varepsilon_1} \|S_1(\varepsilon_2 - s) - S_1(\varepsilon_1 - s)\|_{B(E)} \left[|y^1(\delta)| + \|S_1(T)\| |\varpi_1(0,\delta)|\right] ds$$
$$+ \lambda \int_0^{\varepsilon_1} \|S_1(\varepsilon_2 - s) - S_1(\varepsilon_1 - s)\|_{B(E)}$$
$$\times \int_0^T \|S_1(T - \varepsilon)\|_{B(E)} |f(\varepsilon, y_{\sigma(s,y_s)}, \delta)| d\varepsilon ds$$
$$+ \lambda \int_{\varepsilon_1}^{\varepsilon_2} \|S_1(\varepsilon_2 - s)\|_{B(E)} \left[|y^1(\delta)| + \|S_1(T)\| |\varpi_1(0,\delta)|\right] ds$$
$$+ \lambda \int_{\varepsilon_1}^{\varepsilon_2} \|S_1(\varepsilon_2 - s)\|_{B(E)} \int_0^T \|S_1(T - \varepsilon)\|_{B(E)} |f(\varepsilon, y_{\sigma(\varepsilon,y_\varepsilon)}, \delta)| d\varepsilon ds.$$

Thus,

$$|(T(\delta)y)(\varepsilon_2) - (T(\delta)y)(\varepsilon_1)|$$
$$\leq |S_1(\varepsilon_2) - S_1(\varepsilon_1)| \|\varpi_1\|_{\mathcal{B}} + \|S_2(\varepsilon_2) - S_2(\varepsilon_1)\|_{B(E)} |\varpi_2|$$
$$+ \int_0^{\varepsilon_1} \|S_1(\varepsilon_2 - s) - S_1(\varepsilon_1 - s)\|_{B(E)} f(s, y_{\sigma(s,y_s)}, \delta) ds$$
$$+ \int_{\varepsilon_1}^{\varepsilon_2} \|S_1(\varepsilon_2 - s)\|_{B(E)} f(s, y_{\sigma(s,y_s)}, \delta) ds$$
$$+ \lambda \int_0^{\varepsilon_1} \|S_1(\varepsilon_2 - s) - S_1(\varepsilon_1 - s)\|_{B(E)} ds$$
$$\times \left[\|y^1(\delta)\| + \|S_1(T)\|_{B(E)} |\varpi_1(0,\delta)|\right]$$
$$+ \lambda \int_0^{\varepsilon_1} \|S_1(\varepsilon_2 - s) - S_1(\varepsilon_1 - s)\|_{B(E)} \varkappa((M_T + L^{\varpi_1})\|\varpi_1\|_{\mathcal{B}} + \zeta_T Q(\delta))$$
$$\times \int_0^T p(\varepsilon, \delta) d\varepsilon ds$$
$$+ \lambda M \int_{\varepsilon_1}^{\varepsilon_2} \|y^1\| + \|S_1(T)\|_{B(E)} |\varpi_1(0,\delta)|$$
$$+ M \varkappa((M_T + L^{\varpi_1})\|\varpi_1\|_{\mathcal{B}} + \zeta_T Q(\delta)) \int_0^T p(\varepsilon, \delta) d\varepsilon ds.$$

Hence,

$$|(T(\delta)y)(\varepsilon_2) - (T(\delta)y)(\varepsilon_1)|$$

$$\leq \|S_1(\varepsilon_2) - S_1(\varepsilon_1)\|_{B(E)} \|\varpi_1\|_{\mathcal{B}} + \|S_2(\varepsilon_2) - S_2(\varepsilon_1)\|_{B(E)} |\varpi_2|$$

$$+ \varkappa\left((M_T + L^{\varpi_1})\|\varpi_1\|_{\mathcal{B}} + \zeta_T Q(\delta)\right)$$

$$\times \int_0^{\varepsilon_1} \|S_1(\varepsilon_2 - s) - S_1(\varepsilon_1 - s)\|_{B(E)}\, p(s,\delta)ds$$

$$+ M\varkappa\left((M_T + L^{\varpi_1})\|\varpi_1\|_{\mathcal{B}} + \zeta_T Q(\delta), \delta\right) \int_{\varepsilon_1}^{\varepsilon_2} p(s,\delta)ds$$

$$+ \lambda \int_0^{\varepsilon_1} \|S_1(\varepsilon_2 - s) - S_1(\varepsilon_1 - s)\|_{B(E)}\, ds$$

$$\times \left[\|y^1(\delta)\| + \|S_1(T)\|_{B(E)} |\varpi_1(0,\delta)|\right]$$

$$+ \int_0^{\varepsilon_1} \|S_1(\varepsilon_2 - s) - S_1(\varepsilon_1 - s)\|_{B(E)}\, \varkappa\left((M_T + L^{\varpi_1})\|\varpi_1\|_{\mathcal{B}} + \zeta_T Q(\delta)\right)$$

$$\times \int_0^T p(\varepsilon,\delta)d\varepsilon ds$$

$$+ \lambda M \int_{\varepsilon_1}^{\varepsilon_2} \|y^1(\delta)\| + \|S_1(T)\|_{B(E)} |\varpi_1(0,\delta)|$$

$$+ M\varkappa\left((M_T + L^{\varpi_1})\|\varpi_1\|_{\mathcal{B}} + \zeta_T Q(\delta)\right) \int_0^T p(\varepsilon,\delta)d\varepsilon ds.$$

The right-hand of the above inequality tends to zero as $\varepsilon_2 - \varepsilon_1 \to 0$, since $S_1(t)$ and $S_2(t)$ are strongly continuous compact operators, thus we obtain the continuity in the uniform operator topology.

Further, let $\delta \in \Omega$ be fixed.

(b) Let Φ be a subset of $D(\delta)$ where $\Phi \subset \overline{conv}\,(T(\Phi) \cup \{0\})$. Φ is bounded and equicontinuous, thus the function $t \to v(t) = \alpha(\Phi(t))$ is continuous on $(-\infty, T]$. By (6.10.4), Lemma 2.6 and the properties of the measure α we have for each $t \in (-\infty, T]$

$$v(t) \leq \alpha\left(T(\Phi)\right)(t) \cup \{0\})$$

$$\leq \alpha\left(T(\Phi(t))\right)$$

$$\leq \alpha\left(S_1(t)\varpi_1(0,\delta)\right) + \alpha\left(S_2(t)\varpi_2(\delta)\right)$$

$$+ \alpha\left(\int_0^t S_1(t - s)\, f(s, y_{\sigma(s,y_s)}, \delta), \delta)ds\right)$$

$$+ M\lambda \int_0^t \alpha\left(y^1(\delta) - S_1(T)\varpi_1(0,\delta) - S_2(T)\varpi_2(\delta)\right)$$

$$+ \alpha \left(\int_0^T S_1(T - \varepsilon) f(\varepsilon, y_{\sigma(\varepsilon, y_\varepsilon)}, \delta) d\varepsilon \right) ds$$

$$\leq M \int_0^t \alpha \left(f(s, y_{\sigma(s, y_s)}, \delta), \delta) \right) ds$$

$$+ M\lambda \int_0^t \int_0^T \alpha \left(S_1(T - \varepsilon) f(\varepsilon, y_{\sigma(\varepsilon, y_\varepsilon)}, \delta) \right) d\varepsilon ds$$

$$\leq M \int_0^t \beta(s) \alpha(\{ y_{\sigma(s, y_s)} : y \in \Phi \}) ds$$

$$+ M\lambda \int_0^t \int_0^T \alpha \left(S_1(T - \varepsilon) f(\varepsilon, y_{\sigma(\varepsilon, y_\varepsilon)}, \delta) \right) d\varepsilon ds$$

$$\leq M \int_0^t \beta(s) \zeta(s) \sup_{0 \leq \varepsilon \leq s} \alpha(\Phi(\varepsilon)) ds$$

$$+ M^2\lambda \int_0^t \int_0^T \alpha \left(f(\varepsilon, y_{\sigma(\varepsilon, y_\varepsilon)}, \delta) \right) d\varepsilon ds$$

$$\leq M \int_0^t \beta(s) \zeta(s) \alpha(\Phi(s)) ds$$

$$+ M^2\lambda T \int_0^T \beta(\varepsilon) \alpha(\{ y_{\sigma(\varepsilon, y_\varepsilon)} : y \in \Phi \}) d\varepsilon$$

$$\leq M \int_0^t v(s) \, \beta(s) \zeta(s) ds + M^2\lambda T \int_0^T \beta(\varepsilon) \zeta(\varepsilon) \alpha(\Phi(\varepsilon)) d\varepsilon$$

$$= M \int_0^t \beta(s) \zeta(s) v(s) ds + M^2\lambda T \int_0^T \beta(\varepsilon) \zeta(\varepsilon) v(\varepsilon) d\varepsilon.$$

$$\leq M \int_0^T \beta(s) \zeta(s) v(s) ds + M^2\lambda T \int_0^T \beta(\varepsilon) \zeta(\varepsilon) v(\varepsilon) d\varepsilon.$$

$$\leq M \left(1 + M\lambda T \right) \int_0^T \beta(s) \zeta(s) v(s) ds.$$

$$\leq M \left(1 + M\lambda T \right) \int_0^T \beta(s) \zeta(s) \sup_{0 \leq \varepsilon \leq s} v(\varepsilon) ds.$$

$$\leq M \left(1 + M\lambda T \right) \|v\|_\infty \int_0^T \beta(s) \zeta(s) ds.$$

Thus,

$$\|v\|_\infty \leq M \left(1 + M\lambda T \right) \|v\|_\infty \int_0^T \beta(s) \zeta(s) ds.$$

Then

$$\|v\|_\infty \left(1 - M\left(1 + M\lambda T\right) \int_0^T \beta(s)\zeta(s)ds \right) \le 0.$$

Consequently, $\|v\|_\infty = 0$, thus $v(t) = 0$ for each $t \in J$, and then $\Phi(t)$ is relatively compact in E. As a result of the Ascoli-Arzelà theorem, Φ is relatively compact in $D(\delta)$. By Mönch fixed point theorem, we deduce that T has a fixed point $y(\delta) \in D(\delta)$. As $\bigcap_{\delta \in \Omega} D(\delta) \ne \emptyset$, and a measurable selector of $\mathrm{int}\, D$ exists, by Lemma 2.18, T has a stochastic fixed point $y^*(\delta)$, which is a mild solution of (6.17).

6.4.5   *An Example*

Consider the problem:

$$\frac{\partial^2}{\partial t^2} w(t, y, \delta) = \frac{\partial^2}{\partial y^2} w(t, y, \delta) + f(t, w(t, y, \delta), \delta)$$
$$+ A_2 g\left(t, \delta\right)\ y \in [0, \pi];\ t \in J = [0, T], \tag{6.21}$$

$$w(t, 0, \delta) = w(t, \pi, \delta) = 0;\ t \in [0, T],\ \delta \in \Omega, \tag{6.22}$$

$$w(t, y, \delta) = \varpi_1(t, \delta), \frac{\partial}{\partial t} w(0, y, \delta) = \varpi_2(y, \delta);\ t \in (-\infty, 0],\ \delta \in \Omega, \tag{6.23}$$

where $f : J \times \mathbb{R} \times \Omega \longrightarrow \mathbb{R}$ is a given function. Let $E = L^2[0, \pi]$, and $A_1 : E \longrightarrow E$ given by $A_1\varpi = \varpi''$ with domain $D(A_1) = \{\varpi \in E; \varpi, \varpi''$ are absolutely continuous, $\varpi'' \in E, \varpi(0) = \varpi(\pi) = 0\}$.

The operator A_1 is the infinitesimal generator of a strongly continuous cosine function $(S_1(t))_{t \in \mathbb{R}}$ on E. Furthermore, A_1 has discrete spectrum, the eigenvalues are $-n^2, n \in I\!N$ with corresponding normalized eigenvectors

$$w_n(\varepsilon) := \left(\frac{2}{\pi} \right)^{\frac{1}{2}} \sin(n\varepsilon),$$

and

(a) $\{w_n : n \in I\!N\}$ is an orthonormal basis of E,

(b) If $y \in E$, then $A_1 y = -\sum_{n=1}^{\infty} n^2 \langle y, w_n \rangle w_n$,

(c) For $y \in E, S_1(t)y = \sum\limits_{n=1}^{\infty} \cos(nt) \langle y, w_n \rangle w_n$, and the associated sine family is

$$S_2(t)y = \sum_{n=1}^{\infty} \frac{\sin(nt)}{n} \langle y, w_n \rangle w_n.$$

Consequently, $S_2(t)$ is compact for all $t > 0$ and

$$\|S_1(t)\| = \|S_2(t)\| \leq 1, \text{ for all } t \geq 0.$$

(d) If we denote the group of translations on E by

$$\bar{\Phi}(t)y(w, \delta) = \tilde{y}(w + t, \delta),$$

where $\tilde{y}$ is the extension of y with period 2π, then

$$S_1(t) = \frac{1}{2}(\bar{\Phi}(t) + \bar{\Phi}(-t)); A_1 = D,$$

where D is the infinitesimal generator of the group on

$$X = \{y(\cdot, \delta) \in H^1(0, \pi) : y(0, \delta) = y(\pi, \delta) = 0\}.$$

Suppose that A_2 is a bounded linear operator from U into E and the linear operator $\mathcal{K} : L^2(J, U) \to E$ given by:

$$\mathcal{K}g = \int_0^T S_1(T - s)A_2g(s, \delta)ds,$$

has an inverse operator $\mathcal{K}^{-1}$ in $L^2(J, U)/\mathrm{ker}\mathcal{K}$. We deduce that (6.16) is an abstract formulation of (6.21)–(6.23). If (6.6.1)–(6.6.6) are met. By Theorem 6.3, we conclude that (6.21)–(6.23) is controllable.

6.5   Notes and Remarks

This chapter's results are based on the articles [38, 337, 349]. The monographs [119–121] and the papers [44, 291, 326–328, 331] include additional pertinent results and investigations.

Chapter 7

S-Asymptotically ω-Periodic Mild Solutions for Differential Evolution Equations

7.1 Introduction and Motivations

In this chapter, we study the existence of S-asymptotically ω-periodic mild and almost automorphic solutions to some classes of second order semilinear evolution equation in Banach space, random functional evolution equations with infinite delay and semilinear integro-differential systems with nonlocal conditions via resolvent operators in the sense given by Grimmer. The investigation is based some fixed point theorems. Finally, examples are presented to illustrate the main findings.

The results of our analysis in this chapter can be viewed as a conditional extension of the problems discussed fairly recently in the following:

- Evolution equations arise in many areas of applied mathematics [1, 2]. There are many results concerning the second-order differential equations, see for example [3, 4, 6–8]. In recent years there has been an increasing interest in studying abstract non-autonomous second order initial value problems [260–263, 266].

- While the almost periodic, almost automorphic, and weighted pseudo almost periodic solutions to various evolution equations are investigated by many authors ([75, 79–81]), the notion of S-asymptotically ω-periodic functions have many applications in several problems like functional differential equations, integro-differential equations, fractional differential equations and fractional integro-differential equations. The concept of S-asymptotically ω-periodic function was first introduced in the literature by Henriquez *et al.* in [82, 83]. In the literature, there has been a significant attention devoted this concept in the deterministic case; we refer the reader to [84–91] and the references therein.

- The papers [38, 125–130], where many authors have investigated the existence of solutions for systems of ordinary differential, integral and semi-linear differential equations using the vector version fixed point theorems.
- The several papers in which authors discuss differential problems with various forms of delays, see [38, 39, 44, 107, 109, 273–275, 294, 350] and the references therein.
- In [328], Diop *et al.* considered the following random partial integro-differential equations with unbounded delay:

$$\begin{cases} y'(t,\delta) = Ay(t,\delta) + \displaystyle\int_0^t \beta(t-r)y(r,\delta)dr + F\left(t, y_t(\cdot,\delta), \delta\right), \quad t \in [0,\kappa], \\ y(t,\delta) = \phi(t,\delta), \quad t \in (-\infty, 0], \end{cases}$$

where A is a generator of a C_0 semigroup $(S(t)_{t\geq 0})$ from a Banach space E into E, $\beta(t)$ is a closed linear operator with domain $\mathcal{D}(A) \subset \mathcal{D}(\beta(t))$ and δ is a random variable. The authors used a random fixed point theorem with a stochastic domain combined with Schauder's fixed point theorem and Grimmer's resolvent operator theory in their proofs.

- The authors of [38] studied the existence and controllability results for the second order functional differential equation with delay and random effects that follows:

$$\begin{cases} y''(t,\delta) = \mathcal{F}_1 y(t,\delta) + \psi\left(t, y_t(\cdot,\delta), \delta\right) + \mathcal{F}_2 f(t,\delta), \quad \text{a.e. } t \in [0,\kappa], \\ y(t,\delta) = \varpi_1(t,\delta); t \in (-\infty, 0], \\ y'(0,\delta) = \varpi_2(\delta), \end{cases}$$

with results based on a random fixed point theorem with a stochastic domain.

7.2 Mild Solutions for Second-Order Semilinear Evolution Equations

In this paper, we investigate the existence of S-asymptotically ω-periodic mild solution for second differential equations. More precisely, we will consider the following problem

$$y''(t) - A(t)y(t) = f(t, y(t)), \quad t \in \mathbb{R}^+ := [0, \infty), \tag{7.1}$$

$$y(0) = y_0, \ y'(0) = y_1, \tag{7.2}$$

where $\{A(t)\}_{0\leq t<+\infty}$ is a family of linear closed operators from E into E that generates an evolution system of linear bounded operators $\{\mathcal{U}(t,s)\}_{(t,s)\in\mathbb{R}^+\times\mathbb{R}^+}$ for $0 \leq s \leq t < +\infty$, $f : \mathbb{R}^+ \times E \to E$ is a Carathéodory function, and $(E, \|\cdot\|)$ is a real Banach space.

7.2.1   Existence Results

In this work, the existence of solution the problem (7.1)–(7.2) is related to the existence of an evolution operator $\mathcal{U}(t, s)$ for the following homogeneous problem

$$y''(t) = A(t)y(t) \qquad t \in \mathbb{R}^+. \tag{7.3}$$

This concept of evolution operator has been developed by Kozak [266].

Denote by $\omega^T(y, \varepsilon)$ the modulus of continuity of y on the interval $[0, T]$, i.e.,

$$\omega^T(y, \varepsilon) = \sup \left\{ \|y(t) - y(s)\| \, ; t, s \in [0, T], \|t - s\| \leq \varepsilon \right\}.$$

Moreover, let us put

$$\omega^T(D, \varepsilon) = \sup \left\{ \omega^T(y, \varepsilon); y \in D \right\},$$

$$\omega_0^T(D) = \lim_{\varepsilon \to 0} \omega^T(D, \varepsilon).$$

Definition 7.1. A function $y \in BC(\mathbb{R}^+, E)$ is called a mild solution to the problem (7.1)–(7.2) if y satisfies the integral equation

$$y(t) = -\frac{\partial}{\partial s}\mathcal{U}(t, 0)y_0 + \mathcal{U}(t, 0)y_1 + \int_0^t \mathcal{U}(t, s)f(s, y(s))ds. \tag{7.4}$$

For the proof of our main theorem, we need the following hypotheses:

(7.1.1) There exist constants $M \geq 1$ and $\delta > 0$, such that

$$\|\mathcal{U}(t, s)\|_{B(E)} \leq Me^{-\delta(t-s)} \quad \text{for any } (t, s) \in \Delta$$

and

$$\mathcal{U}(t + \omega, s + \omega) = \mathcal{U}(t, s)(\omega - \text{periodicity}).$$

(7.1.2) There exist constants $\tilde{M} \geq 0$ and $\delta > 0$, such that:

$$\left\| \frac{\partial}{\partial s}\mathcal{U}(t, s) \right\|_{B(E)} \leq \tilde{M}e^{-\delta(t-s)}, (t, s) \in \Delta.$$

(7.1.3) The function $f : \mathbb{R}^+ \times E \to E$ is Carathéodory and satisfies the following:

(a) There exists a constant $\omega > 0$, such that

$$\lim_{t \to +\infty} \|f(t + \omega, u) - f(t, u)\| = 0 \text{ for a.e } t \in \mathbb{R}^+ \text{ and each } u \in E.$$

(b) There exist $p \in L^q(\mathbb{R}^+, \mathbb{R}^+)$, $q \in]1, \infty[$ and a continuous non-decreasing function $\psi : [0, \infty) \to (0, \infty)$ such that:

$$\|f(t, u)\| \le p(t)\psi(\|u\|) \text{ for a.e } t \in \mathbb{R}^+ \text{ and each } u \in E.$$

(7.1.4) There exist a locally integrable function $\eta : \mathbb{R}^+ \to \mathbb{R}^+$ and a continuous non-decreasing function $\varphi : \mathbb{R}^+ \to \mathbb{R}_+$ such that for any non-empty bounded set $D \subset E$ we have:

$$\alpha(f(t, D)) \le \eta(t)\varphi(\alpha(D)) \text{ for a.e. } t \in \mathbb{R}^+.$$

Additionally we assume that

$$\lim_{n \to +\infty} (\psi + \phi)^n(t) = 0 \text{ for a.e } t \in \mathbb{R}^+.$$

(7.1.5) There exists a positive constant R such that

$$\tilde{M}\|y_0\| + M\|y_1\| + \frac{M\psi(R)\|p\|_{L^q}}{\delta^{1-\frac{1}{q}}} \le R.$$

Theorem 7.1. *Assume that hypotheses* (7.1.1)–(7.1.5) *are satisfied. If*

$$M \max\left(4\|\eta\|_{L^1}, \frac{\|p\|_{L^q}}{\delta^{1-\frac{1}{q}}}\right) < 1,$$

then the problem (7.1)–(7.2) *has a unique S-asymptotically ω-periodic mild solution on $\mathbb{R}^+$.*

Proof. Consider the operator $N : SAP_\omega(X) \to SAP_\omega(X)$ defined by

$$(Ny)(t) = -\frac{\partial}{\partial s}\mathcal{U}(t, 0)y_0 + \mathcal{U}(t, 0)y_1 + \int_0^t \mathcal{U}(t, s)f(s, y(s))ds. \qquad (7.5)$$

Let the operators $\overline{N} = N_1 + N_2$, where

$$(N_1 y)(t) = -\frac{\partial}{\partial s}\mathcal{U}(t, 0)y_0,$$

$$(N_2 y)(t) = \mathcal{U}(t, 0)y_1,$$

$$(N_3 y)(t) = \int_0^t \mathcal{U}(t, s)f(s, y(s))ds.$$

Claim 1: $\overline{N} \in SAP_\omega(X)$.

For $t \ge 0$, one has

$$\|(\overline{N}y)(t + \omega) - (\overline{N}y)(t)\|$$
$$= |\frac{\partial}{\partial s}\mathcal{U}(t + \omega, 0)y_0 - \frac{\partial}{\partial s}\mathcal{U}(t, 0)y_0 + \mathcal{U}(t + \omega, 0)y_1 - \mathcal{U}(t, 0)y_1|$$
$$\le |\frac{\partial}{\partial s}\mathcal{U}(t + \omega, 0)y_0 - \frac{\partial}{\partial s}\mathcal{U}(t, 0)y_0| + |\mathcal{U}(t + \omega, 0)y_1 - \mathcal{U}(t, 0)y_1|$$
$$\le \tilde{M}(e^{-\delta(t+\omega)} - e^{-\delta t})\|y_0\| + M(e^{-\delta(t+\omega)} - e^{-\delta t})\|y_1\|. \qquad (7.6)$$

Since $\delta > 0$, we deduce that

$$\lim_{t \to +\infty} \|(\overline{N}y)(t + \omega) - (\overline{N}y)(t)\| = 0.$$

Then $\overline{N} \in SAP_\omega(X)$.

Claim 2. $N_3 \in SAP_\omega(X)$.

For $t \geq 0$, one has

$$
\begin{aligned}
(N_3 y)&(t + \omega) - (N_3 y)(t) \\
&= \int_0^{t+\omega} \mathcal{U}(t + \omega, s) f(s, y(s)) ds - \int_0^t \mathcal{U}(t, s) f(s, y(s)) ds \\
&= \int_0^\omega \mathcal{U}(t + \omega, s) f(s, y(s)) ds \\
&\quad + \int_\omega^{t+\omega} \mathcal{U}(t + \omega, s) f(s, y(s)) ds - \int_0^t \mathcal{U}(t, s) f(s, y(s)) ds \\
&= \Upsilon_1(t) + \Upsilon_2(t),
\end{aligned}
$$

where

$$\Upsilon_1(t) = \int_0^\omega \mathcal{U}(t + \omega, s) f(s, y(s)) ds,$$

and

$$\Upsilon_2(t) = \int_\omega^{t+\omega} \mathcal{U}(t + \omega, s) f(s, y(s)) ds - \int_0^t \mathcal{U}(t, s) f(s, y(s)) ds.$$

By (7.1.3), we have

$$\sup_{s \in [0, \omega]} \|f(s, y(s))\| = f_0 < +\infty.$$

Using the fact that $(U(t, s))_{t \geq s}$ is exponentially stable, we obtain

$$
\begin{aligned}
\|\Upsilon_1(t)\| &\leq M f_0 \int_0^\omega e^{-\delta(t+\omega-s)} ds \\
&= M f_0 e^{-\delta t} \int_0^\omega e^{-\delta(\omega-s)} ds \\
&= M f_0 e^{-\delta t} \frac{1 - e^{-\delta \omega}}{\delta},
\end{aligned}
$$

which shows that

$$\lim_{t \to +\infty} \Upsilon_1(t) = 0.$$

Let us write

$$\Upsilon_2(t) = \int_0^t \mathcal{U}(t + \omega, s + \omega) f(s + \omega, y(s + \omega)) ds - \int_0^t \mathcal{U}(t, s) f(s, y(s)) ds$$

and since the evolution family is ω-periodic, we obtain

$$\Upsilon_2(t) \leq \int_0^t \mathcal{U}(t,s) \|f(s+\omega, y(s+\omega)) - f(s+\omega, y(s))\| \, ds$$

$$+ \int_0^t \mathcal{U}(t,s) \|f(s+\omega, y(s)) - f(s, y(s))\| \, ds$$

$$= I_1(t) + I_2(t).$$

Observing that

$$\|I_1(t)\| \leq \left\| \int_0^{T_\varepsilon} \mathcal{U}(t,s) \|f(s+\omega, y(s+\omega)) - f(s+\omega, y(s))\| \, ds \right\|$$

$$+ \left\| \int_{T_\varepsilon}^t \mathcal{U}(t,s)(f(s+\omega, y(s+\omega)) - f(s+\omega, y(s))) ds \right\|.$$

Since $y \in SAP_\omega$, we know that there is a positive constant $T_\varepsilon > 0$ such that

$$\|y(t+\omega) - y(t)\| \leq \varepsilon \quad \text{for } t \geq T_\varepsilon.$$

By (7.1.3), we have

$$\|f(t, y(s+\omega)) - f(t, y(s))\| \leq \frac{\delta\varepsilon}{M} \quad \text{for } t \geq T_\varepsilon.$$

Thus, we get

$$\left\| \int_{T_\varepsilon}^t \mathcal{U}(t,s)(f(s+\omega, y(s+\omega)) - f(s+\omega, y(s))) ds \right\|$$

$$\leq M \int_{T_\varepsilon}^t e^{-\delta(t-s))} \|f(s+\omega, y(s+\omega)) - f(s+\omega, y(s))\| \, ds$$

$$\leq \int_{T_\varepsilon}^t M e^{-\delta(t-s)} \frac{\delta\varepsilon}{M} ds$$

$$\leq \varepsilon(e^{-\delta t} - e^{-\delta(t-T_\varepsilon)})$$

$$\leq \varepsilon. \tag{7.7}$$

Also, we have

$$\left\| \int_0^{T_\varepsilon} \mathcal{U}(t,s)(f(s+\omega, y(s+\omega)) - f(s+\omega, y(s))) ds \right\|$$

$$\leq M \int_0^{T_\varepsilon} e^{-\delta(t-s)} (\|f(s+\omega, y(s+\omega)) - f(s+\omega, y(s))\|) ds$$

$$\leq M \int_0^{T_\varepsilon} e^{-\delta(t-s)} (\|f(s+\omega, y(s+\omega))\| + \|f(s+\omega, y(s))\|) ds$$

$$\leq 2M \int_0^{T_\varepsilon} e^{-\delta(t-s)} p(s)\psi(\|y(s)\|) ds$$

$$\leq 2M\psi(\|y\|_{BC}) \int_0^{T_\varepsilon} e^{-\delta(t-s)} p(s) ds.$$

For $t \geq 0$, it follows from the Hölder inequality that

$$\left\| \int_0^{T_\varepsilon} \mathcal{U}(t,s)(f(s+\omega, y(s+\omega)) - f(s+\omega, y(s))ds \right\|$$

$$\leq 2M\|p\|_{L^q}\psi(\|y\|_{BC}) \left[\int_0^{T_\varepsilon} e^{-\frac{q\delta}{q-1}(t-\tau)} d\tau \right]^{1-\frac{1}{q}}$$

$$\leq \frac{2M\|p\|_{L^q}\psi(\|y\|_{BC})}{\delta^{1-\frac{1}{q}}} \left(e^{-\frac{q\delta}{q-1}t} - e^{-\frac{q\delta}{q-1}(t-T_\varepsilon)} \right)^{1-\frac{1}{q}}$$

$$\leq \varepsilon. \tag{7.8}$$

By (7.7), (7.8) which implies that $\lim_{t\to\infty} I_2(t) = 0$.

By (7.1.3)(a), we can see that, for every $\varepsilon > 0$, there is a positive constant $T_\varepsilon > 0$ such that

$$\|f(t+\omega, y(t)) - f(t, y(t))\| \leq \varepsilon \quad \text{for } t \geq T_\varepsilon.$$

Similarly, we find

$$I_2(t) \leq M \int_{T_\varepsilon}^t e^{-\delta(t-s))} \|f(s+\omega, y(s+\omega)) - f(s, y(s))\| \, ds$$

$$+ M \int_0^{T_\varepsilon} e^{-\delta(t-s))} \|f(s+\omega, y(s)) - f(s, y(s))\| \, ds$$

$$\leq \int_{T_\varepsilon}^t Me^{-\delta(t-s)} \frac{\delta\varepsilon}{M} \, ds$$

$$+ M \int_0^{T_\varepsilon} e^{-\delta(t-s)} \left(\|f(s+\omega, y(s))\| + \|f(s, y(s))\| \right) ds$$

$$\leq \varepsilon(1 - e^{-\delta(t-T_\varepsilon)})$$

$$+ 2M \int_0^{T_\varepsilon} e^{-\delta(t-s)} p(s)\psi(\|y(s)\|)ds$$

$$\leq \varepsilon(1 - e^{-\delta(t-T_\varepsilon)})$$

$$+ 2M\psi(\|y\|_\infty) \int_0^{T_\varepsilon} e^{-\delta(t-s)} p(s)ds.$$

For $t \geq 0$, it follows from the Hölder inequality that

$$I_2(t) \leq \varepsilon(1 - e^{-\delta(t-T_\varepsilon)})$$

$$+ 2M\|p\|_{L^q}\psi(\|y\|_\infty) \left[\int_0^{T_\varepsilon} e^{-\frac{q\delta}{q-1}(t-\tau)} d\tau \right]^{1-\frac{1}{q}}$$

$$\leq \varepsilon(1 - e^{-\delta(t-T_\varepsilon)})$$

$$+ 2\varepsilon \frac{M\|p\|_{L^q}\psi(\|y\|_\infty)}{\delta^{1-\frac{1}{q}}} \left(e^{-\frac{q\delta}{q-1}(t-T_\varepsilon)} - e^{-\frac{q\delta}{q-1}t} \right)^{1-\frac{1}{q}}$$

$$\leq \varepsilon.$$

Consequently, $\lim_{t \to \infty} I_2(t) = 0$, we conclude that $N_2 \in SAP_\omega$.

From **Claims 1 and 2** we deduce $N \in SAP_\omega(E)$.

Next, we will prove that the operator N satisfies all the assumptions of Theorem 2.8. We will break the proof into several steps.

Let

$$B_R = \{y \in SAP_\omega(E) : \|y\|_\infty \leq R\},$$

where R be any positive constant. Then B_R is a bounded, closed and convex subset of $SAP_\omega(E)$.

Step 1: $N(y) \in B_R$ for any $y \in B_R$.

Now, we have

$$\|Ny(t)\| \leq \left\| \frac{\partial}{\partial s}\mathcal{U}(t,0) \right\|_{B(E)} \|y_0\|$$

$$+ \|\mathcal{U}(t,s)\|_{B(E)} \|y_1\| + \int_0^t \|\mathcal{U}(t,s)\|_{B(E)}\, p(s)\psi(\|y(s)\|)ds$$

$$\leq \tilde{M}\|y_1\| + M\|y_0\| + M\psi(R)\int_0^t e^{-\delta(t-s)}p(s)ds.$$

For $t \geq 0$, it follows from the Hölder inequality that

$$\|Ny(t)\| \leq \tilde{M}\|y_0\| + M\|y_1\| + M\psi(R)\|p\|_{L^q}(1 - e^{-\frac{q\delta}{q-1}t})^{1-\frac{1}{q}}$$

$$\leq \tilde{M}\|y_0\| + M\|y_1\| + \frac{M\psi(R)\|p\|_{L^q}}{\delta^{1-\frac{1}{q}}}$$

$$\leq R. \tag{7.9}$$

The conditions (7.1.5) ensure that the operator N transforms the set B_R into itself.

Step 2. N is continuous.

Let $(y_n)_{n \in N}$ be a sequence in B_R such that $y_n \to y$ in B_R.

Case 1. If $t \in [0,T]$; $T > 0$, then, we have

$$\|(Ny_n)(t) - (Ny)(t)\|$$

$$\leq M\int_0^t \|f(s,y_n(s)) - f(s,y(s))\|\,ds.$$

Since the function f is Carathéodory, the Lebesgue dominated convergence theorem implies that

$$\|Ny_n - Ny\|_\infty \to 0 \quad \text{as } n \to +\infty.$$

Case 2. Since the function f is Carathéodory, we can see that

$$\|f(s, y_n(s)) - f(s, y(s))\| \leq \frac{\delta \varepsilon}{M} \quad \text{for } t \geq T. \tag{7.10}$$

If $t \in (T, \infty)$, $T > 0$, then (7.10) and our hypotheses give us that

$$
\begin{aligned}
&\|Ny_n(t) - Ny(t)\| \\
&\leq \int_0^t \|\mathcal{U}(t,s)\|_{B(E)} \Big\| f(s, y_n(s)) - f(s, y(s)) \Big\| ds \\
&\leq M \frac{\delta \varepsilon}{M} \int_0^t e^{-\delta(t-s)} ds \\
&\leq \frac{M}{\delta} \frac{\delta \varepsilon}{M} (1 - e^{-\delta t}) \\
&\leq \varepsilon.
\end{aligned}
\tag{7.11}
$$

Then the inequality (7.11) reduces to

$$\|N(u_n) - N(u)\|_\infty \to 0 \quad \text{as } n \to \infty.$$

Now, we conclude that N is continuous from B_R to B_R.

Step 3: $N(B_R)$ is equicontinuous.
Let $t_1, t_2 \in [0, T]$ with $t_2 > t_1$ and $y \in B_R$. Then, we have

$$
\begin{aligned}
&\|(Ny)(t_2) - (Ny)(t_1)\| \\
&= \left\| \int_0^{t_1} (\mathcal{U}(t_2, s) - \mathcal{U}(t_1, s)) f(s, y(s)) \right. \\
&\quad \left. + \int_{t_1}^{t_2} \mathcal{U}(t_2, s) f(s, y(s)) ds \right\| \\
&\leq \int_0^{t_1} \|\mathcal{U}(t_2, s) - \mathcal{U}(t_1, s)\|_{B(E)} \ p(\tau) \psi(\|y(s)\|) ds \\
&\quad + M \int_{t_1}^{t_2} e^{-\delta(t-s)} p(s) \psi(\|y(s)\|) ds. \\
&\leq \int_0^{t_1} \|\mathcal{U}(t_2, s) - \mathcal{U}(t_1, s)\|_{B(E)} \ p(s) \psi(\|y(s)\|) ds \\
&\quad + \frac{M \|p\|_{L^q} \psi(R)}{\delta^{1-\frac{1}{q}}} \left(e^{-\frac{q\delta}{q-1}(t-t_2)} - e^{-\frac{q\delta}{q-1}(t-t_2)} \right)^{1-\frac{1}{q}}.
\end{aligned}
$$

The right-hand side of the above inequality tends to zero as $t_2 - t_1 \to 0$, which implies that $N(B_R)$ is equicontinuous.

Consider the measure of noncompactness $\mu(B)$ defined on the family of bounded subsets of the space $BC(\mathbb{R}^+, E)$ (see [198]) by

$$\mu(B) = \omega_0^T(B) + \sup_{t \in \mathbb{R}^+} \alpha(B(t)) + \lim_{T \to +\infty} \sup\{\|y(t)\| : t \geq T, y \in E\}.$$

Step 4: $\mu(N(B)) \leq M \max \left(4\|\eta\|_{L^1}, \frac{\|p\|_{L^q}}{\delta^{1-\frac{1}{q}}} \right)(\varphi + \psi)(\mu(B))$ for all $B \subset B_R$.

For all $B \subset B_R$, $N(B)$ is bounded. Hence, by Lemma 2.4, there exists a countable set $B_1 = \{y\}_{n=1}^{\infty} \subset B$, such that

$$\alpha(N(B)) \leq 2\alpha(N(B_1)). \tag{7.12}$$

Using the properties of α, Lemmas 2.4, 2.6 and assumptions (7.1.1) and (7.1.4), we get

$$\alpha(NB_1(t)) \leq \alpha\left(\left\{\int_0^t \mathcal{U}(t,s)f(s,y_n(s))ds\right\}_{n=0}^{\infty}\right)$$

$$\leq 2M \int_0^t \{\alpha\left(f(s,y_n(s))ds)\right)\}_{n=0}^{\infty} ds$$

$$\leq 2M \int_0^t \eta(s)\varphi\left(\{(\alpha(y_n(s))\}_{n=0}^{\infty}))\right) ds$$

$$\leq 2M \int_0^t \eta(s)\varphi(\alpha(B(s)))ds.$$

Form inequality (7.12), it follows that

$$\alpha(NB(t)) \leq 4M \int_0^t \eta(s)\varphi(\alpha(B(s)))ds,$$

then

$$\alpha(N(B(t)) \leq 4M\|\eta\|_{L^1}\varphi(\sup_{t\in\mathbb{R}^+} \alpha(B(t))).$$

Since

$$\sup_{t\in\mathbb{R}^+} \alpha(B(t)) \leq \sup_{t\in\mathbb{R}^+} \alpha(B(t)) + \lim_{T\to+\infty} \sup\{\|y(t)\| : t \geq T, y \in E\},$$

then

$$\alpha(N(B(t)) \leq 4M\|\eta\|_{L^1}\varphi(\sup_{t\in\mathbb{R}^+} \alpha(B(t)) + \lim_{T\to+\infty} \sup\{\|y(t)\| : t \geq T, y \in E\}).$$

$$\tag{7.13}$$

On the other hand, for $t \geq T$, we have

$$\|(Ny)(t)\| \leq \tilde{M}e^{-\delta t}\|y_1\| + Me^{-\delta t}\|y_0\|$$

$$+ M \int_0^t e^{-\delta(t-s)}p(s)\psi(\|y(s)\|)ds.$$

$$\leq \tilde{M} e^{-\delta t} \|y_1\| + M e^{-\delta t} \|y_0\|$$
$$+ M \int_0^T e^{-\delta(t-s)} p(s) \psi(\|y(s)\|) ds.$$
$$+ M \int_T^t e^{-\delta(t-s)} p(s) \psi(\|y(s)\|) ds.$$
$$\leq \tilde{M} e^{-\delta t} \|y_1\| + M e^{-\delta t} \|y_0\|$$
$$+ M \int_0^T e^{-\delta(t-s)} p(s) ds \psi(\sup_{s \in [0,T]} \|y(s)\|).$$
$$+ M \int_T^t e^{-\delta(t-s)} p(s) ds \psi(\sup\{\|y(t)\| : t \geq T, y \in E\}).$$

Next, applying the Hölder inequality we derive

$$\|(Ny)(t)\| \leq (\tilde{M} e^{-\delta t} \|y_1\| + M e^{-\delta t} \|y_0\|)$$
$$+ \frac{M\|p\|_{L^q}}{\delta^{1-\frac{1}{q}}} (e^{-\frac{q\delta}{q-1}t} - e^{-\frac{q\delta}{q-1}(t-T)})^{1-\frac{1}{q}} \psi(\|y\|_\infty).$$
$$+ \frac{M\|p\|_{L^q}}{\delta^{1-\frac{1}{q}}} (1 - e^{-\frac{q\delta}{q-1}t})^{1-\frac{1}{q}} \psi(\sup\{\|y(t)\| : t \geq T, y \in E\}).$$

Then

$$\|(Ny)(t)\| \leq (\tilde{M} e^{-\delta t} \|y_1\| + M e^{-\delta t} \|y_0\|)$$
$$+ \frac{M\|p\|_{L^q}}{\delta^{1-\frac{1}{q}}} e^{-\delta T} \psi(\|y\|_\infty).$$
$$+ \frac{M\|p\|_{L^q}}{\delta^{1-\frac{1}{q}}} \psi(\sup\{\|y(t)\| : t \geq T, y \in E\}).$$

Since

$$\lim_{T \to +\infty} \sup\{\|y(t)\| : t \geq T, y \in E\}$$
$$\leq \sup_{t \in \mathbb{R}^+} \alpha(B(t)) + \lim_{T \to +\infty} \sup\{\|y(t)\| : t \geq T, y \in E\},$$

then

$$\lim_{T \to +\infty} \sup\{\|(Ny)(t) : t \geq T, y \in E\})$$
$$\leq \frac{M\|p\|_{L^q}}{\delta^{1-\frac{1}{q}}} \psi(\sup_{t \in \mathbb{R}^+} \alpha(B(t)) + \lim_{T \to +\infty} \sup\{\|y(t)\| : t \geq T, y \in E\}). \quad (7.14)$$

Further, combining (7.13) and (7.14), we get

$$\sup_{t \in \mathbb{R}^+} \alpha((NB)(t)) + \lim_{T \to +\infty} \sup\{\|(Ny)(t)\| : t \geq T, y \in E\})$$

$$\leq 4M\|\eta\|_{L^1} \varphi(\sup_{t \in \mathbb{R}^+} \alpha(B(t)) + \lim_{T \to +\infty} \sup\{\|y(t)\| : t \geq T, y \in E\})$$

$$+ \frac{M\|p\|_{L^q}}{\delta^{1-\frac{1}{q}}} \psi(\sup_{t \in \mathbb{R}^+} \alpha(B(t)) + \lim_{T \to +\infty} \sup\{\|y(t)\| : t \geq T, y \in E\}$$

$$\leq M \max\left(4\|\eta\|_{L^1}, \frac{\|p\|_{L^q}}{\delta^{1-\frac{1}{q}}}\right)(\varphi + \psi)$$

$$\times (\sup_{t \in \mathbb{R}^+} \alpha(B(t)) + \lim_{T \to +\infty} \sup\{\|y(t)\| : t \geq T, y \in E\}). \qquad (7.15)$$

From **Step 3** and inequality (7.15), we conclude that

$$\mu(N(B)) \leq M \max\left(4\|\eta\|_{L^1}, \frac{\|p\|_{L^q}}{\delta^{1-\frac{1}{q}}}\right)(\varphi + \psi)(\mu(B)).$$

It follows from Lemma 2.8 that N has at least one fixed point $y \in B_R$, which is mild solution of problem (7.1)–(7.2) on the interval $\mathbb{R}^+$.

7.2.2   *An Example*

Consider the second order differential equation of the form:

$$\begin{cases} \dfrac{\partial^2}{\partial t^2} z(t, \tau) = \dfrac{\partial^2}{\partial \tau^2} z(t, \tau) + (-3 + \cos(\pi t)) \dfrac{\partial}{\partial t} z(t, \tau) \\[2mm] \qquad + \dfrac{t^{\frac{1}{r}}}{2(1+t^2)^{\frac{2}{r}}} \ln\left(1 + \dfrac{2t^{\frac{r-1}{r}}}{(1+t^2)^{\frac{2r-2}{r}}} |z(t, \tau)|\right) \\[2mm] \qquad r > 1, \ t \in \mathbb{R}^+, \ \tau \in [0, \pi], \\[4mm] z(t, 0) = z(t, \pi) = 0, \ t \in \mathbb{R}^+, \\[4mm] z(t+1, \tau) = z(t, \tau), \ t \in \mathbb{R}^+, \ \tau \in [0, \pi], \\[4mm] \dfrac{\partial}{\partial t} z(0, \tau) = \psi(\tau), \ \tau \in [0, \pi]. \end{cases} \qquad (7.16)$$

Let $E = L^2([0, \pi], \mathbb{R})$ be the space of 2-integrable functions from $[0, \pi]$ into $\mathbb{R}$, and let $H^2([0, \pi], \mathbb{R})$ be the Sobolev space of functions $x : [0, \pi] \to \mathbb{R}$, such that $x'' \in L^2([0, \pi], \mathbb{R})$. We consider the operator $A_1 z(\tau) = z''(\tau)$ with domain $D(A_1) = H^2(\mathbb{R}, \mathbb{C})$, which is the infinitesimal generator of strongly continuous cosine function $C(t)$ on E. Moreover, A_1 has discrete spectrum, the spectrum of A_1 consists of eigenvalues $-n^2$ for $n \in \mathbb{Z}$, with associated eigenvector

$$\omega_n(\xi) = \frac{1}{\sqrt{2\pi}} e^{in\xi}, n \in \mathbb{Z},$$

the set $\{\omega_n, n \in \mathbb{Z}\}$ is an orthonormal basis of E. In particular,

$$A_1 x = -\sum_{n=1}^{\infty} n^2 \langle x, w_n \rangle w_n \text{ for } x \in D(A).$$

The cosine function $C(t)$ is given by

$$C(t)x = \sum_{n=1}^{\infty} \cos(nt) \langle x, w_n \rangle w_n \text{ for } x \in D(A), t \in \mathbb{R}.$$

The associated sine function is defined by

$$S(t)x = \sum_{n=1}^{\infty} \frac{\sin(nt)}{n} \langle x, w_n \rangle w_n \text{ for } x \in D(A), t \in \mathbb{R}.$$

From [8], for all $x \in H^2([0, \pi], \mathbb{R}), t \in \mathbb{R}$,

$$\|C(t)\|_{B(E)} \le e^{-t}, \text{ and } \|S(t)\|_{B(E)} \le e^{-t}.$$

Now, we define an operator $A(t) : D(A) \subset H \to H$ by

$$\begin{cases} D(A(t)) = D(A) \\ A(t) = A_1 + b(t, \tau), \end{cases} \tag{7.17}$$

where $b(t, \tau) = -3 + 2 \cos \pi t$.

Note that $A(t)$ generates an evolution system $U(t, s)$ of the form

$$\mathcal{U}(t, s) = S(t - s) e^{\int_s^t b(t,s)ds}.$$

Since $b(t, \tau) = -3 + \cos(\pi t) \le -2$, we have

$$\mathcal{U}(t, s) = S(t - s) e^{-2(t-s)} \tag{7.18}$$

and

$$\|\mathcal{U}\|_{B(E)} \le \|S\|_{B(E)} e^{-2(t-s)} \le e^{-3(t-s)}.$$

Since $b(t + 2, x) = b(t, x)$, we conclude that $\mathcal{U}(t, s)$ is a 2-periodic evolution system exponentially stable. It follows from the estimate (7.18) that $\mathcal{U}(t, s)$ is well defined and satisfies the conditions of Definition 2.10. Hence conditions (7.1.1) and (7.1.2) are satisfied.

Set

$$z(t)(\tau) = w(t)(\tau), \ t \geq 0, \ \tau \in [0, \pi],$$

$$f(t, u)(\tau) = \frac{t^{\frac{1}{r}}}{2(1+t^2)^{\frac{2}{r}}} \ln \left(1 + \frac{2t^{\frac{r-1}{r}}}{(1+t^2)^{\frac{2r-2}{r}}} |z(, \tau)| \right).$$

Further, for arbitrarily fixed $z \in H^2([0, \pi], \mathbb{R})$ and for $t > 0$ we obtain

$$\|f(t+1, z)(\tau) - f(t, z)(\tau)\|$$

$$\leq \frac{t^{\frac{1}{r}}}{2(1+t^2)^{\frac{2}{r}}} \ln \left(1 + \frac{2t^{\frac{r-1}{r}}}{(1+t^2)^{\frac{2r-2}{r}}} |z(t+1, \tau)| \right)$$

$$- \frac{t^{\frac{1}{r}}}{2(1+t^2)^{\frac{2}{r}}} \ln \left(1 + \frac{2t^{\frac{r-1}{r}}}{(1+t^2)^{\frac{2r-2}{r}}} |z(t, \tau)| \right)$$

$$\leq \frac{t^{\frac{1}{r}}}{2(1+t^2)^{\frac{2}{r}}} \ln \left(\frac{1 + 2(t+1)^{\frac{r-1}{r}} \|z(t+1, \tau)\|}{1 + 2t^{\frac{r-1}{r}} \|z(t, \tau)\|} \right)$$

$$\leq \frac{t^{\frac{1}{r}}}{2(1+t^2)^{\frac{2}{r}}} \ln \left(\left(\frac{t+1}{t} \right)^{\frac{r-1}{r}} + \frac{1 - \left(\frac{t+1}{t} \right)^{\frac{r-1}{r}}}{1 + 2t^{\frac{r-1}{r}} \|z(t, \tau)\|} \right)$$

$$\leq \frac{t^{\frac{1}{r}}}{2(1+t^2)^{\frac{2}{r}}} \ln \left(\left(\frac{t+1}{t} \right)^{\frac{r-1}{r}} + 1 \right). \tag{7.19}$$

Now from (7.19), we can see $\|f(t+1, z)(\tau) - f(t, z)(\tau)\|$ tends to zero as $t \to \infty$. Hence condition $(7.1.3)(a)$ is satisfied with $\omega = 1$.

We can estimate the function f:

$$f(t, u)(\tau) \leq \frac{t^{\frac{1}{r}}}{2(1+t^2)^{\frac{2}{r}}} \ln(1 + 2|z(t, \tau)|).$$

Hence condition $(7.1.3)(b)$ is satisfied with

$$p(t) = \frac{t^{\frac{1}{r}}}{2(1+t^2)^{\frac{2}{r}}}, \quad \psi(t) = \frac{1}{4} \ln(1 + 2t).$$

We can also estimate the function f:

$$f(t, u)(\tau) \leq \frac{t}{(t^2+1)^2} |z(t, \tau)|. \tag{7.20}$$

By (7.20), for every $t \in \mathbb{R}^+$, and $B \in D \subset E$, we have

$$\alpha(f(t, D) \leq \frac{t}{2(1+t^2)^2} \alpha(D),$$

Hence condition (7.1.4) is satisfied with

$$\eta(t) = \frac{t}{(1+t^2)^2}, \quad \varphi(t) = \frac{t}{2}.$$

Moreover, we have

$$(\psi + \varphi)(t) = \frac{t}{2} + \frac{1}{4}\ln(1 + 2t) \le t.$$

We conclude that (see Lemma 2.1. [252])

$$\lim_{n \to +\infty} (\psi + \phi)^n(t) = 0 \text{ for a.e } t \in J.$$

Consequently, (7.16) can be written in the abstract form (7.1)–(7.2) with $A(t)$ and f as defined above. The existence of mild solutions can be deduced from an application of Theorem 7.1.

7.3 Coupled Integro-Differential Equations with Nonlocal Conditions

The goal of this paper is to study the existence and attractivity of S-Asymptotic ω-Periodic mild solutions for integrodifferential system with nonlocal conditions via resolvent operators of the form:

$$
\begin{cases}
y'(t) = A_1 y(t) + f_1\left(t, y(t), x(t), \Psi_1(y(t), x(t))\right) \\
\quad + \displaystyle\int_0^t B_1(t-s)y(s)ds, \text{ for } t \in \mathbb{R}^+, \\[2mm]
x'(t) = A_2 x(t) + f_2\left(t, y(t), x(t), \Psi_2(y(t), x(t))\right) \\
\quad + \displaystyle\int_0^t B_2(t-s)x(s)ds, \text{ for } t \in \mathbb{R}^+, \\[2mm]
y(0) = y_0 + \Upsilon_1(y, x), \\[2mm]
x(0) = x_0 + \Upsilon_2(y, x),
\end{cases}
\tag{7.21}
$$

where $\mathbb{R}^+ = [0, +\infty)$, and for $i = 0, 1$, $A_i : D(A_i) \subset E \to E$ are the infinitesimal generators of a strongly continuous semigroup $\{T_i(t)\}_{t \ge 0}$, $B_i(t)$ are a closed linear operators with domain $D(A_i) \subset D(B_i(t))$, the operators Ψ_i are defined by

$$\Psi_i(y, x)(t) = \int_0^a g_i(t, s, y(s), x(s))ds, \quad a > 0.$$

The nonlinear terms $f_i : \mathbb{R}^+ \times E \times E \times E \to E$, $\Upsilon_i \in BC(\mathbb{R}^+, E) \times BC(\mathbb{R}^+, E) \to E$, are given functions, and $(E, \|\cdot\|)$ a Banach space.

7.3.1 Main Results

In this section we discuss the existence of mild solution for the problem (7.21).

First, we consider the following Cauchy problem:

$$\begin{cases} w'(t) = Aw(t) + \int_0^t B(t-s)w(s)ds; & \text{for } t \geq 0, \\ w(0) = w_0 \in E. \end{cases} \qquad (7.22)$$

The existence and properties of a resolvent operator has been discussed in [42, 295, 296]. In what follows, we suppose the following assumptions:

(R1) A is the infinitesimal generator of a uniformly continuous semigroup $\{S(t)\}_{t>0}$,

(R2) For all $t \geq 0$, $B(t)$ is closed linear operator from $D(A)$ to E and $B(t) \in B(D(A), E)$. For any $y \in D(A)$, the map $t \to B(t)y$ is bounded, differentiable and the derivative $t \to B'(t)y$ is bounded uniformly continuous on $\mathbb{R}^+$.

Theorem 7.2 ([295]). *Assume that (R1)–(R2) hold. Then, there exists a unique resolvent operator for the Cauchy problem (7.22).*

Lemma 7.1 ([295]). *Assume that (R1)–(R2) hold. The resolvent operator $(\Phi(t))_{t\geq 0}$ is compact for $t > 0$ if and only if the semigroup $(S(t))_{t\geq 0}$ is compact for $t > 0$.*

Lemma 7.2 ([351]). *Assume that (R1)–(R2) hold. If the resolvent operator $(\Phi(t))_{t\geq 0}$ is compact for $t > 0$, then it is norm continuous (or continuous in the uniform operator topology) for $t > 0$.*

Lemma 7.3 ([351]). *For any $\alpha > 0$ there exists a constant $\gamma = \gamma(\alpha)$ such that*

$$\|\Phi(t+h) - \Phi(h)\Phi(t)\|_{\mathcal{L}(X)} \leq \gamma h, \qquad \text{for} \quad 0 \leq h \leq t \leq \alpha.$$

7.3.1.1 Existence of solutions

We obtain existence and uniqueness results by employing Perov's fixed point theorem.

Definition 7.2. A function $(y, x) \in BC(\mathbb{R}^+, E) \times BC(\mathbb{R}^+, E) = X \times X$ is called a mild solution of problem (7.21) if it satisfies

$$y(t) = \Phi_1(t)(y_0 + \Upsilon_1(y, x))$$

$$+ \int_0^t \Phi_1(t-s) f_1\left(s, y(s), x(s), \Psi_1(y(s), x(s))\right) ds; \quad t \in \mathbb{R}^+,$$

$$x(t) = \Phi_2(t)(x_0 + \Upsilon_2(y, x))$$
$$+ \int_0^t \Phi_2(t - s) f_2 (s, y(s), x(s), \Psi_2(y(s), x(s))) \, ds; \quad t \in \mathbb{R}^+.$$

The following hypotheses are needed in the sequel:

(7.7.1) For $i = 1, 2$, $f_i : \mathbb{R}^+ \times E \times E \times E \to E$ are Carathéodory functions and there exist p_i, $q_i \in L^1(\mathbb{R}^+, \mathbb{R}^+)$ and a continuous non-decreasing functions $\psi_i, \phi_i : \mathbb{R}^+ \to (0, +\infty)$ such that

$$\| f_i(t, u, v, w(u, u)) - f_i(t, \widehat{u}, \widehat{v}, w(\widehat{u}, \widehat{v})) \|$$
$$\leq p_i(t)\psi_i(\|u - \widehat{u}\|) + q_i(t)\phi_i(\|v - \widehat{v}\|),$$

$$\text{for } u, \ \widehat{u}, \ v, \ \widehat{v}, \ w \in E,$$

with

$$\psi_i(t) \leq t, \ \phi_i \leq t, \text{ and } f_i^0 = \| f_i(\cdot)(\cdot, 0, 0, 0) \| \in L^1(\mathbb{R}^+, \mathbb{R}^+).$$

(7.7.2) For $i = 1, 2$, $g_i : D_h \times E \times E \to E$ are continuous and there exist continuous functions $g_{c_i}, \widehat{g}_{c_i} : \mathbb{R}^+ \to (0, +\infty)$ such that

$$\| g_i(t, s, u, v) - g_i(t, s, \widehat{u}, \widehat{v}) \| \leq g_{c_i}(t)\|u - \widehat{u}\| + \widehat{g}_{c_i}(t)\|v - \widehat{v}\|,$$

$$\text{for each } (t, s) \in D_{g_i} \text{ and } u, \ \widehat{u}, \ v, \ \widehat{v} \in E,$$

with

$$\max \Big\{ \sup_{t \in \mathbb{R}^+} \{ g_{c_i}(t) \}, \sup_{t \in \mathbb{R}^+} \{ \widehat{g}_{c_i}(t) \}, \sup_{(t,s) \in D_{g_i}} \{ \| g_i(t, s, 0, 0) \| \} \Big\}$$
$$= \max\{ g_{c_i}^*, \widehat{g}_{c_i}^*, g_i^* \} < +\infty.$$

(7.7.3) For $i = 1, 2$, $\Upsilon_i : X \times X \to E$ are Lipschitz functions, i.e there exist positive constants $L_{\Upsilon_i}, \widehat{L}_{\Upsilon_i}$, such that

$$\| \Upsilon_i(u, v) - \Upsilon_i(\widehat{u}, \widehat{v}) \| \leq L_{\Upsilon_i}\|u - \widehat{u}\|_X + \widehat{L}_{\Upsilon_i}\|v - \widehat{v}\|_X, \text{ for all } u, \widehat{u}, v, \widehat{v} \in X.$$

(7.7.4) Assume that $(R1)$–$(R2)$ hold, and there exist $M_{\Phi_i} \geq 1$ and $\beta_i \geq 0$, such that

$$\| \Phi_i(t) \|_{B(E)} \leq M_{\Phi_i} e^{-\beta_i t}.$$

Theorem 7.3. *Assume that the conditions* $(7.7.1)$–$(7.7.4)$ *are satisfied, and the matrix*

$$\begin{pmatrix} M_{\Phi_1}\left(L_{\Upsilon_1} + \|p_1\|_{L^1}\right) M_{\Phi_1}\left(\widehat{L}_{\Upsilon_1} + \|q_1\|_{L^1}\right) \\ M_{\Phi_2}\left(L_{\Upsilon_2} + \|p_2\|_{L^1}\right) M_{\Phi_2}\left(\widehat{L}_{\Upsilon_2} + \|q_2\|_{L^1}\right) \end{pmatrix}$$

converges to zero. Then, the system (7.21) *has a unique mild solution.*

Proof. Transform the problem (7.21) into a fixed point problem. Consider the operator
$\Theta : BC(\mathbb{R}^+, E) \times BC(\mathbb{R}^+, E) \to BC(\mathbb{R}^+, E) \times BC(\mathbb{R}^+, E)$ define by:

$$\Theta(y(t), x(t)) = (\Theta_1(y(t), x(t)), \Theta_2(y(t), x(t))),$$

where

$$\begin{cases} \Theta_1(y(t), x(t)) = \Phi_1(t)(y_0 + \Upsilon_1(y, x)) \\ \qquad\qquad + \displaystyle\int_0^t \Phi_1(t - s) f_1\left(s, y(s), x(s), \Psi_1(y(s), x(s))\right) ds, \\ \\ \Theta_2(y(t), x(t)) = \Phi_2(t)(x_0 + \Upsilon_2(y, x)) \\ \qquad\qquad + \displaystyle\int_0^t \Phi_2(t - s) f_2\left(s, y(s), x(s), \Psi_2(y(s), x(s))\right) ds. \end{cases}$$

We show that Θ was well defined. Given $(y, x) \in X \times X$, $t \in \mathbb{R}^+$, we have

$$\|\Theta_1(y(t), x(t))\| \leq \|R_1(t)\|(\|y_0\| + \|\Upsilon_1(y, x)\|)$$
$$+ \int_0^t \|\Phi_1(t - s)\| \|f_1\left(s, y(s), x(s), \Psi_1(y(s), x(s))\right)\| ds.$$

From $(7.7.1)$, we have

$$\|f_1(s, y(s), x(s), \Psi_1(y(s), x(s)))\| \leq p_1(s)\psi_1(\|y(s)\|) + q_1(s)\phi_1(\|x(s)\|)$$
$$+ \|f_1(s, 0, 0, 0)\|.$$

Also, we have

$$\|\Upsilon_1(y, x)\| \leq L_{\Upsilon_1}\|y\|_X + \widehat{L}_{\Upsilon_1}\|x\|_X + \Upsilon_1^0.$$

Then, we get

$$\|\Theta_1(y(t), x(t))\| \leq M_{\Phi_1}\left(\|y_0\| + L_{\Upsilon_1}\|y\|_X + \widehat{L}_{\Upsilon_1}\|x\|_X + \Upsilon_1^0\right)$$
$$+ M_{\Phi_1}\left(\|p_1\|_{L^1}\psi_1(\|y\|_X) + \|q_1\|_{L^1}\phi_1(\|x\|_X)\right)$$
$$+ M_{\Phi_1}\int_0^t f_1^0(s) ds.$$

Similarly, we obtain

$$\|\Theta_2(y(t), x(t))\| \leq M_{\Phi_2} \left(\|x_0\| + L_{\Upsilon_2}\|y\|_X + \widehat{L}_{\Upsilon_2}\|x\|_X + \Upsilon_2^0 \right)$$
$$+ M_{\Phi_2} \left(\|p_2\|_{L^1}\psi_2(\|y\|_X) + \|q_2\|_{L^1}\phi_2(\|x\|_X) \right)$$
$$+ M_{\Phi_2} \int_0^t f_2^0(s)ds.$$

Thus,

$$\|\Theta(y, x)\|_{X \times X} < +\infty.$$

Obviously, the fixed point of operator Θ is a mild solution of the problem (7.21).

We shall use the Perov's fixed point theorem to prove that Θ has a fixed point. Let $(y, x), (\widehat{y}, \widehat{x}) \in X \times X$, then (7.7.1) and (7.7.2) imply

$$\|\Theta_1(y(t), x(t)) - \Theta_1(\widehat{y}(t), \widehat{x}(t))\|$$
$$\leq M_{\Phi_1}(L_{\Upsilon_1}\|y - \widehat{y}\|_X + \widehat{L}_{\Upsilon_1}\|x - \widehat{x}\|_X)$$
$$+ M_{\Phi_1} \int_0^t p_1(s)\psi_1(\|y(s) - \widehat{u}(s)\|) + q_1(s)\phi_1(\|v(s) - \widehat{v}(s)\|)ds$$
$$\leq M_{\Phi_1} \left(L_{\Upsilon_1} + \|p_1\|_{L^1} \right) \|y - \widehat{y}\|_X + M_{\Phi_1} \left(\widehat{L}_{\Upsilon_1} + \|q_1\|_{L^1} \right) \|x - \widehat{x}\|_X.$$

Similarly, we get

$$\|\Theta_2(y(t), x(t)) - \Theta_2(\widehat{y}(t), \widehat{x}(t))\|$$
$$\leq M_{\Phi_2} \left(L_{\Upsilon_2} + \|p_2\|_{L^1} \right) \|y - \widehat{y}\|_X + M_{\Phi_2} \left(\widehat{L}_{\Upsilon_2} + \|q_2\|_{L^1} \right) \|x - \widehat{x}\|_X.$$

Then, we have

$$\|\Theta(y, x) - \Theta(\widehat{y}, \widehat{x})\|_{X \times X}$$
$$\leq \begin{pmatrix} M_{\Phi_1}(L_{\Upsilon_1} + \|p_1\|_{L^1}) & M_{\Phi_1}\left(\widehat{L}_{\Upsilon_1} + \|q_1\|_{L^1}\right) \\ M_{\Phi_2}(L_{\Upsilon_2} + \|p_2\|_{L^1}) & M_{\Phi_2}\left(\widehat{L}_{\Upsilon_2} + \|q_2\|_{L^1}\right) \end{pmatrix} \begin{pmatrix} \|y - \widehat{y}\|_X \\ \|x - \widehat{x}\|_X \end{pmatrix}.$$

Applying now Theorem 2.16, we conclude that Θ has a unique fixed point, which is a mild solution of problem (7.21). □

Now, we study the existence of a mild solution by using Schaefer's fixed point theorem.

In the assumption (7.7.1), we suppose that for every $M_1, M_2 \geq 0$, we have

$$\lim_{t \to +\infty} \sup_{t \in \mathbb{R}^+} \int_0^t e^{-\beta(t-s)} \left[M_1 p_i(s) + M_2 q_i(s) \right] ds = 0.$$

Let

$$\widehat{M} = \begin{pmatrix} 1 - M_{\Phi_1}\left(L_{\Upsilon_1} + \|p_1\|_{L^1}\right) & -M_{\Phi_1}\left(\widehat{L}_{\Upsilon_1} + \|q_1\|_{L^1}\right) \\ -M_{\Phi_2}\left(L_{\Upsilon_2} + \|p_2\|_{L^1}\right) & 1 - M_{\Phi_2}\left(\widehat{L}_{\Upsilon_2} + \|q_2\|_{L^1}\right) \end{pmatrix}.$$

Theorem 7.4. *Assume that the conditions* (7.7.1)–(7.7.4) *are satisfied and for* $i = 1, 2,$

$$M_{\Phi_i} \max\{L_{\Upsilon_i} + \|p_i\|_{L^1}, \widehat{L}_{\Upsilon_i} + \|q_i\|_{L^1}\} < 1, \quad and \quad \det\left(\widehat{M}\right) > 0.$$

Then, the system (7.21) *has at least one mild solution.*

Proof. We use Theorem 2.5 to prove that Θ has a fixed point. We have divided the proof into four steps.

Step 1: Θ is continuous.

Let $(y_n, x_n)_{n \in \mathbb{N}}$ be a couple of sequences such that $(y_n, x_n) \to (y^*, x^*)$, then for $t \in \mathbb{R}^+$, we have

$$\|(\Theta_1(y_n, x_n))(t) - (\Theta_1(y^*, x^*))(t)\|$$
$$\leq M_{\Phi_1} \int_0^t \|f_1(s, y_n(s), x_n(s), \Psi_1(y_n(s), x_n(s)))$$
$$- f_1(s, y^*(s), x^*(s), \Psi_1(y^*(s), x^*(s)))\| ds$$
$$+ M_{\Phi_1} \|\Upsilon_1(y_n, x_n) - \Upsilon_1(y^*, x^*)\|.$$

By the continuity of g_1, we get

$$g_1(t, s, y_n(s), x_n(s)) \to g_1(t, s, y^*(s), x^*(s)) \quad as \quad n \to +\infty.$$

And, we have

$$\|g_1(t, s, y_n(s), x_n(s)) - g_1(t, s, y^*(s), x^*(s))\| \leq g_{c_i}^* \|y_n(s) - y^*(s)\|$$
$$+ \widehat{g}_{c_i}^* \|x_n(s) - x^*(s)\|.$$

By the Lebesgue dominated convergence theorem, we obtain

$$\int_0^t g_1(t, s, y_n(s), x_n(s)) ds \to \int_0^t g_1(t, s, y^*(s), x^*(s)) ds, \quad as \quad n \to +\infty.$$

Hence

$$\|\Theta_1(y_n, x_n) - \Theta_1(y^*, x^*)\|_X \to 0, \quad as \quad n \to +\infty.$$

Similarly, we get

$$\|\Theta_2(y_n, x_n) - \Theta_2(y^*, x^*)\|_X \to 0, \quad as \quad n \to +\infty.$$

Thus, Θ is continuous.

Let B_δ be defined by

$$B_\delta = \{(y,x) \in X \times X : (\|y\|_X, \|x\|_X) \leq (\delta_1, \delta_2)\},$$

with $\delta_i > 0$. The set B_δ is bounded, closed and convex.

Step 2: Θ is a completely continuous operator.

Claim 1: $\Theta(B_\delta)$ is bounded.

Let $(y,x) \in B_\delta$ and $t \in \mathbb{R}^+$, from (7.7.1)–(7.7.3), it follows that

$$\|f_1(s,y(s),x(s),\Psi_1(y(s),x(s)))\| \leq p_1(s)\psi_1(\delta_1) + q_1(s)\phi_1(\delta_2) + \|f_1(s,0,0,0)\|,$$

and

$$\|\Upsilon_1(y,x)\| \leq L_{\Upsilon_1}\delta_1 + \widehat{L}_{\Upsilon_1}\delta_2 + \Upsilon_1^0.$$

Then, we get

$$\|\Theta_1(y(t),x(t))\| \leq M_{\Phi_1}\left(\|y_0\| + L_{\Upsilon_1}\delta_1 + \widehat{L}_{\Upsilon_1}\delta_2 + \Upsilon_1^0 + \|p_1\|_{L^1}\psi_1(\delta_1)\right.$$
$$\left. + \|q_1\|_{L^1}\phi_1(\delta_2)\right) + M_{\Phi_1}\int_0^t f_1^0(s)ds$$
$$< +\infty.$$

Similarly, we obtain

$$\|\Theta_2(y(t),x(t))\| \leq M_{\Phi_2}\left(\|x_0\| + L_{\Upsilon_2}\delta_1 + \widehat{L}_{\Upsilon_2}\delta_2 + \Upsilon_2^0\right.$$
$$\left. + \|p_2\|_{L^1}\psi_2(\delta_1) + \|q_2\|_{L^1}\phi_2(\delta_2)\right) + M_{\Phi_1}\int_0^t f_2^0(s)ds$$
$$< +\infty.$$

Thus,

$$\|\Theta(y,x)\|_{X \times X} < +\infty.$$

Claim 2: The set $\Theta(B_\delta)$ is equicontinuous.

For $(y,x) \in B_\delta$ and $\kappa_1, \kappa_2 \in \mathbb{R}^+$, we have

$$\|\Theta_1(y,x)(\kappa_1) - \Theta_1(y,x)(\kappa_2)\|$$
$$\leq \|R(\kappa_1) - R(\kappa_2)\|_{B(E)}(\|y_0\| + L_{\Upsilon_1}\delta_1 + \widehat{L}_{\Upsilon_1}\delta_2 + \Upsilon_1^0)$$
$$+ \int_0^{\kappa_1} \|R(\kappa_1 - s) - R(\kappa_2 - s)\|_{B(E)}\big(p_1(s)\psi_1(\delta_1) + q_1(s)\phi_1(\delta_2)\big)ds$$
$$+ M_{\Phi_i}\int_{\kappa_1}^{\kappa_2} \big(p_1(s)\psi_1(\delta_1) + q_1(s)\phi_1(\delta_2)\big)ds.$$

By the strong continuity of $\Phi_1(\cdot)$ and (7.7.1), we have

$$\|\Theta_1(y,x)(\kappa_1) - \Theta_1(y,x)(\kappa_2)\| \to 0 \ as \ \kappa_1 \to \kappa_2.$$

Similarly, we get

$$\|\Theta_2(y,x)(\kappa_1) - \Theta_2(y,x)(\kappa_2)\| \to 0 \ as \ \kappa_1 \to \kappa_2.$$

Hence, the set $\Theta(B_\delta)$ is equicontinuous.

Claim 3: The set $\Theta(B_\delta)$ is equiconvergent.
Let $(y,x) \in B_\delta$ and $t \in \mathbb{R}^+$, by (7.7.1), (7.7.3), we have

$$\|\Theta_1(y(t),x(t))\| \leq M_{\Phi_1} e^{-\beta_1 t} \left(\|y_0\| + L_{\Upsilon_1}\delta_1 + \widehat{L}_{\Upsilon_1}\delta_2 + \Upsilon_1^0 \right)$$

$$+ M_{\Phi_1} \int_0^t e^{-\beta_1(t-s)} (p_1(s)\psi_1(\delta_1) + q_1(s)\phi_1(\delta_2))$$

$$+ M_{\Phi_1} \int_0^t f_1^0(s)ds$$

$$\xrightarrow[t \to +\infty]{} M_{\Phi_1} \int_0^t f_1^0(s)ds.$$

Then

$$\|\Theta_1(y(t),x(t)) - \Theta_1(y(+\infty),x(+\infty))\| \xrightarrow[t \to +\infty]{} 0.$$

Similarly, we get

$$\|\Theta_2(y(t),x(t)) - \Theta_2(y(+\infty),x(+\infty))\| \xrightarrow[t \to +\infty]{} 0.$$

Claim 4: The set $\Theta(B_\delta(t))$ is relatively compact.
If $t = 0$, $\{\Theta(y(0),x(0)) : (y,x) \in B_\delta\}$ is compact. Let $t > 0$ be fixed, $\eta \in (0,t)$ and $(y,x) \in B_\delta$, we define the operator

$$\Theta_\eta(y(t),x(t)) = (\Theta_1^\eta(y(t),x(t)), \Theta_2^\eta(y(t),x(t))),$$

where

$$\begin{cases} \Theta_1^\eta(y(t),x(t)) = \Phi_1(\eta+t)(y_0 + \Upsilon_1(y,x)) + \Phi_1(\eta) \\ \qquad\qquad \times \int_0^{t-\eta} \Phi_1(t-\eta-s)f_1\left(s,y(s),x(s),\Psi_1(y(s),x(s))\right)ds, \\[2mm] \Theta_2^\eta(y(t),x(t)) = \Phi_2(\eta+t)(x_0 + \Upsilon_2(y,x)) + \Phi_2(\eta) \\ \qquad\qquad \times \int_0^{t-\eta} \Phi_2(t-\eta-s)f_2\left(s,y(s),x(s),\Psi_2(y(s),x(s))\right)ds, \end{cases}$$

and the operator

$$\widehat{\Theta}_\eta(y(t),x(t)) = (\widehat{\Theta}_1^\eta(y(t),x(t)), \widehat{\Theta}_2^\eta(y(t),x(t))),$$

where

$$
\begin{cases}
\widehat{\Theta}_1^\eta(y(t), x(t)) = \Phi_1(\eta + t)(y_0 + \Upsilon_1(y, x)) \\
\qquad\qquad + \displaystyle\int_0^{t-\eta} \Phi_1(t - s)f_1\left(s, y(s), x(s), \Psi_1(y(s), x(s))\right) ds, \\[2mm]
\widehat{\Theta}_2^\eta(y(t), x(t)) = \Phi_2(\eta + t)(x_0 + \Upsilon_2(y, x)) \\
\qquad\qquad + \displaystyle\int_0^{t-\eta} \Phi_2(t - s)f_2\left(s, y(s), x(s), \Psi_2(y(s), x(s))\right) ds.
\end{cases}
$$

Since $\Phi_i(\cdot)$ are compact and by Lemma 7.3, the sets

$$
\{\Theta_i^\eta(y(t), x(t)) : (y, x) \in B_\delta\}_{i=1,2}
$$

are relatively compact. Moreover for $(y, x) \in B_\delta$, we obtain

$$
\|\Theta_1^\eta(y(t), x(t)) - \widehat{\Theta}_1^\eta(y(t), x(t))\|
$$

$$
\leq \int_0^{t-\eta} \|\Phi_1(\eta)\Phi_1(t - \eta - s) - \Phi_1(t - s)\|
$$

$$
\times \|f_1\left(s, y(s), x(s), \Psi_1(y(s), x(s))\right)\| ds.
$$

From Lemma 7.3 and (7.7.1), we get

$$
\|\Theta_1^\eta(y(t), x(t)) - \widehat{\Theta}_1^\eta(y(t), x(t))\| \leq \gamma\eta(\|p_1\|_{L^1}\psi_1(\delta_1) + \|q_1\|_{L^1}\phi_1(\delta_2))
$$

$$
\xrightarrow[\eta \to 0]{} 0.
$$

Similarly, we get

$$
\|\Theta_2^\eta(y(t), x(t)) - \widehat{\Theta}_2^\eta(y(t), x(t))\| \leq \gamma\eta(\|p_2\|_{L^1}\psi_2(\delta_1) + \|q_2\|_{L^1}\phi_2(\delta_2))
$$

$$
\xrightarrow[\eta \to 0]{} 0.
$$

Therefore, the set $\{\widehat{\Theta}^\eta(y(t), x(t)) : (y, x) \in B_\delta\}$ is precompact.
Applying this idea again, we obtain

$$
\|\Theta_1(y(t), x(t)) - \widehat{\Theta}_1^\eta(y(t), x(t))\|
$$

$$
\leq \|R(t) \quad R(t \mid \eta)\| \left(\|y_0\| + L_{\Upsilon_1}\delta_1 + \widehat{L}_{\Upsilon_1}\delta_2 + \Upsilon_1^0\right)
$$

$$
+ M_{\Phi_1}\left(\psi_1(\delta_1)\int_{t-\eta}^t p_1(s) + \phi_1(\delta_2)\int_{t-\eta}^t q_1(s)ds\right)
$$

$$
\xrightarrow[\eta \to 0]{} 0.
$$

Thus, $\Theta(B_\delta(t))$ is precompact. Consequently $\Theta(B_\delta)$ is relatively compact.

Step 3: The following set

$$\Delta = \{(y,x) \in X \times X \ : \ (y,x) = \lambda\Theta(y,x), \text{ for some } \lambda \in (0,1)\},$$

is bounded.

Let $(y,x) \in \Delta$ and $\lambda \in (0,1)$ be such that

$$(y(t), x(t)) = \lambda\Theta(y(t), x(t)).$$

Then, we have

$$\begin{cases} y(t) = \lambda\Theta_1(y(t), x(t)), \\ x(t) = \lambda\Theta_2(y(t), x(t)). \end{cases}$$

Thus

$$\|y(t)\| \le M_{\Phi_1}\left(\|y_0\| + L_{\Upsilon_1}\|y\|_X + \widehat{L}_{\Upsilon_1}\|x\|_X + \Upsilon_1^0\right)$$

$$+ M_{\Phi_1}\left(\|p_1\|_{L^1}\|y\|_X + \|q_1\|_{L^1}\|x\|_X\right) + M_{\Phi_1}\int_0^t f_1^0(s)ds.$$

Similarly, we get

$$\|x(t)\| \le M_{\Phi_2}\left(\|x_0\| + L_{\Upsilon_2}\|y\|_X + \widehat{L}_{\Upsilon_2}\|x\|_X + \Upsilon_2^0\right)$$

$$+ M_{\Phi_2}\left(\|p_2\|_{L^1}\|y\|_X + \|q_2\|_{L^1}\|x\|_X\right) + M_{\Phi_2}\int_0^t f_2^0(s)ds.$$

Therefore

$$\begin{pmatrix} 1 - M_{\Phi_1}\left(L_{\Upsilon_1} + \|p_1\|_{L^1}\right) & -M_{\Phi_1}\left(\widehat{L}_{\Upsilon_1} + \|q_1\|_{L^1}\right) \\ -M_{\Phi_2}\left(L_{\Upsilon_2} + \|p_2\|_{L^1}\right) & 1 - M_{\Phi_2}\left(\widehat{L}_{\Upsilon_2} + \|q_2\|_{L^1}\right) \end{pmatrix} \begin{pmatrix} \|y\|_X \\ \|x\|_X \end{pmatrix}$$

$$\le \begin{pmatrix} M_{\Phi_1}\left(\|y_0\| + \displaystyle\int_0^t f_1^0(s)ds\right) \\ M_{\Phi_2}\left(\|x_0\| + \displaystyle\int_0^t f_2^0(s)ds\right) \end{pmatrix}.$$

From Lemma 2.3, $\widehat{M}^{-1}$ is order preserving, then we get

$$\begin{pmatrix} \|y\|_X \\ \|x\|_X \end{pmatrix} \le \widehat{M}^{-1} \begin{pmatrix} M_{\Phi_1}\left(\|y_0\| + \displaystyle\int_0^t f_1^0(s)ds\right) \\ M_{\Phi_2}\left(\|x_0\| + \displaystyle\int_0^t f_2^0(s)ds\right) \end{pmatrix}.$$

Thus, the set Δ is bounded, hence we deduce from Theorem 2.5, that Θ has a fixed point. Consequently, system (7.21) has at least one mild solution on $\mathbb{R}^+$. $\qquad\square$

7.3.1.2 *Asymptotic periodicity of solutions*

Theorem 7.5. *Assume that the conditions (7.7.1)–(7.7.4) and the following hypothesis are satisfied:*

(7.10.1) *The function f be uniformly S-asymptotically ω-periodic and asymptotically uniformly continuous on bounded sets.*

Then the system (7.21) has an S-asymptotically ω-periodic mild solution.

Proof. Firstly, we will prove that

$$\Theta(SAB_\omega(\mathbb{R}^+, E) \times SAB_\omega(\mathbb{R}^+, E)) \subset SAB_\omega(\mathbb{R}^+, E) \times SAB_\omega(\mathbb{R}^+, E).$$

Lets $(y, x) \in SAB_\omega(\mathbb{R}^+, E)$ and $t \in \mathbb{R}^+$, then we have

$$\|\Theta_1(y(t+\omega), x(t+\omega)) - \Theta_1(y(t), x(t))\|$$

$$\leq M_{\Phi_1}(\|y_0\| + \|\Upsilon_1(y, x)\|)(e^{-\beta_1(t+\omega)} + e^{-\beta_1 t})$$

$$+ \left\| \int_0^{t+\omega} \Phi_1(t+\omega-s) f_1\left(s, y(s), x(s), \Psi_1(y(s), x(s))\right) ds \right.$$

$$\left. - \int_0^t \Phi_1(t-s) f_1\left(s, y(s), x(s), \Psi_1(y(s), x(s))\right) ds \right\|$$

$$\leq M_{\Phi_1}\left(\|y_0\| + L_{\Upsilon_1}\|y\|_Y + \widehat{L}_{\Upsilon_1}\|x\|_Y + \Upsilon_1^0\right)\left(e^{-\beta_1(t+\omega)} + e^{-\beta_1 t}\right)$$

$$+ \int_0^\omega \|\Phi_1(t+\omega-s)\| \|f_1\left(s, y(s), x(s), \Psi_1(y(s), x(s))\right)\| ds$$

$$+ \int_0^t \Phi_1(t-s) f_1\left(s+\omega, y(s+\omega), x(s+\omega), \Psi_1(y, x)(s+\omega)\right)$$

$$- f_1\left(s, y(s), x(s), \Psi_1(y(s), x(s))\right) ds$$

$$\leq M_{\Phi_1}\left(\|y_0\| + L_{\Upsilon_1}\|y\|_Y + \widehat{L}_{\Upsilon_1}\|x\|_Y + \Upsilon_1^0\right)\left(e^{-\beta_1(t+\omega)} + e^{-\beta_1 t}\right)$$

$$+ M_{\Phi_1} \int_0^t e^{-\beta_1(t+\omega-s)}(\psi_1(\|y(s)\|)p_1(s) + \phi_1(\|x(s)\|)q_1(s) + \|f_1^0(s)\|)ds$$

$$+ \int_0^{L_\epsilon} \|\Phi_1(t-s)\| \|f_1\left(s+\omega, y(s+\omega), x(s+\omega), \Psi_1(y, x)(s+\omega)\right)$$

$$- f_1\left(s, y(s), x(s), \Psi_1(y(s), x(s))\right)\| ds$$

$$+ \int_{L_\epsilon}^t \|\Phi_1(t-s)\| \|f_1\left(s+\omega, y(s+\omega), x(s+\omega), \Psi_1(y, x)(s+\omega)\right)$$

$$- f_1\left(s, y(s), x(s), \Psi_1(y(s), x(s))\right)\| ds$$

$$\leq M_{\Phi_1}\left(\|y_0\| + L_{\Upsilon_1}\|y\|_Y + \widehat{L}_{\Upsilon_1}\|x\|_Y + \Upsilon_1^0\right)\left(e^{-\beta_1(t+\omega)} + e^{-\beta_1 t}\right)$$

$$+ M_{\Phi_1} \int_0^t e^{-\beta_1(t+\omega-s)}(2\psi_1(\|y\|_Y)p_1(s) + 2\phi_1(\|x\|_Y)q_1(s) + \|f_1^0(s)\|)ds$$
$$+ \varepsilon M_{\Phi_1} \left(\frac{1 - e^{-\beta_1(t-L_\varepsilon)}}{\beta_1} \right).$$

Then

$$\|\Theta_1(y(t+\omega), x(t+\omega)) - \Theta_1(y(t), x(t))\| \xrightarrow[t\to+\infty]{} 0.$$

Similarly, we get

$$\|\Theta_2(y(t+\omega), x(t+\omega)) - \Theta_2(y(t), x(t))\| \xrightarrow[t\to+\infty]{} 0.$$

Now, let $(y^*, x^*) \in BC(\mathbb{R}^+, E) \times BC(\mathbb{R}^+, E)$ be a solution of (7.21), then from Theorem 7.4 we have

$$(y^*(t), x^*(t)) = \Theta(y^*(t), x^*(t)).$$

Then,

$$\|(y^*(t+\omega), x^*(t+\omega)) - (y^*(t), x^*(t))\|$$
$$= \|\Theta(y^*(t+\omega), x^*(t+\omega)) - \Theta(y^*(t), x^*(t))\|.$$

Thus

$$\|(y^*(t+\omega), x^*(t+\omega)) - (y^*(t), x^*(t))\| \xrightarrow[t\to+\infty]{} 0.$$

Consequently $(y^*, x^*) \in SAB_\omega(\mathbb{R}^+, E) \times SAB_\omega(\mathbb{R}^+, E)$. □

7.3.1.3   *Attractivity of solutions*

In this section we study the local attractivity of solutions for the problem (7.21).

Firstly, we introduce the following concept of attractivity of solutions as in [27].

Definition 7.3. We say that solutions of (7.21) are locally attractive if there exists a closed ball $B(z^*, \gamma)$ in the generalized Banach space $Y \times Y$ for some $z^* = (z_1^*, z_2^*) \in Y \times Y$ such that for arbitrary solutions $z = (z_1, z_2)$ and $\tilde{z} = (\tilde{z}_1, \tilde{z}_2)$ of (7.21) belonging to $B(z^*, \gamma)$, we have that

$$\lim_{t\to+\infty} ((z_1(t), z_2(t)) - (\tilde{z}_1(t), \tilde{z}_2(t))) = 0_{\mathbb{R}^2}.$$

When the last limit is uniform with respect to $B(z^*, \gamma)$, solutions of problem (7.21) are said to be uniformly locally attractive (or equivalently that solutions of (7.21) are locally asymptotically stable).

Lets z^* be a solution of (7.21) and $B_\gamma = B(z^*, \gamma)$ the closed ball in $Y \times Y$, with $\gamma = (\gamma_1, \gamma_2) > 0$, and $\det(M^*) > 0$, where

$$M^* = \begin{pmatrix} 1 - 2M_{\Phi_1}(L_{\Upsilon_1} + \|p_1\|_{L^1}) & -2M_{\Phi_1}\left(\widehat{L}_{\Upsilon_1} + \|q_1\|_{L^1}\right) \\ -2M_{\Phi_2}(L_{\Upsilon_2} + \|p_2\|_{L^1}) & 1 - 2M_{\Phi_2}\left(\widehat{L}_{\Upsilon_2} + \|q_2\|_{L^1}\right). \end{pmatrix}$$

Theorem 7.6. *Suppose that hypotheses (7.7.1)–(7.7.4) hold. Then, the solutions of problem (7.21) are uniformly locally attractive.*

Proof. For $(z_1, z_2) \in B(z^*, \gamma)$, $t \in \mathbb{R}^+$ and by (7.7.1) and (7.7.3), we have

$$\|\Theta(z_1, z_2)(t) - z^*(t)\| = \begin{pmatrix} \|\Theta_1(z_1(t), z_1(t)) - \Theta_1(z_1^*(t), z_2^*(t))\| \\ \|\Theta_2(z_1(t), z_1(t)) - \Theta_2(z_1^*(t), z_2^*(t))\| \end{pmatrix}.$$

Then, we have

$$\|\Theta_1(z_1(t), z_1(t)) - \Theta_1(z_1^*(t), z_2^*(t))\|$$
$$\leq M_{\Phi_1}(L_{\Upsilon_1}\|z_1 - z_1^*\|_X + \widehat{L}_{\Upsilon_1}\|z_2 - z_2^*\|_X)$$
$$\quad + M_{\Phi_1}\int_0^t p_1(s)\psi_1(\|z_1(s) - z_1^*(s)\|) + q_1(s)\phi_1(\|z_2(s) - z_2^*(s)\|)ds$$
$$\leq 2M_{\Phi_1}(L_{\Upsilon_1} + \|p_1\|_{L^1})\gamma_1 + 2M_{\Phi_1}\left(\widehat{L}_{\Upsilon_1} + \|q_1\|_{L^1}\right)\gamma_2.$$

Similarly, we get

$$\|\Theta_1(z_1(t), z_1(t)) - \Theta_1(z_1^*(t), z_2^*(t))\|$$
$$\leq 2M_{\Phi_2}(L_{\Upsilon_2} + \|p_2\|_{L^1})\gamma_1 + 2M_{\Phi_2}\left(\widehat{L}_{\Upsilon_1} + \|q_2\|_{L^1}\right)\gamma_2.$$

Thus

$$\|\Theta(z_1, z_2)(t) - z^*(t)\|$$
$$\leq \begin{pmatrix} 2M_{\Phi_1}(L_{\Upsilon_1} + \|p_1\|_{L^1}) & 2M_{\Phi_1}\left(\widehat{L}_{\Upsilon_1} + \|q_1\|_{L^1}\right) \\ 2M_{\Phi_2}(L_{\Upsilon_2} + \|p_2\|_{L^1}) & 2M_{\Phi_2}\left(\widehat{L}_{\Upsilon_2} + \|q_2\|_{L^1}\right) \end{pmatrix}\begin{pmatrix} \gamma_1 \\ \gamma_2 \end{pmatrix}.$$

From Lemma 2.3, M_1 is order preserving, then we obtain

$$0 < \begin{pmatrix} 1 - 2M_{\Phi_1}(L_{\Upsilon_1} + \|p_1\|_{L^1}) & -2M_{\Phi_1}\left(\widehat{L}_{\Upsilon_1} + \|q_1\|_{L^1}\right) \\ -2M_{\Phi_2}(L_{\Upsilon_2} + \|p_2\|_{L^1}) & 1 - 2M_{\Phi_2}\left(\widehat{L}_{\Upsilon_2} + \|q_2\|_{L^1}\right) \end{pmatrix}\begin{pmatrix} \gamma_1 \\ \gamma_2 \end{pmatrix}.$$

This proves that $\Theta(B_\gamma) \subset B_\gamma$.

So, for each $(z_1, z_2), (\widetilde{z}_1, \widetilde{z}_2) \in B(z^*, \gamma)$ solutions of problem (7.21) and $t \in \mathbb{R}^+$, we have

$$\|(z_1, z_2)(t) - (\widetilde{z}_1, \widetilde{z}_2)(t)\| = \|\Theta((z_1, z_2)(t)) - \Theta((\widetilde{z}_1, \widetilde{z}_2)(t))\|.$$

Then

$$\|\Theta_1((z_1, z_2)(t)) - \Theta_1((\tilde{z}_1, \tilde{z}_2)(t))\|$$

$$\leq M_{\Phi_1}(L_{\Upsilon_1}\|z_1 - \tilde{z}_1\|_X + \widehat{L}_{\Upsilon_1}\|z_2 - \tilde{z}_2\|_X)$$

$$+ M_{\Phi_1} \int_0^t p_1(s)\psi_1(\|z_1(s) - \tilde{z}_1(s)\|) + q_i(s)\phi_1(\|z_2(s) - \tilde{z}_2(s)\|)ds$$

$$\leq 2M_{\Phi_1}(L_{\Upsilon_1}\gamma_1 + \widehat{L}_{\Upsilon_1}\gamma_2)e^{-\beta_1 t}$$

$$+ M_{\Phi_1} \int_0^t e^{-\beta_1(t-s)}\left(\psi_1(\gamma_1)p_1(s) + \phi_1(\gamma_2)q_1(s)\right)ds.$$

Similarly, we obtain

$$\|\Theta_1((z_1, z_2)(t)) - \Theta_1((\tilde{z}_1, \tilde{z}_2)(t))\|$$

$$\leq 2M_{\Phi_2}(L_{\Upsilon_2}\gamma_1 + \widehat{L}_{\Upsilon_2}\gamma_2)e^{-\beta_2 t}$$

$$+ M_{\Phi_2} \int_0^t e^{-\beta_2(t-s)}\left(\psi_2(\gamma_1)p_2(s) + \phi_2(\gamma_2)q_2(s)\right)ds.$$

We conclude that $\|(z_1, z_2)(t) - (\tilde{z}_1, \tilde{z}_2)(t)\| \to 0$, as $t \to +\infty$. $\qquad\square$

7.3.2 *An Example*

Consider the following class of partial integro-differential system:

$$
\begin{cases}
\frac{\partial}{\partial t}\varpi_1(t,x) - \Delta(\theta, \varpi_1(t,x)) - \int_0^t \Gamma_1(t-s)\Delta(\theta, \varpi_1(s,x))ds \\
\quad = \rho_1(t)\frac{\sin(e^{-\sigma t})}{(t^2+1)} \int_0^a \frac{\ln(1+e^{-t^2})(1+\varpi_1(s,x)+\varpi_2(s,x))e^{-\sigma(t-s)}}{1+2t^2+s^2}ds \\
\quad\quad + \frac{(\varpi_1(t,x)+\varpi_2(t,x))\pi\sin(e^{-\sigma t})}{77(t^2+1)}\rho_1(t) \quad \text{if } t \in \mathbb{R}^+ \text{ and } x \in (0,1), \\[2mm]
\frac{\partial}{\partial t}\varpi_2(t,x) - \Delta(\widehat{\theta}, \varpi_2(t,x)) - \int_0^t \Gamma_2(t-s)\Delta(\widehat{\theta}, \varpi_2(s,x))ds \\
\quad = \frac{\rho_2(t)\cos(e^{-\sigma\pi})(\phi_1(t,x)+\phi_2(t,x))}{99(t^2+1)} \\
\quad\quad + e^{-\sigma t}\rho_2(t)\int_0^a e^{s-t-(\pi+1)a}\sqrt{\|\varpi_1(s,x)\| + \|\varpi_2(s,x)\|}ds, \\[2mm]
\varpi_1(t,0) = \varpi_1(t,1) = \varpi_2(t,0) = \varpi_2(t,1) = 0, \quad \text{for } t \geq 0, \\[2mm]
\varpi_1(0,x) = \varpi_1^0(x) + \frac{1}{107+e^t}\sum_{j=1}^3 \ln(1+\|\varpi_1(j,x)\|_{L^2})(1+\|\varpi_2(j,x)\|_{L^2}), \\[2mm]
\varpi_2(0,x) = \varpi_2^0(x) + \frac{e^{-t+\frac{1}{4}}}{333}\sum_{j=1}^4 (\|\varpi_1(2^j,x)\|_{L^2} + \|\varpi_2(2^j,x)\|_{L^2}),
\end{cases}
$$

$$(7.23)$$

where $t \in \mathbb{R}^+$, *and* $x \in (0,1)$, $\mathbb{R}^+ = \mathbb{R}^+$, $\rho_i : \mathbb{R}^+ \mapsto \mathbb{R}$ are continuous, $\theta_1, \theta_2, \widehat{\theta}_1 \widehat{\theta}_2 \in \mathbb{R}$, $\sigma \geq \beta_i$.

The operator Δ defined by

$$\Delta(\theta, \xi) = \frac{\partial}{\partial x} \left(\frac{\partial \xi(t, x)}{\partial x} + \theta_1 \xi(t, x) \right) + \theta_2 \xi(t, x).$$

Let

$$H := L^2(0,1) = \left\{ \xi : (0,1) \longrightarrow \mathbb{R} : \int_0^1 |\xi(x)|^2 dx < +\infty \right\},$$

be the Hilbert space with the scalar product $\langle \xi, v \rangle = \int_0^1 -(x)v(x)dx$.

We define the operators A_i induced on H as follows:

$$A_1 z = z'' + \theta_1 z' + \theta_2 z, \quad \theta_1, \theta_2 \in \mathbb{R} \text{ and } D(A_1) = H^2(0,1) \cap H_0^1(0,1),$$

$$A_2 z = z'' + \widehat{\theta}_1 z' + \widehat{\theta}_2 z, \quad \widehat{\theta}_1, \widehat{\theta}_2 \in \mathbb{R} \text{ and } D(A_2) = H^2(0,1) \cap H_0^1(0,1),$$

which are the infinitesimal generators of an analytic semigroup $(G_1(t))_{t \geq 0}$, $(G_2(t))_{t \geq 0}$ on H. Since the semigroups generated by A_1 are analytic (respectively, A_2), then it is norm continuous for $t > 0$ which implies that resolvent operator is operator-norm continuous for $t > 0$ (see [229]). As in [295], for $i = 1, 2$ and some $\widehat{r}_i > r_i > 1$, we assume that $\|\Gamma_i(t)\| \leq \frac{e^{-\widehat{r}_i t}}{r_i}$, and $\|\Gamma_i'(t)\| \leq \frac{e^{-\widehat{r}_i t}}{r_i^2}$, we get that $\|\Phi_i(t)\| \leq e^{-\widehat{\sigma}_i t}$, where $\widehat{\sigma}_i = 1 - r_i^{-1}$.

We define also the operators $B_i(t) : H \mapsto H$ by:

$$B_i(t)z = \Gamma_i(t)A_i z, \quad for \ t \geq 0, \ z \in D(A_i).$$

More appropriate conditions on operators B_i, (7.7.4) hold with $M_{\Phi_i} = 1$ and $\beta_i = 1 - r_i^{-1}$.

Now, if $\rho(t) = 1$. We put $\varpi(t)(x) = \varpi(t, x)$, for $t \in [0, +\infty)$, and define

$$f_1(t, \phi_1, \phi_2, \Psi_1(\phi_1, \phi_2)(x)) = \frac{\Psi_1(\phi_1, \phi_2)(x)}{77(1 + t\sqrt{\pi t})e^{\gamma t}}$$
$$+ \frac{\pi \rho_1(t)(\phi_1(t, x) + \phi_2(t, x)) \sin(e^{-\sigma t})}{77(t^2 + 1)},$$

$$\Psi_1(\phi_1, \phi_2)(x) = \frac{\rho_1(t) \sin(e^{-\sigma t})}{(t^2 + 1)}$$
$$\times \int_0^a \frac{\ln(1 + e^{-t^2})(1 + (\phi_1(t, x) + \phi_2(t, x)))e^{-\sigma(t-s)}}{1 + 2t^2 + s^2} ds,$$

$$f_2(t, \phi_1, \phi_2, \Psi_2(\phi_1, \phi_2)(x)) = \frac{\rho_2(t)\cos\left(e^{-\sigma\pi}\right)(\phi_1(t,x) + \phi_2(t,x))}{99\,(t^2 + 1)}$$
$$+ e^{-\sigma t}\Psi_2(\phi_1, \phi_2)(x),$$

$$\Psi_2(\phi_1, \phi_2)(x) = \rho_2(t)\int_0^a e^{s-t-(\pi+1)a}\sqrt{\|\phi_1(s,x)\| + \|\phi_2(s,x)\|}ds,$$

$$\Upsilon_1(t, \phi_1, \phi_2) = \frac{1}{707 + e^t}\sum_{j=1}^4 \ln(1 + \|\phi_1(j,x)\|_{L^2})(1 + \|\phi_2(j,x)\|_{L^2}),$$

$$\Upsilon_2(t, \phi_1, \phi_2) = \frac{e^{-t+\frac{1}{4}}}{333}\sum_{j=1}^3 (\|\phi_1(2^j, x)\|_{L^2} + \|\phi_2(2^j, x)\|_{L^2}).$$

We can write (7.23) in the form

$$\begin{cases} y'(t) = A_1 y(t) + f_1\left(t, y(t), y(t), \Psi_1(y(t), x(t))\right) \\ \qquad + \displaystyle\int_0^t B_1(t-s)y(s)ds, \text{ for } t \in I, \\[2mm] x'(t) = A_2 x(t) + f_2\left(t, y(t), x(t), \Psi_1(y(t), x(t))\right) \\ \qquad + \displaystyle\int_0^t B_2(t-s)y(s)ds, \text{ for } t \in I, \\[2mm] y(0) = y_0 + \Upsilon_1(y, x), \\[2mm] x(0) = x_0 + \Upsilon_2(y, x). \end{cases} \qquad (7.24)$$

For $t \in \mathbb{R}^+$, we have

$$|f_1(t, \phi_1(t), \phi_2(t), \Psi_1(\phi_1(t), \phi_2(t))) - f_1(t, \widehat{\phi}_1(t), \widehat{\phi}_2(t), \Psi_1(\widehat{\phi}_1(t), \widehat{\phi}_2(t)))|$$
$$\leq \frac{e^{-\sigma t}}{77}\left(e^{-\sigma t - t^2}\frac{\pi}{2} + \frac{\pi}{t^2 + 1}\right)\left(|\phi_1(t) - \phi_2(t)| + |\widehat{\phi}_1(t) - \widehat{\phi}_2(t)|\right),$$

and

$$|f_2(t, \phi_1(t), \phi_2(t), \Psi_2(\phi_1(t), \phi_2(t))) - f_2(t, \widehat{\phi}_1(t), \widehat{\phi}_2(t), \Psi_2(\widehat{\phi}_1(t), \widehat{\phi}_2(t)))|$$
$$\leq \frac{e^{-\sigma t}}{99(t^2 + 1)}\left(\int_0^a e^{s-t-(\pi+1)a}ds\right)\left(|\phi_1(t) - \phi_2(t)| + |\widehat{\phi}_1(t) - \widehat{\phi}_2(t)|\right).$$

So, $\psi_i(t) = \phi_i(t) = t$; $i = 1, 2$, are continuous non-decreasing functions from $(0, +\infty)$ to $(0, +\infty)$. And, we have

$$p_1(t) = q_1(t) = \frac{e^{-\sigma t}}{77}\left(e^{-\sigma t - t^2}\frac{\pi}{2} + \frac{\pi}{t^2 + 1}\right) \in L^1(\mathbb{R}^+, \mathbb{R}^+),$$

$$p_2(t) = q_2(t) = \frac{e^{-\sigma t}}{99(t^2+1)} \left(\int_0^a e^{s-t-(\pi+1)a} ds \right) \in L^1(\mathbb{R}^+, \mathbb{R}^+).$$

Now, about g_1, g_2, Υ_1 and Υ_2, we have

$$|g_1(t, s, \phi_1, \phi_2) - g_1(t, s, \widehat{\phi}_1, \widehat{\phi}_2)|$$
$$\leq e^{s-t-(\pi+1)a} \left(|\phi_1(t) - \phi_2(t)| + |\widehat{\phi}_1(t) - \widehat{\phi}_2(t)| \right),$$

$$|g_2(t, s, \phi_1, \phi_2) - g_2(t, s, \widehat{\phi}_1, \widehat{\phi}_2)|$$
$$\leq e^{-\sigma t} \left(\int_0^a \frac{\ln(1 + e^{-t^2}) e^{-\sigma(t-s)}}{1 + 2t^2 + s^2} ds + \frac{\pi}{t^2+1} \right)$$
$$\times \left(|\phi_1(t) - \phi_2(t)| + |\widehat{\phi}_1(t) - \widehat{\phi}_2(t)| \right),$$

$$\|\Upsilon_1(t, \phi_1, \phi_2) - \Upsilon_1(t, \widehat{\phi}_1, \widehat{\phi}_2)\| \leq \frac{1}{177} \|\phi_1 - \phi_2\|_H,$$

$$\|\Upsilon_2(t, \phi_1, \phi_2) - \Upsilon_2(t, \widehat{\phi}_1, \widehat{\phi}_2)\| \leq \frac{e^{\frac{1}{4}}}{111} \|\phi_1 - \phi_2\|_H,$$

and

$$g_{c_1}^* = \widehat{g}_{c_1}^* = e^{-(\pi+1)a}, \ g_{c_2}^* = \widehat{g}_{c_2}^* = \ln(2),$$

$$L_{\Upsilon_1} = L_{\widehat{\Upsilon}_1} = \frac{1}{177}, \ L_{\Upsilon_2} = L_{\widehat{\Upsilon}_2} = \frac{e^{\frac{1}{4}}}{111}.$$

Also, we have

$$\frac{1}{177} + \frac{\pi}{308} + \frac{\pi^2}{154} \simeq 0,076$$

$$\frac{e^{\frac{1}{4}}}{111} + \frac{\pi}{198} \simeq 0,027.$$

Then, $\rho(\widetilde{M}) \in (0,1)$, $\det(\widehat{M}) = 0,897$, $\det(M^*) = 0,794$ where

$$\widetilde{M} = \begin{pmatrix} M_{\Phi_1} \left(L_{\Upsilon_1} + \|p_1\|_{L^1} \right) M_{\Phi_1} \left(\widehat{L}_{\Upsilon_1} + \|q_1\|_{L^1} \right) \\ M_{\Phi_2} \left(L_{\Upsilon_2} + \|p_2\|_{L^1} \right) M_{\Phi_2} \left(\widehat{L}_{\Upsilon_2} + \|q_2\|_{L^1} \right) \end{pmatrix}.$$

And $\widehat{M} = I - \widetilde{M}$, $M^* = I - 2\widetilde{M}$. Therefore, $\widetilde{M}$ converges to zero and $\widehat{M}$, M^* are order preserving. By Theorems 7.3 and 7.6, we deduce that (7.23) has a unique mild solution, which is locally attractive.

Now, we assume that $\rho(t) = \sin(\ln(t+1))$. Then, We have

$$\|f_1(t, \phi_1(t), \phi_2(t), \Psi_1(\phi_1(t), \phi_2(t))) - f_1(t, \widehat{\phi}_1(t), \widehat{\phi}_2(t), \Psi_1(\widehat{\phi}_1(t), \widehat{\phi}_2(t)))\|$$
$$\leq \frac{3\pi}{154} \left(\|\phi_1 - \phi_2\| + \|\widehat{\phi}_1 - \widehat{\phi}_2\| \right),$$

and

$$\|f_2(t, \phi_1(t), \phi_2(t), \Psi_2(\phi_1(t), \phi_2(t))) - f_2(t, \widehat{\phi}_1(t), \widehat{\phi}_2(t), \Psi_2(\widehat{\phi}_1(t), \widehat{\phi}_2(t)))\|$$
$$\leq \frac{e^{-\pi a}}{99} \left(\|\phi_1 - \phi_2\| + \|\widehat{\phi}_1 - \widehat{\phi}_2\| \right).$$

Now, for any $\omega > 0$, we have

$$|\sin(\ln(t+\omega+1)) - \sin(\ln(t+1))| \xrightarrow[t\to+\infty]{} 0.$$

In other hand, if we put

$$\Lambda_1(t) = \frac{\Psi_1(\phi_1, \phi_2)(x)}{77(1 + t\sqrt{\pi t})e^{\gamma t}} + \frac{(\phi_1(t,x) + \phi_2(t,x)) \sin(e^{-\sigma t}) \pi}{77(t^2 + 1)}.$$
$$\Lambda_2(t) = \frac{\cos(e^{-\sigma \pi})(\phi_1(t,x) + \phi_2(t,x))}{99(t^2 + 1)} + e^{-\sigma t}\Psi_2(\phi_1, \phi_2)(x).$$

For $\phi_1, \phi_2 \in \widetilde{H} = H \cap SAB_\omega$, Λ_1, Λ_2, are bounded and we have

$$\|\Lambda_1(t)\| \xrightarrow[t\to+\infty]{} 0, \quad \text{and} \quad \|\Lambda_2(t)\| \xrightarrow[t\to+\infty]{} 0.$$

Then

$$\|f_1(t+\omega, \phi_1(t), \phi_2(t), \Psi_1(\phi_1(t), \phi_2(t))) - f_1(t, \phi_1(t), \phi_2(t), \Psi_1(\phi_1(t), \phi_2(t)))\|$$
$$\leq \|\rho(t+\omega) - \rho(t)\| \|\Lambda_1(t+\omega)\| + \|\rho(t)\|(\|\Lambda_1(t)\| + \|\Lambda_1(t+\omega)\|)$$
$$\xrightarrow[t\to+\infty]{} 0.$$

Similarly, we get

$$\|f_2(t+\omega, \phi_1(t), \phi_2(t), \Psi_1(\phi_1(t), \phi_2(t))) - f_1(t, \phi_1(t), \phi_2(t), \Psi_1(\phi_1(t), \phi_2(t)))\|$$
$$\leq \|\rho(t+\omega) - \rho(t)\| \|\Lambda_2(t+\omega)\| + \|\rho(t)\|(\|\Lambda_2(t+\omega)\| + \|\Lambda_2(t)\|)$$
$$\xrightarrow[t\to+\infty]{} 0.$$

Therefore, all conditions of Theorems 7.4, 7.5 and 7.6 are verified. Consequently, the problem (7.23) has at least one S-asymptotically ω-periodic mild solution, which is locally attractive.

7.4 Functional Evolution Equations with Delay and Random Effect

In this section, we will establish a random version of the theory for the almost automorphic solution. More precisely we will verify the existence of mild random almost automorphic solution for evolution equations with delay and random effects of the form:

$$y'(t,\delta) = A(t)y(t,\delta) + f(t, y_t(\cdot,\delta),\delta), \quad \text{a.e.} \ \ t \in \mathbb{R}^+ := [0,\infty), \quad (7.25)$$

$$y(t,\delta) = \phi(t,\delta), \quad t \in (-\infty, 0], \quad (7.26)$$

where (Ω, F, P) is a complete probability space, $\delta \in \Omega$, $f : \mathbb{R}^+ \times \mathcal{B} \times \Omega \to E$, $\phi \in \mathcal{B} \times \Omega$ are given random functions, $\{A(t)\}_{0 \leq t < +\infty}$ is a family of linear closed operators from E into E that generate an evolution system of operators $\{\mathcal{U}(t,s)\}_{(t,s) \in \mathbb{R}^+ \times \mathbb{R}^+}$ for $0 \leq s \leq t < +\infty$, $\mathcal{B}$ is the phase space, and $(E, \|\cdot\|)$ is a Banach space. For y defined on $\mathbb{R} \times \Omega$ and any $t \in \mathbb{R}^+$ we denote by $y_t(\cdot, \delta)$ the element of $\mathcal{B} \times \Omega$ given by $y_t(\kappa, \delta) = y(t + \kappa, \delta), \kappa \in (-\infty, 0]$.

7.4.1 *Main Results*

Throughout the section, we will utilize an axiomatic definition of the phase space $\mathcal{B}$ introduced by Hale and Kato in [218] and follow the terminology used in [219]. Thus, $(\mathcal{B}, \|\cdot\|_{\mathcal{B}})$ will be a seminormed linear space of functions mapping $(-\infty, 0]$ into E, meeting the following axioms:

(A_1) If $y : \mathbb{R} \to E$, is continuous on $\mathbb{R}_+$ and $y_0 \in \mathcal{B}$, then for every $t \in \mathbb{R}_+$ the following requirements hold:
(i) $y_t \in \mathcal{B}$;
(ii) There exists a positive constant β such that $\|y(t)\| \leq \beta \|y_t\|_{\mathcal{B}}$;
(iii) There exist $\beta_1(\cdot), \beta_2(\cdot) : \mathbb{R}_+ \to \mathbb{R}_+$ with β_1 continuous and bounded, and β_2 locally bounded where

$$\|y_t\|_{\mathcal{B}} \leq \beta_1(t) \sup\{ \|y(s)\| : 0 \leq s \leq t\} + \beta_2(t)\|y_0\|_{\mathcal{B}}.$$

(A_2) For the function y in (A_1), y_t is a $\mathcal{B}$−valued continuous function on $\mathbb{R}_+$.
(A_3) The space $\mathcal{B}$ is complete.

Denote

$$\alpha_1 = \sup\{\beta_1(t) : t \in \mathbb{R}_+\},$$

and

$$\alpha_2 = \sup\{\beta_2(t) : t \in \mathbb{R}_+\}.$$

Remark 7.1.

1. (ii) is equivalent to $\|\phi(0)\| \leq \beta\|\phi\|_{\mathcal{B}}$ for every $\phi \in \mathcal{B}$.
2. Since $\|\cdot\|_{\mathcal{B}}$ is a seminorm, two elements $\phi, \ell \in \mathcal{B}$ can verify $\|\phi - \ell\|_{\mathcal{B}} = 0$ without necessarily $\phi(\kappa) = \ell(\kappa)$ for all $\kappa \leq 0$.
3. From the equivalence of in the first remark, we can see that for all $\phi, \ell \in \mathcal{B}$ such that $\|\phi - \ell\|_{\mathcal{B}} = 0$: We necessarily have that $\phi(0) = \ell(0)$.

In this section, we study the main existence result for problem (7.25)–(7.26).

Definition 7.4. A stochastic process $y : \mathbb{R}^+ \times \Omega \to E$ is a random mild solution of (7.25)–(7.26) if $y(t, \delta) = \phi(t, \delta)$, $t \in (-\infty, 0)$ and the restriction of $y(\cdot, \delta)$ to the interval $(0, \infty)$ is continuous and verifies

$$y(t, \delta) = \mathcal{U}(t, 0)\phi(0, \delta) + \int_0^t \mathcal{U}(t, s)f(s, y_s(\cdot, \delta), \delta)ds, \ t \in \mathbb{R}^+ = (0, \infty).$$

$$(7.27)$$

We will need to introduce the following hypotheses which are be assumed thereafter:

(7.14.1) The evolution family $\{\mathcal{U}(t, s)\}_{t \geq s}$ is compact and there exists a constant $\gamma_1 \geq 1$ and $\zeta > 0$ such that

$$\|\mathcal{U}(t, s)\|_{B(E)} \leq \gamma_1 e^{-\zeta(t-s)} \quad \text{for every } (s, t) \in \Delta.$$

and

$$\mathcal{U}(t + \omega, s + \omega) = \mathcal{U}(t, s).$$

(7.14.2) The function $f : \mathbb{R}^+ \times E \times \Omega \to E$ is Carathéodory and satisfies the following:

(a) There exists a constant $\omega > 0$, such that

$$\lim_{t \to +\infty} \|f(t + \omega, u, \delta) - f(t, u, \delta)\| = 0 \text{ for a.e } t, \delta \in \mathbb{R}^+ \times \Omega, \ u \in E.$$

(b) There exist $p \in L^\infty(\mathbb{R}^+ \times \Omega, \mathbb{R}^+)$, and a continuous non-decreasing function $\ell : [0, \infty) \times \Omega \to (0, \infty)$ such that:

$$\|f(t, u, \delta)\| \leq p(t, \delta)\ell(\|u\|_{B}, \delta) \text{ for a.e } t \in \mathbb{R}^+ \times \Omega \text{ and each } u \in E.$$

(7.14.3) There exists a random function $\chi : \Omega \longrightarrow \mathbb{R}^+\backslash\{0\}$ such that:

$$\gamma_1\|\phi\|_{\mathcal{B}} + \frac{\gamma_1\ell(D_1,\delta)\|p\|_{L^\infty}}{\zeta} \leq \chi(\delta),$$

where

$$D_1 := \alpha_1\chi(\delta) + \alpha_2\|\phi\|_{\mathcal{B}}.$$

(7.14.4) For each $\delta \in \Omega, \phi(\cdot,\delta)$ is continuous and for each $t, \phi(t,\cdot)$ is measurable, and $\lim\limits_{t\longrightarrow\infty}(\phi(t+\omega,\delta) - \phi(t,\delta)) = 0.$

(7.14.5) For each $(t,s) \in \Delta$ we have: $\lim\limits_{t\to+\infty}\int_0^t e^{-\zeta(t-s)}p(s,\delta)ds = 0.$

Theorem 7.7. *Assume that* (7.14.1)–(7.14.5) *are met. Then, problem* (7.25)–(7.26) *has at least one S-asymptotically ω-periodic mild random solution on* $(-\infty,\infty)$.

Proof. Consider the operator $\aleph : \Omega \times SAP_\omega(X) \to SAP_\omega(X)$ be the random operator given by

$$(\aleph(\delta)y)(t) = \mathcal{U}(t,0)\ \phi(0,\delta) + \int_0^t \mathcal{U}(t,s)\ f(s,y_s,\delta)\ ds, \quad t \in \mathbb{R}^+, \quad (7.28)$$

and

$$y(t,\delta) = \phi(t,\delta),\ t \in (-\infty, 0).$$

Let the operators $\overline{\aleph} = \aleph_1 + \aleph_2$, where

$$(\aleph_1(\delta)y)(t) = \mathcal{U}(t,0)\phi(0,\delta),$$

$$(\aleph_2(\delta)y)(t) = \int_0^t \mathcal{U}(t,s)f(s,y_s,\delta)ds.$$

Claim 1: $\aleph_1 \in SAP_\omega(E)$.
For $t \geq 0$, one has

$$\|(\aleph_1(\delta)y)(t+\omega) - (\aleph_1(\delta)y)(t)\| = \|\mathcal{U}(t+\omega,0)\phi(0,\delta) - \mathcal{U}(t,0)\phi(0,\delta)\|$$
$$\leq \gamma_1\|\phi\|_{\mathcal{B}}\left(e^{-\zeta(t+\omega)} + e^{-\zeta t}\right).$$
$$(7.29)$$

Since $\zeta > 0$, we deduce that

$$\lim\limits_{t\to+\infty}\|(\aleph_1(\delta)y)(t+\omega) - (\aleph_1(\delta)y)(t)\| = 0.$$

Then $\aleph_1 \in SAP_\omega(X)$.

Claim 2: $\aleph_2 \in SAP_\omega(E)$.

For $t \geq 0$, one has

$$(\aleph_2(\delta)y)(t+\omega) - (\aleph_2(\delta)y)(t)$$

$$= \int_0^{t+\omega} \mathcal{U}(t+\omega,s)f(s,y_s,\delta)ds - \int_0^t \mathcal{U}(t,s)f(s,y_s,\delta)ds$$

$$= \int_0^\omega \mathcal{U}(t+\omega,s)f(s,y_s,\delta)ds$$

$$+ \int_\omega^{t+\omega} \mathcal{U}(t+\omega,s)f(s,y_s,\delta)ds - \int_0^t \mathcal{U}(t,s)f(s,y_s,\delta)ds$$

$$= \Upsilon_1(t) + \Upsilon_2(t),$$

where

$$\Upsilon_1(t) = \int_0^\omega \mathcal{U}(t+\omega,s)f(s,y_s,\delta)ds,$$

and

$$\Upsilon_2(t) = \int_\omega^{t+\omega} \mathcal{U}(t+\omega,s)f(s,y_s,\delta)ds - \int_0^t \mathcal{U}(t,s)f(s,y_s,\delta)ds.$$

Using the fact that $(\mathcal{U}(t,s))_{t \geq s}$ is exponentially stable, we obtain

$$\|\Upsilon_1(t)\| \leq \gamma_1 \int_0^\omega e^{-\zeta(t+\omega-s)}p(s,\delta)\ell(\|y_s\|_\mathcal{B},\delta)ds$$

$$= \gamma_1 e^{-\zeta t}\ell(\|y_s\|_\mathcal{B},\delta)\|p\|_{L^\infty} \int_0^\omega e^{-\zeta(\omega-s)}ds$$

$$= \gamma_1 e^{-\zeta t}\ell(\|y_s\|_\mathcal{B},\delta)\|p\|_{L^\infty} \frac{1 - e^{-\zeta\omega}}{\zeta},$$

which shows that

$$\lim_{t \to +\infty} \Upsilon_1(t) = 0.$$

Let us write

$$\Upsilon_2(t) = \int_0^t \mathcal{U}(t+\omega,s+\omega)f(s+\omega,y_{s+\omega},\delta)ds - \int_0^t \mathcal{U}(t,s)f(s,y_s,\delta)ds$$

and since the evolution family is ω-periodic, we obtain

$$\Upsilon_2(t) \leq \int_0^t \mathcal{U}(t,s)\,\|f(s+\omega,y_{s+\omega},\delta) - f(s+\omega,y_s,\delta)\|\,ds$$

$$+ \int_0^t \mathcal{U}(t,s)\,\|f(s+\omega,y_s,\delta) - f(s,y_s,\delta)\|\,ds$$

$$= I_1(t) + I_2(t),$$

where

$$I_1(t) = \int_0^{T_\varepsilon} \mathcal{U}(t,s) \|f(s+\omega, y_{s+\omega}, \delta) - f(s+\omega, y_s, \delta)\| \, ds$$
$$+ \int_{T_\varepsilon}^t \mathcal{U}(t,s) \|f(s+\omega, y_s, \delta) - f(s, y_s, \delta)\| \, ds,$$

$$I_2(t) = \int_{T_\varepsilon}^t \mathcal{U}(t,s) \|f(s+\omega, y_{s+\omega}, \delta) - f(s+\omega, y_s, \delta)\| \, ds$$
$$+ \int_0^{T_\varepsilon} \mathcal{U}(t,s) \|f(s+\omega, y_s, \delta) - f(s, y_s, \delta)\| \, ds,$$

and $T_\varepsilon >$. Observing that

$$\|I_1(t)\| \le \left\| \int_0^{T_\varepsilon} \mathcal{U}(t,s) \|f(s+\omega, y_s(s+\omega), \delta) - f(s+\omega, y_s, \delta))\| \, ds \right\|$$
$$+ \left\| \int_{T_\varepsilon}^t \mathcal{U}(t,s)(f(s+\omega, y_{s+\omega}, \delta) - f(s+\omega, y_s, \delta) ds \right\|.$$

Since $y \in SAP_\omega$, we know that there is a positive constant $T_\varepsilon > 0$ such that

$$\|y(t+\omega, \delta) - y(t, \delta)\| \le \varepsilon, \quad \text{for } t \ge T_\varepsilon.$$

By (7.14.3), we have

$$\|f(t, y_{s+\omega}, \delta) - f(t, y_s, \delta)\| \le \frac{\zeta \varepsilon}{\gamma_1}, \quad \text{for } t \ge T_\varepsilon.$$

Thus, we get

$$\left\| \int_{T_\varepsilon}^t \mathcal{U}(t,s)(f(s+\omega, y_s(s+\omega), \delta) - f(s+\omega, y_s, \delta) ds \right\|$$
$$\le \gamma_1 \int_{T_\varepsilon}^t e^{-\zeta(t-s)} \|f(s+\omega, y_{s+\omega}, \delta) - f(s+\omega, y_s)\| \, ds$$
$$\le \int_{T_\varepsilon}^t \gamma_1 e^{-\zeta(t-s)} \frac{\zeta \varepsilon}{\gamma_1} \, ds$$
$$\le \varepsilon(e^{-\zeta t} - e^{-\zeta(t-T_\varepsilon)})$$
$$\le \varepsilon. \tag{7.30}$$

Also, we have

$$\left\| \int_0^{T_\varepsilon} \mathcal{U}(t,s)(f(s+\omega, y_{s+\omega}, \delta) - f(s+\omega, y_s)ds \right\|$$

$$\leq \gamma_1 \int_0^{T_\varepsilon} e^{-\zeta(t-s)}(\|f(s+\omega, y_{s+\omega}, \delta) - f(s+\omega, y_s)\|)ds$$

$$\leq \gamma_1 \int_0^{T_\varepsilon} e^{-\zeta(t-s)}\left(\|f(s+\omega, y_{s+\omega}, \delta)\| + \|f(s+\omega, y_s, \delta)\|\right)ds$$

$$\leq 2\gamma_1 \int_0^{T_\varepsilon} e^{-\zeta(t-s)}p(s,\delta)\ell(\|y(s)\|, \delta)ds$$

$$\leq 2\gamma_1 \ell(\|y\|_\infty, \delta) \int_0^{T_\varepsilon} e^{-\zeta(t-s)}p(s,\delta)ds.$$

For $t \geq 0$, it follows from the Hölder inequality that

$$\left\| \int_0^{T_\varepsilon} \mathcal{U}(t,s)(f(s+\omega, y_{s+\omega}, \delta) - f(s+\omega, y_s, \delta)ds \right\|$$

$$\leq 2\gamma_1 \|p\|_{L_\infty} \ell(\|y\|_\infty, \delta) \left[\int_0^{T_\varepsilon} e^{-\zeta(t-\nu)}d\nu \right]$$

$$\leq \frac{2\gamma_1 \|p\|_{L_\infty} \ell(\|y\|_\infty, \delta)}{\zeta} \left(e^{-\zeta t} - e^{-\zeta(t-T_\varepsilon)} \right)$$

$$\leq \varepsilon. \tag{7.31}$$

By (7.30), (7.31) which implies that $\lim_{t\to\infty} I_1(t) = 0$.

By (7.14.2)(a), we can see that, for every $\varepsilon > 0$, there is a positive constant $T_\varepsilon > 0$ such that

$$\|f(t+\omega, y.(t,\delta), \delta) - f(t, y.(t,\delta), \delta)\| \leq \varepsilon \quad \text{for } t \geq T_\varepsilon.$$

Similarly, we find

$$I_2(t) \leq \gamma_1 \int_{T_\varepsilon}^t e^{-\zeta(t-s))} \|f(s+\omega, y_{s+\omega}, \delta) - f(s, y_s, \delta)\| \, ds$$

$$+ \gamma_1 \int_0^{T_\varepsilon} e^{-\zeta(t-s))} \|f(s+\omega, y_s) - f(s, y_s, \delta)\| \, ds$$

$$\leq \int_{T_\varepsilon}^t \gamma_1 e^{-\zeta(t-s)} \frac{\zeta\varepsilon}{\gamma_1} ds$$

$$+ \gamma_1 \int_0^{T_\varepsilon} e^{-\zeta(t-s)} \left(\|f(s+\omega, y_s, \delta)\| + \|f(s, y_s, \delta)\| \right) ds$$

$$\leq \varepsilon(1 - e^{-\zeta(t-T_\varepsilon)})$$

$$+ 2\gamma_1 \int_0^{T_\varepsilon} e^{-\zeta(t-s)}p(s,\delta)\ell(\|y(s)\|, \delta)ds$$

$$\leq \varepsilon(1 - e^{-\zeta(t-T_\varepsilon)})$$
$$+ 2\gamma_1 \ell(\|y\|_\infty, \delta) \int_0^{T_\varepsilon} e^{-\zeta(t-s)} p(s, \delta) ds.$$

For $t \geq 0$, it follows from the Hölder inequality that

$$I_2(t) \leq \varepsilon(1 - e^{-\zeta(t-T_\varepsilon)})$$
$$+ 2\gamma_1 \|p\|_{L_\infty} \ell(\|y\|_\infty, \delta) \left[\int_0^{T_\varepsilon} e^{-\zeta(t-\nu)} d\nu \right]$$
$$\leq \varepsilon(1 - e^{-\zeta(t-T_\varepsilon)})$$
$$+ 2\varepsilon \frac{\gamma_1 \|p\|_{L_\infty} \ell(\|y\|_\infty, \delta)}{\zeta} \left(e^{-\zeta(t-T_\varepsilon)} - e^{-\zeta t} \right)$$
$$\leq \varepsilon.$$

Consequently, $\lim_{t\to\infty} I_2(t) = 0$, we conclude that $\aleph_2 \in SAP_\omega$. From **Claims 1** and **2** we deduce $\aleph \in SAP_\omega(E)$, and $\phi(\cdot, \delta) \in SAP_\omega(E)$.

Next, we will prove that the operator $\aleph$ satisfies all the assumptions of Lemma 2.18.

Let $X = SAP_\omega(E)$. Then we show that the mapping defined by (4) is a random operator. To do this, we need to prove that for any $y \in X$, $\aleph(\cdot)(y) : \Omega \longrightarrow X$ is a random variable. Then we prove that $\aleph(\cdot)(y) : \Omega \longrightarrow X$ is measurable since the mapping $f(t, y, \cdot)$, $t \in \mathbb{R}^+$, $y \in X$ is measurable by assumption (7.14.2) and (7.14.4).

Let $\sigma : \Omega \longrightarrow 2^X$ be defined by:

$$\sigma(\delta) = \{ y \in X : \|y\| \leq \chi(\delta) \}.$$

$\sigma(\delta)$ is bounded, closed, convex and solid for all $\delta \in \Omega$. Then σ is measurable by Lemma 17 (see [243]).

Let $\delta \in \Omega$ be fixed, then for $y \in \sigma(\delta)$, and by (A_1), we have

$$\|y_s\|_{\mathcal{B}} \leq \beta_1(s)\|y(s)\| + \beta_2(s)\|y_0\|_{\mathcal{B}}$$
$$\leq \alpha_1\|y(s)\| + \alpha_2\|\phi\|_{\mathcal{B}}.$$

By (7.14.3) and (7.14.2), we have

$$\|(\aleph(\delta)y)(t)\| \leq \|\mathcal{U}(t,0)\|_{B(E)}\|\phi(0,\delta)\| + \int_0^t \|\mathcal{U}(t,s)\|_{B(E)}\|f(s,y_s,\delta)\| \, ds$$
$$\leq \gamma_1 e^{-\zeta t}\|\phi\|_{\mathcal{B}} + \gamma_1 \int_0^t e^{-\zeta(t-s)} p(s,\delta) \, \ell(\|y_s\|_{\mathcal{B}}, \delta) \, ds.$$

Set

$$D_1 := \alpha_1 \chi(\delta) + \alpha_2 \|\phi\|_{\mathcal{B}}.$$

Then, we have

$$\|(\aleph(\delta)y)(t)\| \leq \gamma_1 \|\phi\|_{\mathcal{B}} + \gamma_1 \ell(D_1, \delta) \|p\|_{L_\infty} (1 - e^{-\zeta t}) \, ds.$$

Thus

$$\|(\aleph(\delta)y)\| \leq \gamma_1 \|\phi\|_{\mathcal{B}} + \frac{\gamma_1 \ell(D_1, \delta) \|p\|_{L_\infty}}{\zeta} \leq \chi(\delta).$$

Thus, $\aleph$ is a random operator with stochastic domain σ and $F(\delta) : \sigma(\delta) \longrightarrow \sigma(\delta)$ for each $\delta \in \Omega$.

Step 1: $\aleph$ is continuous.
Let y^n be a sequence where $y^n \longrightarrow y$ in X. Since the function f is Carathéodory, we can see that

$$\|f(s, y_s^n, \delta) - f(s, y_s, \delta)\| \leq \frac{\zeta \varepsilon}{\gamma_1}. \tag{7.32}$$

If $t \in \mathbb{R}^+$, then (7.32) and our hypotheses give us that

$$
\begin{aligned}
\|\aleph(\delta) y_s^n(t) &- \aleph(\delta) y_s(t)\| \\
&\leq \int_0^t \|\mathcal{U}(t,s)\|_{B(E)} \Big\| f(s, y_n(s), \delta) - f(s, y(s), \delta) \Big\| \, ds \\
&\leq \gamma_1 \frac{\zeta \varepsilon}{\gamma_1} \int_0^t e^{-\zeta(t-s)} \, ds \\
&\leq \frac{\gamma_1}{\zeta} \frac{\zeta \varepsilon}{\gamma_1} (1 - e^{-\zeta t}) \\
&\leq \varepsilon.
\end{aligned}
\tag{7.33}
$$

Then the inequality (7.33) reduces to

$$\|\aleph(\delta)y^n - \aleph(\delta)y\|_\infty \to 0 \quad \text{as } n \to \infty.$$

Now, we conclude that $\aleph$ is continuous.

Step 2: We demonstrate that for $\delta \in \Omega, \{y \in \sigma(\delta) : \aleph(\delta)y = y\} \neq \emptyset$. For prove this we apply Schauder's theorem [214].
$\aleph(\sigma(\delta))$ **relatively compact:** To prove the compactness, we will use Corduneanu's lemma.

(a) Firstly, it is clear that the assumption (i) is holds. Then we will demonstrate that $\aleph(\sigma(\delta))$ is equicontinuous set for each closed bounded interval $[0, A]$ in $\mathbb{R}^+$. Let $\nu_1, \nu_2 \in [0, A]$ with $\nu_2 > \nu_1$, $\sigma(\delta)$ be a bounded set as in Step 2, and $y \in \sigma(\delta)$. Thus

$$\|(\aleph(\delta)y)(\nu_2) - (\aleph(\delta)y)(\nu_1)\|$$
$$\leq \|\mathcal{U}(\nu_2, 0) - \mathcal{U}(\nu_1, 0)\|_{B(E)}\|\phi\|_{\mathcal{B}}$$
$$+ \left\| \int_0^{\nu_1} [\mathcal{U}(\nu_2, s) - \mathcal{U}(\nu_1, s)] f(s, y_s, \delta) \, ds \right\|$$
$$+ \left\| \int_{\nu_1}^{\nu_2} \mathcal{U}(\nu_2, s) f(s, y_s, \delta) \, ds \right\|$$
$$\leq \|\mathcal{U}(\nu_2, 0) - \mathcal{U}(\nu_1, 0)\|_{B(E)}\|\phi\|_{\mathcal{B}}$$
$$+ \int_0^{\nu_1} \|\mathcal{U}(\nu_2, s) - \mathcal{U}(\nu_1, s)\| \|f(s, y_s, \delta)\| \, ds$$
$$+ \int_{\nu_1}^{\nu_2} \|\mathcal{U}(\nu_2, s)\| \|f(s, y_s, \delta)\| \, ds$$
$$\leq \|\mathcal{U}(\nu_2, 0) - \mathcal{U}(\nu_1, 0)\|_{B(E)}\|\phi\|_{\mathcal{B}}$$
$$+ \ell(D_1, \delta) \int_0^{\nu_1} \|\mathcal{U}(\nu_2, s) - \mathcal{U}(\nu_1, s)\|_{B(E)} p(s, \delta) ds$$
$$+ \frac{\gamma_1 \|p\|_{L^\infty} \ell(D_1, \delta)}{\zeta} \left(e^{-\zeta(t-\nu_2)} - e^{-\zeta(t-\nu_1)} \right).$$

The right-hand of the above inequality tends to zero as $\nu_2 - \nu_1 \to 0$, as $\aleph$ is bounded and equicontinuous.

(b) Now we demonstrate that $Q(t, \delta) = \{(\aleph(\delta)y)(t) : y \in \sigma(\delta)\}$ is precompact in E. Let $t \in [0, A]$ be fixed and let ϵ be verifying $0 < \epsilon < t$. For $y \in \sigma(\delta)$ we define

$$(\aleph_\epsilon(\delta)y)(t) = \mathcal{U}(t, 0)\phi(0, \delta) + \mathcal{U}(t, t-\epsilon) \int_0^{t-\epsilon} \mathcal{U}(t-\epsilon, s) f(s, y_s, \delta) \, ds.$$

Since $\mathcal{U}(t, s)$ is a compact operator and the set $Q_\epsilon(t, \delta) = \{(\aleph_\epsilon(\delta)y)(t) : y \in \sigma(\delta)\}$ is the image of bounded set of E then $Q_\epsilon(t, \delta)$ is pre-compact in E for every ϵ, $0 < \epsilon < t$. Moreover

$$\|(\aleph(\delta)y)(t) - (\aleph_\epsilon(\delta)y)(t)\| \leq \int_{t-\epsilon}^t \|\mathcal{U}(t, s)\|_{B(E)} \|f(s, y_s, \delta)\| ds$$

$$\leq \gamma_1 \ell(D_1, \delta) \int_{t-\epsilon}^t e^{-\zeta(t-s)} p(s, \delta) ds.$$

Thus, $Q(t, \delta) = \{(\aleph(\delta)y)(t) : y \in \sigma(\delta)\}$ is precompact in E.

(c) We prove that $\aleph$ is equiconvergent.

Let $y \in \sigma(\delta)$, then from (7.14.1), (7.14.2) we have

$$\|(\aleph(\delta)y)(t)\| \leq \gamma_1 e^{-\zeta t}\|\phi\|_{\mathcal{B}} + \gamma_1 \int_0^t e^{-\zeta(t-s)}p(s,\delta)\, \ell(D_1,\delta)\, ds,$$

it follows immediately by (7.14.5) that $\|(\aleph(\delta)y)(t)\| \longrightarrow 0$ as $t \to +\infty$. Then

$$\lim_{t \to +\infty} \|(\aleph(\delta)y)(t) - (\aleph(\delta)y)(+\infty)\| = 0.$$

Therefore, $\aleph$ is equiconvergent.

Consequently, $\aleph(\delta) : \sigma(\delta) \to \sigma(\delta)$ is continuous and compact. By Schauder's theorem [214], $\aleph(\delta)$ has a fixed point $y(\delta)$ in $\sigma(\delta)$. Since $\bigcap_{\delta \in \Omega} \sigma(\delta) \neq \emptyset$, the hypothesis that a measurable selector of intoσ exists holds. By Lemma 2.18, $\aleph$ has a stochastic fixed point $y^*(\delta)$, which is a random mild solution of (7.25)–(7.26).

7.4.2 *An Example*

Consider the following example:

$$\frac{\partial}{\partial t}\vartheta(t,\alpha,\delta) = \jmath(t,\alpha)\frac{\partial^2}{\partial \alpha^2}\vartheta(t,\alpha,\delta) \;\; + C_0(\delta)K(\delta)\frac{t^{\frac{1}{r}}}{2(1+t^2)^{\frac{2}{r}}}$$

$$\times \ln\left(1 + \frac{2t^{\frac{r-1}{r}}}{(1+t^2)^{\frac{2r-2}{r}}}|\vartheta(t,\alpha,\delta)|\right), \;\; \alpha \in (0,\pi), \; t \in (0,+\infty),$$

$$\tag{7.34}$$

$$\vartheta(t,0,\delta) = \vartheta(t,\pi,\delta) = 0, \; t \in (0,+\infty), \tag{7.35}$$

$$\vartheta(s,\alpha,\delta) = \vartheta_0(s,\alpha,\delta), \; s \in (-\infty,0), \; \alpha \in [0,\pi], \tag{7.36}$$

where $\jmath(t,y)$ is continuous and uniformly Hölder continuous in t, K and C_0 are a real-valued random variable.

Let $E = L^2(0,\pi)$, (Ω, F, P) be a complete probability space, and define $A(t)$ by

$$A(t)\eta = \jmath(t,y)\eta''$$

with domain

$$\sigma(A) = \{\eta \in E, \eta, \eta' \text{are absolutely continuous, } \eta'' \in E, \; \eta(0) = \eta(\pi) = 0\}.$$

Then $A(t)$ generates an evolution system $\mathcal{U}(t,s)$ verifying (7.14.1) (see [352, 353]).

Let $\mathcal{B} = BUC(\mathbb{R}^-; E)$ be the space of bounded uniformly continuous functions endowed with

$$\|y\| = \sup_{s \leq 0} \|y(s)\|, \quad \text{for } y \in \mathcal{B}$$

If $\phi \in BCU(\mathbb{R}^-; E), \alpha \in (0, \pi)$ and $\delta \in \Omega$, then

$$y(t, \alpha, \delta) = \vartheta(t, \alpha, \delta), t \in (0, A)$$

$$\phi(s, \alpha, \delta) = \vartheta_0(s, \alpha, \delta), \quad s \in (-\infty, 0), \quad \alpha \in [0, \pi], \delta \in \Omega.$$

Set

$$f(t, \vartheta, \delta) = C_0(\delta) K(\delta) \frac{t^{\frac{1}{r}}}{2(1+t^2)^{\frac{2}{r}}} \ln \left(1 + \frac{2t^{\frac{r-1}{r}}}{(1+t^2)^{\frac{2r-2}{r}}} |\vartheta(t, \alpha, \delta)| \right),$$

and we have $f(t, \vartheta, \delta) = C_0(\delta) K(\delta) g(t, \vartheta, \delta)$

$$\|g(t+1, \vartheta, \delta)(\nu) - g(t, \vartheta, \delta)(\nu)\|$$

$$\leq \frac{t^{\frac{1}{r}}}{2(1+t^2)^{\frac{2}{r}}} \ln \left(1 + \frac{2t^{\frac{r-1}{r}}}{(1+t^2)^{\frac{2r-2}{r}}} |\vartheta(t+1, \nu, \delta)| \right)$$

$$- \frac{t^{\frac{1}{r}}}{2(1+t^2)^{\frac{2}{r}}} \ln \left(1 + \frac{2t^{\frac{r-1}{r}}}{(1+t^2)^{\frac{2r-2}{r}}} |\vartheta(t, \nu, \delta)| \right)$$

$$\leq \frac{t^{\frac{1}{r}}}{2(1+t^2)^{\frac{2}{r}}} \ln \left(\frac{1 + 2(t+1)^{\frac{r-1}{r}} \|\vartheta(t+1, \nu, \delta)\|}{1 + 2t^{\frac{r-1}{r}} \|\vartheta(t, \nu, \delta)\|} \right)$$

$$\leq \frac{t^{\frac{1}{r}}}{2(1+t^2)^{\frac{2}{r}}} \ln \left(\left(\frac{t+1}{t} \right)^{\frac{r-1}{r}} + \frac{1 - \left(\frac{t+1}{t} \right)^{\frac{r-1}{r}}}{1 + 2t^{\frac{r-1}{r}} \|\vartheta(t, \nu, \delta)\|} \right)$$

$$\leq \frac{t^{\frac{1}{r}}}{2(1+t^2)^{\frac{2}{r}}} \ln \left(\left(\frac{t+1}{t} \right)^{\frac{r-1}{r}} + 1 \right). \tag{7.37}$$

Now from (7.37), we can see $\|g(t+1, \vartheta, \delta)(\nu) - g(t, \vartheta, \delta)(\nu)\|$ tends to zero as $t \to \infty$, then $\|f(t+1, \vartheta)(\nu, \delta) - f(t, z)(\nu, \delta)\|$ tends to zero as $t \to \infty$. Hence condition (7.14.3)(a) is satisfied.

The function $f(t, \varphi(\alpha), \delta)$ is Carathéodory, and verifies (7.14.2) with

$$p(t, \delta) = K(\delta) \frac{t^{\frac{1}{r}}}{2(1+t^2)^{\frac{2}{r}}} \quad \text{and} \quad \ell(\alpha, \delta) = \|C_0(\delta)\| \frac{1}{4} \ln(1 + 2t).$$

Then the problem (7.34)–(7.35) in an abstract formulation of the problem (7.25)–(7.26), and conditions (7.14.1)–(7.14.5) are satisfied. Theorem 7.7 implies that the random problem (7.34)–(7.35) has at least one random mild solutions.

7.5   Notes and Remarks

The conclusions of the present chapter are based on the papers [354–356]. One can see the publications [92–95,97–99], for additional details and results on the subject.

Chapter 8

Impulsive Differential Evolution Equations

8.1 Introduction and Motivations

In this chapter, we study the existence of mild solutions of classes of impulsive integrodifferential equations on bounded and unbounded domains via resolvent operators in Banach space. Our results are based on the fixed point theory and the concept of measure of non-compactness with the help of the resolvent operator. Moreover, we give sufficient conditions that ensure the controllability and attractivity of our problem. Finally, examples are given to validate the theory part.

We explored and demonstrated the results obtained in this chapter by taking into consideration the previously stated publications in the preceding chapters and the works that follow that we mention as references for the reader:

- The papers [357,358], where the existence of mild solutions is developed for some semi-linear functional differential equations. There has been a significant development in functional evolution equations in recent years; see the monographs [359,360], the papers [13,120,361–365], and the references therein.
- In [366], Xue studied the existence of mild solutions for semilinear differential equations with nonlocal initial conditions in separable Banach spaces. Xue [367] discussed the semilinear nonlocal differential equations when the semigroup $T(t)$ generated by the coefficient operator is compact and the nonlocal term g is not compact. Fan and Li [359] discussed the existence for impulsive semilinear differential equations with nonlocal conditions by using Sadovskii's fixed point theorem and Schauder's fixed point theorem.

- Impulsive differential equations have become more important in recent years in some mathematical models of real phenomena, especially in biological or medical domains, and in control theory, see for example the monographs [169–171], and the papers [172–174]. In [162,177,368–370] the authors initially offered to study some classes of impulsive differential equations with non-instantaneous impulses.

- Several researchers obtained other results by application of the technique of measure of non-compactness; see [193,306,371], and the references therein.

- In domains such as population dynamics and optimal control, impulsive integral equations, impulsive integro-differential equations, and impulsive differential equations naturally occur (see the monographs [43,154,175–177,350]). It appears that the earliest discussion of impulsive systems dates back to Krylov and Bogolyubov's work [178].

- In 1982, Grimmer started by utilizing resolvent operators to demonstrate the existence of integro-differential systems in [42,295,296]. The resolvent operator with fixed point technique is the most convenient and appropriate approach for solving integrodifferential equations. Readers may see [42, 296, 372] and the sources given therein for more information on resolvent operators. The existence of nonlocal analytic resolvent operator integro-differential equations in [372, 373] has been demonstrated using facts about resolvent operators and the regularity of evolution integro-differential systems. Because of its applicability in describing numerous issues in physics, fluid dynamics, biological models, and chemical kinetics, integral-differential equations over infinite intervals have sparked a lot of interest see [44, 374–376]. Many authors have examined qualitative properties such as existence, uniqueness, and stability for many integral, differential and integro-differential equations, (see [29,38,39,44,327,350,377–382]), and with nonlocal condition in [200, 298, 383, 384].

- The controllability of linear and nonlinear systems represented by ODEs in finite dimensional space has received a lot of attention, see [38–40]. Several writers have expanded the idea of controllability to infinite-dimensional Banach space systems with unbounded operators, see the monographs [40,64–66]. For more relevant studies on differential equations, see the papers [41–44] and the references therein. Lasiecka and Triggiani [46] set sufficient conditions for controllability.

- The theory and application of integrodifferential equations are important subjects in applied mathematics, see, for example [2,434,435,437] and recent development of the topic. In recent times there has been an increasing interest in studying the abstract autonomous second order, see for example [3, 5–8].

- In [28], Benchohra and Rezoug investigated the existence and local attractivity of the mild solution, defined on a semi-infinite positive real interval $J = [0, \infty)$, for non-autonomous semilinear second order evolution equation of mixed type in a real Banach space. They considered the following problem

$$y''(t) - A(t)y(t) = f\left(t, y(t), \int_0^t K(t, s, y(s))ds\right), \ t \in J,$$

$$y(0) = y_0, \ y'(0) = y_1,$$

where $\{A(t)\}_{0 \le t < +\infty}$ is a family of linear closed operators from E into E, $f : J \times E \times E \to E$ is a Carathéodory function, $K : \Delta \times E \to E$ is a continuous function, $\Delta := \{(t, s) \in J \times J : s \le t\}, y_0, y_1 \in E$ and $(E, |\cdot|)$ is a real Banach space. The results are obtained by using the Mönch fixed point theorem and the Kuratowski measure of non-compactness.

- In [184], Abbas *et al.* discussed the existence of mild solutions for the following nonlocal problem of impulsive integrodifferential equations:

$$\begin{cases} u'(t) = Au(t) + \displaystyle\int_0^t Y(t - s)u(s)ds + f(t, u(t)); t \in I_k, k = 0, \ldots, m, \\ u\left(t_k^+\right) = u\left(t_k^-\right) + L_k\left(u\left(t_k^-\right)\right); \quad k = 1, \ldots, m, \\ u(0) + g(u) = u_0 \in E, \end{cases}$$

where $I_0 = [0, t_1], I_k := (t_k, t_{k+1}]; k = 1, \ldots, m, 0 = t_0 < t_1 < \cdots < t_m < t_{m+1} = T, f : I_k \times E \longrightarrow E; k = 1, \ldots, m, L_k : E \longrightarrow E; k = 1, \ldots, m, g : PC \longrightarrow E$ are given functions, the set PC is given later, E is a real (or complex) Banach space with the norm $\|\cdot\|, u'(t) := du/dt, A : D(A) \subset E \longrightarrow E$ generates a C_0-semigroup on the Banach space E, and $Y(t)$ is a closed linear operator on E with $D(A) \subset D(Y)$.

- Recently in [298], the authors used Schauder's fixed point to prove the existence of mild solutions by considering two cases of the resolvent operator for the following integrodifferential problem:

$$\begin{cases} y'(t) = Ay(t) + \displaystyle\int_0^t B(t - s)y(s)ds + f(t, y(t), (Hy)(t)); \ \text{if } t \in [0, a], \\ y(0) = g(y) + y_0. \end{cases}$$

- Recently, Hernandez *et al.* [385] used non-instantaneous impulsive condition for semi-linear abstract differential equation of the form

$$\begin{cases} y'(t) = Ay(t) + f(t, y(t)), \ t \in (s_i, t_{i+1}], i = 0, \ldots, m, \\ y(t) = g_i(t, y(t)), \qquad\qquad t \in (t_i, s_i], i = 1, \ldots, m, \\ y(0) = y_0, \end{cases} \qquad (8.1)$$

and introduced the concepts of mild and classical solution. To learn more about this kind of problems, we refer [22, 386–393].

8.2 Instantaneous and Non-Instantaneous Impulsive Integro-Differential Equations in Banach Spaces

In this section, we first discuss the existence of mild solutions for the following nonlocal problem of impulsive integro-differential equation:

$$\begin{cases} y'(t) = Ay(t) + \displaystyle\int_0^t \Upsilon(t-s)y(s)ds + f(t, y(t)); \ t \in J_k, \ k = 0, \ldots, m, \\ y(t_k^+) = y(t_k^-) + L_k(y(t_k^-)); \ k = 1, \ldots, m, \\ y(0) + g(y) = y_0 \in E, \end{cases}$$
$$(8.2)$$

where $J_0 = [0, t_1]$, $J_k := (t_k, t_{k+1}]$; $k = 1, \ldots, m$, $0 = t_0 < t_1 < \cdots < t_m < t_{m+1} = T$, $f : J_k \times E \to E$; $k = 1, \ldots, m$, $L_k : E \to E$; $k = 1, \ldots, m$, $g : PC \to E$ are given functions, the set PC is given later, E is a real (or complex) Banach space with norm $\| \cdot \|$, $y'(t) := \frac{dy}{dt}$, $A : D(A) \subset E \to E$ generates a C_0-semigroup on the Banach space E, $\Upsilon(t)$ is a closed linear operator on E with $D(A) \subset D(\Upsilon)$.

We next discuss the existence of mild solutions for the following nonlocal problem of non-instantaneous impulsive integro-differential equations:

$$\begin{cases} y'(t) = Ay(t) + \displaystyle\int_0^t \Upsilon(t-s)y(s)ds + f(t, y(t)); \ t \in J_k, \ k = 0, \ldots, m, \\ y(t) = g_k(t, y(t_k^-)); \ t \in \tilde{J}_k, \ k = 1, \ldots, m, \\ y(s_k) + g(y) = y_k \in E; \ k = 0, \ldots, m, \end{cases}$$
$$(8.3)$$

where $J_0 := [0, t_1]$, $\tilde{J}_k := (t_k, s_k]$, $J_k := (s_k, t_{k+1}]$; $k = 1, \ldots, m$, $f : J_k \times E \to E$, $g_k : \tilde{J}_k \times E \to E$ are given functions such that $g_k(t, y(t_k^-))|_{t=s_k} = y_k \in E$; $k = 1, \ldots, m$, $g : PC \to E$ is a given function, the set PC is given later, and $0 = s_0 < t_1 \le s_1 < t_2 \le s_2 < \cdots \le s_{m-1} < t_m \le s_m < t_{m+1} = T$.

8.2.1   *Mild Solutions with Instantaneous Impulses*

In this section, we are concerned with the existence results of the problem (8.2).

Consider the Banach space

$$PC = \{y : J \to E : y \in C(J_k); \ k = 0, \ldots, m, \ \text{and there exist } y(t_k^-)$$
$$\text{and } y(t_k^+); \ k = 1, \ldots, m, \ \text{with } y(t_k^-) = y(t_k)\},$$

with the norm

$$\|y\|_{PC} = \sup_{t \in J} \|y(t)\|.$$

Definition 8.1 ([364]). A resolvent operator for the Cauchy problem

$$\begin{cases} y'(t) = Ay(t) + \displaystyle\int_0^t \Upsilon(t-s)y(s)ds; \ t \in [0, \infty), \\ y(0) = y_0 \in E, \end{cases} \tag{8.4}$$

is a bounded linear operator-valued function $R(t) \in B(E); \ t \geq 0$, verifying the following conditions:

(i) $R(0) = I$ (the identity map of E) and $\|R(t)\| \leq Ne^{\nu t}$ for some constants $N > 0$, and $\nu \in \mathbb{R}$.

(ii) For each $y \in E$, $R(t)y$ is strongly continuous for $t \geq 0$.

(iii) $R(t)$ is bounded for $t \geq 0$. For $y \in D(A)$, $R(\cdot)y \in C(\mathbb{R}_+, D(A)) \cap C^1(\mathbb{R}_+, E)$ and

$$R'(t)y = AR(t)y + \int_0^t \Upsilon(t-s)R(s)uds$$
$$= R(t)Ay + \int_0^t R(t-s)\Upsilon(s)uds; \ t \in [0, \infty).$$

Let us introduce the following hypotheses:

(8.1.1) The operator A is the infinitesimal generator of a uniformly continuous semigroup $(S(t))_{t \geq 0}$.

(8.1.2) For all $t \geq 0$, $\Upsilon(t)$ is a closed linear operator from $D(A)$ to E and $\Upsilon(t) \in B(E)$. For any $y \subset E$, the map $l \mapsto \Upsilon(t)y$ is bounded, differentiable and the derivative $t \mapsto \Upsilon'(t)y$ is bounded uniformly continuous on $\mathbb{R}_+$.

Theorem 8.1 ([364, 394]). *Assume that (8.1.1) and (8.1.2) hold. Then there exists a unique uniformly continuous resolvent operator for the Cauchy problem (8.4).*

Definition 8.2 ([394]). By a mild solution of the problem (8.2) we mean a function $y \in PC$ that satisfies

$$y(t) = R(t)[y_0 - g(y)] + \int_0^t R(t-s)f(s, y(s))ds + \sum_{0 < t_i < t} R(t-t_i)L_i(y(t_i)); \ t \in J.$$

The following hypotheses will be used in the sequel.

(8.3.1) The function $t \mapsto f(t, y)$ is measurable on J for each $y \in E$, and the function $y \mapsto f(t, y)$ is continuous on E for a.e. $t \in J_k$,

(8.3.2) There exists a function $p \in L^\infty(J)$, such that

$$\|f(t, y)\| \leq p(t)(1 + \|y\|); \ for \ a.e. \ t \in J_k, \ and \ each \ y \in E,$$

(8.3.3) There exist positive constants q^*, l_k^*; $k = 0, \ldots, m$, such that

$$\|g(y)\| \leq q^*(1 + \|y\|_{PC}); \ for \ each \ y \in PC,$$

and

$$\|L_k(y)\| \leq l_k^*(1 + \|y\|); \ k = 0, \ldots, m, \ for \ a.e. \ t \in J, \ and \ each \ y \in E,$$

(8.3.4) For each bounded set $B \subset E$, we have

$$\alpha(f(t, B)) \leq p(t)\alpha(B), \ \alpha(L_k(B)) \leq l_k^*\alpha(B); \ k = 0, \ldots, m,$$

and for each bounded set $B_1 \subset PC$, we have

$$\alpha(g(B_1)) \leq q^* \sup_{t \in J} \alpha(B_1(t)),$$

where $B_1(t) = \{y(t) : y \in B_1\}; \ t \in J.$

Set

$$p^* = \|p\|_{L^\infty}, \ and \ M = \sup_{t \in J} \|R(t)\|_{B(E)}.$$

Theorem 8.2. *Assume that the hypotheses* (8.1.1), (8.1.2), (8.3.1)–(8.3.4) *hold. If*

$$\ell := M\left(q^* + Tp^* + \sum_{k=0}^m l_k^*\right) < 1, \tag{8.5}$$

then the problem (8.2) *has at least one mild solution defined on* J.

Proof. Transform the problem (8.2) into a fixed point problem. Consider the operator $N : PC \to PC$ defined by

$$(Nu)(t) = R(t)[y_0 - g(y)] + \int_0^t R(t-s)f(s, y(s))ds$$

$$+ \sum_{0 < t_i < t} R(t - t_i)L_i(y(t_i)); \ t \in J. \tag{8.6}$$

Let $\rho > 0$, such that

$$\rho \geq \frac{M\left(\|y_0\| + q^* + Tp^* + \sum_{k=0}^{m} l_k^*\right)}{1 - M\left(q^* + Tp^* + \sum_{k=0}^{m} l_k^*\right)},$$

and consider the ball $B_\rho := B(0, \rho) = \{w \in PC : \|w\|_{PC} \leq \rho\}$.
    For any $y \in B_\rho$ and each $t \in J$, we have

$$\|(Nu)(t)\| \leq \|R(t)\|_{B(E)}[\|y_0\| + \|g(y)\|]$$

$$+ \int_0^t \|R(t-s)\|_{B(E)}\|f(s, y(s))\|ds$$

$$+ \sum_{0 < t_i < t} \|R(t - t_i)\|_{B(E)}\|L_i(y(t_i))\|$$

$$\leq M\left[\|y_0\| + (1 + \rho)\left(q^* + Tp^* + \sum_{k=0}^{m} l_k^*\right)\right]$$

$$:\leq \rho.$$

Thus

$$\|N(y)\|_{PC} \leq \rho. \tag{8.7}$$

This proves that N transforms the ball B_ρ into itself. We shall show that the operator $N : B_\rho \to B_\rho$ satisfies all the assumptions of Theorem 2.17. The proof will be given in three steps.

Step 1. $N : B_\rho \to B_\rho$ *is continuous.*
Let $\{y_n\}_{n \in \mathbb{N}}$ be a sequence such that $y_n \to y$ as $n \to \infty$ in B_ρ. Then, for each $t \in J$, we have

$$\|(Ny_n)(t) - (Nu)(t)\| \leq \|R(t)\|_{B(E)} \|g(y_n) - g(y)\|$$

$$+ \int_0^t \|R(t-s)\|_{B(E)} \|f(s, y_n(s)) - f(s, y(s))\| ds$$

$$+ \sum_{0 < t_i < t} \|R(t-t_i)\|_{B(E)} \|L_i(y_n(t_i)) - L_i(y(t_i))\|$$

$$\leq M \|g(y_n) - g(y)\|$$

$$+ M \int_0^T \|f(s, y_n(s)) - f(s, y(s))\| ds$$

$$+ M \sum_{0 < t_i < t} \|L_i(y_n(t_i)) - L_i(y(t_i))\|.$$

Since $y_n \to y$ as $n \to \infty$ and f, g, L_i are continuous, the Lebesgue dominated convergence theorem implies that

$$\|N(y_n) - N(y)\|_{PC} \to 0 \quad as \quad n \to \infty.$$

Step 2. $N(B_\rho)$ *is bounded and equicontinuous.*
Since $N(B_\rho) \subset B_\rho$ and B_ρ is bounded, then $N(B_\rho)$ is bounded.

Next, let $t, \tau \in J$, $\tau < t$ and let $y \in B_\rho$. Thus, we have

$$\|(Nu)(t) - (Nu)(\tau)\| \leq \|R(t) - R(\tau)\|_{B(E)} (\|y_0\| + \|g(y)\|)$$

$$+ \int_0^\tau \|R(t-s) - R(\tau-s)\|_{B(E)} \|f(s, y(s))\| ds$$

$$+ \int_\tau^t \|R(t-s)\|_{B(E)} \|f(s, y(s))\| ds$$

$$+ \sum_{0 < t_i < t} \|R(t-t_i) - R(\tau-t_i)\|_{B(E)} \|L_i(y(t_i))\|.$$

Hence, we get

$$\|(Nu)(t) - (Nu)(\tau)\| \leq (\|y_0\| + q^*(1+\rho)) \|R(t) - R(\tau)\|_{B(E)}$$

$$+ p^*(1+\rho) \int_0^\tau \|R(t-s) - R(\tau-s)\|_{B(E)} ds$$

$$+ Mp^*(1+\rho)(t-\tau)$$

$$+ \sum_{0 < t_i < t} l_i^*(1+\rho) \|R(t-t_i) - R(\tau-t_i)\|_{B(E)}.$$

As the resolvent operator $R(\cdot)$ is uniformly continuous, the right-hand side of the above inequality tends to zero as $\tau \longrightarrow t$.

Step 3. *The implication* (2.2) *holds.*
Now let V be a subset of B_ρ such that $V \subset \overline{N(V)} \cup \{0\}$. V is bounded and equicontinuous and therefore the function $t \to v(t) = \alpha(V(t))$ is continuous

on J. By (8.3.3) and the properties of the measure α, for each $t \in J$, we have

$$
\begin{aligned}
v(t) &\le \alpha((NV)(t) \cup \{0\}) \\
&\le \alpha((NV)(t)) \\
&\le \|R(t)\|_{B(E)} q^* \sup_{t \in J} \alpha(V(t)) + \int_0^t \|R(t-s)\|_{B(E)} p(s)\alpha(V(s))ds \\
&\quad + \sum_{k=0}^m \|R(t-t_k)\|_{B(E)} l_k(t)\alpha(V(t)) \\
&\le Mq^*\|v\|_\infty + Mp^* \int_0^t v(s)ds + M \sum_{k=0}^m l_k^* v(t) \\
&\le M\left(q^* + Tp^* + \sum_{k=0}^m l_k^*\right)\|v\|_\infty.
\end{aligned}
$$

Hence

$$
\|v\|_\infty \le \ell\|v\|_\infty.
$$

From (8.5), we get $\|v\|_\infty = 0$, that is, $v(t) = \alpha(V(t)) = 0$, for each $t \in J$, and then $V(t)$ is relatively compact in E. In view of the Ascoli-Arzelà theorem, V is relatively compact in B_ρ. Applying now Theorem 2.17, we conclude that N has a fixed point which is a mild solution of our problem (8.2).

8.2.2 *Mild Solutions with Not Instantaneous Impulses*

In this section, we are concerned with the existence results of the problem (8.3).

Denote by

$$
\begin{aligned}
\mathcal{PC} = \{ y : J \to E : &\ y \in C([0,t_1] \cup (t_k, s_k] \cup (s_k, t_{k+1}], E),\ k = 1, \ldots, m, \\
&\text{and there exist } y(t_k^-),\ y(t_k^+), y(s_k^-)\text{and } y(s_k^+)\ k = 1, \ldots, m \text{ with} \\
&\ y(t_k^-) = y(t_k)\text{and } y(s_k^-) = y(s_k) \},
\end{aligned}
$$

the Banach space equipped with the standard supremum norm.

Definition 8.3. [394] By a mild solution of the problem (8.3) we mean a function $y \in \mathcal{PC}$ that satisfies

$$
\begin{cases}
y(t) = R(t)[y_k - g(y)] + \displaystyle\int_{s_k}^t R(t-s)f(s,y(s))ds;\ t \in J_k,\ k = 0, \ldots, m, \\
y(t) = g_k(t, y(t_k^-));\ t \in \tilde{J}_k,\ k = 1, \ldots, m,
\end{cases}
$$

The following hypotheses will be used in the sequel.

(8.5.1) The functions $t \mapsto f(t,y)$ and $t \mapsto g_k(t,y)$ are measurable on $J_k, \tilde{J}_k$ respectively for each $y \in E$, and the functions $y \mapsto f(t,y)$ and $y \mapsto g_k(t,y)$ are continuous on E for a.e. t in $J_k, \tilde{J}_k$, respectively.

(8.5.2) There exist functions $p, l_k \in L^\infty(J)$; $k = 0, \ldots, m$, such that

$$\|f(t,y)\| \leq p(t)(1 + \|y\|); \ \ for \ a.e. \ t \in J_k, \ and \ each \ y \in E,$$

and

$$\|g_k(t,y)\| \leq l_k(t)(1+\|y\|); \ k = 1, \ldots, m, \ for \ a.e. \ t \in \tilde{J}_k, \ and \ each \ y \in E.$$

(8.5.3) There exists a positive constant q^*, such that

$$\|g(y)\| \leq q^*(1 + \|y\|_{\mathcal{PC}}); \ for \ a.e. \ t \in J, \ and \ each \ y \in \mathcal{PC}.$$

(8.5.4) For each bounded set $B \subset E$ and for each $t \in J$, we have

$$\alpha(f(t,B)) \leq p(t)\alpha(B), \ \alpha(g_k(t,B)) \leq l_k(t)\alpha(B); \ k = 0, \ldots, m,$$

and for each bounded set $B_0 \subset \mathcal{PC}$, we have

$$\alpha(g(B_0)) \leq q^* \sup_{t \in J} \alpha(B_0(t)),$$

where $B_0(t) = \{y(t) : y \in B_0\}$; $t \in J$.

Set

$$p^* = \|p\|_{L^\infty}, \ l^* = \max_{k=0,\ldots,m} \|l_k\|_{L^\infty}, \ M = \sup_{t \in J} \|R(t)\|_{B(E)}.$$

Theorem 8.3. *Assume that the hypotheses* $(8.1.1), (8.1.2), (8.5.1)–(8.5.4)$ *hold. If*

$$\ell := \max\{l^*, M(q^* + Tp^*)\} < 1, \tag{8.8}$$

then the problem (8.3) *has at least one mild solution defined on* J.

Proof. Transform the problem (8.3) into a fixed point problem. Consider the operator $N : \mathcal{PC} \to \mathcal{PC}$ defined by

$$\begin{cases} (Nu)(t) = R(t)[y_k - g(y)] + \int_{s_k}^{t} R(t-s)f(s,y(s))ds; \ t \in J_k, \ k = 0, \ldots, m, \\ (Nu)(t) = g_k(t, y(t_k^-)); \ t \in \tilde{J}_k, \ k = 1, \ldots, m. \end{cases}$$

$$\tag{8.9}$$

Let $L > 0$, such that

$$L \geq \frac{M(\|y_k\| + q^* + Tp^*)}{1 - M(q^* + Tp^*)}.$$

For any $y \in \mathcal{PC}$ and each $t \in J_k$, we have

$$\|(Nu)(t)\| \leq \|R(t)\|_{B(E)}[\|y_k\| + \|g(y)\|]$$
$$+ \int_{s_k}^{t} \|R(t-s)\|_{B(E)}\|f(s,y(s))\|ds$$
$$\leq M[\|y_k\| + (1+L)(q^* + Tp^*)]$$
$$\leq L.$$

Thus

$$\|N(y)\|_{\mathcal{PC}} \leq L. \tag{8.10}$$

Next, for each $t \in \tilde{J}_k$; $k = 1, \ldots, m$, it is clear that

$$\|(Nu)(t)\|_E \leq l^*.$$

Hence,

$$\|N(y)\|_{\mathcal{PC}} \leq \max\{L, l^*\} := \rho.$$

This proves that N transforms the ball $B_\rho := \{w \in \mathcal{PC} : \|w\|_{\mathcal{PC}} \leq \rho\}$ into itself.

We shall show that the operator $N : B_\rho \to B_\rho$ satisfies all the assumptions of Theorem 2.17. The proof will be given in three steps.

Step 1. $N : B_\rho \to B_\rho$ *is continuous.*
Let $\{y_n\}_{n\in\mathbb{N}}$ be a sequence such that $y_n \to y$ as $n \to \infty$ in B_ρ. Then, for each $t \in \tilde{J}_k$; $k = 1, \ldots, m$, we have

$$\|(Ny_n)(t) - (Nu)(t)\| \leq \|R(t)\|_{B(E)}\|g_k(t, y_n(t_k^-)) - g_k(t, y(t_k^-))\|,$$

and for each $t \in J_k$; $k = 0, \ldots, m$, we have

$$\|(Ny_n)(t) - (Nu)(t)\| \leq \|R(t)\|_{B(E)}\|g(y_n) - g(y)\|$$
$$+ \int_{s_k}^{t} \|R(t-s)\|_{B(E)}\|f(s, y_n(s)) - f(s, y(s))\|ds$$
$$\leq M\|g(y_n) - g(y)\|$$
$$+ M\int_{s_k}^{T} \|f(s, y_n(s)) - f(s, y(s))\|ds.$$

Since $y_n \to y$ as $n \to \infty$ and f, g, g_k are continuous, the Lebesgue dominated convergence theorem implies that

$$\|N(y_n) - N(y)\|_{\mathcal{PC}} \to 0 \text{ as } n \to \infty.$$

Step 2. $N(B_\rho)$ *is bounded and equicontinuous.*
Since $N(B_\rho) \subset B_\rho$ and B_ρ is bounded, then $N(B_\rho)$ is bounded.

Next, let $t, \tau \in J_k$, $\tau < t$ and let $y \in B_\rho$. Thus, we have

$$\|(Nu)(t) - (Nu)(\tau)\| \leq \|R(t) - R(\tau)\|_{B(E)}(\|y_0\| + \|g(y)\|)$$
$$+ \int_0^\tau \|R(t-s) - R(\tau-s)\|_{B(E)}\|f(s, y(s))\|ds$$
$$+ \int_\tau^t \|R(t-s)\|_{B(E)}\|f(s, y(s))\|ds.$$

Hence, we get

$$\|(Nu)(t) - (Nu)(\tau)\| \leq (\|y_0\| + q^*(1+\rho))\|R(t) - R(\tau)\|_{B(E)}$$
$$+ p^*(1+\rho)\int_0^\tau \|R(t-s) - R(\tau-s)\|_{B(E)}ds.$$

As $\tau \longrightarrow t$, the right-hand side of the above inequality tends to zero.

Step 3. *The implication (2.2) holds.*
Now let V be a subset of B_ρ such that $V \subset \overline{N(V)} \cup \{0\}$. V is bounded and equicontinuous and therefore the function $t \to v(t) = \alpha(V(t))$ is continuous on J. By (8.5.3) and the properties of the measure α, for each $t \in J_k$, we have

$$v(t) \leq \alpha((NV)(t) \cup \{0\})$$
$$\leq \alpha((NV)(t))$$
$$\leq l^*\|v\|_\infty$$
$$\leq \ell\|v\|_\infty.$$

Next, for each $t \in J_k$, we have

$$v(t) \leq \alpha((NV)(t) \cup \{0\})$$
$$\leq \alpha((NV)(t))$$
$$\leq \|R(t)\|_{B(E)}q^* \sup_{t \in J} \alpha(V(t)) + \int_0^t \|R(t-s)\|_{B(E)}p(s)\alpha(V(s))ds$$
$$\leq Mq^*\|v\|_\infty + Mp^* \int_0^t v(s)ds$$
$$\leq M(q^* + Tp^*)\|v\|_\infty$$
$$\leq \ell\|v\|_\infty.$$

Thus, for each $t \in J$, we get

$$v(t) \leq \ell\|v\|_\infty.$$

Hence

$$\|v\|_\infty \leq \ell \|v\|_\infty$$

From (8.8), we get $\|v\|_\infty = 0$, that is, $v(t) = \alpha(V(t)) = 0$, for each $t \in J$, and then $V(t)$ is relatively compact in E. In view of the Ascoli-Arzelà theorem, V is relatively compact in B_ρ. Applying now Theorem 2.17, we conclude that N has a fixed point which is a mild solution of problem (8.3).

8.2.3 *Examples*

Let

$$H := L^2([0, \pi]) = \left\{ y : [0, \pi] \to \mathbb{R} : \int_0^\pi |y(x)|^2 dx < \infty \right\},$$

be the Hilbert space with the scalar product $< y, v > = \int_0^\pi y(x)v(x)dx$. It is known that H is a Banach space with the norm

$$\|y\|_2 = \left(\int_0^\pi |y(x)|^2 dx \right)^{\frac{1}{2}}.$$

Example 1. Consider the following problem of impulsive integro-differential equations

$$\begin{cases} \frac{\partial}{\partial t} z(t, x) = \frac{\partial^2}{\partial x^2} z(t, x) + Q(t, z(t, x)) \\ \quad + \int_0^t b(t - s) \frac{\partial^2}{\partial x^2} z(s, x) ds; \ t \in [0, 1] \cup (1, 2], \ x \in [0, \pi], \\ \\ z(1^+, x) = z(1^-, x) + L_1(z(1^-, x)); \ x \in [0, \pi], \\ \\ z(t, 0) = z(t, \pi) = 0; \ t \in [0, 1] \cup (1, 2], \\ \\ z(0, x) + g(z) = 1 + x^2; \ x \in [0, \pi], \ z \in PC, \end{cases} \quad (8.11)$$

where $t \in J = [0, 2]$, $\Gamma C := PC([0, 2], H)$,

$$Q(t, z(t, x)) = \frac{ct^2}{1 + \|z\|_2} \left(e^{-7} + \frac{1}{e^{t+x+5}} \right) (1 + z(t, x)); \ t \in [0, 1] \cup (1, 2],$$

$$L_1(z(1^-), x) = \frac{z(1^-, x)}{3e^4(1 + \|z(1^-, x)\|_2)},$$

and

$$g(z) = \int_0^{\pi} K(x,y) \frac{e^{-y}}{1 + \|z\|_{PC}} dy,$$

with $\int_0^{\pi} \int_0^{\pi} K^2(x,y) dx dy < \infty.$

We define the strongly elliptic operator $A : D(A) \subset H \to H$ by:

$$Ay = \mathcal{A}(x, D)y = \sum_{|\alpha| \le 2m} a_\alpha(x) D^\alpha y,$$

where $a_\alpha \in C^{2m}([0, \pi])$, and $D(A) = H^{2m}([0, \pi]) \cap H_0^m([0, \pi])$.

It is well known (see [11]) that A generates a uniformly continuous semigroup $T(t);\ t \ge 0$ in the Hilbert space H.

For $x \in [0, \pi]$, we have

$$y(t)(x) = z(t, x); \quad t \in [0, 1] \cup (1, 2],$$

$$f(t, y(t))(x) = Q(t, z(t, x)); \quad t \in [0, 1] \cup (1, 2],$$

$$\Upsilon(t) = b(t)A$$

$$y_0(x) = 1 + x^2;\ x \in [0, \pi].$$

Thus, under the above definitions of f, y_0 and A, the system (8.11) can be represented by the problem (8.2). Furthermore, more appropriate conditions on Q ensure the hypotheses $(8.1.1), (8.1.2), (8.3.1)$–$(8.3.4)$. Consequently, Theorem 8.2 implies that the problem (8.11) has at least one mild solution on $[0, 2]$.

Example 2. Consider now the following problem of impulsive integro-differential equations

$$\begin{cases} \frac{\partial}{\partial t} z(t, x) = \frac{\partial^2}{\partial x^2} z(t, x) + Q(t, z(t, x)) \\ \quad + \int_0^t b(t-s) \frac{\partial^2}{\partial x^2} z(s, x) ds;\ t \in [0, 1] \cup (2, 3],\ x \in [0, \pi], \\ \\ z(t, x) = g_1(t, z(1^-, x));\ t \in (1, 2],\ x \in [0, \pi], \\ \\ z(t, 0) = z(t, \pi) = 0;\ t \in [0, 1] \cup (2, 3], \\ \\ z(0, x) + g(z) = 1 + e^x;\ x \in [0, \pi], \\ z(2, x) + g(z) = 2 + e^x;\ x \in [0, \pi], \end{cases} \qquad (8.12)$$

where $t \in [0, 3]$, $\mathcal{PC} := \mathcal{PC}([0, 3], H)$,

$$Q(t, z(t, x)) = \frac{ct^2}{1 + \|z\|_2} \left(e^{-7} + \frac{1}{e^{t+x+5}} \right) (1 + z(t, x)); \ t \in [0, 1] \cup (2, 3],$$

$$g_1(t, z(1^-, x)) = \frac{z(1^-, x)}{(3e^4)(1 + \|z(1^-, x)\|_2)}; \ t \in (1, 2], \ x \in [0, \pi],$$

and

$$g(z) = \int_0^\pi K(x, y) \frac{e^{-y}}{1 + \|z\|_{\mathcal{PC}}} dy,$$

with $\int_0^\pi \int_0^\pi K^2(x, y) dx dy < \infty$.

Again, as the above example; simple computations show that all conditions of Theorem 8.3 are satisfied. It follows that the problem (8.12) has at least one mild solution on $[0, 3]$.

8.3 Existence, Controllability and Attractivity Results on Semi-Infinite Intervals for Integro-Differential Equations with Non-Instantaneous Impulses

Motivated by the above mentioned works, we investigate the existence and attractivity of mild solutions to the following impulsive integrodifferential equations using resolvent operators:

$$\begin{cases} y'(t) = Ay(t) + f\left(t, y(t), (Hy)(t)\right) \\ \qquad + \int_0^t B(t - s) y(s) ds; \ \text{if } t \in J_k; k = 0, 1, \ldots, \\ y(t) = g_k\left(t, y\left(t_k^-\right)\right); \text{if } t \in \tilde{J}_k; k = 1, 2, \ldots, \\ y(0) = y_0, \end{cases} \tag{8.13}$$

where $J_0 = [0, t_1]$, $J_k := (s_k, t_{k+1}]$ and $\tilde{J}_k = (t_k, s_k]$ with $0 = s_0 < t_1 \le s_1 \le t_2 < \ldots < s_{m-1} \le t_m \le s_m \le t_{m+1} \le \ldots \le +\infty$, $A : D(A) \subset E \to E$ is the infinitesimal generator of a strongly continuous semigroup $\{T(t)\}_{t\ge0}$, $B(t)$ is a closed linear operator with domain $D(A) \subset D(B(t))$, the operator H is defined by

$$(Hy)(t) = \int_0^a h(t, s, y(s)) ds,$$

for $a > 0$, $D_h = \{(t, s) \in \mathbb{R}^2 \ ; \ 0 \le s \le t \le a\}$, and $h : D_h \times E \to E$. The nonlinear term $f : J_k \times E \times E \to E$; $k = 0, 1, \ldots$, $g_k : \tilde{J}_k \times E \to E$;

$k = 1, 2, \ldots,$ are a given functions, and $(E, \| \cdot \|)$ is a Banach space. We base their arguments on the fixed point theory and the concept of measure of non-compactness with the help of the resolvent operator. Next, we will investigate the existence and controllability of our problem.

We would like to point out that our work may be viewed as a continuation of the papers [28, 184]. Indeed, unlike the problem in [28], we have added non-instantaneous impulses to our problem, and we investigate our problem on an unbounded domain, as opposed to the problem in [184]. Finally, we investigate our problem's attractivity in a Banach space. The utilization of the assumed hypotheses and the fixed point theorem are novel in the framework of the investigated problem.

8.3.1 *Some Necessary Results*

We consider the following Cauchy problem:

$$\begin{cases} y'(t) = Ay(t) + \displaystyle\int_0^t B(t-s)y(s)ds; & \text{for } t \geq 0, \\ y(0) = y_0 \in E. \end{cases} \tag{8.14}$$

The existence and properties of a resolvent operator have been discussed in [295]. In what follows, we suppose the following assumptions:

(R1) A is the infinitesimal generator of a uniformly continuous semigroup $\{T(t)\}_{t>0}$,

(R2) For all $t \geq 0$, $B(t)$ is closed linear operator from $D(A)$ to E and $B(t) \in B(D(A), E)$. For any $y \in D(A)$, the map $t \to B(t)y$ is bounded, differentiable and the derivative $t \to B'(t)y$ is bounded uniformly continuous on $\mathbb{R}^+$.

Theorem 8.4 ([295]). *Assume that* (R1)–(R2) *hold, then there exists a unique resolvent operator for the Cauchy problem* (8.14).

Consider the space

$$PC(\mathbb{R}^+, E) = \Big\{ y : \mathbb{R}^+ \to E \ : \ y|_{\bar{J}_k} = g_k; \ k = 1, 2, \ldots, y|_{J_k} \ ; \ k = 0, 1, \ldots,$$

$$\text{are continuous, } y\left(s_k^-\right), \ y\left(s_k^+\right), \ y\left(t_k^-\right) \text{ and } y\left(t_k^+\right)$$

$$\text{exist with } y\left(t_k^-\right) = y\left(t_k\right) \Big\},$$

8.3.2 Existence and Attractivity Results

Consider the space

$$X = BPC(\mathbb{R}^+, E)$$
$$= \{y \in PC(\mathbb{R}^+, E) : y \text{ is bounded on } \mathbb{R}^+\},$$

with respect to the norm

$$\|y\|_{BPC} = \sup_{t \in \mathbb{R}^+} \{\|y(t)\|\}.$$

8.3.2.1 *Existence of mild solutions*

Let us recollect the following particular measure of non-compactness that derives from [223], and will be utilized in our main results in order to establish a measure of non-compactness in the space $BPC(\mathbb{R}^+, E)$. Let us fix a non-empty bounded subset H in the space $BPC(\mathbb{R}^+, E)$, for $v \in H$, $T > 0$, $\epsilon > 0$ *and* $\kappa, \tau \in [0, T]$, such that $|\kappa - \tau| \leq \epsilon$. We denote $\omega^T(v, \epsilon)$ the modulus of continuity of the function v on the interval $[0, T]$, namely,

$$\omega^T(v, \epsilon) = \sup\{\|v(\kappa) - v(\tau)\| \; ; \; \kappa, \tau \in [0, T]\},$$
$$\omega^T(H, \epsilon) = \sup\{\omega^T(v, \epsilon) \; ; \; v \in H\},$$
$$\omega_0^T(H) = \lim_{\epsilon \to 0}\{\omega^T(H, \epsilon)\},$$
$$\omega_0(H) = \lim_{T \to +\infty} \omega_0^T(H).$$

If t is fixed from $\mathbb{R}^+$, let us denote $H(t) = \{v(t) \in E \; ; \; v \in H\}$ and

$$d^\Delta(H(t)) = \text{diam}\,(H(t)) = \sup\{\|u(t) - v(t)\| \; ; \; u, \; v \in H\}.$$

Finally, consider the function y_{BPC} defined on the family of subset of $BPC(\mathbb{R}^+, E)$ by the formula

$$y_{BPC}(H) = \omega_0(H) + \lim_{t \to \infty} \sup_{t \in \mathbb{R}^+} d^\Delta(H(t)).$$

It can be shown similar to [223] that the function y_{BPC} is a sublinear measure of non-compactness on the space $BPC(\mathbb{R}^+, E)$.

Definition 8.4. A function $y \in BPC(\mathbb{R}^+, E)$ is called a mild solution of problem (8.13) if it satisfies

$$y(t) = \begin{cases} R(t)y_0 + \displaystyle\int_0^t R(t-s)f(s, y(s), (Hy)(s))ds; & \text{if } t \in J_0, \\[4mm] R(t - s_k)\left[g_k(s_k, y(t_k^-))\right] + \displaystyle\int_{s_k}^t R(t-s)f(s, y(s), (Hy)(s))ds; & t \in J_k, \\[4mm] g_k(t, y(t_k^-)), & \text{if } t \in \tilde{J}_k, \end{cases}$$

where $k = 1, 2, \ldots$

The hypotheses:

(8.8.1) $f : \mathbb{R}^+ \times E \times E \to E$ is a Carathéodory function and there exist a function $\zeta \in L^1(\mathbb{R}^+, \mathbb{R}^+)$ and a continuous non-decreasing function $\varphi : \mathbb{R}^+ \to (0, +\infty)$, such that:

$$\|f(t, u, \bar{u})\| \le \zeta(t)\varphi(\|u\| + \|\bar{u}\|), \quad \text{for } u, \ \bar{u} \in E.$$

(8.8.2) The function $h : D_h \times E \to E$ is continuous and there exists $c_h > 0$, such that

$$\|h(t, s, u) - h(t, s, \bar{u})\| \le c_h\|u - \bar{u}\|, \quad \text{for each } (t, s) \in D_h \text{ and } u, \ \bar{u} \in E,$$

with

$$h^* = \sup\{\|h(t, s, 0)\| \ , \ (t, s) \in D_h\} < \infty.$$

(8.8.3) $g_k : \tilde{J}_k \times E \to E$ are continuous and there exists $\beta_{g_k} \in (0, 1)$; $k \in \mathbb{N}$, such that

$$\|g_k(t, u) - g_k(t, v)\| \le \beta_{g_k}\|u - v\|, \quad \text{for all } u, v \in E, \ k = 1, 2, \ldots$$

and

$$\max_{k \in \mathbb{N}}\{\beta_{g_k}\} = \beta_{g_k}^*.$$

(8.8.4) Assume that there exist $M_R \ge 1$ and $\beta \ge 0$, such that

$$\|R(t)\|_{B(E)} \le M_R e^{-\beta t},$$

and

$$\lim_{t \to +\infty} \sup_{t \in \mathbb{R}^+} \int_0^t e^{-\beta(t-s)}\zeta(s)ds = 0.$$

Theorem 8.5. *Assume that the conditions* (8.8.1)–(8.8.4) *are satisfied. If*

$$M_R \beta_{g_k}^* < 1,$$

then the system (8.13) *has at least one mild solution.*

Proof. Transform the problem (8.13) into a fixed point problem. Consider the operator $\aleph : BPC(\mathbb{R}^+, E) \to BPC(\mathbb{R}^+, E)$ defined by:

$$\aleph y(t) = \begin{cases} R(t)y_0 + \displaystyle\int_0^t R(t - s)f(s, y(s), (Hy)(s))ds; & \text{if } t \in J_0, \\[4mm] R(t - s_k)\left[g_k(s_k, y(t_k^-))\right] \\ \quad + \displaystyle\int_{s_k}^t R(t - s)f(s, y(s), (Hy)(s))ds; & t \in J_k, \\[4mm] g_k(t, y(t_k^-)); & t \in \tilde{J}_k, \end{cases}$$

where $k = 1, 2, \ldots$. Obviously, the fixed points of the operator $\aleph$ are mild solutions of the problem (8.13). Let $D_\rho = \{y \in BPC(\mathbb{R}^+, E) : \|y\| \leq \rho\}$, with

$$\max\left\{ M_R(\|y_0\| + \varphi(K_\rho^*)\|\zeta\|_{L^1}), \frac{M_R(g_0 + \varphi(K_\rho^*)\|\zeta\|_{L^1})}{1 - M_R\beta_{g_k}^*} \right\} \leq \rho,$$

where

$$K_\rho^* = \big((c_h + 1)\rho + ah^*\big),$$

the set D_ρ is bounded, closed and convex.

Step 1: $\aleph(D_\rho) \subset D_\rho$.

- *Case* 1: for $t \in J_0$.
  For each $y \in D_\rho$ and by (8.8.1), we have

$$\|\aleph y(t)\| \leq M_R\|y_0\| + M_R \int_0^t \varphi(\|y\| + \|Hy\|)\zeta(s)ds$$
$$\leq M_R\|y_0\| + M_R\varphi((c_h + 1)\rho + ah^*)\|\zeta\|_{L^1},$$

  then

$$\|\aleph y\|_X \leq M_R\left[\|y_0\| + \varphi((c_h + 1)\rho + ah^*)\|\zeta\|_{L^1}\right].$$

- *Case* 2: for $t \in J_k$.
  For each $y \in D_\rho$ by (8.8.1), (8.8.2) and (8.8.3), we have

$$\|g_k(., u(.))\| \leq \beta_{g_k}^*\|u(\cdot)\| + g_0,$$

  thus

$$\|\aleph y\|_X \leq M_R\left[\beta_{g_k}^*\rho + g_0 + \varphi((c_h + 1)\rho + ah^*)\|\zeta\|_{L^1}\right],$$
$$\leq \rho.$$

- *Case* 3: for $t \in \tilde{J}_k$.
  For each $y \in D_\rho$, we have

$$\|\aleph y(t)\| \leq \beta_{g_k}^*\rho + g_0.$$

Hence

$$\|\aleph y\|_X \leq \rho.$$

Step 2: $\aleph$ is continuous.
Let $(y_n)_{n\in\mathbb{N}}$ be a sequence such that $y_n \to y_*$ in E, then

- *Case* 1: for $t \in J_0$. We have

$$\|(\aleph y_n)(t) - (\aleph y_*)(t)\|$$

$$\leq M_R \int_0^t \|f(s, y_n(s), Hy_n(s)) - f(s, y_*(s), Hy_*(s))\| ds.$$

By the continuity of h and f, we have

$$h(t, s, y_n(s)) \to h(t, s, y_*(s)) \quad as \quad n \to +\infty,$$

and

$$\|h(t, s, y_n(s)) - h(t, s, y_*(s))\| \leq c_h \|y_n - y_*\|.$$

By the Lebesgue dominated convergence theorem

$$\int_0^t h(t, s, y_n(s)) ds \to \int_0^t h(t, s, y_*(s)) ds, \quad as \quad n \to +\infty,$$

then by (8.8.1), we get

$$f(s, y_n(s), Hy_n(s)) \to f(s, y_*(s), Hy_*(s)), \quad as \quad n \to +\infty,$$

consequently

$$\|\aleph y_n - \aleph y_*\|_X \to 0, \quad as \quad n \to +\infty.$$

- *Case* 2: for $t \in J_k$. We have

$$\|(\aleph y_n)(t) - (\aleph y_*)(t)\|$$

$$\leq M_R \|g_k(s_k, y_n(t_k^-)) - g_k((s_k, y_*(t_k^-)))\|$$

$$+ M_R \int_{s_k}^t \|f(s, y_n(s), Hy_n(s)) - f(s, y_*(s), Hy_*(s))\| ds.$$

Similar to Case 1, by the continuity of h, f and g_k, we get

$$\|\aleph y_n - \aleph y_*\|_X \to 0, \quad as \quad n \to +\infty.$$

- *Case* 3: for $t \in \tilde{J}_k$. We have

$$\|(\aleph y_n)(t) - (\aleph y_*)(t)\| \leq \|g_k(t, y_n(t_k^-)) - g_k((t, y_*(t_k^-)))\|.$$

By the continuity of g_k, we obtain

$$\|\aleph y_n - \aleph y_*\|_X \to 0, \quad as \quad n \to +\infty.$$

Thus, $\aleph$ is continuous.

Step 3: In the sequel, we consider the sequence of sets $\{\Omega_n\}_{n=0}^{+\infty}$ defined by induction as follows:

$$\Omega_0 = D_\rho, \quad \Omega_{n+1} = conv(\aleph(\Omega_n)); \text{ for } n = 0, 1, 2 \ldots, \quad \Omega_\infty = \bigcap_{n=0}^{+\infty} \Omega_n,$$

this sequence is non-decreasing, i.e. $\Omega_{n+1} \subset \Omega_n$ for each $n \in \mathbb{N}$.

Now we will prove that $\lim_{n \to +\infty} y_{BPC}(\Omega_n) = 0$, so for $T > 0$ and $k_0 \in \mathbb{N}$, with $T \geq t_{k_0}$ and $y \in \Pi$, we have

- *Case* 1: for $\kappa, \tau \in J_0$.

$$\|\aleph y(\kappa) - \aleph y(\tau)\|$$
$$\leq \|R(\kappa) - R(\tau)\|\|y_0\|$$
$$+ \int_0^\kappa \|R(\kappa - s) - R(\tau - s)\|\zeta(s)\varphi(\|y\|_{BPC} + \|Hy\|_{BPC})ds$$
$$+ \int_\kappa^\tau \|R(\tau - s)\|\zeta(s)\varphi(K_\rho^*)ds,$$
$$\leq \|R(\kappa) - R(\tau)\|\|y_0\|$$
$$+ \varphi(K_\rho^*)\int_0^\kappa \|R(\kappa - s) - R(\tau - s)\|\zeta(s)ds$$
$$+ M_R\varphi(K_\rho^*)\int_\kappa^\tau \zeta(s)ds.$$

By the strong continuity of $R(t)$ and (8.8.1), we get

$$\|\aleph y(\kappa) - \aleph y(\tau)\| \to 0, \ as \ \kappa \to \tau.$$

- *Case* 2: for $\kappa, \tau \in J_k$.

$$\|\aleph y(\kappa) - \aleph y(\tau)\| \leq \|R(\kappa - s_k) - R(\tau - s_k)\|\|g(s_k, y(t_k^-))\|$$
$$+ \int_{s_k}^\kappa \|R(\kappa - s) - R(\tau - s)\|\zeta(s)\varphi(K_\rho^*)ds$$
$$+ \int_\kappa^\tau \|R(\tau - s)\|\zeta(s)\varphi(K_\rho^*)ds$$
$$\leq \|R(\kappa - s_k) - R(\tau - s_k)\|(\beta_{g_k}^*\rho + g_0)$$
$$+ \varphi(K_\rho^*)\int_{s_k}^\kappa \|R(\kappa - s) - R(\tau - s)\|\zeta(s)ds$$
$$+ M_R\varphi(K_\rho^*)\int_\kappa^\tau \zeta(s)ds.$$

Since $R(t)$ is norm continuous and by (8.8.1), we obtain

$$\|\aleph y(\kappa) - \aleph y(\tau)\| \to 0, \ as \ \kappa \to \tau.$$

- *Case* 3: for $\kappa, \tau \in \tilde{J}_k$. We have

$$\|\aleph y(\kappa) - \aleph y(\tau)\| = \|g_k(\kappa, y(t_k^-)) - g_k(\tau, y(t_k^-))\|.$$

From (8.8.3), the set $\{g_k(t, y(t_k^-))\}_{k=1}^{k_0}$ is equicontinuous, then

$$\|\aleph y(\kappa) - \aleph y(\tau)\| \to 0, \ as \ \kappa \to \tau.$$

Finally, the set $\aleph(\Omega_{n+1})$ is equicontinuous, then $\omega_0(\aleph(\Omega_{n+1})) = 0$.

Now for $u, y \in \Omega_n$ and $t \in [0, T]$, we have three cases:

- *Case* 1: for $t \in J_0$.

$$\|(\aleph y)(t) - (\aleph u)(t)\|$$

$$\leq M_R \int_0^t e^{-\beta(t-s)} \|f(s, y(s), Hy(s)) - f(s, u(s), Hu(s))\| ds$$

$$\leq 2M_R \varphi(K_\rho^*) \int_0^t e^{-\beta(t-s)} \zeta(s) ds.$$

Then

$$\sup_{t \in J_0} d^\Delta(\aleph\{\Omega_n(t)\}) \leq 2M_R \varphi(K_\rho^*) \sup_{t \in \mathbb{R}^+} \int_0^t e^{-\beta(t-s)} \zeta(s) ds,$$

when $t \to +\infty$ and by (8.8.4), we get

$$\lim_{n \to +\infty} y_{BPC}(\Omega_n) = 0.$$

- *Case* 2: for $t \in J_k$. We have

$$\|(\aleph y)(t) - (\aleph u)(t)\|$$

$$\leq M_R \|g_k(s_k, u(t_k^-)) - g_k((s_k, y(t_k^-)))\|$$

$$+ M_R \int_{s_k}^t e^{-\beta(t-s)} \|f(s, u(s), Hu(s)) - f(s, y(s), Hy(s))\| ds$$

$$\leq M_R \beta_{g_k}^* \|u(s_k) - y(s_k)\| + 2M_R \varphi(K_\rho^*) \int_{s_k}^t e^{-\beta(t-s)} \zeta(s) ds,$$

when $t \to +\infty$ and by (8.8.4), we get

$$y_{BPC}(\aleph(\Omega_n)) \leq (M_R \beta_{g_k}^*) y_{BPC}(\Omega_n),$$

so

$$y_{BPC}(\Omega_{n+1}) \leq (M_R \beta_{g_k}^*) y_{BPC}(\Omega_n).$$

- *Case* 3: for $t \in \tilde{J}_k$. We have

$$\|(\aleph y)(t) - (\aleph u)(t)\| \leq \|g_k(t, u(t_k^-)) - g_k((t, y(t_k^-)))\|$$

$$\leq \beta_{g_k}^* \|u(t) - y(t)\|.$$

then

$$y_{BPC}(\aleph(\Omega_n)) \leq \beta_{g_k}^* \, y_{BPC}(\Omega_n),$$

Therefore

$$y_{BPC}(\Omega_{n+1}) \leq \beta_{g_k}^* y_{BPC}(\Omega_n).$$

By the method of mathematical induction, we can get

$$y_{BPC}(\Omega_{n+1}) \leq (M_R \beta_{g_k}^*)^{n+1} y_{BPC}(\Omega_0), \quad \text{for all } t \in J_k, \ k = 1, 2, \ldots$$

$$y_{BPC}(\Omega_{n+1}) \leq (\beta_{g_k}^*)^{n+1} y_{BPC}(\Omega_0), \quad \text{for all } t \in \tilde{J}_k, \ k = 0, 1, \ldots$$

Then, we obtain

$$\lim_{n \to +\infty} y_{BPC}(\Omega_n) = 0.$$

Taking into account Lemma 2.7, we infer that $\Omega_\infty = \cap_{n=0}^{+\infty} \Omega_n$ is non-empty, convex and compact. As a consequence of these three steps together with Theorem 2.7, we can conclude that $\aleph : \Omega_\infty \to \Omega_\infty$, has at least one fixed point, which is a mild solution of problem (8.13). □

8.3.2.2 *Attractivity of solutions*

In this section we study the local attractivity of solutions for the problem (8.13).

Firstly, we introduce the following concept of attractivity of solutions.

Definition 8.5 ([27]). We say that the solutions of (8.13) are locally attractive if there exists a closed ball $B(z^*, \gamma)$ in the space X for some $z^* \in X$ such that for arbitrary solutions z and $\tilde{z}$ of (8.13) belonging to $B(z^*, \gamma)$, we have that

$$\lim_{t \to +\infty} (z(t) - \tilde{z}(t)) = 0.$$

When the last limit is uniform with respect to $B(z^*, \gamma)$, solutions of problem (8.13) are said to be uniformly locally attractive (or equivalently that solutions of (8.13) are locally asymptotically stable).

Lets z^* be a solution of (8.13), $B_\gamma = B(z^*, \gamma)$ the closed ball in X and a constant $K_\gamma^* = ((c_h + 1)\gamma + ah^*)$, depends on a positive constant γ.

Theorem 8.6. *Suppose that hypotheses (8.8.1)–(8.8.4) hold, with*

$$\max \left\{ 2M_R(\|y_0\| + \varphi(K_\gamma^*)\|\zeta\|_{L^1}), \ 2M_R(\beta_{g_k}^* \gamma + g_0 + \varphi(K_\gamma^*)\|\zeta\|_{L^1}) \right\} \leq \gamma,$$

and

$$M_R \max\{\beta_{g_k}^*\} < \frac{1}{2},$$

then, the solutions of problem (8.13) are uniformly locally attractive.

Proof. For $z \in B(z^*, \gamma)$ by (8.8.1) and (8.8.3), we get

- *Case 1:* for $t \in J_0$. We have

$$\|(\aleph z)(t) - z^*(t)\| = \|(\aleph z)(t) - (\aleph z^*)(t)\|$$

$$\leq M_R \int_0^t \|f(s, z(s), Hz(s)) - f(s, z^*(s), Hz^*(s))\| ds$$

$$\leq 2M_R(\|y_0\| + \varphi(K_\gamma^*)\|\zeta\|_{L^1})$$

$$\leq \gamma.$$

This proves that $\aleph(B_\gamma) \subset B_\gamma$.
So, for each $z, \tilde{z} \in B(z^*, \gamma)$ solutions of problem (8.13) and $t \in J_0$, we have

$$\|z(t) - \tilde{z}(t)\| = \|(\aleph z)(t) - (\aleph \tilde{z})(t)\|$$

$$\leq 2M_R\varphi(K_\gamma^*) \sup_{t \in J_0} \int_0^t e^{-\beta(t-s)}\zeta(s)ds$$

$$\leq 2M_R\varphi(K_\gamma^*) \sup_{t \in \mathbb{R}^+} \int_0^t e^{-\beta(t-s)}\zeta(s)ds,$$

by (8.8.4), we conclude that

$$\|z(t) - \tilde{z}(t)\| \to 0, \quad as \ t \to +\infty.$$

- *Case 2:* for $t \in J_k$. we have

$$\|(\aleph z)(t) - z^*(t)\|$$

$$\leq M_R\|g_k(s_k, z(t_k^-)) - g_k((s_k, z^*(t_k^-)))\|$$

$$+ M_R \int_{s_k}^t \|f(s, z(s), Hz(s)) - f(s, z^*(s), Hz^*(s))\| ds$$

$$\leq 2M_R(\beta_{g_k}^*\gamma + g_0 + \varphi(K_\gamma^*)\|\zeta\|_{L^1})$$

$$\leq \gamma,$$

therefore, $\aleph(B_\gamma) \subset B_\gamma$.
So, for each $z, \tilde{z} \in B(z^*, \gamma)$ solutions of problem (8.13) and $t \in J_k$, we have

$$\|z(t) - \tilde{z}(t)\| = \|(\aleph z)(t) - (\aleph \tilde{z})(t)\|$$

$$\leq M_R e^{-\beta(t-s_k)}\|g_k(s_k, z(t_k^-)) - g_k((s_k, \tilde{z}(t_k^-)))\|$$

$$+ 2M_R\varphi(K_\gamma^*) \sup_{t \in J_k} \int_{s_k}^t e^{-\beta(t-s)}\zeta(s)ds$$

$$\leq M_R e^{-\beta(t-s_k)}\beta_{g_k}^*\|z(s_k) - \tilde{z}(s_k)\|$$

$$+ 2M_R\varphi(K_\gamma^*) \sup_{t \in \mathbb{R}^+} \int_0^t e^{-\beta(t-s)}\zeta(s)ds,$$

Then, using (8.8.4), we obtain

$$\|z(t) - \widetilde{z}(t)\| \to 0, \quad as \ t \to +\infty.$$

- *Case* 3: for $t \in \tilde{J}_k$. we have

$$
\begin{aligned}
\|(\aleph z)(t) - z^*(t)\| &= \|(\aleph z)(t) - (\aleph z^*)(t)\| \\
&\leq \|g(t, z(t_k^-)) - g((t, z^*(t_k^-)))\| \\
&\leq 2\beta_{g_k}^* \gamma \\
&\leq \gamma.
\end{aligned}
$$

Thus, $\aleph(B_\gamma) \subset B_\gamma$.

So, for each $z, \widetilde{z} \in B(z^*, \gamma)$ solutions of problem (8.13) and $t \in J_k$, we have

$$
\begin{aligned}
\|z(t) - \widetilde{z}(t)\| &= \|(\aleph z)(t) - (\aleph \widetilde{z})(t)\| \\
&\leq \|g(t, z(t_k^-)) - g((t, \widetilde{z}(t_k^-)))\| \\
&\leq \beta_{g_k}^* \|z(t) - \widetilde{z}(t)\|,
\end{aligned}
$$

then

$$(1 - \beta_{g_k}^*)\|z(t) - \widetilde{z}(t)\| \leq 0,$$

hence

$$\|z(t) - \widetilde{z}(t)\| = 0.$$

As a result, the solutions of the problem (8.13) are uniformly locally attractive. $\qquad\square$

8.3.3 *Existence and Controllability Results*

Let the space $PC(\mathbb{R}^+, E)$ be endowed with the family of seminorms

$$\|x\|_n = sup\{\|x(t)\| \ : \ t \in [0, t_n]\}, \ n = 1, 2, \ldots$$

Let $F = C(I, E)$ be the Fréchet space of continuous functions $\Im$ from $\mathbb{R}_+$ into E, with

$$\|\Im\|_n = \sup_{t \in \tilde{J}_n} \|\Im(t)\|, \ \tilde{J}_n := [0, n], \ n \in \mathbb{N},$$

and the distance

$$d(y, \Im) = \sum_{n=1}^{\infty} \frac{2^{-n}\|y - \Im\|_n}{1 + \|y - \Im\|_n}; \quad y, \Im \in C(\mathbb{R}_+, E).$$

Example 8.1. For $\Omega \in \mathcal{M}_{\bar{E}}$, $x \in \Omega$, $n \in \mathbb{N}$ *and* $\epsilon > 0$, let us denote by $\omega^n(x, \epsilon)$ for $n \in \mathbf{N}$, the modulus of continuity of the function x on the interval $\tilde{J}_n$, that is,

$$\omega^n(x, \epsilon) = \sup\{|x(t) - x(s)| \; ; \; t, s \in \tilde{J}_n \; |t - s| < \epsilon\}.$$

Further, let us put

$$\omega^n(\Omega, \epsilon) = \sup\{\omega^n(x, \epsilon) \; ; \; x \in \Omega\}, \quad \omega_0^n(\Omega) = \lim_{\epsilon \to 0^+} \omega^n(\Omega, \epsilon)$$

and

$$\mu_n(\Omega) = \omega_0^n(\Omega) + \sup_{t \in \tilde{J}_n} \alpha\big(\Omega(t)\big).$$

The family of mappings $\{\mu_n\}_{n \in \mathbb{N}}$ where $\mu_n : \mathcal{M}_{\bar{E}} \to \mathbb{R}^+$, satisfies the conditions (a)–(d) of Definition 2.6, then the family of maps $\{\mu_n\}_{n \in \mathbb{N}}$ defined above is a family of measures of non-compactness in the Fréchet space $\bar{E}$.

Definition 8.6 ([395]). A non-empty subset $\Omega \subset \bar{E}$ is bounded if for $n \in \mathbb{N}$, there exists $\mathcal{M}_n > 0$, such that

$$\|y\|_n \le \mathcal{M}_n, \text{ for each } y \in \Omega.$$

8.3.3.1 *Existence of mild solutions*

Definition 8.7. A function $y \in PC(\mathbb{R}^+, E)$ is called a mild solution of problem (8.13) if it satisfies

$$y(t) = \begin{cases} R(t)y_0 + \displaystyle\int_0^t R(t-s)f(s, y(s), (Hy)(s))ds; & \text{if } t \in J_0, \\ R(t - s_k)\big[g_k(s_k, y(t_k^-))\big] + \displaystyle\int_{s_k}^t R(t-s)f(s, y(s), (Hy)(s))ds; t \in J_k, \\ g_k(t, y(t_k^-)); \; t \in \tilde{J}_k, \end{cases}$$

where $j = 1, 2, \ldots$

We will need to introduce the following hypotheses, which will be assumed later on:

(8.14.1) (i) $f : \mathbb{R}^+ \times E \times E \to E$ is a Carathéodory function and there exist a function
$p \in L^1(\mathbb{R}^+, \mathbb{R}^+)$ and a continuous non-decreasing function $\psi : \mathbb{R}^+ \to (0, +\infty)$, such that :

$$\|f(t, y, \bar{y})\| \le p(t)\psi(\|y\| + \|\bar{y}\|), \quad for \; y, \; \bar{y} \in E,$$

(*ii*) There exists a function $l_f \in L^1(\mathbb{R}^+, \mathbb{R}^+)$ such that for any bounded set $B \subset E$ and $t \in \mathbb{R}^+$, we have

$$\alpha(f(t, B, H(B))) \leq l_f(t)\alpha(B).$$

(8.14.2) The function $h : D_h \times E \times E \to E$ is continuous and there exists $c_1 > 0$ such that

$$\|h(t, s, y) - h(t, s, \bar{y})\| \leq c_1 \|y - \bar{y}\|, \quad for\ each\ (t, s) \in D_h\ and\ y,\ \bar{y} \in E,$$

with

$$h^* = \sup\{\|h(t, s, 0)\|\ ,\ (t, s) \in D_h\} < \infty.$$

(8.14.3) $g_k : \tilde{J}_k \times E \to E$ are continuous and there exist positive constants $L_{g_k}, k \in \mathbb{N}$ and $\tau > 1$ such that

$$\|g_k(., y) - g_k(., \Im)\| \leq \frac{L_{g_k}}{\tau}\|y - \Im\|, \quad for\ all\ y, \Im \in E,\ k = 1, 2, \ldots$$

(8.14.4) Assume that $(R1)$–$(R2)$ hold, and there exist $\mathcal{M}_R \geq 1$ and $b \geq 0$, such that

$$\|R(t)\|_{B(E)} \leq \mathcal{M}_R e^{-bt}.$$

Using methods from [396, 397], we can show that the example of family of measures of non-compactness in $PC\,(\mathbb{R}^+, E)$ is

$$\alpha_n(\Pi) = \max_{i=0,\ldots,m} \omega_0\,(\gamma_i^p, \Pi) + \sup_{t \in \tilde{J}_n} \left\{ e^{-\tau \tilde{\zeta}(t)} \alpha(\Pi(t)) \right\};\ p = 0, 1, 2,\ m = 0, 1, \ldots,$$

with γ_i^p is a partition of $\mathbb{R}^+$, in particular

$$\gamma_i^p = \begin{cases} J_0; & \text{if } p = 0,\ m = 0, \\ J_m; & \text{if } p = 1,\ m = 1, 2, \ldots, \\ \tilde{J}_m; & \text{if } p = 2,\ m = 1, 2, \ldots, \end{cases}$$

and $\tilde{\zeta}(t) = \int_0^t \zeta(s)ds,\ \zeta(t) = 4\mathcal{M}_R l(t),\ \tau > 1$, where $\Pi(t) = \{\pi(t) \in \bar{E}\ ;\ \pi \in \Pi\}, t \in \tilde{J}_n.$

Notice that if the set Π is equicontinuous, then $\omega_0\,(\gamma_i^p, \Pi) = 0.$

Theorem 8.7. *Assume that the conditions* (8.14.1)–(8.14.4) *are satisfied, then the system* (8.13) *has at least one mild solution.*

Proof. Transform the problem (8.13) into a fixed point problem. Consider the operator $\aleph : PC\,(\mathbb{R}^+, E) \to PC\,(\mathbb{R}^+, E)$ defined by :

$$\aleph y(t) = \begin{cases} R(t)y_0 + \displaystyle\int_0^t R(t-s)f(s,y(s),(Hy)(s))ds; & \text{if } t \in J_0, \\[2ex] R(t-s_k)\left[g_k(s_k, y(t_k^-))\right] \\[1ex] \quad + \displaystyle\int_{s_k}^t R(t-s)f(s,y(s),(Hy)(s))ds; & \text{if } t \in J_k, \\[2ex] g_k(t, y(t_k^-)); & \text{if } t \in \tilde{J}_k, \end{cases}$$

where $k = 1, 2, \ldots$. Clearly, the fixed points of the operator $\aleph$ are mild solutions of the problem (8.13), so we shall check that operator $\aleph$ satisfies all conditions of Darbo's fixed point theorem [255].

Let $D_{\delta_n} = \{y \in PC\,(\mathbb{R}^+, E)\,;\, \|y\|_n \leq \delta_n\}$, the set D_{δ_n} is bounded, closed and convex.

Step 1: $\aleph(D_{\delta_n}) \subset D_{\delta_n}$.

- **Case 1:** For $t \in J_0 \cap \tilde{J}_n$.
  For any $n \in \mathbb{N}$, $y \in D_{\delta_n}$, $t \in J_0 \cap \tilde{J}_n$ and by (8.14.1), we have

$$\|\aleph y(t)\| \leq \mathcal{M}_R\|y_0\| + \mathcal{M}_R\int_0^t \psi(\|y(s)\| + \|Hy(s)\|)p(s)ds$$
$$\leq \mathcal{M}_R\|y_0\| + \mathcal{M}_R\psi((c_1 + 1)\delta_n + ah^*)\|p\|_{L^1},$$

then

$$\|\aleph y\|_n \leq \mathcal{M}_R\left[\|y_0\| + \psi((c_1 + 1)\delta_n + ah^*)\|p\|_{L^1}\right].$$

- **Case 2:** For $t \in J_k \cap \tilde{J}_n$.
  For each $y \in D_{\delta_n}$, by (8.14.1), (8.14.2) *and* (8.14.3), we have

$$\|g_k(.,y(.))\| \leq \frac{L_{g_k}}{\tau}\|y(.)\| + g_0,$$

then

$$\|\aleph y\|_n \leq \mathcal{M}_R\left[\frac{L_{g_k}}{\tau}\delta_n + g_0 + \psi((c_1 + 1)\delta_n + ah^*)\|p\|_{L^1}\right].$$

- **Case 3:** For $t \in \tilde{J}_k \cap \tilde{J}_n$.
  For each $y \in D_{\delta_n}$ by (8.14.3), we have

$$\|\aleph y\|_n \leq \frac{L_{g_k}}{\tau}\delta_n + g_0,$$

we put

$$K^*_{\delta_n} = \left((c_1 + 1)\delta_n + ah^*\right)\|p\|_{L^1},$$

then

$$\|\aleph y\|_n \leq \delta_n,$$

provided that

$$\max\left\{\mathcal{M}_R(\|y_0\| + \psi(K^*_{\delta_n})), \frac{\mathcal{M}_R(g_0 + \psi(K^*_{\delta_n}))}{1 - \mathcal{M}_R\frac{L_{g_k}}{\tau}}\right\} \leq \delta,$$

and

$$\mathcal{M}_R L_{g_k} < \tau.$$

Step 2: $\aleph$ is continuous.

Let y_m be a sequence such that $y_m \to y_*$ in E, then

- **Case 1:** For $t \in J_0 \cap \tilde{J}_n$.
  We have

$$\|(\aleph y_m)(t) - (\aleph y_*)(t)\|$$
$$\leq \mathcal{M}_R \int_{s_k}^t \|f(s, y_m(s), Hy_m(s)) - f(s, y_*(s), Hy_*(s))\|ds.$$

By the continuity of h and f, we have

$$h(t, s, y_m(s)) \to h(t, s, y_*(s)) \quad as \ \ m \to +\infty,$$

and

$$\|h(t, s, y_m(s)) - h(t, s, y_*(s))\| \leq c_1\|y_m - y_*\|.$$

By the Lebegue dominated theorem,

$$\int_0^t h(t, s, y_m(s))ds \to \int_0^t h(t, s, y_*(s))ds, \quad as \ \ m \to +\infty,$$

then by (8.14.1), we get

$$f(s, y_m(s), Hy_m(s)) \to f(s, y_*(s), Hy_*(s)), \quad as \ \ m \to +\infty,$$

so

$$\|(\aleph y_m) - (\aleph y_*)\|_n \to 0, \quad as \ \ m \to +\infty.$$

- **Case 2:** For $t \in J_k \cap \tilde{J}_n$.
  We have

$$\|(\aleph y_m)(t) - (\aleph y_*)(t)\|$$
$$\leq \mathcal{M}_R \|g_k(s_k, y_m(t_k^-)) - g_k((s_k, y_*(t_k^-)))\|$$
$$+ \mathcal{M}_R \int_0^t \|f(s, y_m(s), Hy_m(s)) - f(s, y_*(s), Hy_*(s))\| ds.$$

  Similar to case 1, by the continuity of h, f and g_k, we get

$$\|(\aleph y_m) - (\aleph y_*)\|_n \to 0, \quad as \ \ m \to +\infty.$$

- **Case 3:** For $t \in \tilde{J}_k \cap \tilde{J}_n$.
  We have

$$\|(\aleph y_m)(t) - (\aleph y_*)(t)\| \leq \|g_k(t_k, y_m(t_k^-)) - g_k((t_k, y_*(t_k^-)))\|,$$

  by the continuity of g_k, we get

$$\|(\aleph y_m) - (\aleph y_*)\|_n \to 0, \quad as \ \ m \to +\infty.$$

Thus, $\aleph$ is continuous.

Step 3: We have $\aleph(D_{\delta_n}) \subset D_{\delta_n}$, that implies that $\aleph(D_{\delta_n})$ is bounded.

Step 4: Let Π be a bounded equicontinuous subset of D_{δ_n}, we have $\{\aleph(\Pi)\}$ is equicontinuous, implies $\omega_0 \ (\gamma_i^p, \aleph(\Pi)) = 0$. Now for any $\varrho > 0$, there exist a sequence $\{y_m\}_{k=0}^\infty \subset \Pi$ such that

- **Case 1:** For $t \in J_0 \cap \tilde{J}_n$.
  We put $O_{fv(s)} = f(s, y(s), Hy(s))$, we have

$$\alpha\left\{ \int_0^t R(t-s)O_{fv(s)} ds \ ; \ y \in \Pi \right\}$$
$$\leq 2\alpha\left\{ \int_0^t R(t-s)O_{fy_m(s)} ds \ ; y \in \Pi \right\} + \varrho$$
$$\leq 4 \int_0^t \mathcal{M}_R l_f(s)\alpha(\{\Pi(s)\}) ds + \varrho$$
$$\leq \int_0^t \zeta(s)\alpha(\Pi(s)) ds + \varrho$$
$$\leq \int_0^t e^{\tau\tilde{\zeta}(s)} e^{-\tau\tilde{\zeta}(s)} \zeta(s)\alpha(\Pi(s)) ds + \varrho$$
$$\leq \int_0^t \zeta(s)e^{\tau\tilde{\zeta}(s)} \sup_{s\in[0,t]} e^{-\tau\tilde{\zeta}(s)}\alpha(\Pi(s)) ds + \varrho$$

$$\leq \alpha_n(\Pi) \int_0^t \left(\frac{e^{\tau\widetilde\zeta(s)}}{\tau}\right)' ds + \varrho$$

$$\leq \frac{e^{\tau\widetilde\zeta(t)}}{\tau}\alpha_n(\Pi) + \varrho,$$

then, we get

$$\alpha(\aleph(\Pi)(t)) \leq \frac{e^{\tau\widetilde\zeta(t)}}{\tau}\alpha_n(\Pi) + \varrho,$$

since ϱ is arbitrary, so

$$\alpha(\aleph(\Pi)(t)) \leq \frac{e^{\tau\widetilde\zeta(t)}}{\tau}\alpha_n(\Pi),$$

therefore

$$\alpha_n(\aleph(\Pi)) \leq \frac{1}{\tau}\alpha_n(\Pi).$$

- **Case 2:** For $t \in J_k \cap \tilde J_n$.
  Similar to Case 1, we get

$$\alpha(\aleph(\Pi)(t)) \leq \mathcal{M}_R\,\alpha(\{g_k(s_k, y_m(t_k^-)); y \in \Pi\}) + \frac{e^{\tau\widetilde\zeta(t)}}{\tau}\alpha_n(\Pi) + \varrho.$$

$$\leq \frac{e^{\tau\widetilde\zeta(t)}(\mathcal{M}_R L_{g_k} + 1)}{\tau}\alpha_n(\Pi) + \varrho,$$

therefore

$$\alpha_n(\aleph(\Pi)) \leq \frac{(\mathcal{M}_R L_{g_k} + 1)}{\tau}\alpha_n(\Pi).$$

- **Case 3:** For $t \in \tilde J_k \cap \tilde J_n$.
  From (8.14.3), the set $\{g_k(t, y_k^-)\}_{k=1}^n$ is equicontinuous, then
$$\omega_0\left(\gamma_i^p, G(\Pi)\right) = 0,$$
with $\{Gv(t)\} = \{g_k(t, y_k^-)\}$.
On the other hand
$$\|g_k(t, y(.)) - g_k(t, \overline y(.))\| \leq \frac{L_{g_k}}{\tau}\|y(.) - \overline y(.)\|,$$
then
$$e^{-\tau\widetilde\zeta(t)}\|g_k(t, y(t_k^-)) - g_k(t, \overline y(t_k^-))\| \leq \frac{L_{g_k}}{\tau}e^{-\tau\widetilde\zeta(t)}\|y(t_k^-) - \overline y(t_k^-)\|,$$
therefore
$$\alpha_n(\aleph(\Pi)) < \frac{L_{g_k}}{\tau}\alpha_n(\Pi).$$

Then $\aleph$ is a contraction (in terms of a measure of non-compactness), provided that
$$\mathcal{M}_R L_{g_k} + 1 < \tau.$$
By Darbo's fixed point theorem [255], we conclude that $\aleph$ has at least one fixed point which is a mild solution of problem (8.13). $\qquad\square$

8.3.3.2   Controllability of Solutions

In this Section, we give controllability result for the system

$$
\begin{cases}
y'(t) = A(t)y(t) + f\left(t, y(t), (Hy)(t)\right) \\
\qquad + \displaystyle\int_0^t B(t-s)y(s)ds + Cu(t); \ \text{if } t \in J_k, k = 0, 1, \ldots, \\
y(t) = g_k\left(t, y\left(t_k^-\right)\right); \text{if } t \in \tilde{J}_k, k = 1, 2, \ldots, \\
y(0) = y_0,
\end{cases}
\tag{8.15}
$$

where the control function u is give function in $L^2(\mathbb{R}^+, S)$, a Banach space of admissible control functions with S as a Banach space. C is a bounded linear operator from S into E. Before this, we introduce the following type of solutions for the problem (8.15).

Definition 8.8. The system (8.15) is said to be controllable on the interval $\mathbb{R}^+$, if for every initial function $y_0 = y(0) \in E$ and $\hat{y} \in E$, there is for some $n > 0$, some control $u \in L^2([0; n]; E)$ such that the mild solution $y(\cdot)$ of this problem satisfies the terminal condition $y(n) = \hat{y}$.

To obtain the controllability of mild solutions of (8.15), we should assume the following conditions:

(8.16.1) There exists a positive constant ρ_n, such that:

$$
\max\left\{\varphi_1^\rho; \varphi_2^\rho; \frac{g_0}{1 - \frac{L_{g_k}}{\tau}}\right\} \leq \rho_n,
$$

with

$$
\varphi_1^\rho = \left\{ \mathcal{M}_R \Big[\|y_0\| + \psi(K_{\rho_n}^*)\|p\|_{L^1} \right.
$$

$$
\left. + c_5 c_6 \left(\rho_n + \mathcal{M}_R\|y_0\| + \mathcal{M}_R\psi(K_{\rho_n}^*)\|p\|_{L^1}\right)\Big] \right\},
$$

$$
\varphi_2^\rho = \left\{ \mathcal{M}_R \Big[\frac{L_{g_k}}{\tau}\rho_n + g_0 + \psi(K_{\rho_n}^*)\|p\|_{L^1} \right.
$$

$$
\left. + c_5 c_6 \left(\rho_n + \mathcal{M}_R\|y_0\| + \mathcal{M}_R\psi(K_{\rho_n}^*)\|p\|_{L^1}\right)\Big] \right\},
$$

and

$$
K_{\rho_n}^* = \left((c_1 + 1)\rho_n + ah^*\right)\|p\|_{L^1},
$$

(8.16.2) (*i*) For each n, the linear operator $W : L^2(\tilde{J}_n, S) \to \bar{E}$, defined by

$$Wu = \int_0^n R(n - s)Cu(s)ds,$$

has a pseudo inverse operator W^{-1}, which takes values in the space $L^2(\tilde{J}_n, S)\backslash Ker(W)$.

(*ii*) There exist positive constants c_5, c_6, such that

$$\|C\| \le c_5 \quad \text{and} \quad \|W^{-1}\| \le c_6.$$

(*iii*) There exist $p_w \in L^1(\mathbb{R}^+, \mathbb{R}^+)$, $k_C \ge 0$, and for any bounded sets $V_1 \subset E$, $V_2 \subset S$,

$$\alpha((W^{-1}V_1)(t)) \le p_w(t)\alpha(V_1), \quad \alpha((CV_2)(t)) \le k_C\alpha(V_2(t)).$$

Theorem 8.8. *Suppose that hypotheses* (8.14.1)–(8.14.4) *and* (8.16.1)–(8.16.2) *are valid, then the problem* (8.15) *is controllable.*

Proof. For $n \in \mathbb{N}$, we define in $PC(\mathbb{R}^+, \bar{E})$ the family of measures of non compactness by

$$\tilde{\alpha}_n(\Pi) = \max_{i=0,\ldots,m} \omega_0 \left(\gamma_i^p, \Pi\right) + \sup_{t \in \tilde{J}_n} \left\{e^{-\tau\tilde{\varkappa}(t)}\alpha(\Pi(t)) :\right\}, p = 0, 1, 2, m = 0, 1, \ldots,$$

where $\tilde{\varkappa}(t) = \int_0^t \varkappa(s)ds$, $\varkappa(t) = 4M_R(l_f(t) + k_C(M_R\|l_f\|^1)p_w(t))$, $\tau > 1$.

Using (8.16.1), we define the control:

$$u_y(t) = \begin{cases} W^{-1}\Big[y(n) - R(n)y_0 \\ \quad - \int_0^n R(n - s)f(s, y(s), (Hy)(s))ds\Big]; \text{ if } t \in J_0, \\ W^{-1}\Big[y(n) - R(n - s_k)\left[g_k(s, y(s_k^-))\right] \\ \quad - \int_{s_k}^n R(t - s)f(s, y(s), (Hy)(s))ds\Big]; \text{ if } t \in J_k, k = 1, 2, \ldots \end{cases}$$

It shall be show that using this control the operator defined by:

$$\Upsilon y(t) = \begin{cases} R(t)y_0 + \int_0^t R(t - s)f(s, y(s), (Hy)(s))ds \\ \quad + \int_0^t R(t - s)Cu_y(s)ds; \text{ if } t \in J_0, \\ R(t - s_k)\left[g_k(s_k, y(t_k^-))\right] + \int_{s_k}^t R(t - s)f(s, y(s), (Hy)(s))ds \\ \quad + \int_{s_k}^t R(t - s)Cu_y(s)ds; \text{ if } t \in J_k, \ k = 1, 2, \ldots, \\ g_k(t, y(t_k^-)); \text{ if } t \in \tilde{J}_k, \ k = 1, 2, \ldots, \end{cases}$$

has a fixed point which is a mild solution of (8.15), and this implies that the system is controllable.

For any $n \in \mathbb{N}$ and by (8.14.4), we define the closed, bounded and convex subset B_{ρ_n} by $B_{\rho_n} = B(0, \rho_n) = \{x \in PC \; : \; \|x\|_n \leq \rho_n\}$.

In the same way that we used previously, we establish the proof in several steps.

Step 1: $\aleph(B_{\rho_n}) \subset B_{\rho_n}$.

For any $y \in B_{\rho_n}$ and by assumptions (8.14.1), (8.14.4) and (8.16.1), we get

- **Case 1:** For $t \in J_0 \cap \tilde{J}_n$.

  For any $n \in \mathbb{N}$, $y \in B_{\rho_n}$, $t \in J_0 \cap \tilde{J}_n$, and by (8.14.1), we have

  $$\|\Upsilon y(t)\| \leq \mathcal{M}_R \Big(\|y_0\| + \psi((c_1 + 1)\rho_n + ah^*)\|p\|_{L^1} $$
  $$+ c_5 c_6 \big(\rho_n + \mathcal{M}_R \|y_0\| + \mathcal{M}_R \psi(K_{\rho_n}^*) \|p\|_{L^1} \big) \Big)$$
  $$\leq \rho_n.$$

- **Case 2:** For $t \in J_k \cap \tilde{J}_n$.

  For each $y \in B_{\rho_n}$ by (8.14.1), (8.14.2) *and* (8.14.3), we get

  $$\|\Upsilon y(t)\| \leq \mathcal{M}_R \bigg[\frac{L_{g_k}}{\tau} \rho_n + g_0 + \psi((c_1 + 1)\delta_n + ah^*)\|p\|_{L^1} $$
  $$+ c_5 c_6 \big(\rho_n + \mathcal{M}_R \|y_0\| + \mathcal{M}_R \psi(K_{\rho_n}^*) \|p\|_{L^1} \big) \bigg]$$
  $$\leq \rho_n.$$

- **Case 3:** For $t \in \tilde{J}_k \cap \tilde{J}_n$.

  For each $y \in B_{\rho_n}$ by (8.14.3), we have

  $$\|\Upsilon y(t)\| \leq \frac{L_{g_k}}{\tau} \rho_n + g_0 \leq \rho_n.$$

Thus

$$\|\Upsilon y\|_n \leq \rho_n,$$

this implies that $\Upsilon(B_{\rho_n}) \subset B_{\rho_n}$ and $\Upsilon(B_{\rho_n})$ is bounded.

Step 2: Υ is continuous on B_{ρ_n}.

Let y_n be a sequence such that $y_n \to y_*$ in B_{ρ_n}.

Since f, h, g_k, C are continuous, and we have by the Lebegue dominated convergence theorem

$$\int_0^t R(t - s) C u_{y_n}(s) ds \to \int_0^t R(t - s) C u_{y_*}(s) ds,$$

then from *Step* 2, we get

$$\|(\Upsilon y_n) - (\Upsilon y_*)\|_n \to 0, \quad as \ n \to +\infty.$$

Thus, we deduce that Υ is continuous.

Step 3: Let Π be a bounded equicontinuous subset of B_{ρ_n}, we have $\{\Upsilon(\Pi)\}$ is equicontinuous, implies $\omega_0 \left(\gamma_i^p, \Upsilon(\Pi)\right) = 0$. Now for any $\varrho > 0$ there exist a sequence $\{y_k\}_{k=0}^{\infty} \subset \Pi$, such that

- **Case 1:** For $t \in J_0 \cap \tilde{J}_n$.
  We have

$$\alpha(\Upsilon(\Pi)(t))$$
$$\leq 2\alpha(\{\int_0^t R(t-s)(f(s, y_k(s), Hy_k(s)) + u_{y_k}(s))ds \ ; \ y \in \Pi\}) + \varrho$$
$$\leq 4\int_0^t \mathcal{M}_R(l_f(s) + k_C(\mathcal{M}_R\|l_f\|_L^1)p_w(s))\alpha(\{\Pi(s)\})ds + \varrho$$
$$\leq \frac{e^{\tau\tilde{\varkappa}(t)}}{\tau}\tilde{\alpha}_n(\Pi) + \varrho.$$

Since ϱ is arbitrary, we have

$$\alpha(\Upsilon(\Pi)(t)) \leq \frac{e^{\tau L(t)}}{\tau}\tilde{\alpha}_n(\Pi).$$

Therefore

$$\tilde{\alpha}_n(\Upsilon(\Pi)) \leq \frac{1}{\tau}\tilde{\alpha}_n(\Pi).$$

- **Case 2:** For $t \in J_k \cap \tilde{J}_n$.
  Similar to Case 1, we get

$$\alpha(\Upsilon(\Pi)(t)) \leq 4\int_0^t \mathcal{M}_R(l_f(s) + k_C(\mathcal{M}_R\|l_f\|_L^1)p_w(s))\alpha(\{\Pi(s)\})ds + \varrho$$
$$+ \frac{\mathcal{M}_R L_{g_k}}{\tau}\alpha(\{\Pi(t)\})$$
$$\leq \frac{e^{\tau\tilde{\varkappa}(t)}(\mathcal{M}_R L_{g_k} + 1)}{\tau}\tilde{\alpha}_n(\Pi) + \varrho,$$

since ϱ is arbitrary, we get

$$\tilde{\alpha}_n(\Upsilon(\Pi)) \leq \frac{\mathcal{M}_R L_{g_k} + 1}{\tau}\tilde{\alpha}_n(\Pi).$$

- **Case 3:** For $t \in \tilde{J}_k \cap \tilde{J}_n$.

  From (8.14.3), the set $\left\{ g_k(t, z(t_k^-)) \right\}_{k=1}^n$ is equicontinuous, then

$$\omega_0 \left(\gamma_i^p, G(\Pi) \right) = 0,$$

with $\{Gz(t)\} = \left\{ g_k(t, z(t_k^-)) \right\}$.

In other hand

$$\| g_k(t, z(.)) - g_k(t, \overline{z}(.)) \| \leq \frac{L_{g_k}}{\tau} \| z(.) - \overline{z}(.) \|,$$

then

$$e^{-\tau \tilde{\zeta}(t)} \| g_k(t, z(t_k^-)) - g_k(t, \overline{z}(t_k^-)) \| \leq \frac{L_{g_k}}{\tau} e^{-\tau \tilde{\zeta}(t)} \| z(t_k^-) - \overline{z}(t_k^-) \|,$$

therefore

$$\tilde{\alpha}_n(\Upsilon(\Pi)) \leq \frac{L_{g_k}}{\tau} \tilde{\alpha}_n(\Pi).$$

We assume that

$$\mathcal{M}_R L_{g_k} + 1 < \tau,$$

then Υ is a contraction. Hence, by Darbo's fixed point theorem [255], we conclude that Υ has a fixed point, implies that the system is controllable. $\qquad \square$

8.3.4 Examples

Example 8.2. Consider the following impulsive integro-differential equation:

$$
\begin{cases}
\frac{\partial}{\partial t} \xi(t, x) = \varepsilon^2 \frac{\partial^2 \xi(t, x)}{\partial x^2} - \int_0^t \Lambda(t - s) \frac{\partial^2 \xi(s, x)}{\partial x^2} ds - \frac{\alpha \| \xi(t, x) \|_{L^2}}{e^{(1 - b^{-1})t}(1 + \sqrt{t})} \\
\qquad + \frac{\alpha e^{-(1 - b^{-1})t}}{1 + e^t} \arctan \left(\int_0^a e^{-st - s} |\xi(s, x)| ds \right), \ t \in J_k, \ x \in (0, \pi), \\[2mm]
\xi(t, x) = \frac{\xi(k^- - 1, x)}{33(1 + |\xi(k^- - 1, x)|)}, \ \text{if } t \in \tilde{J}_k, \ x \in (0, \pi), \\[2mm]
\xi(t, 0) = \xi(t, \pi) = 0, \ t \in \mathbb{R}^+, \\[2mm]
\xi(0, x) = \xi_0(x), \ x \in (0, \pi),
\end{cases}
$$

$$(8.16)$$

where $J_k = (k, k+1], k = 0, 1, \ldots,$ $\tilde{J}_k = (k-1, k], k = 1, 2, \ldots,$ $\varepsilon > 0,$ $b > 1,$ $\alpha \in \left(0, \frac{5}{12}\right)$ and $\Lambda : \mathbb{R}^+ \to \mathbb{R}$ is a C^1-function. Now, we define

$$\xi(t)(x) = \xi(t, x),$$

$$f(t, \xi(t), H\xi(t))(x) = -\frac{\alpha \|\xi(t, x)\|_{L^2}}{e^{(1-b^{-1})t}(1 + \sqrt{t})}$$

$$+ \frac{\alpha e^{-(1-b^{-1})t}}{(1 + e^t)} \arctan\left(\int_0^a e^{-st-s}|\xi(s, x)|ds\right),$$

$$H\xi(t)(x) = \int_0^a e^{-st-s}|\xi(s, x)|ds,$$

$$g_k(t, \xi(t_{k-}, x)) = \frac{\xi(k^- - 1, x)}{33(1 + |\xi(k^- - 1, x)|)},$$

$$B(t) = \Gamma(t)A.$$

To rewrite system (8.16) in the abstract form, we introduce the space $X = L^2(0, \pi)$ and let A be defined by

$$\begin{cases} D(A) = \{\varphi \in L^2(0, \pi) \, / \, \varphi, \, \varphi'' \in L^2(0, \pi) \, , \, \varphi(0) = \varphi(\pi) = 0\}, \\ (A\varphi)(x) = \frac{\partial^2 \varphi(t, x)}{\partial x^2}, \end{cases}$$

It is well known that A generates a strongly continuous semigroup $(S(t))_{t\geq0}$, which is dissipative and compact with $\|S(t)\| \leq e^{-\varepsilon^2 t}$, and for some $b > \frac{1}{\varepsilon^2}$, we assume that $\|\Gamma(t)\| \leq \frac{e^{-\varepsilon^2 t}}{b}$, and $\|\Gamma'(t)\| \leq \frac{e^{-\varepsilon^2 t}}{b^2}$. It follows from [295], that $\|R(t)\| \leq e^{-\varkappa t}$, where $\varkappa = 1 - b^{-1}$.

More appropriate conditions on operator B, (8.8.4) hold with $M_R = 1$ and $\beta = 1 - b^{-1}$.

Then, system (8.16) takes the following abstract form

$$\begin{cases} \xi'(t) = A\xi(t) + f\left(t, \xi(t), (H\xi)(t)\right) + \int_0^t B(t-s)\xi(s)ds, \text{ if } t \in J_k, \\ \xi(t) = g_k\left(t, \xi\left(t_k^-\right)\right), \text{if } t \in \tilde{J}_k, \\ \xi(0) = \xi_0, \end{cases}$$

$$(8.17)$$

where $Y = BPC(\mathbb{R}^+, X)$. With the help of simple computation, we find that

$$\zeta(t) = \frac{\alpha e^{-(1-b^{-1})t}}{1 + e^t}, \quad \varphi(x) = \varepsilon + x, \quad \beta_{g_k}^* = \frac{1}{33},$$

$$L_\varphi = \frac{\pi^2}{24}, \; c_h = a = \frac{12 - \pi^2 - 24\|\zeta\|_{L^1}}{24}.$$

Then

$$|f(t, \xi_1(t), \xi_2(t))| \le \frac{\alpha e^{-(1-b^{-1})t}}{1 + e^t} \left(|\xi_1(t)| + |\xi_2(t)| + \varepsilon \right), \; for \; \varepsilon > 0.$$

The function $\varphi(t) = \varepsilon + x$ is continuous non-decreasing from $\mathbb{R}^+$ to $[\varepsilon, +\infty)$, we have

$$\zeta \in L^1(\mathbb{R}^+, \mathbb{R}^+),$$

$$\lim_{t \to +\infty} \sup_{t \in \mathbb{R}^+} \int_0^t e^{-\beta(t-s)} \zeta(s) ds = \lim_{t \to +\infty} \sup_{t \in \mathbb{R}^+} \int_0^t \frac{\alpha e^{-(1-b^{-1})t}}{1 + e^s} ds = 0,$$

and

$$M_R \max\{\beta_{g_k}^*\} = 0,4112\ldots < \frac{1}{2}.$$

Also, we can choose the values of ρ and γ so that Theorem 8.5 and Theorem 8.6 are applicable. Consequently, the problem (8.16) has at least one mild solution defined on $\mathbb{R}^+$, which is uniformly locally attractive.

Example 8.3. Consider the following integro-differential equation with impulsions:

$$\begin{cases} \frac{\partial}{\partial t}\gamma(t,x) = -\frac{\partial}{\partial x}\gamma(t,x) - \pi\gamma(t,x) - \int_0^t \Gamma(t-s)(\frac{\partial}{\partial x}\gamma(s,x) \\ \quad +\pi\gamma(s,x))ds + \frac{\|\gamma(t,x)\|_{L^2}}{1+t^3\sin^2(t)} \\ \quad +(1+t^3\sin^2(t))^{-1}\sin\left[\int_0^a \cos^2(st)|\gamma(s,x)|ds\right] \\ \quad +Cu(t,x), \quad \text{if } t \in J_k, \; x \in (0,1), \\ \\ \gamma(t,x) = \frac{\|\gamma(2k^- - 1, x)\|_{L^2}}{1 + 17(\|\gamma(2k^- - 1, x)\|_{L^2} + 1)}, \quad \text{if } t \in \tilde{J}_k, \; x \in (0,1), \\ \\ \gamma(t,0) = \gamma(t,1) = 0, \quad t \in \mathbb{R}^+, \\ \\ \gamma(0,x) = e^x, \; x \in (0,1), \end{cases} \tag{8.18}$$

where $J_0 = (0,1]$, $J_k = (2j; 2j+1]$, $k = 0, 1, \ldots$, $\tilde{J}_k = (2j-1; 2j]$, $k = 1, 2, \ldots$.

Set $\bar{E} = L^2(0,1)$ and let A be defined by

$$(A\varphi)(x) = -\left(\frac{d}{dx}\varphi(x) + \pi\varphi(x)\right),$$

And

$$D(A) = \{\varphi \in L^2(0,1) \; / \; \varphi, A\varphi \in L^2(0,1) \; ; \; \varphi(0) = \varphi(1) = 0\}.$$

The operator A is the infinitesimal generator of a C_0-semigroup on $\bar{E}$ with domain $D(A)$, and with more appropriate conditions on operator $B(\cdot) = \Gamma(\cdot)A$, the problem (8.18) has a resolvent operator $(R(t))_{t\geq 0}$ on $\bar{E}$ which is norm continuous. Now, define

$$\gamma(t)(x) = \gamma(t,x),$$

$$f(t, \gamma(t), H\gamma(t))(x) = \frac{\|\gamma(t,x)\|_{L^2}}{1 + t^3 \sin^2(t)} + (1 + t^3 \sin^2(t))^{-1}$$
$$\times \sin\left[\int_0^a \cos^2(st)|\gamma(s,x)|ds\right],$$

$$H\gamma(t)(x) = \int_0^a \cos^2(st)|\gamma(s,x)|ds,$$

$$g_k(t, \gamma(t_{k^-}, x)) = \frac{\|\gamma(2k^- - 1, x)\|_{L^2}}{1 + 17(\|\gamma(2k^- - 1, x)\|_{L^2} + 1)}.$$

Case 1: $Cu = 0$.

With these settings system (8.18) can be written in the abstract form

$$\begin{cases} \gamma'(t) = A\gamma(t) + f\left(t, \gamma(t), (H\gamma)(t)\right) + \int_0^t B(t-s)\gamma(s)ds, \text{ if } t \in J_k, \\ \gamma(t) = g_k\left(t, \gamma\left(t_k^-\right)\right), \text{if } t \in \tilde{J}_k, \\ \gamma(0) = \gamma_0, \end{cases}$$

where $M = PC(\mathbb{R}^+, \bar{E})$. On the other hand

$$|f(t, \gamma_1(t), \gamma_2(t))| \leq (1 + t^3 \sin^2(t))^{-1}\left(|\gamma_1(t)| + |\gamma_2(t)| + 1\right).$$

Also we have for any bounded set $\Sigma \subset \bar{E}$,

$$\alpha(f(t, \Sigma, H(\Sigma))) \leq (1 + t^3 \sin^2(t))^{-1}\alpha(\Sigma).$$

So

$$p(t) = (1 + t^3 \sin^2(t))^{-1}, \text{ which certainly belongs to } L^1(\mathbb{R}^+, \mathbb{R}^+),$$

and the function $\psi(t) = 1 + t$ is continuous non-decreasing function from $\mathbb{R}^+$ to $[1, +\infty)$.

We have also the following estimates,

$$\|h(t, s, \gamma_1) - h(t, s, \gamma_2)\|_n \leq a\|\gamma_1 - \gamma_2\|_n,$$

and

$$\|g_k(\gamma_1) - g_k(\gamma_2)\|_n \le \frac{1}{18}\|\gamma_1 - \gamma_2\|_n.$$

For $\mathcal{M}_R < 3$, all conditions of Theorem 8.7 are satisfied. Hence, the problem (8.18) has at least one mild solution defined on $\mathbb{R}^+$.

Case 2: $Cu = \varkappa u(t, \gamma)$ for $\varkappa > 0$.
The operator $C : L^2(0, 1) \to L^2(0, 1)$ be defined by $Cu = \varkappa u(t, \gamma)$,
And the system (8.18) can be written as follows

$$
\begin{cases}
\gamma'(t) = A\gamma(t) + f\left(t, \gamma(t), (H\gamma)(t)\right) \\
\qquad\quad + \displaystyle\int_0^t B(t - s)\gamma(s)ds + Cu(t), \text{ if } t \in J_k, \\
\gamma(t) = g_k\left(t, \gamma\left(t_k^-\right)\right), \text{ if } t \in \tilde{J}_k, \\
\gamma(0) = \gamma_0.
\end{cases}
$$

Then similar to Case 1, we can easily verify that the assumptions (8.14.1)–(8.16.1) hold and if we assume that the operator W given by $Wu = \displaystyle\int_0^n R(n - s)\varkappa u(s)ds$, satisfies (8.16.2), then all the assumptions given in Theorem 8.8 are verified. Therefore, the problem (8.18) is controllable.

8.4   Second Order Semilinear Volterra-Type Integro-Differential Equations with Non-Instantaneous Impulses

The aim of this section is to establish an result of the existence of mild solution for a class of the non-autonomous second order nonlinear differential equation with non-instantaneous impulses described in the form

$$
\begin{cases}
y''(t) = A(t)y(t) + f\left(t, \displaystyle\int_0^t g(t, s, y(s))ds\right), \quad t \in (s_i, t_{i+1}], i = 0, \dots, m, \\
y(t) = \gamma_i(t, y(t^-)), \quad t \in (t_i, s_i], \quad i = 1, \dots, m, \\
y'(t) = \zeta_i(t, y(t^-)), \quad t \in (t_i, s_i], \quad i = 1, \dots, m, \\
y(0) = y_0, \ y'(0) = y_1,
\end{cases}
$$

$$(8.19)$$

In this text, E is a Banach space endowed with a norm $\|\cdot\|$, $J = [0, a]$, $0 \le s_0 \le t_0 < t_1 < s_1 < t_2, \dots, t_m < s_m < t_{m+1} \le a < \infty$. We consider in problem (8.19) that $y \in C((s_i, t_{i+1}), E)$; $i = 0, 1, \ldots, m$. The functions $\gamma_i(t, y(t))$ and $\zeta_i(t, y(t))$ represent non-instantaneous impulses during the intervals $(t_i, s_i]$ si; $i = 1, \ldots, m$, so impulses at t_i have some duration, namely on intervals $(t_i, s_i]$. $\{A(t)\}_{0 \le t \le a}$ is a family of linear closed operators from E into E, that generate an evolution system of linear bounded operators $\{S(t, s)\}_{(t,s) \in \mathcal{U}}$, f is a given function $J \times E \times E \to E$, $g \in C(\mathcal{U}, E)$, $\mathcal{U} = \{(t, s) \in J \times J : s \le t\}$.

To deal with the above mentioned issues, we investigate the existence results of mild solutions to a system (8.19). By virtue of the theory of measure of non-compactness associated with Darbo's and Darbo-Sadovskii's fixed point theorem. This technique was mainly initiated in the monograph of Banas and Goebel [197] and subsequently developed and used in many papers; see [23, 194, 195, 392, 398].

8.4.1 *Main Results*

To treat the impulsive conditions, we define the space of piecewise continuous functions

$PC(J, E) = \{y : J \to E : y \in C([0, t_1] \cup (t_k, s_k] \cup (s_k, t_{k+1}], E), k = 1, \ldots, m$ and there exist $y(t_k^-)$, $y(t_k^+), y(s_k^-)$ and $y(s_k^+)$ $k = 1, \ldots, m$ with $y(t_k^-) = y(t_k)$ and $y(s_k^-) = y(s_k)\}$.

It can be easily proved that $PC(J, E)$ for all $t \in J$, is a Banach space endowed with the supremum norm

$$\|y\|_{PC} = \sup_{t \in J} \|y(t)\|.$$

Denote α_{PC} by the Kuratowski measure of non-compactness of $PC(J; X)$.

Lemma 8.1 ([197]). *If $W \subset PC(J; E)$ is bounded, then $\alpha(W(t)) \le \alpha_C(W)$, for all $t \in J$; where $W(t) = \{y(t); y \in W \subset E\}$. Furthermore if W is equicontinuous on J, then $\alpha(W(t))$ is continuous on J and*

$$\alpha_{PC}(W) = \sup_{t \in J} \alpha(W(t)).$$

Lemma 8.2 ([254]). *If the map $\Psi : D(\Psi) \subset E \to F$ is Lipschitz continuous with constant k, then $\alpha(\Psi(W)) \le k\alpha(W)$ for any bounded subset $W \subset \mathcal{D}(\Psi)$, where F is another Banach space.*

In this section, we discuss the existence of mild solutions for system (8.19). Firstly, let us propose the definition of the mild solution of system (8.19).

Definition 8.9. A function $y \in PC(J, E)$ is said to be a mild solution to the system (8.19), if it satisfies the following relations:

$y(0) = y_0$, $y'(0) = y_1$, the non-instantaneous conditions

$$y(t) = \gamma_i(t, y(t_i^-)), \quad y'(t) = \zeta_i(t, y(t_i^-)), \quad t \in (t_i, s_i],$$

and y is the solution of the following integral equations

$$y(t) = \begin{cases} -\dfrac{\partial}{\partial s} S(t, 0) y_0 + S(t, 0) y_1 \\ \qquad + \displaystyle\int_0^t S(t, s) f\left(s, y(s), \int_0^s g(s, \tau, y(\tau)) d\tau\right) ds, \ t \in [0, t_1], \\ -\dfrac{\partial}{\partial s} S(t, s_i) \gamma_i(s_i, y(t_i^-)) + S(t, s_i) \zeta_i(s_i, y(t_i^-)) \\ \qquad + \displaystyle\int_{s_i}^t S(t, s) f\left(s, y(s), \int_0^s g(s, \tau, y(\tau)) d\tau\right) ds, \ t \in (s_i, t_{i+1}]. \end{cases}$$

In this section, we list the following hypotheses:

(8.22.1) There exist a pair of constants $M \geq 1$ and $\delta > 0$, such that

$$\|S(t, s)\|_{B(E)} \leq M e^{-\delta(t-s)} \quad \text{for any } (t, s) \in \mathcal{U}.$$

(8.22.2) There exist two constants $\tilde{M}$, $\delta > 0$ such that:

$$\left\| \frac{\partial}{\partial s} S(t, s) \right\|_{B(E)} \leq \tilde{M} e^{-\delta(t-s)}, (t, s) \in \mathcal{U}.$$

(8.22.3) $f : J \times E \to E$ is of Carathéodory type satisfies:

(a) There exist $\Theta_f \in L^r(J, \mathbb{R}^+)$, $r \in [1, \infty)$ and a continuous non-decreasing function $\psi : [0, \infty) \to (0, \infty)$ such that:

$$\|f(t, y, z)\| \leq \Theta_f(t) \psi(\|y\| + \|z\|) \text{ for a.e } t \in J \text{ and each } y, z \in E.$$

(b) There exist integrable functions $\sigma, \varrho : J \to \mathbb{R}^+$, such that:

$$\alpha(f(t, W_1, W_2)) \leq \sigma(t) \alpha(W_1) + \varrho(t) \alpha(W_2)$$

for a.e $t \in J$ and $W_1, W_2 \subset E$.

(8.22.4) $g : \mathcal{U} \times E \to E$ is continuous function satisfies:

(a) There exist $\Theta_g \in L^1(J, \mathbb{R}^+)$, such that:

$$\|g(t, s, y)\| \leq \Theta_g(t) \varphi(\|y\|) \text{ for a.e } (t, s) \in \mathcal{U} \text{ and each } y \in E.$$

(b) There exists constant $K^* > 0$, such that

$$\alpha(g(t, s, W)) \leq K^* \alpha(W) \text{ for a.e } (t, s) \in \mathcal{U} \text{ and } W \subset E.$$

(8.22.5) The functions $\gamma_i : (t_i; s_i] \times E \to E, i = 1, \ldots, m$, are continuous, and they satisfy the following conditions:

(a) there exist positive constants $c_i, i = 1, \ldots, m$ such that

$$\|\gamma_i(t, y_2) - \gamma_i(t, y_1)\| \leq c_i \|y_2 - y_1\| \text{ for a.e } t \in (t_i; s_i], y \in E.$$

(b) there exist positive constants $d_i, i = 1, \ldots, m$ such that

$$d_i = \sup_{t \in [s_i, t_i]} \gamma_i(t, 0).$$

(8.22.6) The functions $\zeta_i : (t_i; s_i] \times E \to E, i = 1, \ldots, m$, are continuous, and satisfy the following conditions:

(a) There exist constants $e_i, l_i > 0, i = 1, \ldots, m$ such that

$$\|\zeta_i(t, y)\| \leq e_i \|y\| + l_i \text{ for a.e } t \in (t_i; s_i] \text{ and each } y \in E.$$

(b) There exists constants $\overline{k}_i > 0$, such that

$$\alpha(\zeta_i(t, W)) \leq \overline{k}_i \alpha(W) \text{ for a.e } t \in (t_i; s_i] \text{ and each } y \in E.$$

(8.22.7) The following inequality is verified

$$\max_{0 \leq i \leq m} (k_i, 1) \left(\max_{0 \leq i \leq m} (\tilde{M}k_i + M\overline{k}_i) + 2M(\|\sigma\|_{L^1} + 2K^* a \|\varrho\|_{L^1})) \right) < 1,$$

Remark 8.1. From Lemma 8.2 and (8.22.4), there exist constants $k_i > 0$, such that

$$\alpha(\gamma_i(t, W)) \leq k_i \alpha(W) \text{ for a.e } t \in (t_i; s_i] \text{ and each } y \in E.$$

Theorem 8.9. *Under the assumption* (8.22.1)–(8.22.7), *the system* (8.19) *has at least one mild solution on* J, *provided that*

$$\int_0^a \max(\delta, \bar{M}\Theta_f(s), \Theta_g(s))ds \leq \int_{m_i}^\infty \frac{ds}{s + \psi(s) + \varphi(s)}, \qquad (8.20)$$

with

$$\bar{M} = \max_{0 \leq i \leq m} \left\{ \frac{M}{1 - L_i}, \frac{Mc_{i+1}}{1 - L_i} \right\},$$

and

$$m_i - \max_{0 \leq i \leq m} \left\{ \tilde{M} \|y_0\| + M \|y_1\|, \frac{\tilde{M}d_i}{1 - L_i} + \frac{Ml_i}{1 - L_i}, \frac{\tilde{M}c_{i+1}d_i}{1 - L_i} + \frac{Mc_i l_i}{1 - L_i} \right\}$$

$$+ \frac{d_i}{1 - L_{i-1}},$$

where

$$L_i = \tilde{M}c_i + Me_i.$$

Proof. Consider the operator $\Lambda : PC(J,E) \to PC(J,E)$, defined by

$$(\Lambda y)(t) = \begin{cases} \gamma_i \left(t, -\dfrac{\partial}{\partial s} g(t, s_i) \gamma_{i-1}(s_{i-1}, y(t_{i-1}^{-})) \right) \\ \quad + S(t, s_{i-1}) \zeta_{i-1}(s_{i-1}, y(t_{i-1}^{-})) \\ \quad + \displaystyle\int_{s_{i-1}}^{t_i} S(t,s) f\left(s, y(s), \displaystyle\int_0^s g(s, \tau, y(\tau)) d\tau \right) ds \Bigg), \quad t \in (t_i, s_i], \\[2ex] -\dfrac{\partial}{\partial s} S(t,0) y_0 + S(t,0) y_1 \\ \quad + \displaystyle\int_0^t S(t,s) f\left(s, y(s), \displaystyle\int_0^s g(s, \tau, y(\tau)) d\tau \right) ds, \quad t \in [0, t_1], \\[2ex] -\dfrac{\partial}{\partial s} S(t, s_i) \gamma_i(s_i, y(t_i^{-})) + S(t, s_i) \zeta_i(s_i, y(t_i^{+})) \\ \quad + \displaystyle\int_{s_i}^t S(t,s) f\left(s, y(s), \displaystyle\int_0^s g(s, \tau, y(\tau)) d\tau \right) ds, \quad t \in (s_i, t_{i+1}]. \end{cases}$$

It is obvious that the fixed point of Λ is the mild solution of (8.19). We shall show that m satisfies the assumptions of Theorem 2.9. The proof will be given in four steps.

Step 1. A priori bounds.

Let $\lambda \in (0,1)$ and let $y \in \partial Y$ be a possible solution of $y = \lambda \Lambda(y)$ for some $0 < \lambda < 1$. Thus,

Case 1. For each $t \in [0, t_1]$, we get

$$y(t) = -\lambda \frac{\partial}{\partial s} S(t,0) y_0 + \lambda S(t,0) y_1$$
$$+ \lambda \int_0^t S(t,s) f\left(s, y(s), \int_0^s g(s, \tau, y(\tau)) d\tau \right) ds.$$

Then

$$\|y(t)\| \leq \left\| \frac{\partial}{\partial s} S(t,0) \right\|_{B(E)} \|y_0\| + \|S(t,0)\|_{B(E)} \|y_1\|$$
$$+ \int_0^t \|S(t,s)\|_{B(E)} \Theta_f(s) \psi \left(\|y(s)\| + \int_0^s \Theta_g(\tau) \varphi(\|y(\tau)\|) d\tau \right) ds$$
$$\leq \tilde{M} \|y_0\| e^{-\delta t} + M \|y_1\| e^{-\delta t}$$
$$+ \int_0^t M e^{-\delta(t-s)} \Theta_f(s) \psi \left(\|y(s)\| + \int_0^s \Theta_g(\tau) \varphi(\|y(\tau)\|) d\tau \right) ds.$$
$$\leq (\tilde{M} \|y_0\| + M \|y_1\|) e^{-\delta t}$$
$$+ \int_0^t M e^{-\delta(t-s)} \Theta_f(s) \psi \left(\|y(s)\| + \int_0^s \Theta_g(\tau) \varphi(\|y(\tau)\|) d\tau \right) ds.$$

Case 2. For each $t \in (s_i, t_{i+1}]$, we have

$$y(t) = -\lambda \frac{\partial}{\partial s} S(t, s_i) \gamma_i(s_i, y(s_i)) + \lambda S(t, s_i) \zeta_i(s_i, y(s_i))$$
$$+ \lambda \int_{s_i}^{t} S(t, s) f(s, y(s), \int_0^s g(s, \tau, y(\tau)) d\tau) ds,$$

then

$$\|y(t)\| \leq \left\| \frac{\partial}{\partial s} S(t, s_i) \right\|_{B(E)} \|\gamma_i(s_i, y(s_i))\| + \|S(t, s_i)\|_{B(E)} \|\zeta_i(s_i, y(s_i))\|$$
$$+ \int_{s_i}^{t} \|S(t, s)\|_{B(E)} \Theta_f \left(\|y(s)\| + \int_0^s \Theta_g(\tau) \varphi(\|y(\tau)\|) d\tau \right) ds$$
$$\leq \tilde{M} c_i \|y(t)\| e^{-\delta(t - s_i)} + \tilde{M} d_i e^{-\delta(t - s_i)}$$
$$+ M e_i \|y(t)\| e^{-\delta(t - s_i)} + M l_i e^{-\delta(t - s_i)}$$
$$+ \int_{s_i}^{t} M e^{-\delta(t - s)} \Theta_f(s) \psi \left(\|y(s)\| + \int_0^s \Theta_g(\tau) \varphi(\|y(\tau)\|) d\tau \right) ds.$$
$$\leq \tilde{M} c_i \|y(t)\| + \tilde{M} d_i e^{-\delta t}$$
$$+ M e_i \|y(t)\| + M l_i e^{-\delta t}$$
$$+ \int_{s_i}^{t} M e^{-\delta(t - s)} \Theta_f(s) \psi \left(\|y(s)\| + \int_0^s \Theta_g(\tau) \varphi(\|y(\tau)\|) d\tau \right) ds.$$

It is easy to see that

$$\|y(t)\| \leq \left(\frac{\tilde{M} d_i}{1 - L_i} + \frac{M l_i}{1 - L_i} \right) e^{-\delta t}$$
$$+ \int_{s_i}^{t} \frac{M}{1 - L_i} e^{-\delta(t - s)} \Theta_f(s) \psi \left(\|y(s)\| + \int_0^s \Theta_g(\tau) \varphi(\|y(\tau)\|) d\tau \right) ds.$$

Case 3. For each $t \in (s_i, t_i]$, we have,

$$y(t) = \lambda \gamma_i \left(t, -\frac{\partial}{\partial s} S(t, s_i) \gamma_{i-1}(s_{i-1}, y(t_{i-1}^-)) + S(t, s_{i-1}) \zeta_{i-1}(s_{i-1}, y(t_{i-1}^-)) \right.$$
$$\left. + \int_{s_{i-1}}^{l_i} S(t, s) f \left(s, y(s), \int_0^s g(s, \tau, y(\tau)) d\tau \right) ds \right).$$

This implies

$$\|y(t)\| \leq \left(\frac{\tilde{M} c_i d_{i-1}}{1 - L_{i-1}} + \frac{M c_i l_{i-1}}{1 - L_{i-1}} \right) e^{-\delta t} + \frac{d_i}{1 - L_{i-1}}$$

$$+ \int_{s_{i-1}}^{t_i} \frac{Mc_i}{1 - L_{i-1}} e^{-\delta(t-s)} \Theta_f(s)\psi$$

$$\times \left(\|y(s)\| + \int_0^s \Theta_g(\tau)\varphi(\|y(\tau)\|)d\tau \right) ds.$$

Then, for all $t \in J$, we have

$$\|y(t)\| \leq M_i^* e^{-\delta t} + \frac{d_i}{1 - L_{i-1}}$$

$$+ e^{-\delta t} \int_0^t \bar{M} e^{\delta s} \Theta_f(s)\psi \left(\|y(s)\| + \int_0^s \Theta_g(\tau)\varphi(\|y(\tau)\|)d\tau \right) ds,$$

where

$$M_i^* = \max_{0 \leq i \leq m} \left\{ \tilde{M}\|y_0\| + M\|y_1\|, \frac{\tilde{M}d_i}{1 - L_i} + \frac{Ml_i}{1 - L_i}, \frac{\tilde{M}c_{i+1}d_i}{1 - L_i} + \frac{Mc_{i+1}l_i}{1 - L_i} \right\}.$$

Let us take the right-hand side of the above inequality as $\mu(t)$. Then $\mu(0) = M_i^* + \frac{d_i}{1 - L_{i-1}}$ and $y(t) \leq \mu(t)$ and

$$\mu'(t) \leq \delta\mu(t) + \bar{M}\Theta_f(t)\psi \left(\mu(t) + \int_0^t \Theta_g(s)\varphi(\mu(s))ds \right).$$

Let $\beta(t) = \mu(t) + \int_0^t \Theta_g(s)\varphi(\mu(s))ds$. Then

$$\beta'(t) = \mu'(t) + \Theta_g(t)\varphi(\mu(t))$$
$$\leq \delta\beta(t) + \bar{M}\Theta_f(t)\psi(\beta(t)) + \Theta_g(t)\varphi(\beta(t)).$$

This implies that

$$\int_{\beta(0)}^{\beta(t)} \frac{ds}{s + \psi(s) + \varphi(s)} \leq \int_{m_i}^a \max(\delta, \bar{M}\Theta_f(s), \Theta_g(s))ds$$

$$< \int_{m_i}^{+\infty} \frac{ds}{s + \psi(s) + \varphi(s)}.$$

This above inequality implies that there exists a constant L such that $\beta(t) \leq L, t \in J$, and hence $\mu(t) \leq L, t \in J$. Since for every $t \in J, \|y(t)\| \leq \mu(t)$, we have $\|y\|_{PC} \leq L$.

Step 2. Λ is continuous.

Let $(y_n)_{n \in \mathbb{N}}$ be a sequence in B_ρ such that $y_n \to y$. By the continuity of nonlinear term γ and ζ with respect to second , for each $s \in J$, we have

$$\sup_{s \in J} \|\gamma_i(s, y_n(s)) - \gamma_i(s, y(s))\| \to 0 \quad \text{as} \quad n \to \infty, \tag{8.21}$$

$$\sup_{s \in J} \|\zeta(s, y_n(s)) - \zeta(s, y(s))\| \to 0 \quad \text{as} \quad n \to \infty. \tag{8.22}$$

By the Carathéodory character of nonlinear term f, for each $s \in J$, we have

$$\left\| f\left(s, y_n(s), \int_0^s g(s, \tau, y_n(\tau))d\tau\right) - f\left(s, y(s), \int_0^s g(s, \tau, y(\tau))d\tau\right) \right\|$$
$$\to 0 \quad \text{as} \quad n \to \infty. \tag{8.23}$$

Case 1. For each $t \in (s_i; t_1]$, we obtain

$$\|(\Lambda y_n)(t) - (\Lambda y)(t)\|$$

$$\leq \gamma_i \left(t, -\frac{\partial}{\partial s}S(t, s_i)\gamma_{i-1}(s_{i-1}, y(t_{i-1}^-)) + \mathcal{U}(t, s_{i-1})\zeta_{i-1}(s_{i-1}, y(t_{i-1}^-))\right.$$
$$\left. + \int_{s_{i-1}}^{t_i} S(t, s)f\left(s, y(s), \int_0^s g(s, \tau, y(\tau))d\tau\right)ds\right)$$

$$- \gamma_i \left(t, -\frac{\partial}{\partial s}S(t, s_i)\gamma_{i-1}(s_{i-1}, y_n(t_{i-1}^-)) + S(t, s_{i-1})\zeta_{i-1}(s_{i-1}, y_n(t_{i-1}^-))\right.$$
$$\left. + \int_{s_{i-1}}^{t_i} S(t, s)f\left(s, y_n(s), \int_0^s g(s, \tau, y_n(\tau))d\tau\right)ds\right)$$

$$\to 0, \quad \text{as} \quad n \to \infty.$$

Case 2. For each $t \in [0; t_1]$, we obtain

$$\|(\Lambda y_n)(t) - (\Lambda y)(t)\| \leq M \int_0^t \left\| f\left(s, y_n(s), \int_0^s g(s, \tau, y_n(\tau))d\tau\right)\right.$$
$$\left. - f\left(s, y(s), \int_0^s g(s, \tau, y(\tau))d\tau\right)\right\| ds$$

$$\to 0, \quad \text{as} \quad n \to \infty.$$

Case 3. For each $t \in (s_i; t_{i+1}]$, we have

$$\|(\Lambda y_n)(t) - (\Lambda y)(t)\| \leq \tilde{M}\|\gamma_i(s_i, y_n(s_i)) - \gamma_i(s_i, y(s_i))\|$$
$$+ M\|\zeta_i(s_i, y_n(s_i)) - \zeta_i(s_i, y(s_i))\|$$
$$+ M \int_{\sigma_i}^t \left\| f\left(s, y_n(s), \int_0^s g(s, \tau, y_n(\tau))d\tau\right)\right.$$
$$\left. - f\left(s, y(s), \int_0^s g(s, \tau, y(\tau))d\tau\right)\right\| ds$$

$$\to 0 \quad \text{as} \quad n \to \infty.$$

As a consequence of **Cases 1–3**, $\Lambda y_n \to \Lambda y$, as $n \to +\infty$. Hence the Λ is continuous.

Step 3. Λ is equicontinuous.

Case 1. For the interval $t \in [0; t_1]$, $0 \leq \tilde{t}_1 \leq \tilde{t}_2 \leq t_1$, any $y \in B_R$, we have

$$\|(\Lambda y)(\tilde{t}_2) - (\Lambda y)(\tilde{t}_1)\|$$

$$= \left\| \int_0^{\tilde{t}_1} (S(\tilde{t}_2, s) - S(\tilde{t}_1, s)) f\left(s, y(s), \int_0^s g(s, \tau, y(\tau)) d\tau\right) ds \right.$$

$$\left. + \int_{\tilde{t}_1}^{\tilde{t}_2} S(\tilde{t}_2, \tau) f\left(s, y(s), \int_0^s g(s, \tau, y(\tau)) d\tau\right) ds \right\|$$

$$\leq \int_0^{\tilde{t}_1} \|S(\tilde{t}_2, \tau) - S(\tilde{t}_1, \tau)\|_{B(E)} \, \Theta_f(\tau)$$

$$\times \psi\left(\|y(s)\| + \int_0^s \Theta_g(\tau) \varphi(\|y(\tau)\|) d\tau\right) ds$$

$$+ M \int_{\tilde{t}_1}^{\tilde{t}_2} \Theta_f(s) \psi\left(\|y(s)\| + \int_0^s \Theta_g(\tau) \varphi(\|y(\tau)\|) d\tau\right) ds.$$

It follows from the Hölder's inequality that

$$\|(\Lambda y)(\tilde{t}_2) - (\Lambda y)(\tilde{t}_1)\|$$

$$\leq \psi\left(R + \varphi(R)\|\Theta_g\|_{L^1}\right) \int_0^{\tilde{t}_1} \|S(\tilde{t}_2, \tau) - S(\tilde{t}_1, \tau)\|_{B(E)} \, \Theta_f(\tau) d\tau$$

$$+ \frac{M \|\Theta_f\|_{L^r} \psi\left(R + \varphi(R)\|\Theta_g\|_{L^1}\right)}{\delta^{1-\frac{1}{r}}} \left(e^{-\frac{r\delta}{r-1}(t-\tilde{t}_2)} - e^{-\frac{r\delta}{r-1}(t-\tilde{t}_1)}\right)^{1-\frac{1}{r}}.$$

Case 2. For each $t \in (s_i; t_{i+1}]$, $s_i \leq \tilde{t}_1 \leq \tilde{t}_2 \leq t_{i+1}$, any $y \in B_R$, then we get

$$\|(\Lambda y)(\tilde{t}_2) - (\Lambda y)(\tilde{t}_1)\|$$

$$\leq \left\| \frac{\partial}{\partial s} S(\tilde{t}_2, s_i) - \frac{\partial}{\partial s} S(\tilde{t}_1, s_i) \right\|_{B(E)} \|\gamma_i(s_i, y(s_i))\|$$

$$+ \|S(\tilde{t}_2, s_i) - S(\tilde{t}_1, s_i)\|_{B(E)} \|\zeta_i(s_i, y(s_i))\|$$

$$+ \left\| \int_{s_i}^{\tilde{t}_1} (S(\tilde{t}_2, s) - S(\tilde{t}_1, s)) f\left(s, y(s), \int_0^s g(s, \tau, y(\tau)) d\tau\right) ds \right.$$

$$\left. + \int_{\tilde{t}_1}^{\tilde{t}_2} S(\tilde{t}_2, \tau) f\left(s, y(s), \int_0^s g(s, \tau, y(\tau)) d\tau\right) ds \right\|$$

$$\leq \int_{s_i}^{\tilde{t}_1} \|S(\tilde{t}_2, \tau) - S(\tilde{t}_1, \tau)\|_{B(E)} \Theta_f(\tau)$$

$$\times \psi\left(\|y(s)\| + \int_0^s \Theta_g(\tau) \varphi(\|y(\tau)\|) d\tau\right) ds$$

$$+ M \int_{\tilde{t}_1}^{\tilde{t}_2} \Theta_f(s)\psi\left(\|y(s)\| + \int_0^s \Theta_g(\tau)\varphi(\|y(\tau)\|)d\tau\right)ds.$$

It follows from the Hölder's inequality that

$$\|(\Lambda y)(\tilde{t}_2) - (\Lambda y)(\tilde{t}_1)\|$$

$$\leq \|\frac{\partial}{\partial s}S(\tilde{t}_2, s_i) - \frac{\partial}{\partial s}S(\tilde{t}_1, s_i)\|_{B(E)}\|\gamma_i(s_i, y(s_i))\|$$

$$+ \|S(\tilde{t}_2, s_i) - S(\tilde{t}_1, s_i)\|_{B(E)}\|\zeta_i(s_i, y(s_i))\|$$

$$+ \psi\left(R + \varphi(R)\|\Theta_g\|_{L^1}\right)\int_{s_i}^{\tilde{t}_1}\|S(\tilde{t}_2, \tau) - S(\tilde{t}_1, \tau)\|_{B(E)}\ p(\tau)d\tau$$

$$+ \frac{M\|\Theta_f\|_{L^r}\psi\left(R + \varphi(R)\|\Theta_g\|_{L^1}\right)}{\delta^{1-\frac{1}{r}}}\left(e^{-\frac{r\delta}{r-1}(t-\tilde{t}_2)} - e^{-\frac{r\delta}{r-1}(t-\tilde{t}_1)}\right)^{1-\frac{1}{r}}.$$

Case 3. For each $t \in (s_i; t_i]$, $s_i \leq \tilde{t}_1 \leq \tilde{t}_2 \leq t_i$, any $y \in B_R$, we have

$$\|(\Lambda y)(\tilde{t}_2) - (\Lambda y)(\tilde{t}_1)\|$$

$$\leq \gamma_i\left(\tilde{t}_2, -\frac{\partial}{\partial s}S(\tilde{t}_2, s_i)\gamma_{i-1}(s_{i-1}, y(t_{i-1}^-)) + S(\tilde{t}_2, s_{i-1})\zeta_{i-1}(s_{i-1}, y(t_{i-1}^-))\right.$$

$$+ \int_{s_{i-1}}^{\tilde{t}_2} S(t, s)f\left(s, y(s), \int_0^s g(s, \tau, y(\tau))d\tau\right)ds\right)$$

$$- \gamma_i\left(\tilde{t}_1, -\frac{\partial}{\partial s}S(\tilde{t}_1, s_i)\gamma_{i-1}(s_{i-1}, y_n(t_{i-1}^-))\right.$$

$$+ S(\tilde{t}_1, s_{i-1})\zeta_{i-1}(s_{i-1}, y_n(t_{i-1}^-))$$

$$+ \int_{s_{i-1}}^{\tilde{t}_1} S(\tilde{t}_1, s)f\left(s, y(s), \int_0^s g(s, \tau, y(\tau))d\tau\right)ds\right).$$

Then,

$$\|(\Lambda y)(\tilde{t}_2) - (\Lambda y)(\tilde{t}_1)\|$$

$$\leq \left\| -\frac{\partial}{\partial s}S(\tilde{t}_2, s_i)\gamma_{i-1}(s_{i-1}, y(t_{i-1}^-)) + S(\tilde{t}_2, s_{i-1})\zeta_{i-1}(s_{i-1}, y(t_{i-1}^-))\right.$$

$$+ \int_{s_{i-1}}^{\tilde{t}_2} S(t, s)f\left(s, y(s), \int_0^s g(s, \tau, y(\tau))d\tau\right)ds)$$

$$+ \frac{\partial}{\partial s}S(\tilde{t}_1, s_i)\gamma_{i-1}(s_{i-1}, y_n(t_{i-1}^-)) - S(\tilde{t}_1, s_{i-1})\zeta_{i-1}(s_{i-1}, y_n(t_{i-1}^-))$$

$$\left. - \int_{s_{i-1}}^{\tilde{t}_1} S(\tilde{t}_1, s)f\left(s, y(s), \int_0^s g(s, \tau, y(\tau))d\tau\right)ds)\right\|.$$

Similarly, one can easily see that

$$\|(\Lambda y)(\tilde{t}_2) - (\Lambda y)(\tilde{t}_1)\|$$
$$\leq \left\| \frac{\partial}{\partial s} S(\tilde{t}_2, s_{i-1}) - \frac{\partial}{\partial s} S(\tilde{t}_1, s_{i-1}) \right\|_{B(E)} \|\gamma_{i-1}(s_{i-1}, y(t_{i-1}^-))\|$$
$$+ \|S(\tilde{t}_2, s_{i-1}) - S(\tilde{t}_1, s_{i-1})\|_{B(E)} \|\zeta_{i-1}(s_{i-1}, y(t_{i-1}^-))\|$$
$$+ \psi\left(R + \varphi(R)\|\Theta_g\|_{L^1}\right) \int_{s_{i-1}}^{\tilde{t}_1} \|S(\tilde{t}_2, \tau) - S(\tilde{t}_1, \tau)\|_{B(E)} \, \Theta_f(\tau) d\tau$$
$$+ \frac{M \|\Theta_f\|_{L^r} \psi\left(R + \varphi(R)\|\Theta_g\|_{L^1}\right)}{\delta^{1-\frac{1}{r}}} \left(e^{-\frac{r\delta}{r-1}(t-\tilde{t}_2)} - e^{-\frac{r\delta}{r-1}(t-\tilde{t}_1)} \right)^{1-\frac{1}{r}}.$$

In view of **Cases 1–3**, as a result, $\|(\Lambda y)(\tilde{t}_2) - (\Lambda y)(\tilde{t}_1)\| \to 0$ as $\tilde{t}_2 \to \tilde{t}_1$, which means that Λ is equicontinuous.

Step 4. Λ is a α_{PC}-contraction operator.
For every bounded subset $B \subset PC(J, E)$, by Lemma 2.4, there exists a countable set $B_1 = \{y\}_{n=1}^{\infty} \subset B$, such that for each $t \in J$, we have

$$\alpha(\Lambda(B)(t)) \leq 2\alpha(\Lambda(B_1)(t)). \tag{8.24}$$

Note that B and ΛB are equicontinuous, we can get from Lemma 2.4, Lemma 2.6, Lemma 8.1 and using the assumptions (8.22.1)–(8.22.6), we obtain
Case 1. For each $t \in (t_i; s_i]$, we have

$$\alpha(\Lambda B_1(t)) \leq \tilde{M} k_i \left\{ \alpha\left(\gamma_{i-1}(s_{i-1}, y_n(t_{i-1}^-))\right) \right\}_{n=0}^{\infty}$$
$$+ M k_i \left\{ \alpha\left(\zeta_{i-1}(s_{i-1}, y_n(t_{i-1}^-))\right) \right\}_{n=0}^{\infty}$$
$$+ k_i \alpha \left(\left\{ \int_{s_{i-1}}^{t} S(t,s) f(s, y_n(s), \int_0^s g(s,\tau, y_n(s)) d\tau) ds \right\}_{n=0}^{\infty} \right)$$
$$\leq \tilde{M} k_i k_{i-1} \left\{ \alpha(y_n(t_i^-)) \right\}_{n=0}^{\infty}) + M k_i \bar{k}_{i-1} \left\{ \alpha(y_n(t_{i-1}^-)) \right\}_{n=0}^{\infty})$$
$$+ 2 M k_i \int_{s_{i-1}}^{t} \left\{ \alpha\left(f(s, y_n(s), \int_0^s K(s,\tau, y_n(\tau)) d\tau) ds \right) \right\}_{n=0}^{\infty} ds$$
$$\leq \tilde{M} k_i k_{i-1} \left\{ \alpha(y_n(t_i^-)) \right\}_{n=0}^{\infty}) + M k_i \bar{k}_{i-1} \left\{ \alpha(y_n(t_{i-1}^-)) \right\}_{n=0}^{\infty})$$
$$+ 2 M k_i \int_{s_{i-1}}^{t} \sigma_1(s) \left\{ \alpha(y_n(s)) \right\}_{n=0}^{\infty}$$
$$+ \varrho_i(s) \left\{ \alpha\left(\int_0^s g(s,\tau, y_n(\tau)) d\tau \right) \right\}_{n=0}^{\infty} \right) ds$$

$$\leq \tilde{M}k_i k_{i-1} \{\alpha(y_n(t_i^-))\}_{n=0}^{\infty}) + Mk_i\bar{k}_{i-1}\{\alpha(y_n(t_{i-1}^-))\}_{n=0}^{\infty})$$

$$+ 2Mk_i \int_{s_{i-1}}^{t} \sigma_i(s)\{\alpha(y_n(s))\}_{n=0}^{\infty})$$

$$+ 2K^*\varrho_i(s)\left\{\int_0^s \alpha(y_n(\tau))d\tau\right\}_{n=0}^{\infty}\right)ds$$

$$\leq (\tilde{M}k_i k_{i-1} + Mk_i\bar{k}_{i-1})\alpha(B_1(t_i^-))$$

$$+ 2Mk_i \int_{s_{i-1}}^{t} \left(\sigma_i(s)\alpha(B_1(s)) + 2K^*\varrho_i(s)\int_0^s \alpha(B_1(\tau))d\tau\right)ds.$$

$$\leq (\tilde{M}k_i k_{i-1} + k_i\bar{k}_{i-1})\sup_{s\in(t_i,s_i]}\alpha(B(t))$$

$$+ 2Mk_i \int_{s_{i-1}}^{t} \left(\sigma(s)\alpha(B_1(s)) + 2K^*\varrho(s)s\sup_{\tau\in[0,s]}\alpha(B_1(\tau))\right)ds.$$

$$\leq (\tilde{M}k_i k_{i-1} + Mk_i\bar{k}_{i-1})\sup_{s\in(s_i,t_{i+1}]}\alpha(B(t))$$

$$+ 2M \int_{s_{i-1}}^{t} \left(\sigma(s)\sup_{s\in(s_i,t_{i+1}]}\alpha(B_1(s))\right.$$

$$\left. + 2K^*\varrho(s)s\sup_{\tau\in(t_i,s_i]}\alpha(B_1(\tau))\right)ds.$$

$$\leq (\tilde{M}k_i k_{i-1} + Mk_i\bar{k}_{i-1})\sup_{s\in(t_i,s_i]}\alpha(B(t))$$

$$+ 2Mk_i \int_{s_{i-1}}^{t} (\sigma(s) + 2K^*s\varrho(s))\sup_{s\in(t_i,s_i]}\alpha(B_1(s))ds$$

$$\leq k_i(\tilde{M}k_{i-1} + M\bar{k}_{i-1} + 2M(\|\sigma\|_{L^1} + 2K^*s_i\|\varrho\|_{L^1}))\sup_{t\in(t_i,s_i]}\alpha(B(t))$$

$$\leq k_i(\tilde{M}k_{i-1} + M\bar{k}_{i-1} + 2M(\|\sigma\|_{L^1} + 2K^*a\|\varrho\|_{L^1})\sup_{t\in(t_i,s_i]}\alpha(B(t)).$$

Then,

$$\alpha(N(B(t)) \leq k_i\left(\tilde{M}k_{i-1} + M\bar{k}_{i-1} + 2M(\|\sigma\|_{L^1} + 2K^*a\|\varrho\|_{L^1}\right)\alpha_{PC}(B(t)), \tag{8.25}$$

Case 2. For each $t \in [0; t_1]$, we have

$$\alpha(\Lambda B_1(t)) \leq \alpha\left(\left\{\int_0^t S(t,s)f(s,y_n(s),\int_0^s g(s,\tau,y_n(s))d\tau)ds\right\}_{n=0}^{\infty}\right)$$

$$\leq 2M \int_0^t \left\{\alpha\left(f(s,y_n(s),\int_0^s g(s,\tau,y_n(\tau))d\tau)ds)\right)\right\}_{n=0}^{\infty}ds$$

$$\leq 2M \int_0^t \sigma_1(s) \left\{\alpha(y_n(s))\right\}_{n=0}^{\infty} ds$$

$$+ \varrho_i(s) \left\{\alpha \left(\int_0^s g(s,\tau,y_n(\tau))d\tau \right) \right\}_{n=0}^{\infty} \Bigg) ds$$

$$\leq 2M \int_0^t \sigma_i(s) \left\{\alpha(y_n(s))\right\}_{n=0}^{\infty}$$

$$+ 2K^* \varrho_i(s) \left\{ \int_0^s \alpha(y_n(\tau))d\tau \right\}_{n=0}^{\infty} \Bigg) ds$$

$$\leq 2M \int_0^t \left(\sigma_i(s)\alpha(B_1(s)) + 2K^* \varrho_i(s) \int_0^s \alpha(B_1(\tau))d\tau \right) ds$$

$$\leq 2M \int_0^t \left(\sigma(s)\alpha(B_1(s)) + 2K^* \varrho(s)s \sup_{\tau \in [0,s]} \alpha(B_1(\tau)) \right) ds$$

$$\leq 2M \int_0^t \left(\sigma(s) \sup_{s \in [0;t_1]} \alpha(B_1(s)) + 2K^* \varrho(s)s \sup_{\tau \in [0;t_1]} \alpha(B_1(\tau)) \right) ds$$

$$\leq 2M \int_0^t \left(\sigma(s) + 2K^* s\varrho(s) \right) \sup_{s \in [0;t_1]} \alpha(B_1(s)) ds.$$

$$\leq 2M(\|\sigma\|_{L^1} + 2K^* t_1 \|\varrho\|_{L^1}) \sup_{t \in [0;t_1]} \alpha(B(t))$$

$$\leq 2M(\|\sigma\|_{L^1} + 2K^* a \|\varrho\|_{L^1}) \sup_{t \in [0;t_1]} \alpha(B(t)).$$

Then,

$$\alpha(\Lambda(B(t)) \leq 2M(\|\sigma\|_{L^1} + 2K^* a \|\varrho\|_{L^1}))\alpha_{PC}(B(t)). \qquad (8.26)$$

Case 3. For each $t \in (s_i; t_{i+1}]$, we have

$$\alpha(\Lambda B_1(t)) \leq \tilde{M} \left\{ \alpha \left(\gamma_i(s, y_n(t_i^-)) \right) \right\}_{n=0}^{\infty} + M \left\{ \alpha \left(\zeta_i(s, y_n(t_i^-)) \right) \right\}_{n=0}^{\infty}$$

$$+ \alpha \left(\left\{ \int_{s_i}^t S(t,s) f(s, y_n(s), \int_0^s g(s,\tau,y_n(s))d\tau) ds \right\}_{n=0}^{\infty} \right)$$

$$\leq \tilde{M} k_i \left\{\alpha(y_n(t_i^-))\right\}_{n=0}^{\infty}) + M \bar{k}_i \left\{\alpha(y_n(t_i^-))\right\}_{n=0}^{\infty})$$

$$+ 2M \int_{s_i}^t \left\{ \alpha \left(f(s, y_n(s), \int_0^s g(s,\tau,y_n(\tau))d\tau) ds \right) \right\}_{n=0}^{\infty} ds$$

$$\leq \tilde{M} k_i \left\{\alpha(y_n(t_i^-))\right\}_{n=0}^{\infty}) + M \bar{k}_i \left\{\alpha(y_n(t_i^-))\right\}_{n=0}^{\infty})$$

$$+ 2M \int_{s_i}^t \sigma_1(s) \left\{\alpha(y_n(s))\right\}_{n=0}^{\infty} ds$$

$$+ \varrho_i(s) \left\{\alpha \left(\int_0^s g(s,\tau,y_n(\tau))d\tau \right) \right\}_{n=0}^{\infty} \Bigg) ds$$

$$\leq \tilde{M}k_i \left\{\alpha(y_n(t_i^-))\right\}_{n=0}^{\infty}) + M\bar{k}_i \left\{\alpha(y_n(t_i^-))\right\}_{n=0}^{\infty})$$

$$+ 2M \int_{s_i}^{t} \sigma_i(s) \left\{\alpha(y_n(s))\right\}_{n=0}^{\infty})$$

$$+ 2K^* \varrho_i(s) \left\{\int_0^s \alpha(y_n(\tau))d\tau\right\}_{n=0}^{\infty}\right) ds$$

$$\leq (\tilde{M}k_i + M\bar{k}_i)\alpha(B(t_i^-))$$

$$+ 2M \int_{s_i}^{t} \left(\sigma(s)\alpha(B_1(s)) + 2K^*\varrho(s)\int_0^s \alpha(B_1(\tau)d\tau\right) ds$$

$$\leq (\tilde{M}k_i + M\bar{k}_i) \sup_{s\in(s_i,t_{i+1}]} \alpha(B_1(s))$$

$$+ 2M \int_{s_i}^{t} \left(\sigma(s)\alpha(B_1(s)) + 2K^*\varrho(s)s \sup_{\tau\in[0,s]} \alpha(B_1(\tau))\right) ds$$

$$\leq (\tilde{M}k_i + M\bar{k}_i) \sup_{s\in(s_i,t_{i+1}]} \alpha(B(s))$$

$$+ 2M \int_{s_i}^{t} \left(\sigma(s) \sup_{s\in(s_i,t_{i+1}]} \alpha(B_1(s))\right.$$

$$\left. + 2K^*\varrho(s)s \sup_{\tau\in(s_i,t_{i+1}]} \alpha(B_1(\tau))\right) ds$$

$$\leq (\tilde{M}k_i + M\bar{k}_i)\alpha(B_1(t_i^-))$$

$$+ 2M \int_{s_i}^{t} (\sigma(s) + 2K^*s\varrho(s)) \sup_{s\in(s_i,t_{i+1}]} \alpha(B_1(s))ds.$$

$$\leq \left(\tilde{M}k_i + M\bar{k}_i + 2M(\|\sigma\|_{L^1} + 2K^*t_{i+1}\|\varrho\|_{L^1})\right) \sup_{t\in(s_i,t_{i+1}]} \alpha(B(t))$$

$$\leq \left(\tilde{M}k_i + M\bar{k}_i + 2M(\|\sigma\|_{L^1} + 2K^*a\|\varrho\|_{L^1})\right) \sup_{t\in(s_i,t_{i+1}]} \alpha(B(t)).$$

Then,

$$\alpha(\Lambda(B(t))) \leq \left(\tilde{M}k_i + M\bar{k}_i + 2M(\|\sigma\|_{L^1} + 2K^*a\|\varrho\|_{L^1})\right) \alpha_{PC}(B(t)). \tag{8.27}$$

From the above cases (8.25), (8.26) and (8.27), for all $t \in J$, we obtain

$$\alpha_{PC}(\Lambda(B))$$

$$\leq \max_{0\leq i\leq N}(k_i, 1) \left(\max_{0\leq i\leq m} (\tilde{M}k_i + M\bar{k}_i) + 2M(\|\sigma\|_{L^1} + 2K^*a\|\varrho\|_{L^1}))\right)$$

$$\times \alpha_{PC}(B).$$

Thus, we find that Λ is α_{PC}-contraction operator. By means of lemma 2.9, as the proof of Theorem 8.9, we conclude that (8.19) has at least a mild solution.

Next, we present another existence result for the mild solution of the system (8.19).

Theorem 8.10. *Under the assumption* (8.22.1)–(8.22.6), *the system* (8.19) *has at least one mild solution on* J. *Furthermore, we suppose*

$$\lim_{R \to +\infty} \inf \frac{\psi(R + \|\Theta_g\|_{L^1}\varphi(R))\|\Theta_f\|_{L^r})}{R} = \rho,$$

and

$$\max_{i=0,\dots,m} (\tilde{M}c_i + Me_i) + \frac{M\rho\|\Theta_f\|_{L^r}}{\delta^{1-\frac{1}{r}}} \le 1. \qquad (8.28)$$

Step 1: $\Lambda(y) \in B_R$ for any $y \in B_R$.
Case 1. For each $t \in (s_i, t_i]$, we have,

$$\|(\Lambda y)(t)\| \le \left\| \frac{\partial}{\partial s} S(t, s_{i-1}) \right\|_{B(E)} \|\gamma_i(s_{i-1}, y(s_{i-1}))\|$$

$$+ \|S(t, s_{i-1})\|_{B(E)} \|\zeta_i(s_{i-1}, y(s_{i-1}))\|$$

$$+ \int_{s_{i-1}}^t \|S(t, s)\|_{B(E)} \Theta_f(s) \left(\|y(s)\| + \int_0^s \Theta_g(\tau)\varphi(\|y(\tau)\|)d\tau \right) ds$$

$$\le \tilde{M}c_{i-1} \|y(t)\| + \tilde{M}d_{i-1}$$

$$+ Me_{i-1} \|y(t)\| + Ml_{i-1}$$

$$+ \int_{s_{i-1}}^t Me^{-\delta(t-s)}\Theta_f(s)\psi \left(\|y(s)\| + \int_0^s \Theta_g(\tau)\varphi(\|y(\tau)\|)d\tau \right) ds.$$

Then,

$$\|(\Lambda y)(t)\| \le Mc_{i-1}R + Ml_{i-1} + Me_{i-1}R + Ml_{i-1}$$

$$+ \int_{s_{i-1}}^t Me^{-\delta(t-s)}\Theta_f(s)\psi \left(\|y(s)\| + \int_0^s \Theta_g(\tau)\varphi(\|y(\tau)\|)d\tau \right) ds.$$

$$\le (Mc_{i-1} + Me_{i-1})R + Ml_{i-1} + Ml_{i-1}$$

$$+ M\psi(R + \|\Theta_g\|_{L^1}\varphi(R)) \int_{s_{i-1}}^t e^{-\delta(t-s)}\Theta_f(s)ds.$$

It follows from the Hölder's inequality that

$$\|(\Lambda y)(t)\| \le (Mc_{i-1} + Me_{i-1})R + Ml_{i-1} + Ml_{i-1}$$

$$+ \frac{M\psi(R + \|\Theta_g\|_{L^1}\varphi(R))\|\Theta_f\|_{L^r}}{\delta^{1-\frac{1}{r}}}.$$

Case 2. For each $t \in [0; t_1]$, we have

$$\|(\Lambda y)(t)\| \leq \left\|\frac{\partial}{\partial s}\mathcal{U}(t,0)\right\|_{B(E)} \|y_0\|$$
$$+ \|S(t,s)\|_{B(E)} \|y_1\|$$
$$+ \int_0^t \|S(t,s)\|_{B(E)} \Theta_f(s)$$
$$\psi\left(\|y(s)\| + \int_0^s \Theta_g(\tau)\varphi(\|y(\tau)\|)d\tau\right) ds$$
$$\leq \tilde{M}\|y_1\| + M\|y_0\| + M\psi(R + \|\Theta_g\|_{L^1}\varphi(R)) \int_0^t e^{-\delta(t-s)}\Theta_f(s)ds.$$

It follows from the Hölder's inequality that

$$\|(\Lambda y)(t)\| \leq \tilde{M}\|y_0\| + M\|y_1\|$$
$$+ M\psi(R + \|\Theta_g\|_{L^1}\varphi(R))\|\Theta_f\|_{L^r}(1 - e^{-\frac{r\delta}{r-1}t})^{1-\frac{1}{r}}$$
$$\leq \tilde{M}\|y_0\| + M\|y_1\| + \frac{M\psi(R + \|\Theta_g\|_{L^1}\varphi(R))\|\Theta_f\|_{L^r}}{\delta^{1-\frac{1}{r}}}.$$

Case 3. For each $t \in (s_i, t_{i+1}]$, we have,

$$R < \|(\Lambda y)(t)\|$$
$$\leq \tilde{M}c_i\|y(t)\| + \tilde{M}d_i$$
$$+ Me_i\|y(t)\| + Ml_i$$
$$+ \int_{s_i}^t \|\mathcal{U}(t,s)\|_{B(E)} \Theta_f(s)\psi\left(\|y(s)\| + \int_0^s \Theta_g(\tau)\varphi(\|y(\tau)\|)d\tau\right) ds.$$
$$\leq \tilde{M}c_iR + \tilde{M}d_i + Me_iR + Ml_i$$
$$+ M\psi(R + \|\Theta_g\|_{L^1}\varphi(R)) \int_{s_i}^t e^{-\delta(t-s)}\Theta_f(s)ds.$$

It follows from the Hölder's inequality that

$$\|(\Lambda y)(t)\| \leq \tilde{M}d_i + Ml_i$$
$$+ (\tilde{M}c_i + Me_i)R$$
$$+ \frac{M\psi(R + \|\Theta_g\|_{L^1}\varphi(R))\|\Theta_f\|_{L^r}}{\delta^{1-\frac{1}{r}}}.$$

In fact, if we assume that the assertion is false, then $R < \|(\Lambda y)(t)\|$. By means of **Steps 1–3**, we have

$$R < \|(\Lambda y)(t)\| \le \max_{i=0,\dots,m} (\tilde{M}c_i + Me_i)R$$
$$+ \max_{i=0,\dots,m} (\tilde{M}d_i + Ml_i, \tilde{M}\|y_0\| + M\|y_1\|)$$
$$+ \frac{M\psi(R + \|\Theta_g\|_{L^1}\varphi(R))\|\Theta_f\|_{L^r}}{\delta^{1-\frac{1}{r}}}.$$

Dividing both sides by R and taking the $\liminf$ as $R \to +\infty$, we have

$$\max_{i=0,\dots,m} (\tilde{M}c_i + Me_i) + \frac{M\rho\|\Theta_f\|_{L^r}}{\delta^{1-\frac{1}{r}}} > 1,$$

which contradicts (8.28). Hence, the operator Λ transforms the set B_R into itself.

The proofs of the other steps are similar to those in Theorem 8.9. Therefore, we omit the details. By means of Theorem 2.11, we conclude that (8.19) has at least a mild solution. The proof is complete.

8.4.2 *An Example*

In this section, we give an example to illustrate the above theoretical result.

Set $E = L^2([0,\pi], \mathbb{R})$ be the space of all square integrable functions from $[0,\pi]$ into $\mathbb{R}$. We denote by $\mathbb{H}^2([0,\pi], \mathbb{R})$ the Sobolev space of functions $u : [0,\pi] \to \mathbb{R}$, such that $u'' \in L^2([0,\pi], \mathbb{R})$. Define the operator $D(\mathbb{A}) \to E$ by

$$\mathbb{A}u(\tau) = u''(\tau),$$

with domain $D(A_1) = \mathbb{H}^2(\mathbb{R}, \mathbb{C})$. It is well known that $\mathbb{A}$ is the infinitesimal generator of a C_0-semigroup and of a strongly continuous cosine function on E, which will be denoted by $(C(t))$. From [8], for all $x \in \mathbb{H}^2([0,\pi], \mathbb{R}), t \in \mathbb{R}$, $\|C(t)\|_{B(E)} \le 1$. Define also the operator $\mathbb{H}^1([0,\pi], \mathbb{R}) \to E$ by

$$\mathbb{B}(t)u(s) = a(t)u'(s),$$

where $a : [0,1] \to \mathbb{R}$ is a Hölder continuous function.

Consider the closed linear operator $\mathcal{A}(t) = \mathbb{B} + \mathbb{A}(t)$. It has been proved by Henríquez in [399] that the family $\{\mathcal{A}(t) : t \in J\}$ generates an evolution operator $\{S(t,s)\}_{t,s\in\Delta}$. Moreover, $S(\cdot,\cdot)$ is well defined and satisfies the conditions (8.22.1) and (8.22.2).

We consider the following system:

$$
\begin{cases}
\dfrac{\partial^2}{\partial t^2} u(t,\tau) = \dfrac{\partial^2}{\partial \tau^2} u(t,\tau) + a(t)\dfrac{\partial}{\partial t} u(t,\tau) + \dfrac{u(t,\tau)}{(\sqrt{t}+1)(1+\|u(t,\tau)\|)} \\
\qquad + \dfrac{e^{-t}}{(\sqrt{t}+1)(t+1)} \displaystyle\int_0^t \dfrac{\sqrt{t}u(s,\tau)}{(1+s^2+t)(1+u^2(s,\tau))}ds, \\
\qquad t \in \left(0,\tfrac{1}{\sqrt{3}}\right] \cup \left(\tfrac{2}{\sqrt{3}},1\right], \\[2mm]
u(t,\tau) = \cos \pi t \, u\left(\tfrac{2}{\sqrt{3}}^{-},\tau\right), \ t \in \left(\tfrac{1}{\sqrt{3}},\tfrac{2}{\sqrt{3}}\right], \tau \in [0,\pi], \\[2mm]
\dfrac{\partial}{\partial t} z\,(t,\tau) = \sin \pi t \, z\left(\dfrac{2}{\sqrt{3}}^{-},\tau\right), \ t \in \left(\dfrac{1}{\sqrt{3}},\dfrac{2}{\sqrt{3}}\right], \tau \in [0,\pi], \\[2mm]
u(t,0) = u(t,\pi) = 0, \ t \in [0,1], \\[2mm]
\dfrac{\partial}{\partial t} u(0,\tau) = v(\tau), \ \tau \in [0,\pi].
\end{cases}
\tag{8.29}
$$

Take $a = t_2 = 1$, $t_0 = s_0 = 0$, $t_1 = \frac{1}{\sqrt{3}}$, $s_1 = \frac{2}{\sqrt{3}}$. By putting

$$u(t)(\tau) = w(t)(\tau), \ t \geq 0, \ \tau \in [0,\pi].$$

Denote

$$
f(t,y,z)(\tau) = \dfrac{\ln(1+\|y(t,\tau)\|)}{(\sqrt{t}+1)(1+\|y(t,\tau)\|)} + \dfrac{e^{-t}}{(\sqrt{t}+1)(t+1)} z(t,\tau),
$$

$$
g(t,s,y)(\tau) = \dfrac{\sqrt{t}y(t,\tau)}{(1+s^2+t)(1+y^2(t,\tau))},
$$

$$
\gamma_i(t,y(t^{-})) = \cos \pi t \, z\left(\dfrac{2}{\sqrt{3}}^{-},\tau\right),
\tag{8.30}
$$

$$
\zeta_i(t,y(t^{-})) = \sin \pi t \, z\left(\dfrac{2}{\sqrt{3}}^{-},\tau\right),
\tag{8.31}
$$

and

$$
\dfrac{\partial}{\partial t} z(0)(\tau) = \dfrac{d}{dt} w(0)(\tau), \ \tau \in [0,\pi].
$$

By a simple computation, we have

$$
\|f(t,y,z)(\tau)\| \leq \dfrac{1}{1+\sqrt{t}} \psi(\|y(t,\tau)\| + \|z(t,\tau)\|),
\tag{8.32}
$$

and

$$|g(t,s,y)(\tau)\| \le \frac{\sqrt{t}}{1+t}\|y(t,\tau)\|. \tag{8.33}$$

From the above discussion, we obtain

$$\psi(t) = \ln(1+t) + t, \quad \Theta_f(t) = \frac{1}{1+\sqrt{t}}, \quad \Theta_g(t) = \frac{\sqrt{t}}{1+t}.$$

For each $t \in J$, and $W_1, W_2 \in W \subset E$, we get

$$\alpha(f(t,W_1,W_2)) \le \frac{1}{(\sqrt{t}+1)}\alpha(D_1) + \frac{e^{-t}}{(\sqrt{t}+1)(t+1)}\alpha(W_2),$$

We shall show that condition (8.22.6) holds with

$$\sigma_1(t) = \frac{1}{(\sqrt{t}+1)}, \quad \sigma_2(t) = \frac{e^{-t}}{(\sqrt{t}+1)(t+1)}.$$

By (8.33), for every $t \in J$ and $D \subset E$, we have

$$\alpha(g(t,s,W)) \le \sup_{t\in[0,1]} \frac{\sqrt{t}}{1+t}\alpha(W),$$

then

$$\alpha(K(t,s,W)) \le \frac{\sqrt{2}}{3}\alpha(W).$$

Hence (8.22.5) is satisfied with $K^* = \dfrac{\sqrt{2}}{3}$.

Next, let us observe that, in view of (8.30) and (8.31), the mapping g and h fulfill the hypotheses (8.22.5) and (8.22.6) with $c_i = e_i = k_i = \bar{k}_i = 1$ and $d_i = l_i = 0$.

Consequently, (8.29) can be written in the abstract form (8.19) with $A(t)$ and f as defined above. The existence of mild solutions can be deduced from an application of Theorem 8.9.

8.5   Notes and Remarks

The results discussed in this chapter are taken from the articles [400–403]. We refer the reader to the monographs [160, 192, 404, 405], the papers [406–410] and the references therein, for some additional and relevant results.

Chapter 9

Periodic Mild Solutions of Evolution Equations with Impulses and Delay

9.1 Introduction and Motivations

This chapter deals with the existence of periodic mild solutions for a class of functional evolution equations with impulses and delay. The techniques used are some fixed point theorems in Banach spaces (Darbo, Kuratowski and Sadovskii fixed point theorems), the Poincaré operator and the measure of non-compactness. Furthermore, illustrative examples are presented to demonstrate the plausibility of our results.

We studied and proved the results in this chapter by considering the previous chapters' publications as well as the publications that follow:

- In [411], an iterative method is used for the existence of mild solutions of evolution equations and inclusions. Using the Tichonov's fixed point theorem, Olszowy and Wędrychowicz [396] considered a class of evolution equations on unbounded intervals. However, in the previous papers, the authors assumed some restrictions like the compactness of the semigroup, the Lipschitz conditions on the nonlinear term or the boundedness of the obtained mild solutions.
- The monographs [2,11,226,269,412] and the papers [12,41,396,411,413], where there has been a significant development in functional evolution equations.
- Most of the research papers deal with the existence of solutions for differential equations with instantaneous impulsive conditions see [154, 414–417].
- In [396], the authors considered a class of evolution equations on unbounded intervals by using the Tichonov's fixed point theorem.

However in the previous papers some restrictions like, the compactness of the semigroup, the Lipschitz conditions on the nonlinear term or the boundedness of the obtained mild solutions, are supposed. Functional differential equations with non-instantaneous impulsive was studied in [177, 418–420].

- In [14, 15, 19], the authors used the Poincaré operator and proved some results concerning the existence of periodic solutions of infinite delay evolution equations.

9.2 Infinite Delay Evolution Equations with Non-Instantaneous Impulses

In the present article, we discuss the existence of periodic mild solutions of the following class of functional differential equations with infinite delay and non-instantaneous impulses

$$
\begin{cases}
y'(t) + A(t)y(t) = f(t, y(t), y_t); & \text{if } t \in I_k; \ k = 0, 1, \ldots, \\
y(t) = g_k(t, y(t_k^-)); & \text{if } t \in J_k; \ k = 1, 2, \ldots, \\
y(t) = \phi(t); & \text{if } t \in \mathbb{R}_- := (-\infty, 0],
\end{cases}
\tag{9.1}
$$

where $I_0 = [0, t_1]$, $I_k := (s_k, t_{k+1}]$, $J_k := (t_k, s_k]$, $0 = s_0 < t_1 \leq s_1 \leq t_2 < \cdots < s_{m-1} \leq t_m \leq s_m \leq t_{m+1} = T \leq s_{m+1} \leq t_{m+2} \leq \ldots < +\infty$, $f : I_k \times E \times \mathcal{B} \to E$; $k = 0, \ldots,$ $g_k : J_k \times E \to E$; $k = 1, 2, \ldots,$ are given functions T-periodic in t, $T > 0$, $\mathcal{B}$ is an abstract phase space to be specified later, $\phi : \mathbb{R}_- \to E$ is a given function, $\{A(t)\}_{t>0}$ is a T-periodic family of unbounded operators from E into E that generate an evolution system of operators $\{U(t, s)\}_{(t,s) \in \mathbb{R}_+ \times \mathbb{R}_+}$; for $(t, s) \in \Lambda := \{(t, s) \in \mathbb{R}_+ \times \mathbb{R}_+ : 0 \leq s \leq t < +\infty\}$, $\mathbb{R}_+ := [0, \infty)$, and $(E, \|\cdot\|_E)$ is a real Banach space.

For any continuous function y and any $t \in \mathbb{R}_+$, we denote by y_t the element of $\mathcal{B}$ defined by $y_t(\theta) = y(t + \theta)$ for $\theta \in \mathbb{R}_- := (-\infty, 0]$. Here, $y_t(\cdot)$ represents the history of the state up to the present time t. We assume that the histories y_t belong to $\mathcal{B}$.

In [14, 15] the authors consider the existence of periodic mild solution of delay evolution equations subject to instantaneous impulse. In this paper we consider the existence of periodic mild solutions for evolution equations with infinite delay and non-instantaneous impulses, which are a more general class of impulsive evolution equations. We use the classical Darbo fixed point theorem, the Poincaré operator and the concept of measure of non-compactness in Banach spaces.

9.2.1   *Necessary Results*

Consider the space

$$\tilde{C}((-\infty,0],E) = \Big\{ y : (-\infty,0] \to E : \ y \text{ is continuous and there exist}$$

$$\tau_k \in (-\infty,0); k = 1,\ldots,m, \text{ such that } y(\tau_k^-) \text{ and } y(\tau_k^+)$$

$$\text{exist with } y(\tau_k^-) = y(\tau_k) \Big\},$$

and the Banach space

$$PC = \Big\{ y : (-\infty,T] \to E : y|_{\mathbb{R}_-} \in \mathcal{B}, \ y|_{J_k} = g_k; \ k = 1,\ldots,m, \ y|_{I_k};$$

$$k = 1,\ldots,m, \text{ is continuous and there exist } y(s_k^-), \ y(s_k^+), \ y(t_k^-)$$

$$\text{and } y(t_k^+) \text{with } y(s_k^+) = g_k(s_k, y(s_k^-)) \text{ and } y(t_k^-) = g_k(t_k, y(t_k^-)) \Big\},$$

with the norm

$$\|y\|_{PC} = \max\{\|y\|_C, \|\phi\|_{\mathcal{B}}\}.$$

In what follows, for the family $\{A(t), \ t \geq 0\}$ of closed densely defined linear unbounded operators on the Banach space E we assume that it satisfies the following assumptions (see [226], p. 158).

(P_1) The domain $D(A(t))$ is independent of t and is dense in E,

(P_2) For $t \geq 0$, the resolvent $R(\lambda, A(t)) = (\lambda I - A(t))^{-1}$ exists for all λ with $Re\lambda \leq 0$, and there is a constant K independent of λ and t such that

$$\|R(t, A(t))\| \leq K(1 + |\lambda|)^{-1}, \text{ for } Re\lambda \leq 0,$$

(P_3) There exist constants $L > 0$ and $0 < \alpha \leq 1$ such that

$$\|(A(t) - A(\theta))A^{-1}(\tau)\| \leq L|t - \tau|^{\alpha}, \text{ for } t, \theta, \tau \in J.$$

Lemma 9.1 ([11, 226]). *Under assumptions $(P_1) - (P_3)$, the Cauchy problem*

$$y'(t) - A(t)y(t) = 0, \ t \in J, \ y(0) = y_0,$$

has a unique evolution system $U(t,s)$, $(t,s) \in \Delta := \{(t,s) \in J \times J : 0 \leq s \leq t \leq T\}$ satisfying the following properties:

(1) $U(t,t) = I$ where I is the identity operator in E,
(2) $U(t,s)\,U(s,\tau) = U(t,\tau)$ for $0 \le \tau \le s \le t \le T$,
(3) $U(t,s) \in B(E)$ the space of bounded linear operators on E, where for every $(t,s) \in \Delta$ and for each $y \in E$, the mapping $(t,s) \to U(t,s)y$ is continuous.

More details on evolution systems and their properties can be found in the books of Ahmed [226] and Pazy [11].

In this paper, we assume that the phase space $(\mathcal{B}, \|\cdot\|_{\mathcal{B}})$ is a seminormed linear space of functions mapping $\mathbb{R}_-$ into E and satisfying the following fundamental axioms introduced by Hale and Kato in [308].

(A_1) If $y \in PC$ and $y_0 \in \mathcal{B}$, then for every $t \in J$ the following conditions hold:

   (i) $y_t \in \mathcal{B}$

   (ii) $\|y_t\|_{\mathcal{B}} \le K(t) \displaystyle\int_0^t \|y(s)\|ds + M(t)\|\phi\|_{\mathcal{B}}$,

   (iii) $\|y(t)\| \le H\|y_t\|_{\mathcal{B}}$, where $H \ge 0$ is a constant, $K : J \to \mathbb{R}_+$ is continuous; $M : \mathbb{R}_+ \to \mathbb{R}_+$ is locally bounded and H, K, M, are independent of $y(\cdot)$.

(A_2) For the function $y(\cdot)$ in (A_1), y_t is a $\mathcal{B}$-valued continuous function on J.

(A_3) The space $\mathcal{B}$ is complete.

Denote

$$K_T = \sup\{K(t) : t \in J\}, \quad M_T = \sup\{M(t) : t \in J\}.$$

Remark 9.1. Axiom (A_1)(ii) is equivalent to $\|\phi(0)\| \le H\|\phi\|_{\mathcal{B}}$; for every $\phi \in \mathcal{B}$. From this equivalence; we can see that for all $\phi, \psi \in \mathcal{B}$ such that $\|\phi - \psi\|_{\mathcal{B}} = 0$, we necessarily have $\phi(0) = \psi(0)$.

Lemma 9.2 (Lemma 2.1 in [14]). *There exists an integer $k_0 > 1$ such that*

$$\left(\frac{1}{2}\right)^{k_0-1} M < 1,$$

where $M = \displaystyle\sup_{(t,s)\in\Delta} \|U(t,s)\|_{B(E)}$ *is finite, and there exists a function h on* $(-\infty, 0]$ *such that* $h(0) = 1$, $h(-\infty) = +\infty$, h *is decreasing on* $(-\infty, 0]$, *and for* $d \ge w_0 := \frac{T}{K_0}$ *one has* $\displaystyle\sup_{s\in(-\infty,0]} \frac{h(s)}{h(s-d)} \le \frac{1}{2}$.

Now we indicate two examples of phase spaces.

Example 9.1. For the function h given in Lemma 9.2, we define the spaces:
$$C_h := \left\{ \phi \in \tilde{C}(\mathbb{R}_-, E) : \frac{\phi(\theta)}{h(\theta)} \text{ is bounded on } \mathbb{R}_- \right\},$$
and $C_h^0 := \left\{ \phi \in C_h : \lim_{\theta \to -\infty} \frac{\phi(\theta)}{h(\theta)} = 0 \right\}$, endowed with the uniform norm

$$\|\phi\| = \sup \left\{ \frac{\|\phi(\theta)\|}{h(\theta)} : \theta \leq 0 \right\}.$$

Then we have that the spaces C_h and C_h^0 satisfy condition (A_3).
Also; C_h and C_h^0 satisfy conditions (A_1) and (A_2) if

$$\sup_{t \in J} \sup_{-\infty < \theta \leq -t} \frac{\phi(t + \theta)}{h(\theta)} < \infty.$$

Example 9.2. For any real positive constant γ, we define the space

$$C_\gamma := \{ \phi \in \tilde{C}((-\infty, 0]), E) : \lim_{\theta \to -\infty} e^{\gamma\theta} \phi(\theta) \text{ exist in } E \},$$

endowed with the norm

$$\|\phi\| = \sup \{ e^{\gamma\theta} \|\phi(\theta)\| : \theta \leq 0 \}.$$

Then in the space C_γ the axioms $(A_1) - (A_3)$ are satisfied.

In all what follows, we consider the phase space

$$\mathcal{B} := \left\{ \phi \in \tilde{C}((-\infty, 0]), E) : \sup_{s \in (-\infty, 0]} \frac{\|\phi(s)\|}{h(s)} < \infty \right\},$$

where $h : \mathbb{R}_- \to \mathbb{R}_+$ is the function given in Lemma 9.2.
Then $\mathcal{B}$ endowed with the norm

$$\|\phi\|_{\mathcal{B}} = \sup_{s \in (-\infty, 0]} \frac{\|\phi(s)\|}{h(s)},$$

is a Banach space [16].

9.2.2 Existence Results

Definition 9.1. By a periodic mild solution of problem (9.1) we mean a measurable and T-periodic function y that satisfies

$$y(t) = \begin{cases} U(t,0)\phi(0) + \displaystyle\int_0^t U(t,s)\, f(s,y(s),y_s)ds.; & \text{if } t \in I_0, \\[4mm] U(t,s_k)g_k(s_k,y(s_k^-)) \\ \quad + \displaystyle\int_{s_k}^t U(t,s)\, f(s,y(s),y_s)ds.; & \text{if } t \in I_k;\ k=1,\ldots,m, \\[4mm] g_k(t,y(t_k^-));\ \text{if } t \in J_k;\ k=1,\ldots,m, \\[4mm] \phi(t);\ \text{if } t \in \mathbb{R}_-. \end{cases}$$

Now, we shall prove the following theorem concerning the existence of periodic mild solutions of problem (9.1).

Theorem 9.1. *Assume that the following hypotheses hold:*

(9.7.1) *The functions f and g_k are continuous in their variables, and they map bounded sets into bounded sets.*

(9.7.2) *The function $t \mapsto f(t,y,v)$ is measurable on I_k, $k = 0,\ldots,m$, for each $y,v \in E \times \mathcal{B}$, and the functions $y \mapsto f(t,y,v)$ and $v \mapsto f(t,y,v)$ are continuous on $E \times \mathcal{B}$ for a.e. $t \in I_k$; $k = 0,\ldots,m$.*

(9.7.3) *For a constant $T > 0$, $f(t+T,y,v) = f(t,y,v)$, $A(t+T) = A(t)$; $t \in I_k$; $k = 0,\ldots,m$, $y,v \in E \times \mathcal{B}$, and $g_k(t+T,z) = g_k(t,z)$ $t \in J_k$; $k = 1,\ldots,m$, $z \in E$.*

(9.7.4) *There exist continuous functions $p : I_k \to \mathbb{R}_+$, $q : J_k \to \mathbb{R}_+$, such that*

$$\|f(t,y,v)\| \le p(t),\ \text{for a.e. } t \in I_k;\ k=0,\ldots,m,\ \text{and each } y,v \in E \times \mathcal{B},$$

and

$$\|g_k(t,z)\| \le q(t),\ \text{for a.e. } t \in J_k,\ \text{and each } z \in E,\ k=0,\ldots,m.$$

(9.7.5) *For each bounded and measurable sets $B(t) \subset E$; $t \in \mathbb{R}_+$ and $B_t \subset \mathcal{B}$; $t \in \mathbb{R}_+$, such that*

$$B(t) = \{y(t) : y \in C(J)\},\ \text{and } B_t = \{y_t : y_t \in \mathcal{B}\},$$

we have

$$\alpha(f(t, B(t), B_t)) \leq p(t)\alpha(B); \ for \ a.e. \ t \in I_k; \ k = 0, \ldots, m,$$

and

$$\alpha(g_k(t, B)) \leq q(t)\alpha(B); \ for \ a.e. \ t \in J_k; \ k = 1, \ldots, m,$$

where α is a measure of non-compactness on the Banach space E, and

$$M = \sup_{(t,s) \in \Delta} \|U(t, s)\|_{B(E)}, \ \ p^* = \sup_{t \in I_k} p(t), \ \ and \ \ q^* = \sup_{t \in J_k} q(t).$$

If $4MTp^* < 1$, then the problem (9.1) has at least one T-periodic mild solution defined on $\mathbb{R}$.

Proof. The proof will be given in two parts. Consider the problem

$$\begin{cases} y'(t) + A(t)y(t) = f(t, y(t), y_t); \ \text{if } t \in I_k, \ k = 0, \ldots, m, \\ y(t) = g_k(t, y(t_k^-)); \ \text{if } t \in J_k, \ k = 1, \ldots, m, \\ y(t) = \phi(t); \ \text{if } t \in \mathbb{R}_- := (-\infty, 0], \end{cases} \tag{9.2}$$

Part 1. We begin by showing that (9.2) has a mild solution $y \in PC$, with $\|y\|_{PC} \leq R$ where

$$R \geq \max\{\|\phi\|_{\mathcal{B}}, q^*, M(\|\phi(0)\| + p^*), M(q^* + p^*)\}.$$

Consider the operator $N : PC \to PC$ defined by:

$$(Ny)(t) = \begin{cases} U(t, 0)\phi(0) + \displaystyle\int_0^t U(t, s) \ f(s, y(s), y_s)ds.; \ \text{if } t \in I_0, \\ U(t, s_k)g_k(s_k, y(s_k^-)) \\ \quad + \displaystyle\int_{s_k}^t U(t, s) \ f(s, y(s), y_s)ds.; \ \text{if } t \in I_k; \ k = 1, \ldots, m, \\ g_k(t, y(t_k^-)); \ \text{if } t \in J_k; \ k = 1, \ldots, m, \\ \phi(t); \ \text{if } t \in \mathbb{R}_-. \end{cases}$$

$$\tag{9.3}$$

Clearly, the fixed points of the operator N are mild solution of the problem (9.2).

For any $y \in PC$ and each $t \in I_0$, we have

$$\|(Ny)(t)\| \leq M\|\phi(0)\| + M \int_0^t \|f(s, y(s), y_s)\| ds$$
$$\leq M\|\phi(0)\| + Mp^* \leq R.$$

Next, for any $y \in PC$ and each $t \in I_k$; $k = 1, \ldots, m$, we have

$$\|(Ny)(t)\| \leq Mq^* + M \int_{s_k}^t \|f(s, y(s), y_s)\| ds$$
$$\leq Mq^* + Mp^* \leq R.$$

Also, for any $y \in PC$ and each $t \in J_k$; $k = 1, \ldots, m$, we have

$$\|(Ny)(t)\| \leq q^* \leq R,$$

and for any $y \in PC$ and each $t \in \mathbb{R}_-$, we have

$$\|(Ny)(t)\| = \|\phi\|_{\mathcal{B}} \leq R.$$

This proves that N transforms the ball $B_R := \{w \in PC : \|w\|_{PC} \leq R\}$ into itself. We shall show that the operator $N : B_R \to B_R$ satisfies all the assumptions of Theorem 2.6. The proof will be given in two steps.

Step 1. $N : B_R \to B_R$ *is continuous.*
Let $\{y_n\}_{n \in \mathbb{N}}$ be a sequence such that $y_n \to y$ in B_R. For each $t \in \mathbb{R}_- \cup J_k$; $k = 1, \ldots, m$, we have

$$\|(Ny_n)(t) - (Ny)(t)\| = 0 \to 0 \quad \text{as } n \to \infty,$$

and, for each $t \in I_k$; $k = 0, \ldots, m$, we have

$$\|(Ny_n)(t) - (Ny)(t)\| \leq M \int_0^t \|f(s, y_n(s), y_{sn}) - f(s, y(s), y_s)\| ds. \quad (9.4)$$

Since $y_n \to y$ as $n \to \infty$ and f is continuous, then by the Lebesgue dominated convergence theorem, equation (9.4) implies

$$\|(Ny_n)(t) - (Ny)(t)\| \to 0 \quad \text{as } n \to \infty.$$

Hence

$$\|N(y_n) - N(y)\|_{PC} \to 0 \quad \text{as } n \to \infty.$$

Step 2. *For each closed subset D of $C(J)$, $\alpha(N(D)) \leq \ell\alpha(D)$.*
From Lemmas 2.5 and 2.6, for any $D \subset B_R$ and any $\epsilon > 0$, there exists a sequence $\{y_k\}_{k=0}^{\infty} \subset D$, such that for all $t \in I_k$; $k = 0, \ldots, m$, we have

$$\alpha((ND)(t)) = \alpha\left(\left\{U(t,0)\phi(0) + \int_0^t U(t,s)\ f(s,y(s),y_s)ds;\ y \in D\right\}\right)$$

$$\leq 2\alpha\left(\left\{\int_0^t U(t,s)f(s,y_k(s),y_{ks})ds\right\}_{k=1}^{\infty}\right) + \epsilon$$

$$\leq 4\int_0^t \alpha\left(\|U(t,s)\|_{B(E)}\{f(s,y_k(s),y_{ks})\}_{k=1}^{\infty}\right)ds + \epsilon$$

$$\leq 4M\int_0^t \alpha\left(\{f(s,y_k(s),y_{ks})\}_{k=1}^{\infty}\right)ds + \epsilon$$

$$\leq 4M\int_0^t p(s)\alpha\left(\{y_k(s)\}_{k=1}^{\infty}\right)ds + \epsilon$$

$$\leq 4Mp^*\int_0^t \alpha\left(\{y_k(s)\}_{k=1}^{\infty}\right)ds + \epsilon,$$

$$\leq 4MTp^*\alpha(D) + \epsilon,$$

and, for all $t \in I_k$; $k = 1, \ldots, m$, we get

$$\alpha((ND)(t))$$

$$= \alpha\left(\left\{U(t,s_k)g_k(s_k,y(s_k^-)) + \int_{s_k}^t U(t,s)\ f(s,y(s),y_s)ds;\ y \in D\right\}\right)$$

$$\leq 2\alpha\left(\left\{\int_0^t U(t,s)f(s,y_k(s),y_{ks})ds\right\}_{k=1}^{\infty}\right) + \epsilon$$

$$\leq 4MTp^*\alpha(D) + \epsilon.$$

Since $\epsilon > 0$ is arbitrary, then

$$\alpha((ND)(t)) \leq 4MTp^*\alpha(D).$$

As a consequence of these two steps together with Theorem 2.6, we can conclude that N has a fixed point in $y \in B_R$ which is a mild solution of problem (9.1).

Part 2. *Periodic mild solutions.*
A standard approach in deriving T-periodic solutions is to define the Poincaré operator $P : \mathcal{B} \to \mathcal{B}$ given by $P(\phi) = y_T(\phi)$ such that

$$(P\phi)(s) = y_T(s,\phi) = y(T+s,\phi);\ s \in \mathbb{R}_-,$$

which maps an initial function (or value) ϕ along the unique mild solution $y(\phi)$ to our problem (9.1) by $T-$ units (i.e., T units along the unique solution $y(\cdot, \phi)$ determined by the initial function ϕ). We show that P is a condensing operator with respect to Kuratowski's measure of non-compactness in the phase space $\mathcal{B}$, then the given conditions such that the fixed point theorem (Theorem 2.10) can be applied to get fixed points for the Poincaré operator, which give rise to periodic solutions. We do this in two steps.

Step 1. *The fixed points of P give rise to periodic mild solutions of* (9.1). Let $\phi \in \mathcal{B}$ be such that $p(\phi) = \phi$. Then for the solution $y(\cdot) = y(\cdot, \phi)$ with $y_0(\cdot, \phi) = \phi$, we can define $v(t) = y(t + T)$. Now, for $t > 0$, we can use the known properties of $U(t, s)$ in Lemma 9.1, and the fact that $A(t), f$ and g_k are T-periodic functions in t, to obtain that v is also a solution with $v_0(\cdot, \phi) = y_T(\phi) = y(\cdot, \phi)$. Indeed; we can obtain that

$$
v(t) =
\begin{cases}
U(t,0)\phi(0) + \displaystyle\int_0^t U(t,s)\, f(s, v(s), v_s)ds.; & \text{if } t \in I_0, \\[1.5em]
U(t, s_k)g_k(s_k, v(s_k^-)) \\[0.3em]
\quad + \displaystyle\int_{s_k}^t U(t,s)\, f(s, v(s), v_s)ds.; & \text{if } t \in I_k; \ k = 1, \dots, m, \\[1.5em]
g_k(t, v(t_k^-)); & \text{if } t \in J_k; \ k = 1, \dots, m, \\[1.5em]
\phi(t); & \text{if } t \in \mathbb{R}_-.
\end{cases}
$$

Then the uniqueness of $\{U(t, s)\}_{(t,s) \in \Lambda}$ implies that $v(t) = y(t)$, so that $y(t) = y(t + T)$ is a T-periodic solution.

Step 2. *P is condensing.*
Now, we prove that the operator $P : \mathcal{B} \to \mathcal{B}$ is condensing. Let $D \subset \mathcal{B}$ be bounded with $\alpha(D) > 0$. From Theorem 4.1 in [14], we get

$$
\alpha(P(D)) \leq \left(\frac{1}{2}\right)^{k_0 - 1} M\alpha(D) < \alpha(D).
$$

Thus from Theorem 2.10, P has a fixed point which gives rise to a periodic mild solution of problem (9.1).

9.2.3   *An Example*

We consider the following functional evolution problem

$$
\begin{cases}
\dfrac{\partial z}{\partial t}(t,x) = a(t,x)\dfrac{\partial^2 z}{\partial x^2}(t,x) \\[2mm]
\qquad +Q(t, z(t,x), z_t(\cdot,x)); x \in [0,\pi],\ t \in I_k;\ k = 0,\ldots, \\[4mm]
z(t,x) = g_k(t,x); t \in \mathbb{R}_+,\ x \in [0,\pi],\ t \in J_k;\ k = 1,\ldots, \\[4mm]
z(t,0) = z(t,\pi) = 0;\ t \in \mathbb{R}_+. \\[4mm]
z(0,x) = \Phi(x);\ x \in [0,\pi], \\[4mm]
z(t,x) = \phi(t,x);\ t \in \mathbb{R}_-,\ x \in [0,\pi],
\end{cases}
\tag{9.5}
$$

where $a(t,x) : \mathbb{R}_+ \times [0,\pi] \to \mathbb{R}$ is a continuous function and is uniformly Hölder continuous in t, $Q : \mathbb{R}_+ \times \mathbb{R} \times C_h \to \mathbb{R}$, $\Phi : [0,\pi] \to \mathbb{R}$ and $\phi : \mathbb{R}_- \times [0,\pi] \to \mathbb{R}$ are continuous functions such that $\Phi(x) = \phi(0,x);\ x \in [0,\pi]$, and C_h is the phase space defined in Example 9.1.

Consider $E = L^2([0,\pi],\mathbb{R})$ and define $A(t)$ by $A(t)w = a(t,x)w''$ with domain

$$D(A) = \{w \in E : w, w' \text{ are absolutely continuous}, w'' \in E, w(0) = w(\pi) = 0\}.$$

Then $A(t)$ generates an evolution system $U(t,s)$ (see [352]).

For $x \in [0,\pi]$, we have

$$y(t)(x) = z(t,x); \quad t \in \mathbb{R}_+,$$

$$f(t, y(t), y_t, x) = Q(t, z(t,x), z_t(\cdot,x)); \quad t \in \mathbb{R}_+,$$

$$y_0(x) = \Phi(t,x); \quad x \in [0,\pi],$$

and

$$y(t,x) = \phi(t,x);\ x \in [0,\pi].$$

Thus, under the above definitions of f, y_0 and $A(\cdot)$, the system (9.5) can be represented by the functional evolution problem (9.1). Furthermore, more appropriate conditions on Q ensure the hypotheses (9.7.1)–(9.7.5). Consequently, Theorem 9.1 implies that the evolution problem (9.5) has at least one periodic mild solution.

9.3   Infinite Delay Second Order Evolution Equations with Impulses

In this paper, we discuss the existence of periodic mild solutions of the following class of second order evolution equations with infinite delay and non-instantaneous impulses

$$
\begin{cases}
y''(t) + A(t)y(t) = f(t, y(t), y_t); & \text{if } t \in I_k; \ k = 0, 1, \ldots, \\
y(t) = g_k(t, y(t_k^-)); & \text{if } t \in J_k; \ k = 1, 2, \ldots, \\
y(t) = \phi(t); & \text{if } t \in \mathbb{R}_- := (-\infty, 0], \\
y'(s_k) = \psi_k \in E; & k = 0, \ldots, m, \ldots,
\end{cases}
\tag{9.6}
$$

where $I_0 = [0, t_1]$, $I_k := (s_k, t_{k+1}]$, $J_k := (t_k, s_k]$, $0 = s_0 < t_1 \leq s_1 \leq t_2 < \cdots < s_{m-1} \leq t_m \leq s_m \leq t_{m+1} = T \leq s_{m+1} \leq t_{m+2} \leq \cdots < +\infty$, $f : I_k \times E \times \mathcal{B} \to E$; $k = 0, \ldots,$ $g_k : J_k \times E \to E$; $k = 1, 2, \ldots,$ are given functions T-periodic in t, $T > 0$, $\mathcal{B}$ is an abstract phase space to be specified later, $\phi : \mathbb{R}_- \to E$ is a given function, $\{A(t)\}_{t>0}$ is a T-periodic family of unbounded operators from E into E that generate an evolution system of operators $\{U(t, s)\}_{(t,s) \in \mathbb{R}_+ \times \mathbb{R}_+}$; for $(t, s) \in \Lambda := \{(t, s) \in \mathbb{R}_+ \times \mathbb{R}_+ : 0 \leq s \leq t < +\infty\}$, $\mathbb{R}_+ := [0, \infty)$, and $(E, \|\cdot\|_E)$ is a real Banach space.

For any continuous function y and any $t \in \mathbb{R}_+$, we denote by y_t the element of $\mathcal{B}$ defined by $y_t(\theta) = y(t + \theta)$ for $\theta \in \mathbb{R}_- := (-\infty, 0]$. Here, $y_t(\cdot)$ represents the history of the state up to the present time t. We assume that the histories y_t belong to $\mathcal{B}$.

9.3.1   *Necessary Results*

Consider the space

$$
\tilde{C}((-\infty, 0], E) = \Big\{ y : (-\infty, 0] \to E : \ y \text{ is continuous and there exist}
$$
$$
\tau_k \in (-\infty, 0); \ k = 1, \ldots, m, \text{ such that } y(\tau_k^-) \text{ and}
$$
$$
y(\tau_k^+) \text{ exist with } y(\tau_k^-) = y(\tau_k) \Big\},
$$

and the Banach space

$$
PC = \Big\{ y : (-\infty, T] \to E : y|_{\mathbb{R}_-} \in \mathcal{B}, \ y|_{J_k} = g_k; \ k = 1, \ldots, m,
$$
$$
y|_{I_k}; \ k = 1, \ldots, m, \text{ is continuous and there exist } y(s_k^-), \ y(s_k^+), \ y(t_k^-)
$$
$$
\text{and } y(t_k^+) \text{ with } y(s_k^+) = g_k(s_k, y(s_k^-)) \text{ and } y(t_k^-) = g_k(t_k, y(t_k^-)) \Big\},
$$

with the norm

$$\|y\|_{PC} = \max\{\|y\|_C, \|\phi\|_\mathcal{B}\}.$$

Lemma 9.3. *(Lemma 2.1 in [14]) There exists an integer $k_0 > 1$ such that*

$$\left(\frac{1}{2}\right)^{k_0-1} M < 1,$$

where $M = \sup\limits_{(t,s)\in\Lambda} \|U(t,s)\|_{B(E)}$ is finite, and there exists a function h on $\mathbb{R}_-$ such that $h(0) = 1$, $h(-\infty) = +\infty$, h is decreasing on $\mathbb{R}_-$, and for $d \geq w_0 := \frac{T}{K_0}$ one has $\sup\limits_{s\in(-\infty,0]} \dfrac{h(s)}{h(s-d)} \leq \dfrac{1}{2}.$

In all what follows, we consider the phase space

$$\mathcal{B} := \left\{\phi \in \tilde{C}((-\infty,0]), E) : \sup\limits_{s\in(-\infty,0]} \frac{\|\phi(s)\|}{h(s)} < \infty\right\},$$

where $h : \mathbb{R}_- \to \mathbb{R}_+$ is the function given in Lemma 9.3. We have that the space $\mathcal{B}$ satisfies the condition (A_3). Also; $\mathcal{B}$ satisfies conditions (A_1) and (A_2) if

$$\sup\limits_{t\in J} \sup\limits_{-\infty<\theta\leq-t} \frac{\phi(t+\theta)}{h(\theta)} < \infty.$$

The space $\mathcal{B}$ endowed with the norm

$$\|\phi\|_\mathcal{B} = \sup\limits_{s\in(-\infty,0]} \frac{\|\phi(s)\|}{h(s)},$$

is a Banach space [16].

9.3.2 *Main Results*

Definition 9.2. By a periodic mild solution of problem (9.6) we mean a measurable and T-periodic function y that satisfies

$$y(t) = \begin{cases} -\frac{\partial}{\partial s}U(t,0)\phi(0) + U(t,0)\psi_0 + \displaystyle\int_0^t U(t,s)\, f(s,y(s),y_s)ds.; & \text{if } t \subset I_0, \\ -\frac{\partial}{\partial s}U(t,s_k)g_k(s_k,y(s_k^-)) + U(t,s_k)\psi_k \\ + \displaystyle\int_{s_k}^t U(t,s)\, f(s,y(s),y_s)ds.; & \text{if } t \in I_k;\ k=1,\ldots,m, \\ g_k(t,y(t_k^-));\ \text{if } t \in J_k;\ k=1,\ldots,m, \\ \phi(t);\ \text{if } t \in \mathbb{R}_-. \end{cases}$$

Now, we shall prove the following theorem concerning the existence of periodic mild solutions of problem (9.6).

Theorem 9.2. *Assume that the following hypotheses hold:*

(9.10.1) *The functions f and g_k are continuous in their variables, and they map bounded sets into bounded sets,*

(9.10.2) *The function $t \mapsto f(t, y, v)$ is measurable on I_k, $k = 0, \ldots, m$, for each $y, v \in E \times \mathcal{B}$, and the functions $y \mapsto f(t, y, v)$ and $v \mapsto f(t, y, v)$ are continuous on $E \times \mathcal{B}$ for a.e. $t \in I_k$; $k = 0, \ldots, m$,*

(9.10.3) *For a constant $T > 0$, $f(t + T, y, v) = f(t, y, v)$, $A(t + T) = A(t)$; $t \in I_k$; $k = 0, \ldots, m$, $(y, v) \in E \times \mathcal{B}$, and $g_k(t + T, z) = g_k(t, z)$ $t \in J_k$; $k = 1, \ldots, m, z \in E$,*

(9.10.4) *There exist continuous functions $p : I_k \to \mathbb{R}_+$, $q : J_k \to \mathbb{R}_+$, such that*

$$\|f(t, y, v)\| \le p(t), \ \text{for a.e. } t \in I_k; \ k = 0, \ldots, m, \ \text{and each } y, v \in E \times \mathcal{B},$$

and

$$\|g_k(t, z)\| \le q(t), \ \text{for a.e. } t \in J_k, \ \text{and each } z \in E, \ k = 0, \ldots, m,$$

(9.10.5) *For each bounded sets $B(t) \subset E$, and $B_t \subset \mathcal{B}$; $t \in \mathbb{R}_+$, such that*

$$B(t) = \{y(t) : y \in C(J)\}, \ \text{and} \ B_t = \{y_t : y_t \in \mathcal{B}\},$$

we have

$$\alpha(f(t, B(t), B_t)) \le p(t)\alpha(B); \ \text{for a.e. } t \in I_k; \ k = 0, \ldots, m,$$

$$\alpha(g_k(t, B)) \le q(t)\alpha(B); \ \text{for a.e. } t \in J_k; \ k = 1, \ldots, m,$$

where

$$M_0 = \sup_{(t,s) \in \Lambda} \left\| \frac{\partial}{\partial s} U(t, s) \right\|_{B(E)}, \quad p^* = \sup_{t \in I_k} p(t), \quad \text{and} \quad q^* = \sup_{t \in J_k} q(t).$$

If $\ell := 4MTp^* < 1$, then the problem (9.6) has at least one T-periodic mild solution defined on $\mathbb{R}$.

Proof. The proof will be given in two parts. Consider the problem

$$\begin{cases} y''(t) + A(t)y(t) = f(t, y(t), y_t); & \text{if } t \in I_k; \ k = 0, \ldots, m, \\ y(t) = g_k(t, y(t_k^-)); & \text{if } t \in J_k; \ k = 1, \ldots, m, \\ y(t) = \phi(t); & \text{if } t \in \mathbb{R}_- := (-\infty, 0], \\ y'(s_k) = \psi_k; & k = 0, \ldots, m. \end{cases} \quad (9.7)$$

Part 1. *Existence of mild solutions.*

We prove that problem (9.7) has a mild solution $y \in PC$, with $\|y\|_{PC} \le R$ where

$$R \ge \max\{\|\phi\|_{\mathcal{B}}, q^*, M_0\|\phi(0)\| + M\|\psi_0\| + Mp^*, M_0 q^* + M\|\psi_k\| + Mp^* + Mp^*\}.$$

Consider the operator $N : PC \to PC$ defined by:

$$(Ny)(t) = \begin{cases} -\frac{\partial}{\partial s}U(t,0)\phi(0) + U(t,0)\psi_0 + \displaystyle\int_0^t U(t,s)\ f(s,y(s),y_s)ds; t \in I_0, \\ -\frac{\partial}{\partial s}U(t,s_k)g_k(s_k, y(s_k^-)) + U(t,s_k)\psi_k \\ + \displaystyle\int_{s_k}^t U(t,s)\ f(s,y(s),y_s)ds.; \quad \text{if } t \in I_k; \ k = 1,\ldots,m, \\ g_k(t, y(t_k^-)); \text{ if } t \in J_k; \ k = 1,\ldots,m, \\ \phi(t); \text{ if } t \in \mathbb{R}_-. \end{cases}$$

$$(9.8)$$

Clearly, the fixed points of the operator N are mild solutions of problem (9.7).

For any $y \in PC$ and each $t \in I_0$, we have

$$\|(Ny)(t)\| \le M_0\|\phi(0)\| + M\|\psi_0\| + M\int_0^t \|f(s,y(s),y_s)\|ds$$
$$\le M_0\|\phi(0)\| + M\|\psi_0\| + Mp^*$$
$$\le R.$$

Next, for any $y \in PC$ and each $t \in I_k; \ k = 1,\ldots,m$, we have

$$\|(Ny)(t)\| \le M_0 q^* + M\|\psi_k\| + M\int_{s_k}^t \|f(s,y(s),y_s)\|ds$$
$$\le M_0 q^* + M\|\psi_k\| + Mp^*$$
$$\le R.$$

Also, for any $y \in PC$ and each $t \in J_k; \ k = 1,\ldots,m$, we have

$$\|(Ny)(t)\| \le q^* \le R,$$

and for any $y \in PC$ and each $t \in \mathbb{R}_-$, we have

$$\|(Ny)(t)\| = \|\phi\|_{\mathcal{B}} \le R.$$

This proves that N transforms the ball $B_R := \{w \in PC : \|w\|_{PC} \le R\}$ into itself. We shall show that the operator $N : B_R \to B_R$ satisfies all the assumptions of Theorem 2.6. The proof will be given in two steps.

Step 1. $N : B_R \to B_R$ *is continuous.*

Let $\{y_n\}_{n \in \mathbb{N}}$ be a sequence such that $y_n \to y$ in B_R.

For each $t \in \mathbb{R}_- \cup J_k$; $k = 1, \dots, m$, we have

$$\|(Ny_n)(t) - (Ny)(t)\| = 0 \to 0 \quad \text{as } n \to \infty,$$

and for each $t \in I_k$; $k = 0, \dots, m$, we have

$$\|(Ny_n)(t) - (Ny)(t)\| \leq M \int_0^t \|f(s, y_n(s), y_{sn}) - f(s, y(s), y_s)\| ds. \quad (9.9)$$

Since $y_n \to y$ as $n \to \infty$ and f is continuous, then by the Lebesgue dominated convergence theorem, equation (9.9) implies

$$\|(Ny_n)(t) - (Ny)(t)\| \to 0 \quad \text{as } n \to \infty.$$

Hence

$$\|N(y_n) - N(y)\|_{PC} \to 0 \quad \text{as } n \to \infty.$$

Step 2. *For each closed subset D of $C(J)$, $\alpha(N(D)) \leq \ell\alpha(D)$.*

From Lemmas 2.5 and 2.6, for any $D \subset B_R$ and any $\epsilon > 0$, there exists a sequence $\{y_k\}_{k=0}^\infty \subset D$, such that for all $t \in I_k$; $k = 0, \dots, m$, we have

$$\alpha((ND)(t)) = \alpha\left(\left\{-\frac{\partial}{\partial s} U(t,0)\phi(0) + U(t,0)\psi_0\right.\right.$$

$$+ \int_0^t U(t,s)\, f(s, y(s), y_s) ds;\ y \in D\Big\}\Big)$$

$$\leq 2\alpha\left(\left\{\int_0^t U(t,s) f(s, y_k(s), y_{ks}) ds\right\}_{k=1}^\infty\right) + \epsilon$$

$$\leq 4\int_0^t \alpha\left(\|U(t,s)\|_{B(E)}\{f(s, y_k(s), y_{ks})\}_{k=1}^\infty\right) ds + \epsilon$$

$$\leq 4M \int_0^t \alpha\left(\{f(s, y_k(s), y_{ks})\}_{k=1}^\infty\right) ds + \epsilon$$

$$\leq 4M \int_0^t p(s)\alpha\left(\{y_k(s)\}_{k=1}^\infty\right) ds + \epsilon$$

$$\leq 4Mp^* \int_0^t \alpha\left(\{y_k(s)\}_{k=1}^\infty\right) ds + \epsilon,$$

$$\leq 4MTp^*\alpha_c(D) + \epsilon,$$

and, for all $t \in I_k$; $k = 1, \ldots, m$, we get

$$
\begin{aligned}
\alpha((ND)(t)) &= \alpha \left(\left\{ -\frac{\partial}{\partial s} U(t, s_k) g_k(s_k, y(s_k^-)) + U(t, s_k) \psi_k \right. \right. \\
&\qquad \left. \left. + \int_{s_k}^{t} U(t, s) \ f(s, y(s), y_s) ds; \ y \in D \right\} \right) \\
&\leq 2\alpha \left(\left\{ \int_{0}^{t} U(t, s) f(s, y_k(s), y_{ks}) ds \right\}_{k=1}^{\infty} \right) + \epsilon \\
&\leq 4MTp^* \alpha_c(D) + \epsilon.
\end{aligned}
$$

Since $\epsilon > 0$ is arbitrary, then

$$
\alpha_c(ND) \leq \ell \alpha_c(D).
$$

As a consequence of these two steps together with Theorem 2.6, we can conclude that N has a fixed point in $y \in B_R$ which is a mild solution of problem (9.6).

Part 2. *Periodic mild solutions.*
A standard approach in deriving T-periodic solutions is to define the Poincaré operator $P : \mathcal{B} \to \mathcal{B}$ given by $P(\phi) = y_T(\phi)$ such that

$$
(P\phi)(s) = y_T(s, \phi) = y(T + s, \phi); \ s \in \mathbb{R}_-,
$$

which maps an initial function (or value) ϕ along the unique mild solution $y(\phi)$ to our problem (9.6) by $T-$ units (i.e., T units along the unique solution $y(\cdot, \phi)$ determined by the initial function ϕ).

We show that P is a condensing operator with respect to Kuratowski's measure of non-compactness in the phase space $\mathcal{B}$, then the given conditions such that the fixed point theorem (Theorem 2.10) can be applied to get fixed points for the Poincaré operator, which give rise to periodic solutions. We do this in two steps.

Step 1. *The fixed points of P give rise to periodic mild solutions of* (9.6).
Let $\phi \in \mathcal{B}$ be such that $p(\phi) = \phi$. Then for the solution $y(\cdot) = y(\cdot, \phi)$ with $y_0(\cdot, \phi) = \phi$, we can define $v(t) = y(t + T)$. Now, for $t > 0$, we can use the known properties of $U(t, s)$, and the fact that $A(t), f$ and g_k are T-periodic functions in t, to obtain that v is also a solution with

$v_0(\cdot, \phi) = y_T(\phi) = y(\cdot, \phi)$. Indeed; we can obtain that

$$
v(t) = \begin{cases}
-\frac{\partial}{\partial s}U(t,0)\phi(0) + U(t,0)\psi_0 + \int_0^t U(t,s)\, f(s,v(s),v_s)ds.; & \text{if } t \in I_0, \\[2mm]
-\frac{\partial}{\partial s}U(t,s_k)g_k(s_k, y(s_k^-)) + U(t,s_k)\psi_k \\[1mm]
\quad + \int_{s_k}^t U(t,s)\, f(s,v(s),v_s)ds.; & \text{if } t \in I_k;\ k = 1, \ldots, m, \\[2mm]
g_k(t, v(t_k^-));\ \text{if } t \in J_k;\ k = 1, \ldots, m, \\[2mm]
\phi(t);\ \text{if } t \in \mathbb{R}_-.
\end{cases}
$$

Then the uniqueness of $\{U(t,s)\}_{(t,s)\in\Lambda}$ implies that $v(t) = y(t)$, so that $y(t) = y(t + T)$ is a T-periodic solution.

Step 2. *P is condensing.*
Now, we prove that the operator $P : \mathcal{B} \to \mathcal{B}$ is condensing. Let $D \subset \mathcal{B}$ be bounded with $\alpha_c(D) > 0$. From Theorem 4.1 in [14], we get

$$
\alpha_c(P(D)) \leq \left(\frac{1}{2}\right)^{k_0-1} M\alpha_c(D) < \alpha_c(D).
$$

Thus from Theorem 2.10, P has a fixed point which gives rise to a periodic mild solution of our problem (9.6).

9.3.3 *An Example*

Consider the following functional evolution problem:

$$
\begin{cases}
\dfrac{\partial^2 z}{\partial t^2}(t,x) = a(t,x)\dfrac{\partial^2 z}{\partial x^2}(t,x) \\
\qquad\qquad + Q(t, z(t,x), z_t(\cdot, x));\ x \in [0,\pi],\ t \in I_k;\ k = 0, \ldots, \\
z(t,x) = g_k(t,x);\ x \in [0,\pi],\ t \in J_k;\ k = 1, \ldots, \\
z(t,0) = z(t,\pi) = 0;\ t \in \mathbb{R}_+, \\
z(0,x) = \Phi(x);\ x \in [0,\pi], \\
z(t,x) = \phi(t,x);\ t \in \mathbb{R}_-,\ x \in [0,\pi],
\end{cases}
\tag{9.10}
$$

where $a(t,x) : \mathbb{R}_+ \times [0,\pi] \to \mathbb{R}$ is a continuous function and is uniformly Hölder continuous in t, $Q : \mathbb{R}_+ \times \mathbb{R} \times \mathcal{B} \to \mathbb{R}$, $\Phi : [0,\pi] \to \mathbb{R}$ and $\phi : \mathbb{R}_- \times [0,\pi] \to \mathbb{R}$ are continuous functions such that $\Phi(x) = \phi(0,x);\ x \in [0,\pi]$.

Consider $E = L^2([0, \pi], \mathbb{R})$ and define $A(t)$ by $A(t)w = a(t, x)w''$ with domain

$$D(A) = \{w \in E : w, w' \text{ are absolutely continuous, } w'' \in E, \ w(0) = w(\pi) = 0\}.$$

Then $A(t)$ generates an evolution system $U(t, s)$ (see [352]).

For $x \in [0, \pi]$, we have

$$y(t)(x) = z(t, x); \quad t \in \mathbb{R}_+,$$

$$f(t, y(t), y_t, x) = Q(t, z(t, x), z_t(\cdot, x)); \quad t \in \mathbb{R}_+,$$

$$y_0(x) = \Phi(x); \quad x \in [0, \pi],$$

$$y(t, x) = \phi(t, x); \ x \in [0, \pi], \ t \in \mathbb{R}_-.$$

Thus, under the above definitions of f, y_0 and $A(\cdot)$, the system (9.10) can be represented by the functional evolution problem (9.6). Furthermore, more appropriate conditions on Q ensure the hypotheses (9.10.1)–(9.10.5). Consequently, Theorem 9.2 implies that the evolution problem (9.10) has at least one periodic mild solution.

9.4   Notes and Remarks

The results of Chapter 9 are taken from the paper [421, 422]. For more relevant information, we recommend the books [160, 192, 404, 423] and the articles [424, 425].

Chapter 10

Periodic Mild Solutions of Evolution Inclusions with Impulses and Delay

10.1 Introduction and Motivations

This chapter deals with the existence of periodic mild solutions for a class of functional impulsive evolution equations and inclusions. We base our arguments on fixed point theory paired with the approach of measure of non-compactness using the resolvent operator. We show that the Poincaré operator is a condensing operator with respect to Kuratowski's measure of non-compactness in a determined phase space, and then derive periodic solutions from bounded solutions by using Sadovskii's fixed point theorem. Furthermore, illustrative examples are presented to demonstrate the plausibility of our results.

We studied and proved the results in this chapter by considering the previous chapters' publications as well as the publications that follow:

- The monographs [2, 226, 412] and the papers [12, 41, 413], where the authors keep developing functional evolution equations and inclusions.
- In [411], an iterative method is used for the existence of mild solutions of evolution equations and inclusions. Using the Tichonov's fixed point theorem, Olszowy and Wędrychowicz [396] considered a class of evolution equations on unbounded intervals.
- Liang *et al.* [15] considered the existence of periodic mild solutions of a class of impulsive differential equations with infinite delay in Banach spaces. However in the previous papers some restrictions like, the compactness of the semigroup, the Lipschitz conditions on the nonlinear term or the boundedness of the obtained mild solutions, are supposed.

- Various researchers have recently obtained further results by using the approach of measure of non-compactness; see [193, 306, 371], and the sources within.
- Impulsive differential inclusions have gained much importance in several mathematical models of real phenomena, particularly in biological or medical fields, as well as in control theory. Recent research and results on impulsive differential inclusions and equations can be found in the monographs [170, 171], and the papers [172–174, 426].
- In [162, 368–370, 427, 428], the authors investigated several types of impulsive differential equations with non-instantaneous impulses.
- For basic results and recent development on evolution equations with instantaneous and non-instantaneous impulses, one can refer to [39, 43, 109, 167, 183, 185, 187, 426–429].

10.2 Infinite Delay Non-Instantaneous Impulsive Evolution Inclusions

In the present section, we discuss the existence of periodic mild solutions of the following class of first order differential inclusions with infinite delay and not instantaneous impulses

$$
\begin{cases}
y'(t) + A(t)y(t) \in F(t, y(t), y_t); & \text{if } t \in I_k; \ k = 0, 1, \ldots, \\
y(t) = g_k(t, y(t_k^-)); & \text{if } t \in J_k; \ k = 1, 2, \ldots, \\
y(t) = \phi(t); & \text{if } t \in \mathbb{R}_- := (-\infty, 0],
\end{cases}
\tag{10.1}
$$

where $I_0 = [0, t_1]$, $I_k := (s_k, t_{k+1}]$, $J_k := (t_k, s_k]$, $0 = s_0 < t_1 \leq s_1 \leq t_2 < \cdots < s_{m-1} \leq t_m \leq s_m \leq t_{m+1} = T \leq s_{m+1} \leq t_{m+2} \leq \ldots < +\infty$, $F : I_k \times E \times \mathcal{B} \to \mathcal{P}(E)$; $k = 0, \ldots,$ is compact valued multivalued map, and T-periodic in t, $T > 0$, $\mathcal{P}(E)$ is the family of all subsets of E, $\mathcal{B}$ is an abstract phase space to be specified later, $\phi : \mathbb{R}_- \to E$, $g_k : J_k \times E \to E$; $k = 1, 2, \ldots,$ are given functions, $\{A(t)\}_{t>0}$ is a T-periodic family of unbounded operators from E into E that generate an evolution system of operators $\{U(t, s)\}_{(t,s)\in\mathbb{R}_+\times\mathbb{R}_+}$; for $(t, s) \in \Lambda := \{(t, s) \in \mathbb{R}_+ \times \mathbb{R}_+ : 0 \leq s \leq t < +\infty\}$, $\mathbb{R}_+ := [0, \infty)$, and $(E, \|\cdot\|_E)$ is a real separable Banach space.

For any continuous function y and any $t \in \mathbb{R}_+$, we denote by y_t the element of $\mathcal{B}$ defined by $y_t(\theta) = y(t + \theta)$ for $\theta \in \mathbb{R}_- := (-\infty, 0]$. Here, $y_t(\cdot)$ represents the history of the state up to the present time t. We assume that the histories y_t belong to $\mathcal{B}$.

10.2.1 Necessary Results

Consider the space

$$\tilde{C}((-\infty,0],E) = \Big\{y:(-\infty,0] \to E: \ y \text{ is continuous and there exist}$$
$$\tau_k \in (-\infty,0); \ k=1,\ldots,m, \text{ such that } y(\tau_k^-) \text{ and } y(\tau_k^+)$$
$$\text{exist with } y(\tau_k^-) = y(\tau_k)\Big\},$$

and the Banach space

$$PC = \Big\{y:(-\infty,T] \to E: y|_{\mathbb{R}_-} \in \mathcal{B}, \ y|_{J_k} = g_k; \ k=1,\ldots,m,$$
$$y|_{I_k}; \ k=1,\ldots,m, \text{ is continuous and there exist } y(s_k^-), \ y(s_k^+),$$
$$y(t_k^-) \text{ and } y(t_k^+) \text{ with } y(s_k^+) = g_k(s_k,y(s_k^-))$$
$$\text{and } y(t_k^-) = g_k(t_k,y(t_k^-))\Big\},$$

with the norm

$$\|y\|_{PC} = \max\{\|y\|_C, \|\phi\|_{\mathcal{B}}\}.$$

Also, we define the space

$$\widetilde{PC} = \{y:\mathbb{R} \to E: y|_{\mathbb{R}_-} \in \mathcal{B}, \ y|_{J_k} = g_k; \ k=1,\ldots; \ y|_{I_k}; \ k=1,\ldots;$$
$$\text{is continuous and there exist } y(s_k^-), \ y(s_k^+), \ y(t_k^-) \text{ and } y(t_k^+)$$
$$\text{with } y(s_k^+) = g_k(s_k,y(s_k^-)) \text{ and } y(t_k^-) = g_k(t_k,y(t_k^-))\}.$$

We need also some properties of multivalued maps. Let $(X,\|\cdot\|)$ be a Banach space. let $P_{cl}(X) = \{Y \in \mathcal{P}(X) : Y \text{ closed}\}$, $P_b(X) = \{Y \in \mathcal{P}(X) : Y \text{ bounded}\}$, $P_{cp}(X) = \{Y \in \mathcal{P}(X) : Y \text{ compact}\}$, $P_{cp,cv}(X) = \{Y \in \mathcal{P}(X) : Y \text{ compact and convex}\}$.

Definition 10.1. A multivalued map $T : X \to P(X)$ is convex(closed) valued if $T(x)$ is convex (closed) for all $x \in X$. T is bounded on bounded sets if $T(B) = \cup_{x\in B}T(x)$ is bounded in X for all $B \in P_b(X)$ (i.e. $\sup\{\sup\{|y| : y \in T(x)\}\} < \infty$). T is called upper semi-continuous (u.s.c.) on X if for each $x_0 \in X$, the set $T(x_0)$ is a non-empty closed subset of X, and if for each open set N of X containing $T(x_0)$, there exists an open neighborhood N_0 of x_0 such that $T(N_0) \subseteq N$. T is said to be completely continuous if $T(B)$ is relatively compact for every $B \in P_b(X)$. A multivalued map $T : X \to P(X)$ has a fixed point if there is $x \in X$ such that $x \in T(x)$. The fixed point set of the multivalued operator T will be denoted by $FixT$.

Definition 10.2. A multivalued map $T : I_k \to P_{cl}(E)$; $k = 0, 1, \ldots$, is said to be measurable if for every $y \in E$, the function

$$t \longmapsto d(y, T(t)) = \inf\{\|y - z\| : z \in T(t)\}$$

is measurable.

Definition 10.3. A multivalued map $F : I_k \times E \times \mathcal{B} \to \mathcal{P}(E)$; $k = 0, 1, \ldots$; is said to be L^1-Carathéodory if

(i) $t \longmapsto F(t, y, v)$ is measurable for each $(y, v) \in E \times \mathcal{B}$;
(ii) $y \longmapsto F(t, y, v)$ and $v \longmapsto F(t, y, v)$ are upper semicontinuous for almost all $t \in I_k$.
(iii) for each $\ell > 0$, there exists $p_\ell \in L^1(I_k, \mathbb{R}_+)$ such that

$$\|F(t, y)\|_{\mathcal{P}} = \sup\{\|v\| : v \in F(t, y)\}$$

$$\leq p_\ell(t) \quad \text{for all } \|y\| \leq \ell, \ \|v\|_{\mathcal{B}} \leq \ell \text{ and for a.e. } t \in I_k.$$

F is said to be Carathéodory if (i) and (ii) hold.

For each $y \in C(I_k, E)$; $k = 0, 1, \ldots$, we define the set of selections of F by

$$S_{F \circ y} = \{v \in L^1(I_k, E) : v(t) \in F(t, y(t), y_t) \text{ a.e. } t \in I_k\}.$$

Let (X, d) be a metric space induced from the normed space X. Consider $H_d : \mathcal{P}(X) \times \mathcal{P}(X) \longrightarrow \mathbb{R}_+ \cup \{\infty\}$ given by

$$H_d(A, B) = \max\left\{\sup_{a \in A} d(a, B), \sup_{b \in B} d(A, b)\right\},$$

where $d(A, b) = \inf_{a \in A} d(a, b)$, $d(a, B) = \inf_{b \in B} d(a, b)$.

Lemma 10.1 ([226], p. 159). *Under assumptions (P_1)–(P_3), the Cauchy problem*

$$y'(t) + A(t)y(t) = 0, \ t \in J, \ y(0) = y_0,$$

has a unique evolution system $U(t, s)$, $(t, s) \in \Delta := \{(t, s) \in J \times J : 0 \leq s \leq t \leq T\}$ satisfying the following properties:

(1) $U(t, t) = I$ where I is the identity operator in E,
(2) $U(t, s) U(s, \tau) = U(t, \tau)$ for $0 \leq \tau \leq s \leq t \leq T$,
(3) $U(t, s) \in B(E)$ the space of bounded linear operators on E, where for every $(t, s) \in \Delta$ and for each $y \in E$, the mapping $(t, s) \to U(t, s)y$ is continuous.

More details on evolution systems and their properties can be found in the books of Ahmed [226] and Pazy [11].

Lemma 10.2 (Lemma 2.1 in [14]). *There exists an integer $k_0 > 1$ such that*

$$\left(\frac{1}{2}\right)^{k_0-1} M < 1,$$

where $M = \sup_{(t,s)\in\Delta} \|U(t,s)\|_{B(E)}$ is finite, and there exists a function h on $(-\infty, 0]$ such that $h(0) = 1$, $h(-\infty) = +\infty$, h is decreasing on $(-\infty, 0]$, and for $d \geq w_0 := \frac{T}{K_0}$ one has $\sup_{s\in(-\infty,0]} \frac{h(s)}{h(s-d)} \leq \frac{1}{2}$.

Now we indicate two examples of phase spaces.

Example 10.1. For the function h given in Lemma 10.2, we define the spaces:
$C_h := \left\{\phi \in \tilde{C}(\mathbb{R}_-, E) : \frac{\phi(\theta)}{h(\theta)} \text{ is bounded on } \mathbb{R}_-\right\}$,
and $C_h^0 := \left\{\phi \in C_h : \lim_{\theta\to-\infty} \frac{\phi(\theta)}{h(\theta)} = 0\right\}$, endowed with the uniform norm

$$\|\phi\| = \sup\left\{\frac{|\phi(\theta)|}{h(\theta)} : \theta \leq 0\right\}.$$

Then we have that the spaces C_h and C_h^0 satisfy condition (A_3).
    Also; C_h and C_h^0 satisfy conditions (A_1) and (A_2) if

$$\sup_{t\in J} \sup_{-\infty<\theta\leq-t} \frac{\phi(t+\theta)}{h(\theta)} < \infty.$$

Example 10.2. For any real positive constant γ, we define the space

$$C_\gamma := \{\phi \in \tilde{C}((-\infty, 0]), E) : \lim_{\theta\to-\infty} e^{\gamma\theta}\phi(\theta) \text{ exist in } E,$$

endowed with the norm

$$\|\phi\| = \sup\{e^{\gamma\theta}|\phi(\theta)| : \theta \leq 0\}.$$

Then in the space C_γ the axioms (A_1)–(A_3) are satisfied.

In all what follows, we consider the phase space

$$\mathcal{B} := \left\{\phi \in \tilde{C}((-\infty, 0]), E) : \sup_{s\in(-\infty,0]} \frac{\|\phi(s)\|}{h(s)} < \infty\right\},$$

where $h : \mathbb{R}_- \to \mathbb{R}_+$ is the function given in Lemma 10.2.

Then $\mathcal{B}$ endowed with the norm

$$\|\phi\|_{\mathcal{B}} = \sup_{s \in (-\infty, 0]} \frac{\|\phi(s)\|}{h(s)},$$

is Banach space [16].

Definition 10.4. A mapping $\phi : \mathbb{R}_+ \to \mathbb{R}_+$, is called a dominating function or; in short, D-function if it is an upper semi-continuous and monotonic non-decreasing function satisfying $\phi(0) = 0$.

Definition 10.5. Let X be a Banach space. A multivalued mapping $T : X \in \mathcal{P}_{bd,cl}(X)$ is called D−set-Lipschitz if there exists a continuous non-decreasing function $\phi : \mathbb{R}_+ \to \mathbb{R}_+$, such that $\alpha(T(B)) \leq \phi(\alpha(B))$, for all $B \in \mathcal{P}_{bd,cl}(X)$ and $\phi(0) = 0$.

Remark 10.1. If $\phi(r) = kr$; $k > 0$, then T is called a k−set-Lipschitz mapping. Moreover, if $k < 1$, then T is called a k−set-contraction on E. If $\phi(r) < r$, for $r > 0$, then T is called a nonlinear D−set-contraction on X.

Definition 10.6. Let X be a Banach space and α be a measure of non-compactness on X. An operator $P : X \to X$ is said to be condensing if P is continuous and takes bounded sets into bounded sets, and $\alpha(P(B)) < \alpha(B)$ for every bounded set B of X with $\alpha(B) > 0$.

10.2.2   *Existence Results*

Definition 10.7. A measurable and T-periodic function $y \in \widetilde{PC}$ is said to be a periodic mild solution of problem (10.1) if there exists $f \in S_{Fo y}$, for a.e. $t \in \mathbb{R}_+$, such that

$$y(t) = \begin{cases} U(t,0)\phi(0) + \displaystyle\int_0^t U(t,s)\, f(s)ds.; & \text{if } t \in I_0, \\[2mm] U(t,s_k)g_k(s_k, y(s_k^-)) + \displaystyle\int_{s_k}^t U(t,s)\, f(s)ds.; & \text{if } t \in I_k;\ k = 1, \ldots, \\[2mm] g_k(t, y(t_k^-));\ \text{if } t \in J_k;\ k = 1, \ldots, \\[2mm] \phi(t);\ \text{if } t \in \mathbb{R}_-, \end{cases}$$

Set

$$M = \sup_{(t,s) \in \Delta} \|U(t,s)\|_{B(E)},$$

$$p_\ell^* := \|p_\ell\|_{L^1(I_k)} = \int_{I_k} p_\ell(t)dt,$$

$$q^* := \|q\|_C = \sup_{t \in J_k} q(t).$$

Now, we shall prove the following theorem concerning the existence of periodic mild solutions of problem (10.1).

Theorem 10.1. *Assume that the following hypotheses hold:*

(10.13.1) *The multivalued map $F(t, y, v)$ is L^1-Carathéodory, and has compact and convex values and maps bounded sets into bounded sets.*

(10.13.2) *The functions g_k; $k = 1, 2, \ldots$; are continuous in their variables, and they map bounded sets into bounded sets.*

(10.13.3) *For a constant $T > 0$, $F(t + T, y, v) = F(t, y, v)$, $A(t + T) = A(t)$; $t \in I_k$; $k = 0, \ldots, m$, $y, v \in E \times \mathcal{B}$, and $g_k(t + T, z) = g_k(t, z)$ $t \in J_k$; $k = 1, \ldots, m, z \in E$.*

(10.13.4) *There exists a continuous function $q \in C(J_k, \mathbb{R}_+)$, such that*

$$\|g_k(t, z)\| \le q(t), \text{ for a.e. } t \in J_k, \text{ and each } z \in E, \ k = 0, \ldots, m,$$

(10.13.5) *For each bounded sets $B \subset E$ and $B_t \subset \mathcal{B}$; $t \in \mathbb{R}_+$, such that*

$$B_t = \{y_t : y_t \in \mathcal{B}\},$$

we have

$$\alpha(F(t, B, B_t)) \le p_\ell(t)\alpha(B); \text{ for a.e. } t \in I_k; \ k = 0, \ldots, m,$$

and

$$\alpha(g_k(t, B)) \le q(t)\alpha(B); \text{ for a.e. } t \in J_k; \ k = 1, \ldots, m,$$

where $p_\ell \in L^1(I_k)$, and α is a measure of non-compactness on the Banach space E.

If $\max\{4MT p_\ell^*, q^*\} < 1$, then the problem (10.1) has at least one T-periodic mild solution defined on $\mathbb{R}$.

Proof. The proof will be given in two parts. Consider the problem

$$\begin{cases} y'(t) + A(t)y(t) \in F(t, y(t), y_t); & \text{if } t \in I_k, \ k = 0, \ldots, m, \\ y(t) = g_k(t, y(t_k^-)); & \text{if } t \in J_k, \ k = 1, \ldots, m, \\ y(t) = \phi(t); & \text{if } t \in \mathbb{R}_- := (-\infty, 0]. \end{cases} \quad (10.2)$$

Part 1. We begin by showing that (10.2) has a mild solution $y \in PC$, with $\|y\|_{PC} \le R$ where

$$R \ge \max\{\|\phi\|_{\mathcal{B}}, q^*, M(\|\phi(0)\| + p_\ell^*), M(q^* + p_\ell^*)\}.$$

Define the multivalued map $N : PC \to \mathcal{P}(PC)$ defined by

$$
N(y) = \left\{ h \in PC : h(t) = \begin{cases} U(t,0)\phi(0) \\ \quad + \displaystyle\int_0^t U(t,s)\, f(s)ds; \;\; \text{if } t \in I_0,\; f \in S_{F \circ y} \\ U(t,s_k)g_k(s_k,y(s_k^-)) + \displaystyle\int_{s_k}^t U(t,s) \\ \quad \times f(s)ds.; \;\; \text{if } t \in I_k;\; k=1,\dots,m, \\ g_k(t,y(t_k^-)); \;\; \text{if } t \in J_k;\; k=1,\dots,m \\ \phi(t); \;\; \text{if } t \in \mathbb{R}_-. \end{cases} \right\}
$$

(10.3)

Clearly, the fixed points of N are mild solution of the problem (10.2). Let $h \in N(y)$. Then there exists $f \in S_{F \circ y}$ such that

$$
h(t) = \begin{cases} U(t,0)\phi(0) + \displaystyle\int_0^t U(t,s)\, f(s)ds; \;\; \text{if } t \in I_0, \\ U(t,s_k)g_k(s_k,y(s_k^-)) + \displaystyle\int_{s_k}^t U(t,s)\, f(s)ds.; \;\; \text{if } t \in I_k;\; k=1,\dots,m, \\ g_k(t,y(t_k^-)); \;\; \text{if } t \in J_k;\; k=1,\dots,m, \\ \phi(t); \;\; \text{if } t \in \mathbb{R}_-. \end{cases}
$$

For each $t \in I_0$, we have

$$
\|h(t)\| \leq M\|\phi(0)\| + M \int_0^t \|f(s)\|ds
$$
$$
\leq M\|\phi(0)\| + Mp_\ell^*
$$
$$
\leq R.
$$

Next, for each $t \in I_k;\; k=1,\dots,m$, we have

$$
\|h(t)\| \leq Mq^* + M \int_{s_k}^t \|f(s)\|ds
$$
$$
\leq Mq^* + Mp_\ell^*
$$
$$
\leq R.
$$

Also, for each $t \in J_k;\; k=1,\dots,m$, we have

$$
\|h(t)\| \leq q^* \leq R,
$$

and for each $t \in \mathbb{R}_-$, we have

$$
\|h(t)\| = \|\phi\|_{\mathcal{B}} \leq R.
$$

This proves that N transforms the ball $B_R := \{w \in PC : \|w\|_{PC} \leq R\}$ into $\mathcal{P}(B_R)$. We shall show that the operator $N : B_R \to \mathcal{P}(B_R)$ satisfies all the assumptions of Theorem 2.24. The proof will be given in two steps.

Step 1. $N(y) \in \mathcal{P}_{cl}(PC)$ *for each* $y \in B_R$.
Let $(y_n)_{n \geq 0} \in N(y)$ such that $y_n \longrightarrow \tilde{y}$ in PC. Then, $\tilde{y} \in PC$ and there exists $f_n \in S_{F \circ y}$ such that

$$
y_n(t) = \begin{cases}
U(t,0)\phi(0) + \displaystyle\int_0^t U(t,s) \, f_n(s)ds; & \text{if } t \in I_0, \\[2mm]
U(t,s_k)g_k(s_k, y_n(s_k^-)) + \displaystyle\int_{s_k}^t U(t,s) \, f_n(s)ds.; & \text{if } t \in I_k; \; k = 1,\ldots,m, \\[2mm]
g_k(t, y_n(t_k^-)); & \text{if } t \in J_k; \; k = 1,\ldots,m, \\[2mm]
\phi(t); & \text{if } t \in \mathbb{R}_-.
\end{cases}
$$

Using the fact that F has compact values and from (10.13.1), we may pass to a subsequence if necessary to get that f_n converges to f in $L^1(J,E)$, and hence $f \in S_{F \circ y}$. Then

$$
y_n(t) \to y(t) = \begin{cases}
U(t,0)\phi(0) + \displaystyle\int_0^t U(t,s) \, f(s)ds; & \text{if } t \in I_0, \\[2mm]
U(t,s_k)g_k(s_k, y(s_k^-)) \\[1mm]
+ \displaystyle\int_{s_k}^t U(t,s) \, f(s)ds.; & \text{if } t \in I_k; \; k = 1,\ldots,m, \\[2mm]
g_k(t, y(t_k^-)); & \text{if } t \in J_k; \; k = 1,\ldots,m, \\[2mm]
\phi(t); & \text{if } t \in \mathbb{R}_-.
\end{cases}
$$

So, $\tilde{y} \in N(y)$.

Step 2. $N : B_R \to \mathcal{P}_{cl,cv}(B_R)$ *is* $D-$*set-contraction*.
From Lemmas 2.5 and 2.6, for any $B \subset B_R$ and any $\epsilon > 0$, there exists a sequence $\{y_k\}_{k=0}^{\infty} \subset B$, such that for all $t \in I_0$, we have

$$
\alpha((NB)(t)) = \alpha\left(\left\{ U(t,0)\phi(0) + \int_0^t U(t,s) \, f(s)ds; \; f \in S_{F \circ y}, \; y \in B \right\} \right)
$$

$$
\leq 2\alpha\left(\left\{ \int_0^t U(t,s)f(s)ds; \; f \in S_{F \circ y_k} \right\}_{k=1}^{\infty} \right) + \epsilon
$$

$$\leq 4 \int_0^t \alpha \left(\|U(t,s)\|_{B(E)} \{f(s); \ f \in S_{F \circ y_k}\}_{k=1}^\infty \right) ds + \epsilon$$

$$\leq 4M \int_0^t \alpha \left(\{f(s); \ f \in S_{F \circ y_k}\}_{k=1}^\infty \right) ds + \epsilon$$

$$\leq 4M \int_0^t p_\ell(s) \alpha \left(\{y_k(s)\}_{k=1}^\infty \right) ds + \epsilon$$

$$\leq 4M p_\ell^* \int_0^t \alpha \left(\{y_k(s)\}_{k=1}^\infty \right) ds + \epsilon,$$

$$\leq 4MT p_\ell^* \alpha_{PC}(B) + \epsilon,$$

and, for all $t \in I_k; \ k = 1, \ldots, m$, we get

$$\alpha((NB)(t))$$
$$= \alpha \left(\left\{ U(t,s_k) g_k(s_k, y(s_k^-)) + \int_{s_k}^t U(t,s) \ f(s) ds; \ f \in S_{F \circ y}, \ y \in B \right\} \right)$$
$$\leq 2\alpha \left(\left\{ \int_0^t U(t,s) f(s) ds; \ f \in S_{F \circ y_k} \right\}_{k=1}^\infty \right) + \epsilon$$
$$\leq 4MT p_\ell^* \alpha_{PC}(B) + \epsilon.$$

Since $\epsilon > 0$ is arbitrary, then for all $t \in I_k; \ k = 0, \ldots, m$, we get

$$\alpha((NB)(t)) \leq \phi_1(\alpha_{PC}(B)).$$

where $\phi_1 : \mathbb{R}_+ \to \mathbb{R}_+, \ \phi_1(x) = 4MT p_\ell^* x$.

Also, for all $t \in J_k; \ k = 1, \ldots, m$, we get

$$\alpha((NB)(t)) \leq \phi_2(\alpha_{PC}(B)).$$

where $\phi_2 : \mathbb{R}_+ \to \mathbb{R}_+, \ \phi_2(x) = q^* x$.

Hence, for all $t \in (-\infty, T]$ we obtain

$$\alpha((NB)(t)) \leq \phi(\alpha_{PC}(B)),$$

where $\phi : \mathbb{R}_+ \to \mathbb{R}_+$ defined by

$$\begin{cases} \phi(t) = \phi_1(t); & \text{if } t \in I_k; \ k = 0, 1, \ldots, \\ \phi(t) = \phi_2(t); & \text{if } t \in J_k; \ k = 1, 2, \ldots. \end{cases}$$

As a consequence of these two steps together with Theorem 2.24, we can conclude that N has a fixed point in $y \in B_R$ which is a mild solution of problem (10.1).

Part 2. *Periodic mild solutions.*

A standard approach in deriving T-periodic solutions is to define the Poincaré operator $P : \mathcal{B} \to \mathcal{B}$ given by

$$P(\phi) = y_T(\phi): \ (P\phi)(s) = y_T(s, \phi) = y(T + s, \phi); \ s \in \mathbb{R}_-,$$

which maps an initial function (or value) ϕ along the unique mild solution $y(\phi)$ to our problem (10.1) by $T-$ units (i.e., T units along the unique solution $y(\cdot, \phi)$ determined by the initial function ϕ). We show that P is a condensing operator with respect to Kuratowski's measure of non-compactness in the phase space $\mathcal{B}$, then the given conditions such that the fixed point theorem (Theorem 2.10) can be applied to get fixed points for the Poincaré operator, which give rise to periodic solutions. We do this in two steps.

Step 1. *The fixed points of P give rise to a periodic mild solutions of* (10.1).

Let $\phi \in \mathcal{B}$ be such that $P(\phi) = \phi$. Then for the solution $y(\cdot) = y(\cdot, \phi)$ with $y_0(\cdot, \phi) = \phi$, we can define $v(t) = y(t + T)$. Now, for $t > 0$, we can use the known properties of $U(t, s)$ in Lemma 10.1, and the fact that $A(t), f$ and g_k are T-periodic functions in t, to obtain that v is also a solution with $v_0(\cdot, \phi) = y_T(\phi) = y(\cdot, \phi)$. Indeed, let $h \in N(y)$. Then there exists $f \in S_{F \circ y}$ such that

$$
v(t) =
\begin{cases}
U(t, 0)\phi(0) + \displaystyle\int_0^t U(t, s)\, f(s)ds; & \text{if } t \in I_0, \ f \in S_{F \circ v}, \\[2mm]
U(t, s_k)g_k(s_k, y(s_k^-)) \\[1mm]
\quad + \displaystyle\int_{s_k}^t U(t, s)\, f(s)ds.; & \text{if } t \in I_k; \ k = 1, \dots, m, \ f \in S_{F \circ v}, \\[2mm]
g_k(t, v(t_k^-)); & \text{if } t \in J_k; \ k = 1, \dots, m, \\[2mm]
\phi(t); & \text{if } t \in \mathbb{R}_-.
\end{cases}
$$

Then the uniqueness of $\{U(t, s)\}_{(t, s) \in \Lambda}$ implies that $v(t) = y(t)$, so that $y(t) = y(t + T)$ is a T-periodic solution.

Step 2. *P is condensing.*

Now, we prove that the operator $P : \mathcal{B} \to \mathcal{B}$ is condensing. Let $B \subset \mathcal{B}$ be bounded with $\alpha_{\mathcal{B}}(B) > 0$. If y_0 is the unique solution with $y_0(\phi) = \phi$, we define $W_s(B) = \{y_s(\phi) : \phi \in B\}$ and $W_{[h,r]}(B) = \{y_{[h,r]}(\phi) : \phi \in B\}$, where

$y_{[h,r]}(\phi)$ means the restriction of y on $[h, r]$. As in the proof of Theorem 4.1 in [14], we obtain

$$\alpha_{\mathcal{B}}(P(B)) \leq \left(\frac{1}{2}\right)^{k_0-1} M\alpha_{\mathcal{B}}(B) < \alpha_{\mathcal{B}}(B).$$

Thus from Theorem 2.10, P has a fixed point which gives rise to a periodic mild solution of problem (10.1).

10.2.3   *An Example*

We consider the following functional evolution problem:

$$
\begin{cases}
\dfrac{\partial z}{\partial t}(t, x) + a(t, x)\dfrac{\partial^2 z}{\partial x^2}(t, x) \in Q(t, z(t, x), z_t(\cdot, x)); \\
\qquad\qquad x \in [0, \pi], \ t \in I_k; \ k = 0, \ldots, \\[2mm]
z(t, x) = g_k(t, x); \ t \in \mathbb{R}_+, \ x \in [0, \pi], \ t \in J_k; \ k = 1, \ldots, \\[2mm]
z(t, 0) = z(t, \pi) = 0; \ t \in \mathbb{R}_+. \\[2mm]
z(0, x) = \Phi(x); \ x \in [0, \pi], \\[2mm]
z(t, x) = \phi(t, x); \ t \in \mathbb{R}_-, \ x \in [0, \pi],
\end{cases}
\tag{10.4}
$$

where $a(t, x) : \mathbb{R}_+ \times [0, \pi] \to \mathbb{R}$ is a continuous function in x and is uniformly Hölder continuous in t, $Q : \mathbb{R}_+ \times \mathbb{R} \times C_h \to \mathcal{P}(\mathbb{R})$,

$$
\begin{aligned}
Q(t, &z(t, x), z_t(\cdot, x)) \\
&= \{v \in \mathbb{R} : |f_1(t, z(t, x), z_t(\cdot, x))| \leq |v| \leq |f_2(t, z(t, x), z_t(\cdot, x))|\},
\end{aligned}
$$

$\Phi : [0, \pi] \to \mathbb{R}$ and $\phi : \mathbb{R}_- \times [0, \pi] \to \mathbb{R}$ are continuous functions such that $\Phi(x) = \phi(0, x)$; $x \in [0, \pi]$, $f_1, f_2 : \mathbb{R}_+ \times \mathbb{R} \times C_h \to \mathbb{R}$ are continuous functions, and C_h is the phase space defined in Example 10.1.

Consider $E = L^2([0, \pi], \mathbb{R})$ and define $A(t)$ by $A(t)w = a(t, x)w''$ with domain

$$D(A) = \{w \in E : w, w' \text{ are absolutely continuous}, w'' \in E, w(0) = w(\pi) = 0\}.$$

Then $A(t)$ generates an evolution system $U(t, s)$ (see [352]).

For $x \in [0, \pi]$, we have

$$y(t)(x) = z(t, x); \quad t \in \mathbb{R}_+,$$

$$F(t, y(t), y_t, x) = Q(t, z(t, x), z_t(\cdot, x)); \quad t \in \mathbb{R}_+,$$

$$y_0(x) = \Phi(t, x); \quad x \in [0, \pi],$$

$$y(t, x) = \phi(t, x); \; x \in [0, \pi].$$

Thus, under the above definitions of F, y_0 and $A(\cdot)$, the system (10.4) can be represented by the functional evolution problem (10.1). Furthermore, more appropriate conditions on Q ensure the hypotheses (10.13.1)–(10.13.5). Consequently, Theorem 10.1 implies that the evolution problem (10.4) has at least one periodic mild solution.

10.3   Infinite Delay Integro-Differential Inclusions with Non-Instantaneous Impulses

Motivated by the preceding articles, in this section, we consider the following problem:

$$\begin{cases} y'(t) - Ay(t) - \displaystyle\int_0^t \Psi(t-s)y(s)ds \in f(t, y(t), y_t); \; t \in J_j, \; j = 0, \ldots, \\ y(t) = g_j(t, y(t_j^-)); \; t \in \tilde{J}_j, \; j = 1, \ldots, \\ y(t) = \psi(t); \; \text{if } t \in \mathbb{R}_- := (-\infty, 0], \end{cases}$$

$$(10.5)$$

where $J_0 := [0, t_1]$, $\tilde{J}_j := (t_j, s_j]$, $J_j := (s_j, t_{j+1}]$; $j = 1, \ldots, 0 = s_0 < t_1 \le s_1 < t_2 \le s_2 < \cdots \le s_{m-1} < t_m \le s_m < t_{m+1} = T \le s_{m+1} < t_{m+2} \le \cdots < +\infty, f : J_j \times E \times \mathcal{B} \to \mu(E); \; j = 0, \ldots,$ is T-periodic compact multivalued map, t, $T > 0$, is the family of subsets of E is denoted by $\mu(E)$, $\mathcal{B}$ is a phase space given in the sequel, $g_j : \tilde{J}_j \times E \to E$ are given functions, and T-periodic in t, $T > 0, \psi : \mathbb{R}_- \to E$ is a given function, and $(E, \| \cdot \|)$ is a Banach space, $y'(t) := \frac{dy}{dt}$, $A : D(A) \subset E \to E$ generates a C_0-semigroup on the Banach space E, $\Psi(t)$ is a closed linear operator on E, and T-periodic in t, $T > 0$, with $D(A) \subset D(\Psi)$. For each continuous function y and any $t \in \mathbb{R}_+$, y_t is the element of $\mathcal{B}$ given by $y_t(\varepsilon) = y(t + \varepsilon)$ for $\varepsilon \in \mathbb{R}_-$.

10.3.1   *Necessary Results*

Consider the space

$$\tilde{F}((-\infty, 0], E) = \Big\{ y : (-\infty, 0] \to E : \; y \text{ is continuous and there exist}$$

$$\tau_j \in (-\infty, 0); \; j = 1, \ldots, m, \text{ where } \; y(\tau_j^-) \text{ and } y(\tau_j^+)$$

$$\text{exist with } y(\tau_j^-) = y(\tau_j) \Big\}.$$

Consider the space

$$\mathcal{P}_c = \Big\{ y : (-\infty, T] \to E : y|_{\mathbb{R}_-} \in \mathcal{B},\ y|_{\bar{J}_j} = g_j;\ j = 1, \dots, m,$$

$$y|_{J_j};\ j = 1, \dots, m,\ \text{is continuous and there exist } y(s_j^-),\ y(s_j^+),\ y(t_j^-)$$

$$\text{and } y(t_j^+) \text{ with } y(s_j^+) = g_j(s_j, y(s_j^-)) \text{ and } y(t_j^-) = g_j(t_j, y(t_j^-)) \Big\},$$

with the norm

$$\|y\|_{\mathcal{P}_c} = \max\{\|y\|_C, \|\psi\|_{\mathcal{B}}\}.$$

Consider the space

$$\widetilde{\mathcal{P}_c} = \Big\{ y : \mathbb{R} \to E : y|_{\mathbb{R}_-} \in \mathcal{B},\ y|_{\bar{J}_j} = g_j;\ j = 1, \dots;\ y|_{J_j};\ j = 1, \dots;$$

$$\text{is continuous and there exist } y(s_j^-),\ y(s_j^+),\ y(t_j^-) \text{ and } y(t_j^+)$$

$$\text{with } y(s_j^+) = g_j(s_j, y(s_j^-)) \text{ and } y(t_j^-) = g_j(t_j, y(t_j^-)) \Big\}.$$

A semigroup of bounded linear operators $S(t)$ is uniformly continuous if

$$\lim_{t \to 0} \|S(t) - I\|_E = 0,$$

where I is the identity operator in E.

Note that if a semigroup $S(t)$ is of class (C_0) then it verifies the growth condition

$$\|S(t)\|_{B(E)} \le \kappa e^{\bar{\kappa} t}, \text{ for } 0 \le t < \infty \text{ with some constants } \kappa > 0 \text{ and } \bar{\kappa} \ge 0.$$

If, for instance $\kappa = 1$ and $\bar{\kappa} = 0$, i.e; $\|S(t)\|_{B(E)} \le 1$, for $t \ge 0$, then the semigroup $S(t)$ is called a *contraction semigroup*.

Let $(W, \|\cdot\|)$ be a Banach space. let $\mu_{cl}(W) = \{\Lambda \in \mu(W) : \Lambda \text{ closed}\}$, $\mu_b(W) = \{\Lambda \in \mu(W) : \Lambda \text{ bounded}\}$, $\mu_{cp}(W) = \{\Lambda \in \mu(W) : \Lambda \text{ compact}\}$, $\mu_{cp,cv}(W) = \{\Lambda \in \mu(W) : \Lambda \text{ compact and convex}\}$.

Let (W, d) be a metric space induced from the normed space W. Consider $\Omega_d : \mu(W) \times \mu(W) \longrightarrow \mathbb{R}_+ \cup \{\infty\}$ given by

$$\Omega_d(Q, \bar{Q}) = \max \left\{ \sup_{\iota \in Q} d(\iota, \bar{Q}), \sup_{\bar{\iota} \in \bar{Q}} d(Q, \bar{\iota}) \right\},$$

where $d(Q, \bar{\iota}) = \inf_{\iota \in Q} d(\iota, \bar{\iota}),\ d(\iota, \bar{Q}) = \inf_{\bar{\iota} \in \bar{Q}} d(\iota, \bar{\iota})$.

Let $(\mathcal{B}, \|\cdot\|_{\mathcal{B}})$ be a seminormed linear space of functions from $\mathbb{R}_-$ into E, and meeting the following essential assumptions:

$(\mathcal{C}d_{A_1})$ If $y \in \mathcal{P}_c$ and $y_0 \in \mathcal{B}$, then for all $t \in J$ the requirements that follows are met:

(i) $y_t \in \mathcal{B}$;

(ii) $\|y_t\|_\mathcal{B} \le E(t) \sup\limits_{s \in [0,t]} \|y(s)\| + \tilde{E}(t)\|\psi\|_\mathcal{B}$;

(iii) $\|y(t)\| \le \bar{E}\|y_t\|_\mathcal{B}$;

where $\bar{E} \ge 0$, $E : J \to \mathbb{R}_+$ is continuous; $\tilde{E} : \mathbb{R}_+ \to \mathbb{R}_+$ is locally bounded, and $\bar{E}, E, \tilde{E}$, are independent of $y(\cdot)$.

$(\mathcal{C}d_{A_2})$ For the function $y(\cdot)$ in $(\mathcal{C}d_{A_1})$, y_t is a $\mathcal{B}$-valued continuous function on J.

$(\mathcal{C}d_{A_3})$ The space $\mathcal{B}$ is complete.

Denote $E_\beta = \sup\{E(t) : t \in J\}$ and $\tilde{E}_\beta = \sup\{\tilde{E}(t) : t \in J\}$. (See [308] for more details).

Remark 10.2. Axiom $(\mathcal{C}d_{A_1})$(ii) is equivalent to $\|\psi(0)\| \le \bar{E}\|\psi\|_\mathcal{B}$, for all $\psi \in \mathcal{B}$. As a consequence, we have for all ψ, $\bar{\psi} \in \mathcal{B}$ such that $\|\psi - \bar{\psi}\|_\mathcal{B} = 0$, thus $\psi(0) = \bar{\psi}(0)$.

Lemma 10.3. *[14] Let $\alpha_0 > 1$ such that*

$$\left(\frac{1}{2}\right)^{\alpha_0 - 1} \kappa < 1,$$

where $\kappa = \sup\limits_{(t) \in J} \|R(t)\|_{B(E)}$ *and there exists a function ξ on $\mathbb{R}_-$ where* $\xi(0) = 1$, $\xi(-\infty) = +\infty$, *ξ is decreasing on $\mathbb{R}_-$, and for $d \ge \beta_0 := \frac{T}{\alpha_0}$ one has* $\sup\limits_{s \in (-\infty,0]} \frac{\xi(s)}{\xi(s-d)} \le \frac{1}{2}$.

Example 10.3. We define the spaces:

$$C_\xi := \left\{\psi \in \tilde{F}(\mathbb{R}_-, E) : \frac{\psi(\varepsilon)}{\xi(\varepsilon)} \text{ is bounded on } \mathbb{R}_-\right\},$$

and

$$C_\xi^0 := \left\{\psi \in C_\xi : \lim\limits_{\varepsilon \to -\infty} \frac{\psi(\varepsilon)}{\xi(\varepsilon)} = 0\right\},$$

with the norm

$$\|\psi\| = \sup\left\{\frac{|\psi(\varepsilon)|}{\xi(\varepsilon)} : \varepsilon \le 0\right\}.$$

Therefore, the spaces C_ξ and C_ξ^0 verify the condition $(\mathcal{C}d_{A_3})$. As well as conditions $(\mathcal{C}d_{A_1})$ and $(\mathcal{C}d_{A_2})$ if

$$\sup_{t\in J}\ \sup_{-\infty<\varepsilon\le -t}\ \frac{\psi(t+\varepsilon)}{\xi(\varepsilon)}<\infty.$$

Example 10.4. For any $\sigma\in(0,\infty)$, we define the space

$$C_\sigma:=\{\psi\in\tilde{F}((-\infty,0]),E):\ \lim_{\varepsilon\to-\infty}e^{\sigma\varepsilon}\psi(\varepsilon)\ exist\ in\ E\},$$

with the norm

$$\|\psi\|=\sup\{e^{\sigma\varepsilon}|\psi(\varepsilon)|:\varepsilon\le 0\}.$$

The axioms $(\mathcal{C}d_{A_1})-(\mathcal{C}d_{A_3})$ are therefore satisfied in the space C_σ.

In everything that follows, we take into account the phase space

$$\mathcal{B}:=\Big\{\psi\in\tilde{F}((-\infty,0]),E):\ \sup_{s\in(-\infty,0]}\frac{\|\psi(s)\|}{\xi(s)}<\infty\Big\}.$$

Therefore, $\mathcal{B}$ verifies the assumption $(\mathcal{C}d_{A_3})$. As well as $(\mathcal{C}d_{A_1})$ and $(\mathcal{C}d_{A_2})$ if

$$\sup_{t\in J}\ \sup_{-\infty<\varepsilon\le -t}\ \frac{\psi(t+\varepsilon)}{\xi(\varepsilon)}<\infty.$$

The space $\mathcal{B}$ equipped with the norm

$$\|\psi\|_\mathcal{B}=\sup_{s\in(-\infty,0]}\frac{\|\psi(s)\|}{\xi(s)},$$

is a Banach space [16].

10.3.2   *Main Results*

Definition 10.8 ([364, 394]). A resolvent operator for the problem

$$\begin{cases} y'(t)=Ay(t)+\displaystyle\int_0^t\Psi(t-s)y(s)ds,\ t\in[0,\infty),\\ y(0)=y_0\in E,\end{cases}\tag{10.6}$$

is a bounded linear operator-valued function $R(t)\in B(E);\ t\ge 0$, verifying the following assumptions:

(i) $R(0)=I$ and $\|R(t)\|\le Ne^{mt}$ for some constants $N>0$, and $m\in\mathbb{R}$.
(ii) For each $y\in E$, $R(t)y$ is strongly continuous for $t\ge 0$.

(iii) $R(t)$ is bounded for $t \geq 0$. For $y \in D(A)$, $R(\cdot)y \in C(\mathbb{R}_+, D(A)) \cap C^1(\mathbb{R}_+, E)$ and

$$R'(t)y = AR(t)y + \int_0^t \Psi(t-s)R(s)yds$$

$$= R(t)Ay + \int_0^t R(t-s)\Psi(s)yds; \ t \in [0, \infty).$$

Theorem 10.2 ([364,394]). *Assume that the following requirements hold:*

(10.19.1) *The operator A is the infinitesimal generator of a uniformly continuous semigroup $(S(t))_{t\geq 0}$.*

(10.19.2) *For all $t \geq 0$, $\Psi(t)$ is a closed linear operator from $D(A)$ to E and $\Psi(t) \in B(E)$. For any $y \in E$, the map $t \mapsto \Psi(t)y$ is bounded, differentiable and the derivative $t \mapsto \Psi'(t)y$ is bounded and uniformly continuous on $\mathbb{R}_+$.*

Then there exists a unique uniformly continuous resolvent operator for the problem (10.6).

Definition 10.9. A T-periodic function $y \in \widetilde{\mathcal{P}_c}$ is a periodic mild solution of problem (10.5) if there exists $\bar{\xi} \in S_{f \circ y}$, for a.e. $t \in \mathbb{R}_+$, such that

$$y(t) = \begin{cases} R(t)\psi(0) + \int_0^t R(t-s)\bar{\xi}(s)ds, \ t \in J_0, \\ R(t-s_j)g_j(s_j, y(s_j^-)) + \int_{s_j}^t R(t-s)\bar{\xi}(s)ds, \ t \in J_j, \ j = 1, \ldots, \\ g_j(t, y(t_j^-)), \ t \in \tilde{J}_j, \ j = 1, \ldots, \\ \psi(t), \ \text{if } t \in \mathbb{R}_-. \end{cases}$$

Theorem 10.3. *Assume that the hypotheses* (10.19.1), (10.19.2), *and the following requirements are met:*

(10.21.1) *The multivalued map $f(t, y, \bar{y})$ is $L^1 - Carathéodory$, and has compact and convex values and maps bounded sets into bounded sets.*

(10.21.2) *The continuous functions g_j, $j = 1, 2, \ldots$; map bounded sets into bounded sets.*

(10.21.3) *For $T > 0$, $f(t+T, y, \bar{y}) = f(t, y, \bar{y})$, $A(t+T) = A(t)$, $t \in J_j$, $j = 0, \ldots, m$, $y, \bar{y} \in E \times \mathcal{B}$, $g_j(t+T, z) = g_j(t, z)$, $t \in \tilde{J}_j$, $j = 1, \ldots, m, z \in E$, and $\psi(s+T) = \psi(s)$, $s \in (-\infty, 0]$.*

(10.21.4) *There exist continuous functions $l_j \in L^\infty(J)$, such that*

$$\|g_j(t, z)\|_E \leq l_j(t)(1 + \|z\|), \ \text{for a.e. } t \in \tilde{J}_j, \ z \in E, \ j = 0, \ldots, m.$$

(10.21.5) *For bounded sets* $\Omega \subset E$ *and* $\Omega_t \subset \mathcal{B}$, $t \in \mathbb{R}_+$, *such that*

$$\Omega_t = \{y_t : y_t \in \mathcal{B}\},$$

we have

$$\alpha_E(f(t, \Omega, \Omega_t)) \leq \eta_\delta(t)\alpha_E(B), \;\; for \; a.e. \; t \in J_j, \; j = 0, \ldots, m,$$

and

$$\alpha_E(g_j(t, \Omega)) \leq l_j(t)\alpha_E(\Omega), \;\; for \; a.e. \; t \in \tilde{J}_j, \; j = 1, \ldots, m,$$

where $\eta_\delta \in L^1(J_j)$, *and* α_E *is a measure of non-compactness on the Banach space* E, *and*

$$\eta_\delta^* := \|\eta_\delta\|_{L^1(J_j)} = \int_{J_j} \eta_\delta(t)dt,$$

$$l^* = \max_{j=0,\ldots,m} \|l_j\|_{L^\infty}, \;\;\; \kappa = \sup_{t \in J} \|R(t)\|_{B(E)}.$$

If

$$\delta := \max\{l^*, 4\kappa T\eta_\delta^*, \kappa(l^* + T\eta_\delta^*)\} < 1, \tag{10.7}$$

then the problem (10.5) has at least one mild solution.

Proof. Consider the problem

$$\begin{cases} y'(t) - Ay(t) - \displaystyle\int_0^t \Psi(t-s)y(s)ds \in f(t, y(t), y_t), \; t \in J_j, \; j = 0, \ldots, m, \\ y(t) = g_j(t, y(t_j^-)), \; t \in \tilde{J}_j, \; j = 1, \ldots, m, \\ y(t) = \psi(t), \;\; \text{if } t \in \mathbb{R}_- := (-\infty, 0], \end{cases}$$

$$\tag{10.8}$$

Part 1. We start by demonstrating that (10.8) has a mild solution $y \in \mathcal{P}_c$. Transform the problem (10.8) into a fixed point problem. Consider the multivalued map $T_1 : \mathcal{P}_c \to \mu(\mathcal{P}_c)$ defined by

$$T_1(y) = \left\{ \xi \in \mathcal{P}_c : \xi(t) = \begin{cases} R(t)\psi(0) + \displaystyle\int_0^t R(t-s)\bar{\xi}(s)ds; \; t \in J_0, \\ R(t-s_j)g_j(s_j, y(s_j^-)) \\ \quad + \displaystyle\int_{s_j}^t R(t-s)\bar{\xi}(s)ds, \; t \in J_j, \; j = 1, \ldots, m, \\ g_j(t, y(t_j^-)), \; t \in \tilde{J}_j, \; j = 1, \ldots, m, \\ \psi(t), \;\; \text{if } t \in \mathbb{R}_- := (-\infty, 0], \end{cases} \right\}$$

$$\tag{10.9}$$

where $\bar{\xi} \in S_{f \circ y}$.

Let $\xi \in T_1(y)$. Thus, there exists $\bar{\xi} \in S_{f \circ y}$ where

$$
\xi(t) = \begin{cases}
R(t)\psi(0) + \displaystyle\int_0^t R(t-s)\bar{\xi}(s)ds, \ t \in J_0, \\
R(t-s_j)g_j(s_j, y(s_j^-)) + \displaystyle\int_{s_j}^t R(t-s)\bar{\xi}(s)ds, \ t \in J_j, \ j = 1, \ldots, m, \\
g_j(t, y(t_j^-)), \ t \in \tilde{J}_j, \ j = 1, \ldots, m, \\
\psi(t), \ \text{if } t \in \mathbb{R}_- := (-\infty, 0],
\end{cases}
$$

Let $\gamma > 0$ be, such that

$$
\gamma \geq \max\left\{ \|\psi\|_{\mathcal{B}}, \frac{l^*}{1 - l^*}, \kappa(\|\psi(0)\| + \eta_\delta^*), \frac{\kappa(l^* + \eta_\delta^*)}{1 - \kappa l^*} \right\},
$$

and consider the ball $\Omega_\gamma := \Omega(0, \gamma) = \{\ell \in \mathcal{P}_c : \|\ell\|_{\mathcal{P}_c} \leq \gamma\}$.
For any $y \in \Omega_\gamma$ and each $t \in J_0$, we have

$$
\begin{aligned}
\|\xi(t)\| &\leq \kappa\|\psi(0)\| + \kappa \int_0^t \|\bar{\xi}(s)\|ds \\
&\leq \kappa\|\psi(0)\| + \kappa\eta_\delta^* \\
&\leq \gamma.
\end{aligned}
$$

Next, for each $t \in J_j$, $j = 1, \ldots, m$, we have

$$
\begin{aligned}
\|\xi(t)\| &\leq \kappa l^*(1 + \gamma) + \kappa \int_{s_j}^t \|\bar{\xi}(s)\|ds \\
&\leq \kappa l^*(1 + \gamma) + \kappa\eta_\delta^* \\
&\leq \gamma.
\end{aligned}
$$

Also, for each $t \in \tilde{J}_j$, $j = 1, \ldots, m$, we have

$$
\|\xi(t)\| \leq l^*(1 + \gamma) \leq \gamma,
$$

and for each $t \in \mathbb{R}_-$, we have

$$
\|\xi(t)\| = \|\psi\|_{\mathcal{B}} \leq \gamma.
$$

Hence,

$$
\|T_1(y)\|_{\mathcal{P}_c} \leq \gamma.
$$

As a consequence, the operator T_1 transforms the ball $\Omega_\gamma := \{p \in \mathcal{P}_c : \|p\|_{\mathcal{P}_c} \leq \gamma\}$ into $\mu(\Omega_\gamma)$. Now, we will demonstrate that the operator $T_1 : \Omega_\gamma \to \mu(\Omega_\gamma)$ verifies all the requirements of Theorem 2.24.

Step 1. $T_1(y) \in \mu_{cl}(\mathcal{P}_c)$ *for each $y \in \Omega_\gamma$.*
Let $(y_n)_{n \geq 0} \in T_1(y)$ such that $y_n \longrightarrow \tilde{y}$ in $\mathcal{P}_c$. Then, $\tilde{y} \in \mathcal{P}_c$ and there exists $\bar{\xi}_n \in S_{foy}$ such that

$$
y_n(t) = \begin{cases}
R(t)\psi(0) + \displaystyle\int_0^t R(t-s)\,\bar{\xi}_n(s)ds, & \text{if } t \in J_0, \\[2ex]
R(t-s_j)g_j(s_j, y_n(s_j^-)) \\
\quad + \displaystyle\int_{s_j}^t R(t-s)\,\bar{\xi}_n(s)ds, & \text{if } t \in J_j, \ j = 1,\dots,m \\[2ex]
g_j(t, y_n(t_j^-)), & \text{if } t \in \tilde{J}_j, \ j = 1,\dots,m \\[2ex]
\psi(t), & \text{if } t \in \mathbb{R}_-.
\end{cases}
$$

Given that f contains compact values and by (10.21.1), we can pass to a subsequence if required to get that $\bar{\xi}_n$ converges to $\bar{\xi}$ in $L^1(J, E)$, and therefore $\bar{\xi} \in S_{foy}$. Then,

$$
y_n(t) \to y(t) = \begin{cases}
R(t)\psi(0) + \displaystyle\int_0^t R(t-s)\,\bar{\xi}(s)ds, & \text{if } t \in J_0, \\[2ex]
R(t-s_j)g_j(s_j, y(s_j^-)) \\
\quad + \displaystyle\int_{s_j}^t R(t-s)\,\bar{\xi}(s)ds, & \text{if } t \in J_j, \ j = 1,\dots,m \\[2ex]
g_j(t, y(t_j^-)), & \text{if } t \in \tilde{J}_j, \ j = 1,\dots,m \\[2ex]
\psi(t), & \text{if } t \in \mathbb{R}_-.
\end{cases}
$$

So, $\tilde{y} \in T_1(y)$.

Step 2. $T_1 : \Omega_\gamma \to \mu_{cl,cv}(\Omega_\gamma)$ *is D-set-contraction.*
Using Lemmas 2.5 and 2.6, for any $\Omega \subset \Omega_\gamma$ and any $\epsilon > 0$, there exists a sequence $\{y_j\}_{j=0}^\infty \subset \Omega$, where for every $t \in J_0$, we obtain

$$
\alpha_E((T_1\Omega)(t)) = \alpha_E\left(\left\{ R(t)\psi(0) + \int_0^t R(t-s)\,\bar{\xi}(s)ds;\ \bar{\xi} \in S_{foy},\ y \in \Omega \right\}\right)
$$

$$
\leq 2\alpha_E\left(\left\{ \int_0^t R(t-s)\bar{\xi}(s)ds;\ \bar{\xi} \in S_{foy_j} \right\}_{j=1}^\infty\right) + \epsilon
$$

$$\leq 4 \int_0^t \alpha_E \left(\|R(t-s)\|_{B(E)} \{\bar{\xi}(s); \ \bar{\xi} \in S_{f \circ y_j}\}_{j=1}^\infty \right) ds + \epsilon$$

$$\leq 4\kappa \int_0^t \alpha_E \left(\{\bar{\xi}(s); \ \bar{\xi} \in S_{f \circ y_j}\}_{j=1}^\infty \right) ds + \epsilon$$

$$\leq 4\kappa \int_0^t \eta_\delta(s) \alpha \left(\{y_j(s)\}_{j=1}^\infty \right) ds + \epsilon$$

$$\leq 4\kappa \eta_\delta^* \int_0^t \alpha_E \left(\{y_j(s)\}_{j=1}^\infty \right) ds + \epsilon$$

$$\leq 4\kappa T \eta_\delta^* \alpha_{\mathcal{P}_c}(\Omega) + \epsilon,$$

and, for all $t \in J_j$, $j = 1, \ldots, m$, we get

$$\alpha_E((T_1\Omega)(t)) = \alpha_E \left(\{R(t-s_j)g_j(s_j, y(s_j^-)) \right.$$

$$\left. + \int_{s_j}^t R(t-s) \ \bar{\xi}(s)ds; \ \bar{\xi} \in S_{f \circ y}, \ y \in \Omega \right\} \right)$$

$$\leq 2\alpha_E \left(\left\{ \int_0^t R(t-s)\bar{\xi}(s)ds; \ \bar{\xi} \in S_{f \circ y_j} \right\}_{j=1}^\infty \right) + \epsilon$$

$$\leq 4\kappa T \eta_\delta^* \alpha_{\mathcal{P}_c}(\Omega) + \epsilon.$$

Since $\epsilon > 0$ is arbitrary, then for all $t \in J_j$, $j = 0, \ldots, m$, we get

$$\alpha_E((T_1\Omega)(t)) \leq \zeta_1(\alpha_{\mathcal{P}_c}(\Omega)),$$

where $\zeta_1 : \mathbb{R}_+ \to \mathbb{R}_+$, $\zeta_1(w) = 4\kappa T \eta_\delta^* w$.

Also, for all $t \in \tilde{J}_j$, $j = 1, \ldots, m$, we get

$$\alpha_E((T_1\Omega)(t)) \leq l_j(t)\alpha_{\mathcal{P}_c}(\Omega),$$

$$\leq l^*\alpha_{\mathcal{P}_c}(\Omega)$$

$$\leq \zeta_2(\alpha_{\mathcal{P}_c}(\Omega)),$$

where $\zeta_2 : \mathbb{R}_+ \to \mathbb{R}_+$, $\zeta_2(w) = l^* w$.

Hence, for all $t \in (-\infty, T]$ we obtain

$$\alpha_E((T_1\Omega)(t)) \leq \zeta(\alpha_{\mathcal{P}_c}(\Omega)),$$

where $\zeta : \mathbb{R}_+ \to \mathbb{R}_+$ defined by

$$\begin{cases} \zeta(t) = \zeta_1(t), & \text{if } t \in J_j; \ j = 0, 1, \ldots, \\ \zeta(t) = \zeta_2(t), & \text{if } t \in \tilde{J}_j; \ j = 1, 2, \ldots. \end{cases}$$

As a result, we may deduce that T_1 admit a fixed point in $y \in \Omega_\gamma$.

Part 2. *Periodic mild solutions.*

A common method for obtaining T-periodic solutions is to define the Poincaré operator $T_2 : \mathcal{B} \to \mathcal{B}$ by

$$T_2(\psi) = y_T(\psi) : \text{ where } (T_2\psi)(s) = y_T(s, \psi) = y(T + s, \psi), \ s \in \mathbb{R}_-,$$

which translates a starting function ψ along the single mild solution $y(\psi)$ to our problem (10.5) by $T-$ units. We prove that T_2 is condensing in $\mathcal{B}$. Thus, the provided assumptions indicate that Theorem 2.25 may be used to obtain fixed points for the Poincaré operator, resulting in periodic solutions.

Step 1. *The fixed points of T_2 provide a periodic mild solutions of* (10.5).

Let $\psi \in \mathcal{B}$ where $T_2(\psi) = \psi$. Then for the solution $y(\cdot) = y(\cdot, \psi)$ with $y_0(\cdot, \psi) = \psi$, we can define $\bar{y}(t) = y(t + T)$. And, for $t > 0$, We can make advantage of the properties of $R(t)$, and the knowledge that $\bar{\xi}$ and g_j are T-periodic functions in t, to get that $\bar{y}$ is also a solution with $\bar{y}_0(\cdot, \psi) = y_T(\psi) = y(\cdot, \psi)$. Indeed, let $\xi \in T_1(y)$. Then there exists $\bar{\xi} \in S_{foy}$ such that

$$\bar{y}(t) = \begin{cases} R(t)\psi(0) + \displaystyle\int_0^t R(t - s)\,\bar{\xi}(s)ds, & \text{if } t \in J_0, \ \bar{\xi} \in S_{foy} \\[2ex] R(t - s_j)g_j(s_j, \bar{y}(s_j^-)) \\[1ex] \quad + \displaystyle\int_{s_j}^t R(t - s)\,\bar{\xi}(s)ds, & \text{if } t \in J_j, \ j = 1, \ldots, m, \ \bar{\xi} \in S_{foy} \\[2ex] g_j(t, \bar{y}(t_j^-)), & \text{if } t \in \tilde{J}_j, \ j = 1, \ldots, m \\[1ex] \psi(t), & \text{if } t \in \mathbb{R}_-. \end{cases}$$

Then the uniqueness of $R(t)$ implies that $\bar{y}(t) = y(t)$, so that $y(t) = y(t+T)$ is a T-periodic solution.

Step 2. *T_2 is condensing.*

Let $\Omega \subset \mathcal{B}$ be bounded with $\alpha_{\mathcal{B}}(\Omega) > 0$. If y_0 is the unique solution with $y_0(\psi) = \psi$, we define $Js(\Omega) = \{y_s(\psi) : \psi \in \Omega\}$ and $J_{[\beta_1, \beta_2]}(\Omega) = \{y_{[\beta_1, \beta_2]}(\psi) : \psi \in \Omega\}$. Same as the proof of Theorem 4.1 in [14], we get

$$\alpha_{\mathcal{B}}(T_2(\Omega)) \le \left(\frac{1}{2}\right)^{\alpha_0 - 1} \kappa \alpha_{\mathcal{B}}(\Omega) < \alpha_{\mathcal{B}}(\Omega).$$

Therefore, Theorem 2.25 implies that T_2 admit a fixed point.

10.3.3   *An Example*

Let us investigate the following problem of impulsive integro-differential inclusions:

$$
\begin{cases}
\dfrac{\partial \bar{p}}{\partial t}(t,\theta) - \dfrac{\partial^2 \bar{p}}{\partial \theta^2}(t,\theta) - \displaystyle\int_0^t b(t-s)\dfrac{\partial^2}{\partial \theta^2}\bar{p}(s,\theta)ds \in f(t,\bar{p}(t,\theta),\bar{p}_t(\cdot,\theta)), \\[2mm]
\qquad \theta \in J := [0,\pi], \ t \in J_j, \ j = 0,\ldots, \\[2mm]
\bar{p}(t,\theta) = g_j(t,\theta), \ \ \theta \in J, \ t \in \tilde{J}_j, \ j = 1,\ldots, \\[2mm]
\bar{p}(t,0) = \bar{p}(t,\pi) = 0, \ t \in \mathbb{R}_+. \\[2mm]
\bar{p}(0,\theta) = \mathcal{G}(\theta), \ \theta \in J, \\[2mm]
\bar{p}(t,\theta) = \psi(t,\theta); \ t \in \mathbb{R}_-, \ \theta \in J,
\end{cases}
$$

$$(10.10)$$

where $f : \mathbb{R}_+ \times \mathbb{R} \times C_{\xi} \to \mathbb{R}$, $\mathcal{G} : J \to \mathbb{R}$ and $\psi : \mathbb{R}_- \times J \to \mathbb{R}$ are continuous functions such that $\mathcal{G}(\theta) = \psi(0,\theta)$, $\theta \in J$, and C_{ξ} is the phase space given in Example 10.3.

Take Let

$$
E := L^2(J) = \left\{ y : J \to \mathbb{R} : \int_0^{\pi} |y(\theta)|^2 dx < \infty \right\}
$$

be the Hilbert space with the scalar product $< y, \bar{y} >= \displaystyle\int_0^{\pi} y(\theta)\bar{y}(\theta)dx$. It is known that E is a Banach space with the norm

$$
\|y\|_2 = \left(\int_0^{\pi} |y(\theta)|^2 dx \right)^{\frac{1}{2}},
$$

and define $A : D(A) \subset E \to E$ by $A\lambda = \lambda''$ with domain

$$D(A) = \{\lambda \in E, \lambda, \lambda' \ \text{are absolutely continuous, } \lambda'' \in E, \lambda(0) = \lambda(\pi) = 0\}.$$

Then

$$
A\lambda = \sum_{n=1}^{\infty} n^2(\lambda, \lambda_n)\lambda_n, \ \lambda \in D(A)
$$

where $(\cdot, \cdot)$ is the inner product in L^2 and $\lambda_n(s) = \sqrt{\dfrac{2}{\pi}} \sin ns$, $n = 1,2,\ldots$ is the orthogonal set of eigenvectors in A. It is well known (see [11]) that

A is the infinitesimal generator of an analytic semigroup $S(t)$, $t \geq 0$ in E and is defined by

$$S(t)\lambda = \sum_{n=1}^{\infty} exp(-n^2 t)(\lambda, \lambda_n)\lambda_n, \ \lambda \in E.$$

Since the analytic semigroup $S(t)$ is compact, there exists a constant $\kappa \geq 1$ such that

$$\|S(t)\|_{B(E)} \leq \kappa.$$

For $\theta \in J$, we have

$$y(t)(\theta) = \bar{p}(t, \theta); \quad t \in \mathbb{R}_+,$$

$$f(t, y(t), y_t) = f(t, \bar{p}(t, \theta), \bar{p}_t(\cdot, \theta)); \quad t \in \mathbb{R}_+,$$

$$\Psi(t) = b(t)A$$

$$y_0(\theta) = \mathcal{G}(\theta); \quad \theta \in J,$$

$$y(t)(\theta) = \psi(t, \theta); \ \theta \in J, t \in \mathbb{R}_-.$$

Therefore, by the definitions of $\bar{\xi}$, y_0 and A, the system (10.10) can be represented by (10.5). Moreover, more relevant assumptions on f guarantee that (10.19.1), (10.19.2) and (10.21.1)–(10.21.5) are met. As a consequence, By Theorem 10.3, we can deduce that problem (10.10) has at least one mild solution on $\mathbb{R}$.

10.4   Infinite Delay Second Order Evolution Inclusions with Non-Instantaneous Impulses

In this section, we consider the following problem:

$$\begin{cases} y''(t) + A(t)y(t) \in f(t, y(t), y_t); & \text{if } t \in J_k; \ k = 0, 1, \ldots, \\ y(t) = g_k(t, y(t_k^-)); & \text{if } t \in \tilde{J}_k; \ k = 1, 2, \ldots, \\ y(t) = \varkappa(t); & \text{if } t \in \mathbb{R}_-, \\ y'(s_k) = \wp_k \in E; & k = 0, \ldots, m, \ldots, \end{cases} \quad (10.11)$$

where $J_0 = [0, t_1]$, $J_k := (s_k, t_{k+1}]$, $\tilde{J}_k := (t_k, s_k]$, $0 = s_0 < t_1 \leq s_1 \leq t_2 < \cdots < s_{m-1} \leq t_m \leq s_m \leq t_{m+1} = T \leq s_{m+1} \leq t_{m+2} \leq \cdots < +\infty$, $f : J_k \times E \times \mathcal{B} \to \mu(E)$; $k = 0, \ldots$, is compact multivalued map, and T-periodic in t, $T > 0$, $\mu(E)$ is the family of subsets of E, $\mathcal{B}$ is an abstract phase space, $\varkappa : \mathbb{R}_- \to E$, $g_k : \tilde{J}_k \times E \to E$; $k = 1, 2, \ldots$, are given function, and T-periodic in t, $T > 0$, $\{A(t)\}_{t>0}$ is a T-periodic family of unbounded operators from E into E that generate an evolution system of operators $\{\mathcal{U}(t,s)\}_{(t,s)\in\mathbb{R}_+\times\mathbb{R}_+}$; for $(t,s) \in \nabla := \{(t,s) \in \mathbb{R}_+ \times \mathbb{R}_+ : 0 \le s \le t < +\infty\}$, $\mathbb{R}_+ := [0, \infty)$, and $(E, \|\cdot\|_E)$ is a Banach space. y_t is the element of $\mathcal{B}$ given by $y_t(\iota) = y(t + \iota)$ for $\iota \in \mathbb{R}_- := \mathbb{R}_-$.

10.4.1 Necessary Results

Consider the space

$$\tilde{K}(\mathbb{R}_-, E) = \Big\{ y : \mathbb{R}_- \to E : \ y \text{ such that there exist } \gamma_k \in (-\infty, 0);$$
$$k = 1, \ldots, m, \text{ such that } y(\gamma_k^-) \text{ and } y(\gamma_k^+) \text{ exist with}$$
$$y(\gamma_k^-) = y(\gamma_k) \Big\},$$

and the Banach space

$$PC = \Big\{ y : (-\infty, T] \to E : y|_{\mathbb{R}_-} \in \mathcal{B}, \ y|_{\tilde{J}_k} = g_k; \ k = 1, \ldots, m,$$
$$y|_{J_k}; \ k = 1, \ldots, m, \text{ is continuous and there exist } y(s_k^-), \ y(s_k^+),$$
$$y(t_k^-) \text{ and } y(t_k^+) \text{ with } y(s_k^+) = g_k(s_k, y(s_k^-))$$
$$\text{and } y(t_k^-) = g_k(t_k, y(t_k^-)) \Big\},$$

with the norm

$$\|y\|_{PC} = \max\{\|y\|_C, \|\varkappa\|_{\mathcal{B}}\}.$$

Consider

$$\widetilde{PC} = \{ y : \mathbb{R} \to E : y|_{\mathbb{R}_-} \in \mathcal{B}, \ y|_{\tilde{J}_k} = g_k; \ k = 1, \ldots; \ y|_{J_k}; \ k = 1, \ldots;$$
$$\text{is continuous and there exist } y(s_k^-), \ y(s_k^+), \ y(t_k^-) \text{ and } y(t_k^+)$$
$$\text{with } y(s_k^+) = g_k(s_k, y(s_k^-)) \text{ and } y(t_k^-) = g_k(t_k, y(t_k^-)) \}.$$

Let (E, δ) be a metric space induced from the normed space E. Consider $Q_\delta : \mu(E) \times \mu(E) \longrightarrow \mathbb{R}_+ \cup \{\infty\}$ given by

$$Q_\delta(\Omega_2, \Omega_1) = \max \Big\{ \sup_{a\in\Omega_2} \delta(a, \Omega_1), \sup_{b\in\Omega_1} \delta(\Omega_2, b) \Big\},$$

where $\delta(\Omega_2, b) = \inf_{a\in\Omega_2} \delta(a, b)$, $\delta(a, \Omega_1) = \inf_{b\in\Omega_1} \delta(a, b)$.

Let $(\mathcal{B}, \|\cdot\|_{\mathcal{B}})$ be a seminormed linear space of functions mapping $\mathbb{R}_-$ into E, and versifying (see [308], for more details):

(A_1) If $y \in PC$ and $y_0 \in \mathcal{B}$, then for $t \in J$, we have:

  (i) $y_t \in \mathcal{B}$

  (ii) $\|y_t\|_{\mathcal{B}} \leq \xi_1(t) \sup_{s \in [0,t]} \|y(s)\| + \xi_2(t)\|\varkappa\|_{\mathcal{B}}$,

  (iii) $\|y(t)\| \leq \widehat{\xi}\|y_t\|_{\mathcal{B}}$, where $\widehat{\xi} \geq 0$ is a constant, $\xi_1 : J \to \mathbb{R}_+$ is continuous; $\xi_2 : \mathbb{R}_+ \to \mathbb{R}_+$ is locally bounded and $\widehat{\xi}, \xi_1, \xi_2$, are independent of $y(\cdot)$.

(A_2) For the function $y(\cdot)$ in (A_1), y_t is a $\mathcal{B}$-valued continuous function on J.

(A_3) The space $\mathcal{B}$ is complete.

Denote $\xi_b = \sup\{\xi_1(t) : t \in J\}$ and $\bar{\xi}_b = \sup\{\xi_2(t) : t \in J\}$.

Remark 10.3. Axiom (A_1)(ii) is equivalent to $\|\rho(0)\| \leq \widehat{\xi}\|\rho\|_{\mathcal{B}}$; for $\rho \in \mathcal{B}$. Thus, ρ, $\wp \in \mathcal{B}$ such that $\|\rho - \wp\|_{\mathcal{B}} = 0$, we necessarily have $\rho(0) = \wp(0)$.

Lemma 10.4 ([14]). *There exists an integer $\eta_0 > 1$ where*

$$\left(\frac{1}{2}\right)^{\eta_0 - 1} \sigma < 1,$$

and $\sigma = \sup_{(t,s) \in \nabla} \|\mathcal{U}(t,s)\|_{B(E)}$ is finite, and there exists a function $\beth$ on $\mathbb{R}_-$ where $\beth(0) = 1$, $\beth(-\infty) = +\infty$, $\beth$ is decreasing on $\mathbb{R}_-$, and for $\delta \geq \varpi_0 := \frac{T}{\eta_0}$, we have $\sup_{s \in \mathbb{R}_-} \dfrac{\beth(s)}{\beth(s - \delta)} \leq \dfrac{1}{2}$.

Example 10.5. We consider the spaces:

$C_\beth := \left\{\varkappa \in \tilde{K}(\mathbb{R}_-, E) : \frac{\varkappa(\iota)}{\beth(\iota)} \text{ is bounded on } \mathbb{R}_-\right\}$,

and $C_\beth^0 := \left\{\varkappa \in C_\beth : \lim_{\iota \to -\infty} \frac{\varkappa(\iota)}{\beth(\iota)} = 0\right\}$, with

$$\|\varkappa\| = \sup\left\{\frac{|\varkappa(\iota)|}{\beth(\iota)} : \iota \leq 0\right\}.$$

Thus, $C_\beth$ and $C_\beth^0$ verify (A_3). Also, $C_\beth$ and $C_\beth^0$ verify (A_1)–(A_2) if

$$\sup_{\kappa \in J} \sup_{-\infty < \iota \leq -\kappa} \frac{\varkappa(\kappa + \iota)}{\beth(\iota)} < \infty.$$

Example 10.6. For $\varpi > 0$, we define the space

$$C_\varpi := \{\varkappa \in \tilde{K}(\mathbb{R}_-), E) : \lim_{\iota \to -\infty} e^{\varpi \iota}\varkappa(\iota) \text{ exist in } E,$$

with

$$\|\varkappa\| = \sup\{e^{\varpi \iota}|\varkappa(\iota)| : \iota \leq 0\}.$$

Then in the space C_ϖ the axioms (A_1)–(A_3) are satisfied.

Consider

$$\mathcal{B} := \left\{ \varkappa \in \tilde{K}(\mathbb{R}_-), E) : \sup_{s \in \mathbb{R}_-} \frac{\|\varkappa(s)\|}{\beth(s)} < \infty \right\},$$

Thus, $\mathcal{B}$ verifies (A_3). Also, $\mathcal{B}$ verifies (A_1)–(A_2) if

$$\sup_{\kappa \in J} \sup_{-\infty < \iota \le -\kappa} \frac{\varkappa(\kappa + \iota)}{\beth(\iota)} < \infty.$$

The space $\mathcal{B}$ endowed with the norm

$$\|\varkappa\|_{\mathcal{B}} = \sup_{s \in \mathbb{R}_-} \frac{\|\varkappa(s)\|}{\beth(s)},$$

is a Banach space.

10.4.2   *Existence Results*

Definition 10.10. A measurable and T-periodic function $y \in \widetilde{PC}$ is a periodic mild solution of (10.11) if there exists $f \in S_{f \circ y}$, for a.e. $t \in \mathbb{R}_+$, such that

$$y(t) = \begin{cases} -\frac{\partial}{\partial s}\mathcal{U}(t,0)\varkappa(0) + \mathcal{U}(t,0)\wp_0 + \displaystyle\int_0^t \mathcal{U}(t,s)\, f(s)ds; & \text{if } t \in J_0, \\[2mm] -\frac{\partial}{\partial s}\mathcal{U}(t,s_k)g_k(s_k, y(s_k^-)) + \mathcal{U}(t,s_k)\wp_k \\[2mm] \quad + \displaystyle\int_{s_k}^t \mathcal{U}(t,s)\, f(s)ds; & \text{if } t \in J_k;\ k = 1,\ldots, \\[2mm] g_k(t, y(t_k^-)); & \text{if } t \in \tilde{J}_k;\ k = 1,\ldots, \\[2mm] \varkappa(t); & \text{if } t \in \mathbb{R}_-. \end{cases}$$

Theorem 10.4. *Assume that the following hypotheses hold:*

(10.27.1) $f(t, y, \vartheta)$ *is L^1–Carathéodory, and has compact and convex values and maps bounded sets into bounded sets.*

(10.27.2) g_k; $k = 1, 2, \ldots$; *are continuous, and they map bounded sets into bounded sets.*

(10.27.3) *For $T > 0$, $f(t + T, y, \vartheta) = f(t, y, \vartheta)$, $A(t + T) = A(t)$; $t \in J_k$; $k = 0, \ldots, m$, $y, \vartheta \in E \times \mathcal{B}$, and $g_k(t + T, \vartheta) = g_k(t, \vartheta)$ $t \in \tilde{J}_k$; $k = 1, \ldots, m, \vartheta \in E$.*

(10.27.4) *There exist continuous function $q : \tilde{J}_k \to \mathbb{R}_+$, such that*

$$\|g_k(t, \vartheta)\|_E \le q(t), \quad t \in \tilde{J}_k, \quad \vartheta \in E, \quad k = 0, \ldots, m,$$

(10.27.5) *For each bounded sets* $\overline{\Omega} \subset E$ *and* $D \subset \mathcal{B}$ *we have*

$$\alpha_E(f(t,\overline{\Omega},D)) \le \ell_\omega(t)\alpha_E(\overline{\Omega}); \quad t \in J_k; \ k = 0,\ldots,m,$$

and

$$\alpha_E(g_k(t,\overline{\Omega})) \le q(t)\alpha_E(\overline{\Omega}); \quad t \in \widetilde{J}_k; \ k = 1,\ldots,m,$$

where $\ell_\omega \in L^1(J_k)$, α_E *is a measure of non-compactness on the Banach space* E, *and*

$$\sigma_0 = \sup_{(t,s)\in\nabla} \left\| \frac{\partial}{\partial s}\mathcal{U}(t,s) \right\|_{B(E)},$$

$$\ell_\omega^* := \|\ell_\omega\|_{L^1(J_k)} = \int_{J_k} \ell_\omega(t)dt,$$

$$q^* := \|q\|_C = \sup_{t\in\widetilde{J}_k} q(t).$$

If $\max\{4\sigma T\ell_\omega^*, q^*\} < 1$, *then* (10.11) *admit at least one* T-*periodic mild solution.*

Proof. Consider the problem:

$$\begin{cases} y''(t) + A(t)y(t) = f(t,y(t),y_t); & \text{if } t \in J_k; \ k = 0,\ldots,m \\ y(t) = g_k(t,y(t_k^-)); & \text{if } t \in \widetilde{J}_k; \ k = 1,\ldots,m \\ y(t) = \varkappa(t); & \text{if } t \in \mathbb{R}_- := \mathbb{R}_-, \\ y'(s_k) = \wp_k \in E; & k = 0,\ldots,m. \end{cases} \quad (10.12)$$

Part 1. We begin by showing that (10.12) has a mild solution $y \in PC$, with $\|y\|_{PC} \le \chi$ where

$$\chi \ge \max\{\|\varkappa\|_{\mathcal{B}}, q^*, \sigma_0\|\varkappa(0)\| + \sigma(\|\wp_0\| + \ell_\omega^*), \sigma_0 q^* + \sigma(\|\wp_k\| + \ell_\omega^*)\}.$$

Let $\aleph : PC \to \mu(PC)$ given by

$$\aleph(y) = \left\{ \varphi \in PC : \varphi(t) = \begin{cases} -\frac{\partial}{\partial s}\mathcal{U}(t,0)\varkappa(0) + \mathcal{U}(t,0)\wp_0 \\ \quad + \int_0^t \mathcal{U}(t,s)\,f(s)ds; & \text{if } t \in J_0, \\ -\frac{\partial}{\partial s}\mathcal{U}(t,s_k)g_k(s_k,y(s_k^-)) + \mathcal{U}(t,s_k)\wp_k \\ \quad + \int_{s_k}^t \mathcal{U}(t,s)\,f(s)ds; & \text{if } t \in J_k; \ k = 1,\ldots,m \\ g_k(t,y(t_k^-)); & \text{if } t \in \widetilde{J}_k; \ k = 1,\ldots,m \\ \varkappa(t); & \text{if } t \in \mathbb{R}_-, \end{cases} \right\} \quad (10.13)$$

where $f \in S_{f\circ y}$.

Let $\varphi \in \aleph(y)$. Then there exists $f \in S_{f \circ y}$ where

$$\varphi(t) = \begin{cases} -\frac{\partial}{\partial s}\mathcal{U}(t,0)\varkappa(0) + \mathcal{U}(t,0)\wp_0 + \displaystyle\int_0^t \mathcal{U}(t,s)\ f(s)ds; & \text{if } t \in J_0, \\[2mm] -\frac{\partial}{\partial s}\mathcal{U}(t,s_k)g_k(s_k, y(s_k^-)) + \mathcal{U}(t,s_k)\wp_k \\[2mm] + \displaystyle\int_{s_k}^t \mathcal{U}(t,s)\ f(s)ds; & \text{if } t \in J_k;\ k = 1,\ldots,m \\[2mm] g_k(t, y(t_k^-)); & \text{if } t \in \tilde{J}_k;\ k = 1,\ldots,m \\[2mm] \varkappa(t); & \text{if } t \in \mathbb{R}_-. \end{cases}$$

For each $t \in J_0$, we have

$$\begin{aligned} \|\varphi(t)\| &\le \sigma_0\|\varkappa(0)\| + \sigma\|\wp_0\| + \sigma\int_0^t \|f(s)\|ds \\ &\le \sigma_0\|\varkappa(0)\| + \sigma\|\wp_0\| + \sigma\ell_\omega^* \\ &\le \chi. \end{aligned}$$

Next, for each $t \in J_k;\ k = 1,\ldots,m$, we have

$$\begin{aligned} \|\varphi(t)\| &\le \sigma_0 q^* + \sigma\|\wp_k\| + \sigma\int_{s_k}^t \|f(s)\|ds \\ &\le \sigma_0 q^* + \sigma\|\wp_k\| + \sigma\ell_\omega^* \\ &\le \chi. \end{aligned}$$

Also, for each $t \in \tilde{J}_k;\ k = 1,\ldots,m$, we have

$$\|\varphi(t)\| \le q^* \le \chi,$$

and for each $t \in \mathbb{R}_-$, we have

$$\|\varphi(t)\| = \|\varkappa\|_{\mathcal{B}} \le \chi.$$

Thus, $\aleph$ transforms the ball $\overline{\Omega}_\chi := \{\varpi \in PC : \|\varpi\|_{PC} \le \chi\}$ into $\mu(\overline{\Omega}_\chi)$.

Step 1. $\aleph(y) \in \mu_{cl}(PC)$ *for each* $y \in \overline{\Omega}_\chi$.
Let $(y_n)_{n\ge 0} \in \aleph(y)$ such that $y_n \longrightarrow \tilde{y}$ in PC. Then, $\tilde{y} \in PC$ and there exists $f_n \in S_{f \circ y}$ where

$$y_n(t) = \begin{cases} -\frac{\partial}{\partial s}\mathcal{U}(t,0)\varkappa(0) + \mathcal{U}(t,0)\wp_0 + \displaystyle\int_0^t \mathcal{U}(t,s)\ f_n(s)ds; & \text{if } t \in J_0, \\[2mm] -\frac{\partial}{\partial s}\mathcal{U}(t,s_k)g_k(s_k, y_n(s_k^-)) + \mathcal{U}(t,s_k)\wp_k \\[2mm] + \displaystyle\int_{s_k}^t \mathcal{U}(t,s)\ f_n(s)ds; & \text{if } t \in J_k;\ k = 1,\ldots,m \\[2mm] g_k(t, y_n(t_k^-)); & \text{if } t \in \tilde{J}_k;\ k = 1,\ldots,m \\[2mm] \varkappa(t); & \text{if } t \in \mathbb{R}_-. \end{cases}$$

Since f has compact values and by (10.27.1), we have that f_n converges to f in $L^1(J, E)$, and thus $f \in S_{f \circ y}$. Then

$$
y_n(t) \to y(t) = \begin{cases}
-\frac{\partial}{\partial s}\mathcal{U}(t,0)\varkappa(0) + \mathcal{U}(t,0)\wp_0 + \int_0^t \mathcal{U}(t,s)\, f(s)ds; & \text{if } t \in J_0, \\[2ex]
-\frac{\partial}{\partial s}\mathcal{U}(t,s_k)g_k(s_k, y(s_k^-)) + \mathcal{U}(t,s_k)\wp_k \\[1ex]
\quad + \int_{s_k}^t \mathcal{U}(t,s)\, f(s)ds; & \text{if } t \in J_k; \ k=1,\dots,m \\[2ex]
g_k(t, y(t_k^-)); & \text{if } t \in \tilde{J}_k; \ k=1,\dots,m \\[1ex]
\varkappa(t); & \text{if } t \in \mathbb{R}_-.
\end{cases}
$$

So, $\tilde{y} \in \aleph(y)$.

Step 2. $\aleph : \overline{\Omega}_\chi \to \mu_{cl,cv}(\overline{\Omega}_\chi)$ *is* $D-$*set-contraction.*
From Lemmas 2.5 and 2.6, for any $\overline{\Omega} \subset \overline{\Omega}_\chi$ and any $\epsilon > 0$, there exists a sequence $\{y_k\}_{k=0}^\infty \subset \overline{\Omega}$, such that for all $t \in J_0$, we have

$$
\begin{aligned}
\alpha_E((NB)(t)) &= \alpha_E \left(\left\{ -\frac{\partial}{\partial s}\mathcal{U}(t,0)\varkappa(0) + \mathcal{U}(t,0)\wp_0 \right. \right. \\
&\qquad \left. \left. + \int_0^t \mathcal{U}(t,s)\, f(s)ds; \ f \in S_{f \circ y}, \ y \in \overline{\Omega} \right\} \right) \\
&\leq 2\alpha_E \left(\left\{ \int_0^t \mathcal{U}(t,s)f(s)ds; \ f \in S_{f \circ y_k} \right\}_{k=1}^\infty \right) + \epsilon \\
&\leq 4 \int_0^t \alpha_E \left(\|\mathcal{U}(t,s)\|_{B(E)}\{f(s); \ f \in S_{f \circ y_k}\}_{k=1}^\infty \right) ds + \epsilon \\
&\leq 4\sigma \int_0^t \alpha_E \left(\{f(s); \ f \in S_{f \circ y_k}\}_{k=1}^\infty \right) ds + \epsilon \\
&\leq 4\sigma \int_0^t \ell_\omega(s)\alpha \left(\{y_k(s)\}_{k=1}^\infty \right) ds + \epsilon \\
&\leq 4\sigma\ell_\omega^* \int_0^t \alpha_E \left(\{y_k(s)\}_{k=1}^\infty \right) ds + \epsilon \\
&\leq 4\sigma T\ell_\omega^* \alpha_{PC}(\overline{\Omega}) + \epsilon,
\end{aligned}
$$

and, for all $t \in J_k; \ k=1,\dots,m$, we get

$$
\begin{aligned}
\alpha_E((NB)(t)) &= \alpha_E \left(\left\{ -\frac{\partial}{\partial s}\mathcal{U}(t,s_k)g_k(s_k, y(s_k^-)) + \mathcal{U}(t,s_k)\wp_k \right. \right. \\
&\qquad \left. \left. + \int_{s_k}^t \mathcal{U}(t,s)\, f(s)ds; \ f \in S_{f \circ y}, \ y \in \overline{\Omega} \right\} \right)
\end{aligned}
$$

$$\leq 2\alpha_E \left(\left\{ \left[\int_0^t \mathcal{U}(t,s) f(s) ds; \ f \in S_{f \circ y_k} \right\}_{k=1}^\infty \right) + \epsilon \right.$$

$$\leq 4\sigma T \ell_\omega^* \alpha_{PC}(\overline{\Omega}) + \epsilon.$$

Since $\epsilon > 0$ is arbitrary, then for all $t \in J_k; \ k = 0, \ldots, m$, we get

$$\alpha_E((NB)(t)) \leq \varkappa_1(\alpha_{PC}(\overline{\Omega})),$$

where $\varkappa_1 : \mathbb{R}_+ \to \mathbb{R}_+, \ \varkappa_1(y) = 4\sigma T \ell_\omega^* y$.

Also, for all $t \in \tilde{J}_k; \ k = 1, \ldots, m$, we get

$$\alpha_E((NB)(t)) \leq \varkappa_2(\alpha_{PC}(\overline{\Omega})),$$

where $\varkappa_2 : \mathbb{R}_+ \to \mathbb{R}_+, \ \varkappa_2(y) = q^* y$.

Hence, for all $t \in (-\infty, T]$ we obtain

$$\alpha_E((NB)(t)) \leq \varkappa(\alpha_{PC}(\overline{\Omega})),$$

where $\varkappa : \mathbb{R}_+ \to \mathbb{R}_+$ defined by

$$\begin{cases} \varkappa(t) = \varkappa_1(t); & \text{if } t \in J_k; \ k = 0, 1, \ldots, \\ \varkappa(t) = \varkappa_2(t); & \text{if } t \in \tilde{J}_k; \ k = 1, 2, \ldots. \end{cases}$$

Consequently, by the Generalized Schauder's fixed point theorem [259], $\aleph$ has a fixed point in $y \in \overline{\Omega}_\chi$.

Part 2. *Periodic mild solutions.*

We define the Poincaré operator $\Theta : \mathcal{B} \to \mathcal{B}$ by

$$\Theta(\varkappa) = y_T(\varkappa) : \ (\Theta \varkappa)(s) = y_T(s, \varkappa) = y(T + s, \varkappa); \ s \in \mathbb{R}_-.$$

Step 1. *The fixed points of Θ give rise to a periodic mild solutions of* (10.11).

Let $\varkappa \in \mathcal{B}$ be where $\Theta(\varkappa) = \varkappa$. Then for $y(\cdot) = y(\cdot, \varkappa)$ with $y_0(\cdot, \varkappa) = \varkappa$, we can give $\vartheta(t) = y(t + T)$. Now, for $t > 0$, by Definition 2.10, and since $A(t)$, f and g_k are T-periodic functions in t, we have that ϑ is also a solution with $\vartheta_0(\cdot, \varkappa) = y_T(\varkappa) = y(\cdot, \varkappa)$. Indeed, let $\varphi \in \aleph(y)$. Then there exists $f \in S_{f \circ y}$ such that

$$\vartheta(t) = \begin{cases} -\frac{\partial}{\partial s} \mathcal{U}(t,0) \varkappa(0) + \mathcal{U}(t,0) \wp_0 + \int_0^t \mathcal{U}(t,s) \, f(s) ds; & \text{if } t \in J_0, \ f \in S_{F \circ y} \\ -\frac{\partial}{\partial s} \mathcal{U}(t,s_k) g_k(s_k, \vartheta(s_k^-)) + \mathcal{U}(t,s_k) \wp_k \\ + \int_{s_k}^t \mathcal{U}(t,s) \, f(s) ds; & \text{if } t \in J_k; \ k = 1, \ldots, m, \ f \in S_{F \circ y} \\ g_k(t, \vartheta(t_k^-)); & \text{if } t \in \tilde{J}_k; \ k = 1, \ldots, m \\ \varkappa(t); & \text{if } t \in \mathbb{R}_-. \end{cases}$$

Then the uniqueness of $\{\mathcal{U}(t,s)\}_{(t,s)\in\nabla}$ implies that $\vartheta(t) = y(t)$, so that $y(t) = y(t+T)$ is a T-periodic solution.

Step 2. Θ *is condensing.*
Let $\overline{\Omega} \subset \mathcal{B}$ be bounded with $\alpha_{\mathcal{B}}(\overline{\Omega}) > 0$. If y_0 is the unique solution with $y_0(\varkappa) = \varkappa$, we define $W_s(\overline{\Omega}) = \{y_s(\varkappa) : \varkappa \in \overline{\Omega}\}$ and $W_{[\varphi,r]}(\overline{\Omega}) = \{y_{[\varphi,r]}(\varkappa) : \varkappa \in \overline{\Omega}\}$, where $y_{[\varphi,r]}(\varkappa)$ means the restriction of y on $[\varphi,r]$. As in [14], we get

$$\alpha_{\mathcal{B}}(\Theta(\overline{\Omega})) \le \left(\frac{1}{2}\right)^{\eta_0 - 1} \sigma\alpha_{\mathcal{B}}(\overline{\Omega}) < \alpha_{\mathcal{B}}(\overline{\Omega}).$$

Thus from Sadovskii's fixed point theorem [14], Θ has a fixed point which gives rise to a periodic mild solution of problem (10.11). $\qquad\square$

10.4.3 *An Example*

Consider the following problem:

$$\begin{cases} \dfrac{\partial^2\vartheta}{\partial t}(t,\gamma) + a(t,\gamma)\dfrac{\partial^2\vartheta}{\partial\gamma^2}(t,\gamma) \\ \quad \in \widehat{f}(t,\vartheta(t,\gamma),\vartheta_t(\cdot,\gamma)); \quad \gamma \in J := [0,\pi],\ t \in J_k;\ k = 0,\ldots, \\[1mm] \vartheta(t,\gamma) = g_k(t,\gamma); \quad\quad\quad\quad t \in \mathbb{R}_+,\ \gamma \in J,\ t \in \widetilde{J}_k;\ k = 1,\ldots, \\[1mm] \vartheta(t,0) = \vartheta(t,\pi) = 0; \quad\quad\quad\quad t \in \mathbb{R}_+, \\[1mm] \vartheta(0,\gamma) = \Omega(\gamma); \quad\quad\quad\quad\quad\quad \gamma \in J, \\ \vartheta(t,\gamma) = \varkappa(t,\gamma); \quad\quad\quad\quad\quad t \in \mathbb{R}_-,\ \gamma \in J, \end{cases}$$

$$(10.14)$$

where $a(t,\gamma) : \mathbb{R}_+ \times J \to \mathbb{R}$ is a continuous function in γ and uniformly Hölder continuous in t, $\widehat{f} : \mathbb{R}_+ \times \mathbb{R} \times C_\varphi \to \mu(\mathbb{R})$,

$$\begin{aligned} \widehat{f}(t,\vartheta(t,\gamma),\vartheta_t(\cdot,\gamma)) = \{\vartheta \in \mathbb{R} : |f_1(t,\vartheta(t,\gamma),\vartheta_t(\cdot,\gamma))| \\ \le |\vartheta| \\ \le |f_2(t,\vartheta(t,\gamma),\vartheta_t(\cdot,\gamma))|\}, \end{aligned}$$

$\Omega : J \to \mathbb{R}$ and $\varkappa : \mathbb{R}_- \times J \to \mathbb{R}$ are continuous functions where $\Omega(\gamma) = \varkappa(0,\gamma)$; $\gamma \in J$, $f_1, f_2 : \mathbb{R}_+ \times \mathbb{R} \times C_\varphi \to \mathbb{R}$ are continuous functions, and C_φ is the phase space given in Example 10.5.

Let $E = L^2(J,\mathbb{R})$ and define $A(t)$ by $A(t)\varpi = a(t,\gamma)\varpi''$ with domain

$$D(A) = \{\varpi \in E : \varpi, \varpi'\ \text{are absolutely continuous},$$
$$\varpi'' \in E,\ \varpi(0) = \varpi(\pi) = 0\}.$$

Then $A(t)$ generates an evolution system $\mathcal{U}(t, s)$ (see [352]).

For $\gamma \in J$, we have

$$y(t)(\gamma) = \vartheta(t, \gamma); \quad t \in \mathbb{R}_+,$$

$$f(t, y(t), y_t, \gamma) = \widehat{f}(t, \vartheta(t, \gamma), \vartheta_t(\cdot, \gamma)); \quad t \in \mathbb{R}_+,$$

$$y_0(\gamma) = \Omega(t, \gamma); \quad \gamma \in J,$$

$$y(t)(\gamma) = \varkappa(t, \gamma); \quad \gamma \in J; \quad t \in \mathbb{R}_-.$$

As a consequence, by the definitions of f, y_0 and $A(\cdot)$, the system (10.14) can be represented by the functional evolution problem (10.11). And, some conditions on $\widehat{f}$ make certain that (10.27.1)–(10.27.5) hold. Finally, Theorem 10.4 implies that the evolution problem (10.14) has at least one periodic mild solution.

10.5   Notes and Remarks

The results of Chapter 12 are taken from the paper [430–432]. For more relevant information, we recommend the books [160, 192, 404, 423] and the articles [406, 407, 433].

Chapter 11

Qualitative and Quantitative Analysis of Dynamical Models

11.1 Non-Autonomous Physiologically Structured Model with Nonlocal Diffusion

11.1.1 *Introduction and Motivations*

Lobesia botrana stands out as a significant vineyard pest, causing economic losses globally. Various strategies, including chemical tools, mating disruption, and integrated pest management (refer to [438,439]), are employed to combat this insect. Despite these efforts, insecticides remain the primary method of control. Notably, in formulating optimal control problems for Lobesia botrana, dispersal ability has often been overlooked. Recognizing the importance of nonlocal diffusion, it is acknowledged as a mechanism to describe the movement of pests over longer distances. Let d be a positive constant. The nonlocal logistic equation

$$\frac{\partial u}{\partial t} = dAu + u(1 - u)$$

has been investigated in [440] where the diffusion process is described by the nonlocal dispersal operator A given by

$$A(u)(x) := \int_D J(x - y)(u(y) - u(x))dy.$$

The function $J(x - y)$ is the rate at which individuals are dispersing from position y to x. The integro-differential equation

$$\frac{\partial u}{\partial t} = dAu + f(x, u)$$

has been analyzed in [441,442]. Recently, age structured population models with nonlocal diffusion have been studied in [443], the authors investigated the model

$$\frac{\partial u}{\partial t} + \frac{\partial u}{\partial a} = -\mu(a, t)u + dAu.$$

Mutation is another integral process shaping the population dynamics of Lobesia botrana. Structured models have given less emphasis to this aspect. When subjected to a low insecticide rate, individuals experience heightened stress levels. In response, the survivors undergo mutation, leading to the development of insecticide resistance. A continuous phenotype $\omega \in \Omega$ describing the resistance level to insecticides is introduced where $\Omega \subset \mathbb{R}^n$ is a bounded domain. At time t, size or physiological age a, position x, and resistance level ω', the pest population with density $u(t, a, \omega', x)$ gives birth to population with resistance level ω at a rate

$$\beta(P(t, x), t, a)\gamma(\omega, \omega') u(t, a, \omega', x).$$

Here β is the birth rate, and

$$P(t, x) := \int_{\Omega} \int_0^L u(t, a, \omega, x) da d\omega,$$

is the total population at time t and position x. The quantity $\gamma(\omega, \omega')$, represents the probability that individuals with trait ω' emerge with a new trait ω.

In this section, we consider the following nondimensionalized form of reaction-diffusion model:

$$\begin{cases} \dfrac{\partial}{\partial t} u(t, a, \omega, x) + \dfrac{\partial}{\partial a}[v(t, a, \omega)u(t, a, \omega, x)] \\ = -\mu(P(t, x), t, a)u(t, a, \omega, x) + dA u(t, a, \omega, x), \\[2mm] v(t, a = 0, \omega)u(t, 0, \omega, x) \\ = \displaystyle\int_0^L \int_{\Omega} \beta(P(t, x), t, s)\gamma(\omega, \omega') u(t, s, \omega', x) d\omega' ds, \\[2mm] u(0, a, \omega, x) = u_0(a, \omega, x), (a, \omega, x) \in (0, L) \times \Omega \times D, \end{cases} \quad (11.1)$$

where $D \subset \mathbb{R}^m$ is a bounded domain with smooth boundary ∂D, $u(t, a, \omega, x)$ is the population density at time $t \in [0, T]$, size $a \in [0, L]$, position $x \in D$, an a continuous phenotype $\omega \in \Omega$, $T > 0$ is a given time, and L is the maximal size. We assume that the death μ and the fertility β rates depend on the total population P. The growth rate v depends on time t, size a and ω. The boundary conditions are of nonlocal Neumann type, since in the definition of the operator A, the integral is defined only on D, the individuals may not enter or leave D see for instance [444], and the references therein. The dispersal rate d is constant and strictly positive.

Almost all size structured models have been studied under the assumption that $d = 0$, and the newborn individuals do not mutate, see for instance [444, 445] and the references therein.

11.1.2 Some Necessary Results

Let $L^1 := L^1((0, L) \times D; \mathbb{R})$ be the Banach space of Lebesgue integrable functions with the norm

$$\|u\|_{L^1} := \int_D \int_0^L |u(a, x)| dadx.$$

Let $Q = [0, T] \times \Omega$ and let B be the Banach space

$$B = L^\infty(Q; L^1),$$

endowed with the norm

$$\|u\|_T = \sup_Q \|u\|_{L^1}.$$

We also define the Banach space B_0 by

$$B_0 = L^\infty(\Omega; L^1),$$

endowed with the norm

$$\|u\| = \sup_\Omega \|u\|_{L^1}.$$

Throughout this section, we require the following assumptions:

(11.1.1) The function $J(.) \in C(\bar{D})$, J is bounded, $J \geq 0$, $J \neq 0$, $\int_{\mathbb{R}^N} J(x) dx = 1$, and $J(x) = J(-x)$.

(11.1.2) The function $\gamma : \Omega \times \Omega \to \mathbb{R}^n$ is continuous, bounded, and $\gamma \geq 0$. In addition, for all $\omega, \omega' \in \Omega$, we assume that

$$\gamma(\omega, \omega') = \gamma(\omega', \omega).$$

(11.1.3) The function $u_0 \in L^\infty((0, L) \times \Omega \times D)$ is everywhere positive.

(11.1.4) The function $(t, a, \omega) \to v(t, a, \omega)$ is bounded, continuous with respect to its arguments, strictly positive, continuously differentiable with respect to a. In addition, there exists a positive constant L_v such that

$$L_v = \sup_{t,a,w} \left| \frac{\partial v}{\partial a}(t, a, \omega) \right| < \infty.$$

Further, we assume that $v(., L,) =.0$ and

$$V_0 = \min_{0 \leq t \leq T, w \in \Omega} \{v(t, 0, \omega)\} > 0.$$

(11.1.5) $\beta(P, t, a)$ is bounded, and nonnegative measurable function on $\mathbb{R} \times [0, T] \times [0, L]$. In addition, the function β is locally Lipschitz with respect to the first variables P.

(11.1.6) $\mu(P, t, a)$ is bounded, and nonnegative measurable function on $\mathbb{R} \times [0, T] \times [0, L]$. In addition, the function μ is locally Lipschitz with respect to the first variable P.

Remark 11.1. The assumptions (11.1.1) are biologically relevant. The assumption $\int_{\mathbb{R}^N} J(x) dx = 1$ means that individuals are neither created nor destroyed during the movement. The symmetry of J implies that an individual at position x has the same probability of dispersion to position y as an individual in location y has of jumping to x. Let

$$
\begin{cases}
G(u(t, ., \omega, .))(a, x) = -\mu(P(t, x), t, a) u(t, a, \omega, x), \\
F(u(t, ., \omega, .))(x) = \displaystyle\int_0^L \int_\Omega \beta(P(t, x), t, s) \gamma(\omega, \omega') u(t, s, \omega', x) \, d\omega' ds.
\end{cases}
$$

Lemma 11.1.

(i) *There exists a constant $c_1 > 0$ such that for $\varphi_1, \varphi_2 \in B$, the function*

$$ F : B \to L^1(D) $$

satisfies

$$ |F(\varphi_1) - F(\varphi_2)|_{L^1(D)} \le c_1 \|\varphi_1 - \varphi_2\|_T . $$

(ii) *There exists a positive constant $c_2 > 0$ such that for $\varphi_1, \varphi_2 \in B$, the function*

$$ G : B \to L^1 $$

satisfies

$$ \|G(\varphi_1) - G(\varphi_2)\|_{L^1} \le c_2 \|\varphi_1 - \varphi_2\|_T . $$

The concept of a strong solution to system (11.1) requires the differentiability of the functions u and vu respectively with respect to the variables t, and a. This is quite restrictive. Along characteristics the system (11.1) behaves like a Cauchy problem. Let $T(t)$ be the semigroup generated by the bounded operator dA, where d is the dispersal rate. We have

$$ T(t) = e^{tdA} := \sum_{n \ge 0} (tdA)^n . $$

Lemma 11.2.

(a) $\|T(t)\| \le e^{2td}$.

(b) *the positive cone $L_+^1(D)$ is positively invariant by the semigroup defined by $T(t), t \ge 0$.*

Proof.

(a) Since A is an integral operator, it follows that

$$\|Au\|_{L^1(D)} \le 2\|u\|_{L^1(D)}.$$

This implies that for all $n \ge 1$

$$\|A^n u\|_{L^1(D)} \le 2^n \|u\|_{L^1(D)}$$

and

$$\|T(t)\| \le e^{2td}.$$

(b) We need to show that if $f \in L_+^1(D)$, then

$$T(t)f \in L_+^1(D), \forall t \ge 0.$$

Indeed, let $f \in L_+^1((D)$, then there exists a sequence of positive continuous functions with compact support

$$f_n \in C_c(D),$$

such that

$$f_n \to f \quad \text{in } L^1(D).$$

We will show that $T(t)f_n \ge 0$ for $t \ge 0$. Let $u_n(t, x) = (T(t)f_n)(x)$, then $u_n(t, x)$ satisfies

$$\begin{cases} \dfrac{\partial u_n(t, x)}{\partial t} = dAu_n(t, x) \\ u_n(0, x) = f_n(x). \end{cases}$$

Note that $v_n(t, x) = e^{\lambda t} u_n(t, x)$ verifies

$$\frac{\partial v_n(t, x)}{\partial t} = d \int_D J(x - y) \left(v_n(t, y) - v_n(t, x) \right) dy + \lambda v_n(t, x). \quad (11.2)$$

For λ positive and large enough, we have $p_0 := \lambda - d > 0$. Let

$$J_0 = d \max_{x \in D} \int_D J(x - y) dy$$

and

$$\tau = \frac{1}{p_0 + J_0}$$

Suppose that for some $x \in D$, and $t \in [0, \tau]$,

$$v_n(t, x) < 0,$$

then there exist $x_1 \in D$, and $t_1 \in [0, \tau]$ such that

$$\min_{x \in D, t \in [0, \tau]} v_n(t, x) = v_n(t_1, x_1) < 0.$$

Integrating the equation (11.2) over $[0, t_1]$, we obtain

$$v_n(t_1, x_1) - v_n(0, x_1) \geq (p_0 + J_0) v_n(t_1, x_1) t_1.$$

Since $v_n(0, x_1) = f_n(x_1) \geq 0$ and $t_1 \leq \tau$, then

$$v_n(t_1, x_1) \geq 0,$$

which is a contradiction. It follows that $v_n(t, x) \geq 0$ for $x \in D$ and $t \in [0, \tau]$. Hence $u_n(t, x) \geq 0$ for $x \in D$ and $t \in [0, \tau]$. Then we repeat the same arguments on the interval $[k\tau, (k+1)\tau]$ for $k = 1, 2 \ldots$. We conclude that

$$T(t)f_n \geq 0, \forall t \geq 0.$$

Since the linear operator $T(t)$ is bounded, then

$$T(t)f_n \to T(t)f \quad \text{in } L^1(D),$$

and there exists a subsequence, such that

$$T(t)f_{n_j} \to T(t)f \quad \text{a.e in} \quad D.$$

It follows that

$$T(t)f \geq 0 \text{ a.e in } D. \qquad \square$$

We define a characteristic curve $\phi(t; \tau, \eta, \omega)$ through the point (τ, η) as the solution of the equation

$$\begin{cases} \dfrac{da}{dt} = v(t, a, \omega), \\ \quad a(\tau) = \eta. \end{cases}$$

The function $a = \phi(t; \tau, \eta, \omega)$, is differentiable with respect to τ, and η, see for instance ([446], Chap 2., p. 116, Th. 9.2.). We have

$$\frac{da}{d\tau} = -v(\tau, \eta, \omega) \exp\left(\int_\tau^t \frac{\partial v}{\partial a}(\sigma, \phi(\sigma; \tau, \eta, \omega), \omega) d\sigma \right), \qquad (11.3)$$

and

$$\frac{da}{d\eta} = \exp\left(\int_\tau^t \frac{\partial v}{\partial a}(\sigma, \phi(\sigma; \tau, \eta, \omega), \omega)d\sigma\right). \qquad (11.4)$$

We define $z(t,\omega) = \phi(t; 0, 0, \omega)$, and $\tau = \tau(t, a, \omega)$, implicitely by the relation

$$\phi(t; \tau, 0, \omega) = a. \qquad (11.5)$$

Definition 11.1 ([447]). By a mild solution to system (11.1), we mean a function $u \in B$ such that $u = K(u)$, where

$$K(u)(t,a,\omega,x) = \begin{cases} \int_\tau^t T(t-s)\widetilde{G}(s, u(s,\cdot,\omega,\cdot))(\phi(s,\tau,0,\omega),x)ds \\ +T(t-\tau)\dfrac{F(u(\tau,\cdot,\omega,\cdot))(x)}{v(\tau,0,\omega)}, \quad \text{if } a < z(t,\omega), \\ \int_0^t T(t-s)\widetilde{G}(s, u(s,\cdot,\omega,\cdot))(\phi(s,t,a,\omega),x)ds \\ +T(t)u_0(\phi(0,t,a,\omega),\omega,x), \quad \text{if } a \geq z(t,\omega). \end{cases}$$

The definition of the mild solution is justified as follows. Let u be a solution of system (11.1), and define

$$U_{t_0,a_0}(t,\omega,x) := u\left(t, \phi\left(t; t_0, a_0, \omega\right), \omega, x\right),$$

then U_{t_0,a_0} satisfies

$$\frac{dU_{t_0,a_0}(t,\omega,x)}{dt} = \frac{\partial u\left(t, \phi\left(t; t_0, a_0, \omega\right), \omega, x\right)}{\partial t}$$
$$+ \frac{\partial u\left(t, \phi\left(t; t_0, a_0, \omega\right), \omega, x\right)}{\partial a} v\left(t, \phi\left(t; t_0, a_0, \omega\right), \omega\right),$$

and

$$\frac{dU_{t_0,a_0}(t,\omega,x)}{dt} = \widetilde{G}(t, u(t,.,\omega,.))\left(\phi\left(t; t_0, a_0, \omega\right), x\right) + dAU_{t_0,a_0}(t,\omega,x),$$

where

$$\widetilde{G}(t, u(t,\cdot,\omega,\cdot))(a,x) = G(u(t,\cdot,\omega,\cdot))(a,x) - \frac{\partial v(t,a,\omega)}{\partial a}u(t,a,\omega,x).$$

Let $T(t)$ be the semigroup generated by the dispersal operator dA, then

$$U_{t_0,a_0}(t,\omega,x) = T\left(t - \tau_0^*\right) U_{t_0,a_0}\left(\tau_0^*, \omega, x\right)$$
$$+ \int_{\tau_0^*}^t T(t-s)\widetilde{G}(s, u(s,.,\omega,.))\left(\phi\left(s; t_0, a_0, \omega\right), x\right) ds,$$

$$(11.6)$$

where $\tau_0^* \in [0, T]$ is an initial time defined by

$$\tau_0^* := \tau_0^* (t_0, a_0, \omega) = \begin{cases} \tau (t_0, a_0, \omega), & \text{if } a_0 < z (t_0, \omega), \\ 0, & \text{if } a_0 \geq z (t_0, \omega). \end{cases}$$

We distinguish two cases:

(i) If $a_0 < z (t_0, \omega)$, then

$$\tau_0 := \tau (t_0, a_0, \omega) > 0,$$

and we consider equation (11.6) with initial time $\tau_0^* = \tau_0$. Using equation (11.5), we obtain

$$\begin{aligned} U_{t_0, a_0} (t, \omega, x) &= U_{\tau_0, 0} (t, \omega, x) \\ &= T (t - \tau_0) U_{\tau_0, 0} (\tau_0, \omega, x) \\ &\quad + \int_{\tau_0}^t T(t - s) \widetilde{G}(s, u(s, \omega, \cdot)) (\phi (s; \tau_0, 0, \omega), x) \, ds. \end{aligned}$$

Since $U_{\tau_0, 0} (\tau_0, \omega, x) = u (\tau_0, \phi (\tau_0; \tau_0, 0, \omega), \omega, x) = u (\tau_0, 0, \omega, x)$, this gives

$$\begin{aligned} U_{\tau_0, 0} (t, \omega, x) &= T (t - \tau_0) u (\tau_0, 0, \omega, x) \\ &\quad + \int_{\tau_0}^t T(t - s) \widetilde{G}(s, u(s, \cdot, \omega, \cdot)) (\phi (s; \tau_0, 0, \omega), x) \, ds. \end{aligned}$$

In particular, we have

$$\begin{aligned} U_{t, a} (t, \omega, x) &= u(t, \phi(t; t, a, \omega), \omega, x) \\ &= u(t, a, \omega, x) \\ &= T(t - \tau) u(\tau, 0, \omega, x) \\ &\quad + \int_{\tau}^t T(t - s) \widetilde{G}(s, u(s, \cdot, \cdot, \omega, \cdot))(\phi(s; \tau, 0, \omega), x) ds. \end{aligned}$$

(ii) If $a_0 \geq z (t_0, \omega)$, then $\tau_0 \leq 0$, and we consider the equation (11.6) with initial time $\tau_0^* = 0$. Hence,

$$\begin{aligned} U_{t_0, a_0} (t, \omega, x) &= T(t) U_{t_0, a_0} (0, \omega, x) \\ &\quad + \int_0^t T(t - s) \widetilde{G}(s, u(s, \cdot, \omega, \cdot)) (\phi (s; t_0, a_0, \omega), x) \, ds. \end{aligned}$$

This leads to

$$\begin{aligned} U_{t_0, a_0} (t, \omega, x) &= T(t) u_0 (\phi (0, t_0, a_0, \omega), \omega, x) \\ &\quad + \int_0^t T(t - s) \widetilde{G}(s, u(s, \cdot, \omega, \cdot)) (\phi (s; t_0, a_0, \omega), x) \, ds. \end{aligned}$$

In particular, we have

$$U_{t,a}(t,\omega,x) = u(t,\phi(t;t,a,\omega),\omega,x) = u(t,a,\omega,x)$$
$$= T(t)u_0(\phi(0,t,a,\omega),\omega,x)$$
$$+ \int_0^t T(t-s)\widetilde{G}(s,u(s,\cdot,\cdot,\omega,\cdot))(\phi(s;t,a,\omega),x)ds.$$

This justifies the definition of a mild solution via characteristics and semi-group.

11.1.3 *Existence of Positive Solutions*

It is important to ensure that the model (11.1) is well posed. As in [448], the following result shows that the mild solutions satisfy system (11.1) along characteristic curves. Let u be a mild solution of the system (11.1), i.e $u \in B$ and $u = K(u)$.

Lemma 11.3. *For fixed* $(t,a,\omega,x) \in (0,T) \times (0,L) \times \Omega \times D$, *the function*
$$U(s) = u(s,\phi(s;t,a,\omega),\omega,x),$$
is differentiable a.e. on (τ^*,T) *and satisfies*
$$\frac{dU(s)}{ds} = \widetilde{G}(s,u(s,\cdot,\omega,\cdot))(\phi(s;t,a,\omega),x) + dAU(s),$$
where $\tau^* := \tau_0^*(t,a,\omega)$.

Proof. We distinguish two cases. If $\phi(s,t,a,\omega) \in (0,z(t,\omega))$, then
$$\frac{1}{h}[Ku(s+h,\phi(s+h,t,a,\omega),\omega,x) - Ku(s,\phi(s,t,a,\omega),\omega,x)] = Q_1^h + Q_2^h,$$
where
$$Q_1^h = \frac{1}{h}[T(s+h-\tau) - T(s-\tau)]\frac{F(u(\tau,\cdot,\omega,\cdot))(x)}{v(\tau,0,\omega)}$$
$$= \frac{1}{h}[T(h) - 1]T(s-\tau)\frac{F(u(\tau,\cdot,\omega,\cdot))(x)}{v(\tau,0,\omega)}.$$
Since the semigroup $T(t)$ is differentiable, then
$$Q_1^h \to dAT(s-\tau)\frac{F(u(\tau,\cdot,\omega,\cdot))(x)}{v(\tau,0,\omega)},$$
and
$$Q_2^h = \frac{1}{h}\left[\int_\tau^{s+h} T(s+h-\eta)(\widetilde{G}(\eta,u(\eta,\cdot,\omega,\cdot))(\phi(\eta,\tau,0,\omega),x))d\eta\right.$$
$$\left. - \int_\tau^s T(s-\eta)(\widetilde{G}(\eta,u(\eta,\cdot,\omega,\cdot))(\phi(\eta,\tau,0,\omega),x))d\eta\right]$$

which implies that,

$$Q_2^h = \frac{1}{h}[T(h) - 1] \int_\tau^s T(s-\eta)(\widetilde{G}(\eta, u(\eta, \cdot, \omega, \cdot))(\phi(\eta, \tau, 0, \omega)), x)d\eta$$
$$+ \left. \frac{T(h)}{h} \int_s^{s+h} T(s-\eta)(\widetilde{G}(\eta, u(\eta, \cdot, \omega, \cdot))(\phi(\eta, \tau, 0, \omega)), x) \right) d\eta,$$

then

$$Q_2^h \to \widetilde{G}(s, u(s, \cdot, \omega, \cdot))(\phi(s, \tau, 0, \omega), x)$$
$$+ dA \int_\tau^s T(s-\eta)(\widetilde{G}(\eta, u(\eta, \cdot, \omega, \cdot))(\phi(\eta, \tau, 0, \omega), x))d\eta.$$

Since $\phi(s, \tau, 0, \omega) = \phi(s, t, a, \omega)$, it follows that when h goes to zero

$$Q_1^h + Q_2^h \to \widetilde{G}(s, u(s, \cdot, \omega, \cdot))(\phi(s, t, a, \omega), x)$$
$$+ dAKu(s, \phi(s, t, a, \omega), \omega, x)$$
$$= \widetilde{G}(s, u(s, \cdot, .\omega, \cdot))(\phi(s, t, a, \omega), x) + dAU(s).$$

In a similar manner, we consider the case where $\phi(s, t, a, \omega) \in (z(t, \omega), L)$. $\qquad\square$

For $\alpha \in \mathbb{R}$, and $u \in B$, we define the operators $K_\alpha(u)$ as follows

$$K_\alpha(u)(t, a, \omega, x)$$
$$= \begin{cases} \int_\tau^t T(t-s)\left(\widetilde{G}(s, u(s, \cdot, \omega, \cdot) + \alpha I)(\phi(s, \tau, .0, \omega), x)e^{-\alpha(t-s)}\right) ds \\ +T(t-\tau)e^{-\alpha(t-\tau)}\dfrac{F(u(\tau, \cdot, \cdot))(x)}{v(\tau, 0, \omega)}, \quad \text{if } a < z(t, \omega), \\[2mm] \int_0^t T(t-s)\left(\widetilde{G}(s, u(s, \cdot, \omega, \cdot) + \alpha I)(\phi(s, t, a, \omega), x)e^{-\alpha(t-s)}\right) ds \\ +T(t)e^{-\alpha t}u_0(\phi(0, t, a, x), \omega, x), \quad \text{if } a \geq z(t, \omega). \end{cases}$$

Lemma 11.4. *Let* $u \in B$. *For fixed* $(t, a, \omega, x) \in (0, T) \times (0, L) \times \Omega \times D$, *the function*

$$w_\alpha(s) := K_\alpha(u)(s, \phi(s, t, a, \omega), \omega, x),$$

is differentiable a.e. on (τ^*, T), *and satisfies*

$$\frac{d}{ds}w_\alpha(s) = -\alpha w_\alpha(s) + (\widetilde{G}(s, u(s, ., \omega) + \alpha I))(\phi(s, t, a, x), x) + dAw_\alpha(s).$$

Lemma 11.5. *Let* $\alpha, \beta \in \mathbb{R}$, *and* $u \in B$. *Then*

$$K_\beta u(s, \phi(s, t, a, w), w, x)$$
$$= K_\alpha u(s, \phi(s, t, a, w), w, x)$$
$$+ (\alpha - \beta) \int_{\tau^*}^t T(t - \eta) e^{-\beta(t-\eta)} \left(K_\alpha u - u \right) (\eta, \phi(\eta, t, a, w), w, x) d\eta.$$

Proof. We have

$$\frac{d}{ds} (w_\beta - w_\alpha) = -\beta (w_\beta - w_\alpha) + (\alpha - \beta) [w_\alpha - u(s, \phi(s, t, a, x), w, x)]$$
$$+ dA (w_\beta - w_\alpha).$$

This gives

$$\frac{d}{ds} \left(e^{\beta s} (w_\beta - w_\alpha) \right) = e^{\beta s} (\alpha - \beta) [w_\alpha - u(s, \phi(s, t, a, w), w, \phi)]$$
$$+ dA e^{\beta s} (w_\beta - w_\alpha),$$

and

$$(w_\beta - w_\alpha)(t)$$
$$= (\alpha - \beta) \int_{\tau^*}^t T(s - \eta) e^{-\beta(t-\eta)} [w_\alpha(\eta) - u(\eta, \phi(\eta, t, a, w), w, x)] d\eta.$$

$\square$

Let $\alpha, \beta \in \mathbb{R}$ and $u \in B$, then $K_\alpha(u) = u$ implies that $K_\beta(u) = u$.

Let L^1_+ be the positive cone of L^1, and let

$$B^+ = L^\infty \left(Q, L^1_+ \right).$$

Theorem 11.1. *Let* $u_0 \in B_0^+$. *Under conditions* (11.1.1)–(11.1.6), *the problem* (11.1) *has a unique solution* $u \in B^+$.

Proof. Let

$$\alpha = \|\mu\|_\infty + L_v.$$

(a) First step: We show that K_α is a map from B to itself. For simplicity of notations, we put

$$\widetilde{G}_\alpha = \widetilde{G} + \alpha.$$

Let $u \in B$. Following [449], we have

$$\int_D \int_0^L |K_\alpha(u)(t, a, w, x)| \, da dx \le J_1 + J_2 + J_3 + J_4,$$

where

$$J_1 = \int_D \int_0^{z(t,\omega)} \left| T(t-\tau) \frac{F(u(\tau,\cdot,\omega,\cdot))(x)\,|}{v(\tau,0,\omega)} \right| da\,dx,$$

$$J_2 = \int_D \int_0^{z(t,\omega)} \int_\tau^t \left| T(t-s)\widetilde{G}_\alpha(u(s,\cdot,\omega,\cdot))(\phi(s,\tau,0,\omega),x) \right| ds\,da\,dx,$$

$$J_3 = \int_D \int_{z(t,\omega)}^L |T(t)u_0(\phi(0,t,a,\omega),\omega,x)|\,da\,dx,$$

$$J_4 = \int_D \int_{z(t,\omega)}^L \int_0^t \left| T(t-s)\widetilde{G}_\alpha(u(s,\cdot,\omega,\cdot))(\phi(s,t,a,\omega),x) \right| ds\,da\,dx.$$

To estimate J_1, we make the change of variables from a to τ by the relation $\tau = \tau(t,a,\omega)$. It follows from (11.3) that

$$J_1 = \int_D^{z(t,\omega)} \int_0^{z(t,\omega)} T(t-\tau) \frac{F(u(\tau,\cdot,\omega,\cdot))(x)}{v(\tau,0,\omega)} \mid da\,dx$$

$$= \int_0^t \|T(t-\tau)F(u(\tau,\cdot,\omega,\cdot))\|_{L^1(D)} \frac{1}{v(\tau,0,\omega)} da$$

$$\leq \int_0^t \|T(t-\tau)F(u(\tau,\cdot,\omega,\cdot))\|_{L^1(D)}$$

$$\times \left(\exp \int_\tau^t \frac{\partial v}{\partial a}(s,\phi(s,\tau,0,\omega),\omega)ds \right) d\tau$$

$$\leq e^{TL_v} \int_0^t \|T(t-\tau)F(u(\tau,\cdot,\omega,\cdot))\|_{L^1(D)} d\tau$$

$$\leq e^{T(2d+L_v)} \int_0^t \|F(u(\tau,\cdot,\omega,\cdot))\|_{L^1(D)} d\tau.$$

By Lemma 11.1, since $F(0) = 0$ it follows that

$$J_1 \leq e^{T(2d+L_v)} \int_0^t \|F(u(\tau,\cdot,\omega,\cdot)) - F(0)\|_{L^1(D)} d\tau$$

$$\leq e^{T(2d+L_v)} \int_0^t c_1\|u\|_T d\tau \leq e^{T(2d+L_v)} c_1\|u\|_T T.$$

Similarly to estimate $J_2 + J_4$, we make the change of variables

$$\eta = \phi(s,t,a,\omega) = \phi(s,\tau,0,\omega),$$

then

$$J_2 + J_4 \leq e^{TL_V} \left\{ \int_D \int_0^t \int_\tau^{z(t,\omega)} \left| T(t-s)\widetilde{G}_\alpha(u(s,.,\omega,.))(\eta,x) \right| d\eta ds dx \right.$$

$$\left. + \int_D \int_0^t \int_{z(t,\omega)}^L \left| T(t-s)\widetilde{G}_\alpha(u(s,\cdot,\cdot,\omega,\cdot))(\eta,x) \right| d\eta ds dx \right\}$$

$$\leq e^{T(2d+L_v)} \left\{ \int_0^t \int_0^L \int_D |G(u(s,\cdot,\cdot,\omega,\cdot))(\eta,x)| dx d\eta ds \right.$$

$$\left. + \int_0^t \int_0^L \int_D \left[\left| \frac{\partial v}{\partial a}(s,\eta,\omega) + \alpha \right] u(s,\eta,\omega,x) \right| dx d\eta ds \right\}.$$

From Lemma 11.1, it follows that

$$\int_D \int_0^L |G(u(s,\cdot,\omega,\cdot))(\eta,x)| d\eta dx = \|G(u(s,\cdot,\omega,\cdot))\|_{L^1}$$

$$= \|G(u(s,\cdot,\omega,\cdot)) - G(0)\|_{L^1}$$

$$\leq \|u\|_T c_2$$

and

$$\int_D \int_0^L \left| \frac{\partial v}{\partial a}(s,\eta,\omega)u(s,\eta,\omega,x) \right| d\eta dx \leq \|u\|_T L_v.$$

Therefore, we obtain that

$$J_2 + J_4 \leq e^{T(2d+L_v)} \left\{ c_2 + \alpha + L_v \right\} T \|u\|_T.$$

To estimate J_3, we use the change of variables

$$\zeta = \phi(0,t,a,\omega),$$

this gives

$$J_3 \leq e^{T(2d+L_v)} \int_0^L \int_D |u_0(\zeta,\omega,x)| dx d\zeta \leq \|u_0\| e^{T(2d+L_v)}.$$

Hence

$$J_1 + J_2 + J_3 + J_4 \leq \|u_0\| e^{T(2d+L_v)} + e^{T(2d+L_v)} \left\{ \alpha + c_2 + L_v + c_1 \right\} \|u\|_T T.$$

This shows that

$$K_\alpha(u) \in B.$$

(b) The second step: We introduce an equivalent norm

$$\|u\|_\lambda = \sup_{t\in[0,T]} e^{-\lambda t}\|u\|_T,$$

where $\lambda > 0$ is a constant. Let $u_1, u_2 \in B$, then

$$\int_D \int_0^L |K_\alpha(u_1) - K_\alpha(u_2)|\, dadx \le P_1 + P_2 + P_3,$$

where

$$P_1 = \int_0^{z(t,\omega)} \int_D \frac{|T(t-\tau)\,(F(u_1(\tau,.,\omega,.)))(x) - F(u_2(\tau,.,\omega,.))(x))|}{v(\tau,0,\omega)}\, dxda$$

and

$$P_2 = \int_0^{z(t,\omega)} \int_\tau^t \int_D |S_1(s,t,x,u_1,u_2)|\, dxdsda,$$

$$P_3 = \int_{z(t,\omega)}^L \int_0^t \int_D |S_2(s,t,x,u_1,u_2)|\, dxdsda.$$

Here

$$S_1(s,t,x,u_1,u_2) = T(t-s)\left(\widetilde{G}_\alpha(s,u_1(s,\cdot,\omega,\cdot))\,(\phi(s,\tau,0,\omega),x)\right)$$
$$- T(t-s)\left(\widetilde{G}_\alpha(s,u_2(s,\cdot,\omega,\cdot))\,(\phi(s,\tau,0,\omega),x)\right),$$

and

$$S_2(s,t,x,u_1,u_2) = T(t-s)\left(\widetilde{G}_\alpha(s,u_1(s,\cdot,\omega,\cdot))\,(\phi(s,t,a,\omega),x)\right)$$
$$- T(t-s)\left(\widetilde{G}_\alpha(s,u_2(s,\cdot,\cdot,\omega,\cdot))\,(\phi(s,t,a,\omega),x)\right).$$

Note that

$$P_1 \le e^{TL_v} \int_0^t \|T(t-\tau)\,(F(u_1(\tau,\cdot,\omega,\cdot))(x) - F(u_2(\tau,\cdot,\omega,\cdot))(x))\|_{L^1(D)}\, d\tau$$

$$\le e^{T(2d+L_v)}\|\gamma\|_\infty\|\beta\|_\infty \int_0^t \int_\Omega \|u_1(\tau,\cdot,\cdot,\omega') - u_2(\tau,\cdot,\cdot,\omega')\|_{L^1}\, d\omega'd\tau.$$

This gives

$$e^{-\lambda t}P_1 \le e^{T(2d+L_v)}\|\gamma\|_\infty\|\beta\|_\infty e^{-\lambda t}$$
$$\times \int_\Omega \int_0^t e^{-\lambda\tau}e^{\lambda\tau}\|u_1(\tau,\cdot,\cdot,\omega') - u_2(\tau,\cdot,\cdot,\omega')\|_{L^1}\, d\tau d\omega',$$

and

$$e^{-\lambda t} P_1 \leq e^{T(2d+L_v)} |\Omega| \|\gamma\|_\infty \|\beta\|_\infty \frac{1}{\lambda} \|u_1 - u_2\|_\lambda,$$

where $|\Omega|$ denotes the measure of the set Ω. Using the change of variables

$$\eta = \phi(s, t, a, \omega) = \phi(s, \tau, 0, \omega),$$

we obtain that

$$
\begin{aligned}
P_2 + P_3 \leq \int_0^L \int_0^t \int_D & |T(t-s) G\left(s, u_1(s, \cdot, \omega, \cdot)\right)(\eta, x) \\
& - G\left(s, u_2(s, \cdot, \omega, \cdot)\right)(\eta, x)| dx d\eta da \\
& + \int_0^L \int_0^t \int_D \left| T(t-s) \left[\frac{\partial v}{\partial a}(t, \eta, \omega) + \alpha \right] \right. \\
& \times \left. (u_1(s, \eta, \omega, x) - u_2(s, \eta, \omega, x)) \right| dx d\eta da.
\end{aligned}
$$

This implies that

$$e^{-\lambda t} (P_2 + P_3) \leq e^{T(2d+L_v)} (\alpha + L_v + \|\mu\|_\infty) \frac{1}{\lambda} \|u_1 - u_2\|_\lambda.$$

It follows that

$$\|K_\alpha(u_1) - K_\alpha(u_2)\|_\lambda \leq C_T \frac{1}{\lambda} \|u_1 - u_2\|_\lambda,$$

where

$$C_T = e^{T(2d+L_v)} (\alpha + L_v + \|\mu\|_\infty + |\Omega| \|\gamma\|_\infty \|\beta\|_\infty).$$

Choosing $\lambda > C_T$ gives that K_α is a contraction on the Banach space $(B, \|\cdot\|_\lambda)$, and K_α has a unique fixed point $u \in B$.

c) Third step. It is clear that for $u \in B^+$, $K_\alpha u \in B^+$. By corollary 1, it follows that

$$K_0(u) = u,$$

and u is a positive solution of system (11.1). □

Remark 11.2. Theorem 11.1 remains valid for function F and G having more general forms, provided that they satisfy some general properties such as Lipschitz conditions and assumptions to ensure that $K_\alpha(B^+) \subset B^+$.

11.1.4 *Comparison Principle and Continuous Dependence*

We prove the following auxiliary theorems which are useful in proving our results of optimal control. Let

$$\tilde{D}_T = [0, T] \times [0, L] \times \Omega \times D.$$

Consider the following problem:

$$
\begin{cases}
\dfrac{\partial}{\partial t} u(t, a, \omega, x) + \dfrac{\partial}{\partial a}[v(t, a, \omega)u(t, a, \omega, x)] \\
\quad = G(u(t, \cdot, \omega, \cdot))(a, x) + dAu, \\
v(t, a = 0, \omega)u(t, 0, \omega, x) = F(u(t, \cdot, \omega, \cdot))(x), \\
u(0, a, \omega, x) = u_0(a, \omega, x), (a, \omega, x) \in (0, L) \times \Omega \times D,
\end{cases}
\tag{11.7}
$$

where

$$
\begin{cases}
G(u(t, \cdot, \omega, \cdot))(a, x) = -\mu(P(t, \omega), t, a)u(t, a, \omega, x) + f(t, a, \omega, x), \\
F(u(t, \cdot, \omega, \cdot))(x) = \displaystyle\int_\Omega \int_0^L \beta(P(t, \omega), t, s)\gamma(\omega, \omega')\, u(t, s, \omega', x)\, ds d\omega'.
\end{cases}
$$

Theorem 11.2. *If* $\mu_i, \beta_i, \gamma_i, u_{i0}$ *and* $f_i \in B$ *satisfy* (11.1.6), (11.1.5), (11.1.2), (11.1.3) *respectively with* $f_1 \leq f_2, \beta_1 \leq \beta_2, \gamma_1 \leq \gamma_2, u_{10} \leq u_{20}$, *and* $\mu_1 \geq \mu_2$ *then*

$$u^1(t, a, \omega, x) \leq u^2(t, a, \omega, x) \ \text{a.e on } \tilde{D}_T,$$

where u^i *is the solution to* (11.7) *corresponding to* $\beta = \beta_i, \gamma = \gamma_i \mu = \mu_i, f = f_i,$ *and* $u_0 = u_{i0}.$

Proof. Let

$$\alpha = \|\mu_1\|_\infty + \|\mu_2\|_\infty + L_v,$$

and

$$K_{k,\alpha}(u)(t, a, \omega, x)$$

$$
= \begin{cases}
T(t - \tau)e^{-\alpha(t-\tau)} \dfrac{F_k(u(\tau, \cdot, \cdot, \cdot))(x)}{v(\tau, 0, \omega)} \\
\quad + \displaystyle\int_\tau^t T(t - s)\left(\tilde{G}_k(s, u(s, \cdot, \omega, \cdot) + \alpha I)(\phi(s, \tau, 0, \omega), x)e^{-\alpha(t-s)}\right) ds \\
\hfill \text{if } a < z(t, \omega), \\
T(t)e^{-\alpha t} u_{k0}(\phi(0, t, a, x), \omega, x) \\
\quad + \displaystyle\int_0^t T(t - s)\left(\tilde{G}_k(s, u(s, \cdot, \omega, \cdot) + \alpha I)(\phi(s, t, a, \omega), x)e^{-\alpha(t-s)}\right) ds \\
\hfill \text{if } a \geq z(t, \omega),
\end{cases}
$$

Similarly as in the proof of Theorem 11.1, for $k = 1, 2$, the operator $K_{k,\alpha}$ has a fixed point u_k in B^+. Hence

$$K_{1,\alpha}(u_1) = u_1, \text{ and } K_{2,\alpha}(u_2) = u_2.$$

Since for $u \in B^+$, we have

$$\begin{aligned}
G_1(u(t, \cdot, \omega, \cdot))(a, x) &= -\mu_1(P(t, \omega), t, a)u(t, a, \omega, x) + f_1(t, a, \omega, x) \\
&\leq -\mu_2(P(t, \omega), t, a)u(t, a, \omega, x) + f_2(t, a, \omega, x) \\
&= G_2(u(t, \cdot, \omega, \cdot))(a, x),
\end{aligned}$$

and

$$F_1(u(t, \cdot, \omega, \cdot))(x) \leq F_2(u(t, \cdot, \cdot, \omega, \cdot))(x),$$

then

$$u_1 = K_{1,\alpha}(u_1) \leq K_{2,\alpha}(u_1).$$

Moreover, the monotony of $K_{2,\alpha}$ with respect to $u \in B^+$ implies that

$$u_1 \leq K_{2,\alpha}(u_1) \leq K_{2,\alpha}(K_{2,\alpha}(u_1)),$$

this leads to

$$u_1 \leq K_{2,\alpha}^2(u_1).$$

By induction, we obtain for each $j \in \mathbb{N}$, and $j \geq 2$

$$u_1 \leq K_{2,\alpha}^2(u_1) \leq \ldots \leq K_{2,\alpha}^j(u_1).$$

Since B^+ is a normal cone, then

$$u_1 \leq \lim_{j \to \infty} K_{2,\alpha}^j(u_1) = u_2. \qquad \square$$

We will analyze the continuous dependence of the solution of the system (11.7) with respect to the function f. We assume that $f_n(t, a, \omega, x)$, $f(t, a, \omega, x) \in B$ and

$$\|f_n - f\|_T = \sup_Q \|f_n - f\|_{L^1} \to 0, \text{ as } n \to \infty.$$

Theorem 11.3. *Let u be a solution of system (11.7), and let u_n be a solution a system (11.7) with f replaced by f_n. If $f_n, f \in B$, then*

$$u_n \to u \text{ as } n \to \infty \text{ in } B.$$

Proof. Let G_n be the function G with f replaced by f_n. We have

$$\|u_n(t, \cdot, \omega, \cdot) - u(t, \cdot, \omega, \cdot)\|_{L^1} \leq \Sigma_1 + \Sigma_2 + \Sigma_3,$$

where

$$\Sigma_1 = \int_0^{z(t,\omega)} \int_D \left| \frac{T(t-\tau)F\left(u_n(\tau, \cdot, \omega, \cdot)\right)(x)}{v(\tau, 0, \omega)} \right. $$
$$\left. - \frac{T(t-\tau)F(u(\tau, \cdot, \omega, \cdot))(x)}{v(\tau, 0, \omega)} \right| dxda,$$

$$\Sigma_2 = \int_D \int_0^{z(t,\omega)} \int_\tau^t |S_1(s,t,\omega,x)| \, dxdsda,$$

$$\Sigma_3 = \int_D \int_{z(t,\omega)}^L \int_0^t |S_2(s,t,\omega,x)| \, dxdsda,$$

with

$$S_1 = T(t-s)\left(\widetilde{G}_n\left(s, u_n(s, \cdot, \omega, \cdot)\right)(\phi(s,\tau,0,\omega), x)\right)$$
$$- T(t-s)(\widetilde{G}(s, u(s, \cdot, \omega, \cdot))(\phi(s,\tau,0,\omega), x)),$$

and

$$S_2 = T(t-s)\left(\widetilde{G}_n\left(s, u_n(s, \cdot, \omega, \cdot)\right)(\phi(s,t,a,\omega), x)\right)$$
$$- T(t-s)(\widetilde{G}(s, u(s, \cdot, \omega, \cdot))(\phi(s,t,a,\omega), x)).$$

By the change of variables $\tau = \tau(t, a, \omega)$, it follows that

$$\Sigma_1 \leq \frac{1}{V_0} e^{T(L_v + 2d)} \|\beta\|_\infty \|\gamma\|_\infty |\Omega| \int_0^t \|u(\tau, \cdot, \omega, \cdot) - u_n(\tau, \cdot, \omega, \cdot)\|_{L^1} \, d\tau.$$

where $|\Omega|$ denotes the measure of the set Ω. Similarly, we divide Σ_2 into two parts, we obtain

$$\Sigma_2 \leq \Sigma_2^1 + \Sigma_2^2,$$

where Σ_2^1 satisfies

$$\Sigma_2^1 \leq \int_D \int_0^{z(t,\omega)} \int_\tau^t \left| S_1^1(s,t,\omega,x) \right| dxdsda$$

with

$$S_1^1(s,t,\omega,x) = T(t-s)(\widetilde{G}(s, u(s, \cdot, \omega, \cdot))(\phi(s,\tau,0,\omega), x)$$
$$- \widetilde{G}_n(s, u(s, \cdot, \omega, \cdot))(\phi(s,\tau,0,\omega), x)\Big),$$

this implies that

$$\Sigma_2^1 \leq e^{2dT} \int_0^T \|f_n(s, \cdot\omega, \cdot) - f(s, \cdot\omega, \cdot)\|_{L^1} \, ds$$
$$\leq e^{2dT} T \|f_n - f\|_T \to 0 \text{ as } n \to \infty.$$

The second term Σ_2^2 satisfies

$$\Sigma_2^2 \leq \int_D \int_0^{z(t,\omega)} \int_\tau^t \left| S_1^2(s,t,\omega,x) \right| dxdsda,$$

where

$$S_1^2(s,t,\omega,x) = T(t-s)\left(\widetilde{G}_n(s,u(s,\cdot,\omega,\cdot))(\phi(s,\tau,0,\omega),x) \right)$$
$$- \widetilde{G}_n\left(s,u_n(s,\cdot,\omega,\cdot)\right)(\phi(s,\tau,0,\omega),x) \right),$$

with the change of variables $\phi(s,\tau,0,\omega) = \eta$, we obtain that

$$\Sigma_2^2 \leq e^{2dT}\left(\|\mu\|_\infty + L_v \right) \int_0^t \|u_n(\eta,.,\omega,.) - u(\eta,.,\omega,.)\|_{L^1} \, d\eta,$$

hence

$$\Sigma_2 \leq \delta_n^1 + e^{2dT}\left(\|\mu\|_\infty + L_v \right) \int_0^t \|u_n(\eta,\cdot,\omega,\cdot) - u(\eta,\cdot,\omega,\cdot)\|_{L^1} \, d\eta.$$

We divide Σ_3 into two parts. We have

$$\Sigma_3 \leq \Sigma_3^1 + \Sigma_3^2,$$

where

$$\Sigma_3^1 \leq \int_D \int_{z(t,\omega)}^L \int_0^t \left| S_2^1(s,t,\omega,x) \right| dsdadx$$

with

$$S_2^1(s,t,\omega,x) = T(t-s)(\widetilde{G}(s,u(s,\cdot,\omega,\cdot))(\phi(s,t,a,\omega),x))$$
$$- T(t-s)\left(\widetilde{G}_n(s,u(s,\cdot,\omega,\cdot))(\phi(s,t,a,\omega),x) \right),$$

hence

$$\Sigma_3^1 \leq e^{2dT} \int_0^T \|f_n(s,\cdot,\omega,\cdot) - f(s,\cdot,\omega,\cdot)\|_{L^1} \, ds$$
$$\leq Te^{2dT} \|f_n - f\|_T \to 0 \text{ as } n \to \infty.$$

The quantity Σ_3^2 verifies

$$\Sigma_3^2 \leq \int_D \int_{z(t,\omega)}^L \int_0^t \left| S_2^2(s,t,\omega,x) \right| dsdadx,$$

where

$$S_2^2(s,t,\omega,x) = T(t-s)\left(\widetilde{G}_n(s,u(s,\cdot,\omega,\cdot))(\phi(s,t,a,\omega),x) \right)$$
$$- T(t-s)\left(\widetilde{G}_n\left(s,u_n(s,\cdot,\omega,\cdot)\right)(\phi(s,t,a,\omega),x) \right).$$

By using the change of variables $\phi(s, t, a, \omega) = \eta$, we will have

$$\Sigma_3^2 \leq e^{2dT} \left(L_v + \|\mu\|_\infty\right) \int_0^t \|u_n(\eta, \cdot, \omega, \cdot) - u(\eta, \cdot, \omega, \cdot)\|_{L^1} \, d\eta.$$

We summarize the estimates as follows

$$\|u_n(t, \cdot, \omega, \cdot) - u(t, \cdot, \omega, \cdot)\|_{L^1} \leq 2e^{2dT} T \|f_n - f\|_T$$
$$+ C \int_0^t \|u(\eta, \cdot, \omega, \cdot) - u_n(\eta, \cdot, \omega, \cdot)\|_{L^1} \, d\eta,$$

for some positive constant C independent of n. Gronwall's Lemma implies that

$$\|u(t, \cdot, \omega, \cdot) - u_n(t, \cdot, \omega, \cdot)\|_{L^1} \leq \left(2e^{2dT} T \|f_n - f\|_T\right) e^{CT}.$$

The right-hand side of the previous inequality is independent of t and ω, this yields that

$$\|u - u_n\|_T \leq \left(2e^{2dT} T \|f_n - f\|_T\right) e^{CT},$$

and by letting n goes to ∞, we obtain the desired result. $\qquad\square$

11.1.5 *Application to Optimal Control*

Let $\Omega_0 \subset \Omega$, and let

$$D_T = [0, T] \times [0, L] \times \Omega_0 \times D.$$

In this section, we are concerned with the optimal control problem:

$$\text{maximize } J(I), \qquad (11.8)$$

where

$$J(I) = \int_{D_T} [\eta(t, a, \omega, x) I(t, a, \omega, x)) u^I(t, a, \omega, x)] \, dx d\omega da dt,$$

subject to the control

$$I \in A = \{I \in L^\infty(D_T) : 0 \leq I \leq I_{\max} \text{ a.e on } D_T\}.$$

Here u^I is the solution to

$$\begin{cases} \dfrac{\partial}{\partial t} u(t, a, \omega, x) + \dfrac{\partial}{\partial a}[v(t, a, \omega) u(t, a, \omega, x)] \\ = -\mu(t, a) u(t, a, \omega, x) - \chi_{\Omega_0}(\omega) I(t, a, \omega, x) u(t, a, \omega, x) + dAu, \\ v(t, 0, \omega) u(t, 0, \omega, x) = \displaystyle\int_\Omega \int_0^L \beta(t, s) \gamma(\omega, \omega') u(t, s, \omega', x) \, ds d\omega', \\ u(0, a, \omega, x) = u_0(a, \omega, x), (a, \omega, x) \in (0, L) \times \Omega \times D. \end{cases} \qquad (11.9)$$

The control I represents the insecticide effort, and plays the role of additional mortality. Note that I is acting only on a non-empty open subset $\Omega_0 \subset \Omega$. That means that individuals with level of resistance $\omega \in \Omega_0$ are more vulnerable. The quantity χ_{Ω_0} is the characteristic function. The function η is the net profit generated by the elimination of an individual with size a, and phenotype ω, at position x and time t. We assume that η is non-negative and bounded by $\eta_{\max}$. Our aim is to maximize J, the benefits from eliminating the pest. By Mazur's Theorem, we establish the existence of the optimal solution, and by the concept of normal cone, we give conditions for optimality.

Definition 11.2. A pair $\left(I^*, u^{I^*}\right)$ is said to be optimal for the control problem if $I^* \in A$, maximizes the functional J, and the pair $\left(I^*, u^{I^*}\right)$ solves the problem 11.9.

In a similar way as in Theorem 11.1, we prove that for any $I \in A$, problem 11.9 has a unique nonnegative solution u^I. Let w be the nonnegative solution of the problem 11.9 corresponding to to $I = 0, \mu = 0, \beta = \|\beta\|_\infty$, and $\gamma = \|\gamma\|_\infty$.

Lemma 11.6. *There exists $C_w > 0$ such that*

$$0 \leq w(t, a, \omega, x) \leq C_w \ a.e \ on \ D_T.$$

Proof. Let P^w be the total population corresponding to w. Integrating the equation of w, we obtain

$$\frac{\partial P^w}{\partial t} + \int_\Omega v(t, L, \omega)u(t, L, \omega, x)d\omega - \int_\Omega v(t, 0, \omega)u(t, 0, \omega, x)d\omega = dAP^w,$$

this implies that

$$\frac{\partial P^w}{\partial t} = dAP^w + \|\beta\|_\infty \|\gamma\|_\infty |\Omega| P^w.$$

Define $M : L^\infty(D) \to L^\infty(D)$ by

$$Mu = dAu + \|\beta\|_\infty \|\gamma\|_\infty |\Omega| u.$$

Then

$$P^w(t, x) = e^{Mt} P^w(0, x),$$

and

$$\sup_{0 \leq t \leq T} \|P^w(t, \cdot)\|_\infty \leq \|P^w(0, \cdot)\|_\infty e^{[2d + \|\beta\|_\infty \|\gamma\|_\infty |\Omega|]T}.$$

For $\rho \in L^\infty([0,T] \times [0,L] \times \Omega)$, we define

$$\Pi_\rho(t,\tau,t_0,a_0,\omega) = e^{\int_\tau^t \rho(\sigma,X(\sigma,t_0,a_0,\omega),\omega)d\sigma}.$$

It is easy to see that the solution w is given by

$$w(t,a,\omega,x) = \begin{cases} T(t-\tau)\Pi_\rho(t,\tau,t,a,\omega)\frac{F(w(\tau,\cdot,\cdot))(x)}{v(\tau,0,\omega)}, & \text{if } a < z(t,\omega), \\ T(t)\Pi_\rho(t,0,t,a,\omega)u_0(\phi(0,t,a,\omega),\omega,x), & \text{if } a \geq z(t,\omega), \end{cases}$$

with

$$\rho(t,a,w) = -\frac{\partial v}{\partial a}(t,a,\omega).$$

Since

$$F(w(\tau,\cdot,\cdot))(x) = \|\beta\|_\infty \|\gamma\|_\infty P^w(t,x),$$

and u_0 are bounded, assumption (11.1.4) implies that there exists a positive constant C_w such that

$$0 \leq w(t,a,\omega,x) \leq C_w \text{ a.e on } D_T. \qquad \square$$

Theorem 11.4. *The problem* (11.8) *has at least one optimal solution.*

Proof. Let $d = \max_{(I)\in A} J(I)$. Using comparaison Theorem 11.2, we obtain

$$0 \leq J(I) \leq \int_{D_T} [\eta_{\max} I_{\max} w(t,a,\omega,x)]\, dx\,d\omega\,da\,dt.$$

Hence $0 \leq d < \infty$. Let (I_n) be a maximizing sequence satisfying

$$d - \frac{1}{n} < J(I_n) \leq d$$

The same comparaison theorem implies that

$$0 \leq u^{I_n}(t,a,\omega,x) \leq w(t,a,\omega,x) \text{ a.e in } D_T,$$

and so $\|u^{I_n}\|_{L^2(D_T)}$ is bounded. It follows that there exists a subsequence denoted again by u^{I_n} such that

$$u^{I_n} \to u^* \text{ weakly in } L^2(D_T).$$

By Mazur's Theorem, see [450] we obtain a sequence $\tilde{u}_n$ verifying

$$\tilde{u}_n \to u^* \text{ in } L^2(D_T),$$

where $\tilde{u}_n$ is given by the following convex combination

$$\tilde{u}_n = \sum_{i=n+1}^{k_n} \lambda_i u^{I_i}, \lambda_i \geq 0, \sum_{i=n+1}^{k_n} \lambda_i = 1, k_n \geq n+1.$$

Let the control $\tilde{I}_n$ be defined as follows

$$\tilde{I}_n(t,a,\omega,x)$$

$$= \begin{cases} \dfrac{\displaystyle\sum_{i=n+1}^{k_n} \lambda^i u^{I_i}(t,a,\omega,x) I_i(t,a,\omega,x)}{\displaystyle\sum_{i=n+1}^{k_n} \lambda^i u^{I_i}(t,a,\omega,x)}, & \text{if } \displaystyle\sum_{i=n+1}^{k_n} \lambda^i u^{I_i}(t,a,\omega,x) \neq 0, \\[4ex] 0, & \text{if } \displaystyle\sum_{i=n+1}^{k_n} \lambda^i u^{I_i}(t,a,\omega,x) = 0. \end{cases}$$

It is clear that $\tilde{I}_n \in A$. The sequence $\tilde{I}_n$ is bounded in $L^2(D_T)$ space, as a consequence there exists a subsequence, denoted again by $\tilde{I}_n$, such that $\tilde{I}_n$ converges weakly in L^2 to I^*. The system (11.9) is linear, then $\tilde{u}_n \in B^+$ is a solution of system (11.9) corresponding to $I = \tilde{I}_n \in A$, i.e

$$\tilde{u}_n = u^{\tilde{I}_n},$$

and

$$u^{\tilde{I}_n}(t,a,\omega,x) = \begin{cases} \displaystyle\int_{\tau}^{t} T(t-s)\widetilde{G}\left(s, u^{\tilde{I}_n}(s,\cdot,\cdot\omega,\cdot)\right)(\phi(s,\tau,0,\omega),x)ds \\ +T(t-\tau)\dfrac{F\left(u^{I_n}(\tau,\cdot,\omega,\cdot)\right)}{v(\tau,0,\omega)}(x), \quad \text{if } a < z(t,\omega), \\ \displaystyle\int_{0}^{t} T(t-s)\widetilde{G}\left(s, u^{\tilde{I}_n}(s,\cdot,\omega,\cdot)\right)(\phi(s,t,a,\omega),x)ds \\ +T(t)u_0(\phi(0,t,a,\omega),\omega,x), \quad \text{if } a \geq z(t,\omega). \end{cases}$$

$$(11.10)$$

Since $u^{\tilde{I}_n}$ converges strongly to u^*, we find that $u^* \in B^+$, and passing to the limit in (11.10), we find that u^* is a mild solution of the problem (11.9) corresponding to $I = I^*$. Next, we show that the control I^* is optimal. On the one hand, we have

$$J\left(\tilde{I}_n\right) = \int_{D_T} \left[\eta(t,a,\omega,x)\tilde{I}_n(t,a,\omega,x)\right) u^{\tilde{I}_n}(t,a,\omega,x)\right] dx\,d\omega\,da\,dt$$

$$= \sum_{i=n+1}^{k_n} \lambda_i J(I_i).$$

Since

$$d - \frac{1}{i} < J\left(I_i\right) \leq d,$$

and $i \geq n$, we have

$$d - \frac{1}{n} < J\left(I_i\right) \leq d.$$

Using $\lambda_i \geq 0$ and $\sum_{i=n+1}^{k_n} \lambda_i = 1$, we obtain

$$d - \frac{1}{n} < \sum_{i=n+1}^{k_n} \lambda_i J\left(I_i\right) \leq d.$$

This implies that

$$\sum_{i=n+1}^{k_n} \lambda_i J\left(I_i\right) \to d,$$

as $n \to +\infty$. We conclude that

$$J\left(\tilde{I}_n\right) \to d,$$

as $n \to +\infty$. This means that $\tilde{I}_n$ is a maximizing sequence. On the other hand, we have

$$J\left(\tilde{I}_n\right) \to \int_{D_T} \left[\eta(t, a, \omega, x)I^*(t, a, \omega, x)\right) u^{I^*}(t, a, \omega, x)\right] dx d\omega da dt,$$

as $n \to +\infty$. By uniqueness of the limit, we obtain

$$J\left(I^*\right) = \int_{D_T} \left[\eta(t, a, \omega, x)I^*(t, a, \omega, x)\right) u^{I^*}(t, a, \omega, x)\right] dx d\omega da dt = d,$$

and we conclude that $\left(I^*, u^{I^*}\right)$ is optimal. □

Let $\left(I^*, u^{I^*}\right)$ be an optimal pair for the problem. Then for any $\varepsilon > 0$ small enough, and for any $h \in L^\infty\left(D_T\right)$ such that $I^* + \varepsilon h \in A$, the solution u^{I^*} is differentiable with respect to the control I^* in the following sense.

Lemma 11.7.

$$\frac{u^{I^*+\varepsilon h} - u^{I^*}}{\varepsilon} \to z \text{ in } B, \text{ as } \varepsilon \to 0,$$

where $u^{I^*+\varepsilon h}$, and u^{I^*} are the solutions of system corresponding to controls $I^* + \varepsilon h$ and I^* respectively. The sensitivity function z, is a solution of the system

$$\begin{cases} \dfrac{\partial z}{\partial t} + \dfrac{\partial(v(t,a,\omega)z)}{\partial a} = dAz(t,a,\omega,x) \\ \qquad\qquad -\mu(t,a)z(t,a,\omega,x) \\ \qquad\qquad -h(t,a,\omega,x)\chi_{\Omega_0}(\omega)u^{I^*}(t,a,\omega,x) \\ \qquad\qquad -I^*(t,a,\omega,x)\chi_{\Omega_0}(\omega)z(t,a,\omega,x), \\ z(t,0,\omega,x) = \displaystyle\int_0^L \int_\Omega \beta(t,s)\gamma(\omega,\omega')\, z(t,s,\omega',x)\, d\omega'ds, \\ z(0,a,\omega,x) = 0. \end{cases} \qquad (11.11)$$

Proof. The existence and uniqueness of solution to (11.11) can be proved in a similar way as that in Theorem 11.1. By Lemma 11.3, we have

$$u^{I^*+\varepsilon h} - u^{I^*} \to 0 \text{ in } B, \text{ as } \varepsilon \to 0.$$

Define

$$w_\varepsilon = \left[\frac{u^{I^*+\varepsilon h} - u^{I^*}}{\varepsilon} \right],$$

then w_ε is a solution of the system

$$Dw(t,a,\omega,x) = dAw(t,a,\omega,x)$$
$$\qquad - \mu(t,a)w(t,a,\omega,x)$$
$$\qquad - h(t,a,\omega,x)\chi_{\Omega_0}(\omega)u^{I^*+\varepsilon h} - I^*(t,a,\omega,x)\chi_{\Omega_0}(\omega)w(t,a,\omega,x),$$
$$w(t,0,\omega,x) = \int_\Omega \int_0^L \beta(t,s)\gamma(\omega,\omega')\, w(l,s,\omega',x)\, dsd\omega',$$
$$w(0,a,\omega,x) = 0.$$

Passing to the limit $\varepsilon \to 0$, and using Lemma 11.3, we obtain

$$w_\varepsilon(t,\cdot,\omega,\cdot) \to z \text{ in } B, \text{ as } \varepsilon \to 0. \qquad \square$$

Let $N_A (I^*)$ be the normal cone of A at I^* in X. To characterize the optimal strategy, we define the dual problem

$$
\begin{cases}
\dfrac{\partial q}{\partial t} + v(t,a,\omega)\dfrac{\partial q}{\partial a} = -\, dAq(t,a,\omega,x) + \mu(t,a)q(t,a,\omega,x) \\
\qquad\qquad + \eta(t,a,\omega,x)I^*(t,a,\omega,x)\chi_{\Omega_0}(\omega) \\
\qquad\qquad + I^*(t,a,\omega,x)\chi_{\Omega_0}(\omega)q(t,a,\omega,x) \\
\qquad\qquad - \beta(t,a)\displaystyle\int_\Omega \gamma\,(w,w')\,q\,(t,0,\omega',x)\,d\omega', \\
q(t,L,\omega,x) = 0, \\
q(T,a,\omega,x) = 0.
\end{cases}
\tag{11.12}
$$

Under the change of variables $v := T - t$, $s := L - a$, and $\tilde{q}(v,s,\omega,x) := q(T - v, L - s, \omega, x)$, the above problem becomes

$$
\begin{cases}
\dfrac{\partial \tilde{q}}{\partial v} + v(T - v, L - s, \omega)\dfrac{\partial \tilde{q}}{\partial s} = dA\tilde{q}(v,s,\omega,x) \\
\qquad\qquad - \mu(T - v, L - s)\tilde{q}(v,s,\omega,x) \\
\qquad\qquad - \eta(t,a,\omega,x)I^*(t,a,\omega,x)\chi_{\Omega_0}(\omega) \\
\qquad\qquad - I^*\chi_{\Omega_0}(\omega)\tilde{q}(v,s,\omega,x) \\
\qquad\qquad + \beta(T - v, L - s) \\
\qquad\qquad \times \displaystyle\int_\Omega \gamma\,(w,w')\,\tilde{q}\,(v,0,\omega',x)\,d\omega', \\
\tilde{q}(v,0,\omega,x) = 0, \\
\tilde{q}(0,s,\omega,x) =.
\end{cases}
\tag{11.13}
$$

Treating the system (11.13) in the same manner as in Theorem 11.1, we get existence and uniqueness. The main result of this section is the following result

Theorem 11.5. *Assume that $\left(I^*, u^{I^*}\right)$ is optimal, q is the solution of the dual problem, and z is a solution of system (11.11), then*

$$
I^* = \begin{cases}
0 & \text{if } \eta(t,a,\omega,x) + q(t,a,\omega,x) < 0, \\
I_{\max} & \text{if } \eta(t,a,\omega,x) + q(t,a,\omega,x) > 0.
\end{cases}
$$

Proof. Let $T_A (I^*)$ be the tangent cone to A at I^*. For any element $h \in T_A (I^*)$, and for any $\varepsilon > 0$ small enough, we have $I^* + \varepsilon h \in A$. Since I^* is optimal, we obtain

$$
J(I^*) \ge \int_{D_T}\left[\eta u^{I^* + \varepsilon h}\,(I^* + \varepsilon h)\right](t,a,\omega,x)dt\,da\,d\omega\,dx
$$

this gives

$$\int_{D_T} \left[\eta I^* w_\varepsilon + \eta h u^{I^* + \varepsilon h} \right] (t, a, \omega, x) dt da d\omega dx \leq 0.$$

By Lemma 11.7, passing to the limit, we obtain

$$\int_{D_T} \left[\eta I^* z + \eta h u^{I^*} \right] (t, a, \omega, x) dt da d\omega dx \leq 0.$$

Multiplying the dual problem by z and integrating over D_T, we get

$$\int_{D_T} \left[\frac{\partial q}{\partial t} + v(t, a, \omega) \frac{\partial q}{\partial a} + dAq - \mu(t, a) q(t, a, \omega, x) \right] z(t, a, \omega, x) dt da d\omega dx$$

$$= \int_{D_T} \left[I^* q(t, a, \omega, x) + \eta I^* - \beta(t, a) \int_\Omega \gamma(\omega, \omega') q(t, 0, \omega', x) d\omega' \right] z dt da d\omega dx.$$

Note that the operator A is self-adjoint on $L^2(D)$, see appendix for a proof. Using the equation of z, we obtain

$$- \int_{D_T} \left[\frac{\partial z}{\partial t} + \frac{\partial (v(t, a, \omega) z)}{\partial a} - dAz + \mu(t, a) z(t, a, \omega, x) \right] q dt da d\omega dx$$

$$- \int_{D_T} q(t, 0, \omega, x) \beta(t, a) \int_\Omega \gamma(\omega, \omega') z(t, a, \omega', x) d\omega' dt da d\omega dx$$

$$= \int_{D_T} \left[I^* q(t, a, \omega, x) + \eta I^* - \beta(t, a) \int_\Omega \gamma(\omega, \omega') q(t, 0, \omega', x) d\omega' \right] z dt da d\omega dx,$$

this implies that

$$\int_{D_T} \left[h u^{I^*} + I^* z \right] q dt da d\omega dx$$

$$- \int_{D_T} q(t, 0, \omega, x) \beta(t, a) \int_\Omega \gamma(\omega, \omega') z(t, a, \omega', x) d\omega' dt da d\omega dx$$

$$= \int_{D_T} \left[I^* q(t, a, \omega, x) + \eta I^* - \beta(t, a) \int_\Omega \gamma(\omega, \omega') q(t, 0, \omega', x) d\omega' \right] z dt da d\omega dx.$$

Changing ω' by ω and using assumption (11.1.2), we have

$$\int_{D_T} q(t, 0, \omega, x) \beta(t, a) \int_\Omega \gamma(\omega, \omega') z(t, a, \omega', x) d\omega' dt da d\omega dx$$

$$= \int_{D_T} \int_\Omega q(t, 0, \omega', x) \beta(t, a) \gamma(\omega, \omega') z(t, a, \omega, x) d\omega' dt da d\omega dx,$$

hence

$$\int_{D_T} h u^{I^*} q dt da d\omega dx = \int_{D_T} \eta I^* z dt da d\omega dx.$$

It follows that

$$\int_{D_T} hu^{I^*}(\eta + q)dtdad\omega dx \geq 0.$$

for any element of tangent cone $h \in T_A(I^*)$, that is

$$u^{I^*}(\eta + q) \in N_A(I^*). \tag{11.14}$$

which implies that

$$I^* = \begin{cases} 0 & \text{if } \eta(t,a,\omega,x) + q(t,a,\omega,x) < 0, \\ I_{\max} & \text{if } \eta(t,a,\omega,x) + q(t,a,\omega,x) > 0. \end{cases} \qquad \square$$

Remark 11.3.

- The first equations of (11.9) and (11.12) with (11.14) represent Pontryagin's principle. The first equation of (11.12) and (11.14) constitute the first order necessary conditions of optimality, see for instance [450].
- The optimal policity depends on $(\eta + q)$. If $(\eta + q) < 0$, that is η is small enough then there is no need to apply insecticides; otherwise, one needs to apply the insecticides at the maximum control $I_{\max}$.

11.2 Stock-Effort Parabolic Model Incorporating Marine Reserve

11.2.1 *Introduction and Motivations*

The global demand for fisheries products has witnessed a notable increase in recent decades, and this upward trajectory is anticipated to persist owing to population growth, rising affluence, and a growing inclination toward healthy dietary choices. However, this heightened demand has led to a depletion of marine fish stocks and the deterioration of their habitats. Consequently, effective fisheries management has emerged as a significant economic and environmental challenge.

Marine reserves, areas where fishing is either controlled or prohibited, present promising opportunities for the restoration of overexploited stocks and the enhancement of fisheries [451–453]. The authors in reference [454] propose that marine reserves can serve as insurance against scientific uncertainties, errors in stock assessment, or regulatory effectiveness. Reference [455] explores the impacts of establishing a marine reserve on both economic and biological aspects, demonstrating the existence of an optimal reserve size that maximizes catch at equilibrium.

In [456], the authors introduce and analyze two models comparing strategies: several small reserves versus a single large reserve. They investigate whether multiple smaller reserves, collectively covering the same area as a single large reserve, can conserve more species and yield a greater catch. The findings suggest that, in certain cases, having multiple small marine reserves may be preferable to a single large reserve.

This section delves into the impact of marine reserves and harvesting in a reaction-diffusion model derived from fisheries management. The examination involves a diffusive stock-effort model featuring a single reserve zone, where fish are free to move in and out, but fishing effort is restricted. The rate of change in fish density is governed by three processes: population growth, movement, and harvesting. The model comprises two equations, one describing fish density dynamics and the other capturing variations in fishing effort. For ease of presentation and theoretical analysis, we assume that all parameters in the model are constant, except for the catchability function, which is position-dependent on x. Thus, we consider the following nondimensionalized form of reaction-diffusion model:

$$
\begin{cases}
\dfrac{\partial u}{\partial t} - d_1 \Delta u = ru(1-u) - \chi(x)ue, & x \in \Omega, t > 0, \\[2mm]
\dfrac{\partial e}{\partial t} - d_2 \Delta e = -ce + p\chi(x)uedx, & x \in \Omega_*, t > 0, \\[2mm]
\dfrac{\partial u}{\partial \eta} = 0, & x \in \partial\Omega, t > 0, \\[2mm]
\dfrac{\partial e}{\partial \eta} = 0, & x \in \partial\Omega_*, t > 0, \\[2mm]
u(x,0) = u_0(x) \geq 0, & x \in \Omega, \\[2mm]
e(x,0) = e_0(x) \geq 0, & x \in \Omega_*,
\end{cases}
\tag{11.15}
$$

where Ω is a bounded region in $\mathbb{R}^N$ for $N \geq 1$ with $C^{2,\alpha}$ smooth boundary $\partial\Omega$, the reserve zone (or no-harvesting zone) Ω_0 is a subdomain of Ω whose boundary $\partial\Omega_0$ is also smooth, thus the real space for the fishing effort is $\Omega_* = \Omega/\bar{\Omega}_0$, $\dfrac{\partial}{\partial \eta}$ is the outward normal derivative on the boundary, $u(x,t)$ is the density of fish and $e(x,t)$ is the fishing effort at position x and time t. The Laplace operator Δ represents the spatial diffusion with d_1 and d_2 are the diffusion rates of the fish population and fishing effort respectively. The periphery of the reserve zone has no impact on the fish's dispersal;

however, it functions as a barrier, preventing fishing effort from entering Ω_0. In the absence of harvesting, we posit a logistic growth model for the fish population with a maximal intrinsic growth rate of $r > 0$ and a carrying capacity of $K = 1$. The assumption is that the catch is directly proportional to both the fish density and the fishing effort, as indicated in references [457, 458]. Here, c represents the unit cost of fishing effort, and p denotes the per-unit price for fish. χ is the characteristic function of Ω_*, that is

$$\chi(x) = \chi_{\Omega_*}(x) = \begin{cases} 0 & \text{if } x \in \bar{\Omega}_0, \\ q(x) \geq 0 & \text{if } x \in \Omega_*, \end{cases} \tag{11.16}$$

where $q(x)$ is a catchability function. The fact of $\chi(x) = 0$ in $\bar{\Omega}_0$ implies that no fishing could take place there.

An evident outcome of this contribution is to underscore the significance of selecting the size of the reserve. The obtained results can be biologically interpreted as follows: when presented with a habitat Ω for the harvested species, the size of the reserve $|\Omega_0|$ can be chosen as a control variable to influence the sustainability or extinction of the fishery. The determination of these outcomes involves a combination of factors, including the reserve zone, the cost, and the market price of the resource.

11.2.2 *Global Existence and Boundedness*

In this section, we show that the model (11.15) is well-posed in the sense that for any pair of positive initial functions $(u_0(x), e_0(x))$, the system (11.15) possess a unique componentwise nonnegative solution that exists globally in time. For notational convenience, we denote by $F_1(u, e)$ and $F_2(u, e)$ the two functions on the right-hand side of (11.15), that is,

$$F_1(u, e) := ru(1 - u) - \chi(x)ue,$$
$$F_2(u, e) := -ce + p\chi(x)ue.$$

In order to consider classic solutions of (11.15), we introduce the space $\mathbb{X} = L^\infty(\Omega) \times L^\infty(\Omega_*)$. The following theorem confirms the well-posedness of (11.15), including existence, uniqueness, positivity and boundedness of a solution to (11.15).

Theorem 11.6. *Let (u_0, e_0) be nonnegative initial data in $\mathbb{X}$ with $\|u_0\|_{L^\infty(\Omega)} < 1$ (the normalized carrying capacity). Then, there exists a unique nonnegative classical solution of the stock effort model (11.15) for all*

$x \in \Omega$ and $t \geq 0$. Moreover, if $u_0 \not\equiv 0$ and $e_0 \not\equiv 0$, then $u(x,t) > 0$ and $e(x,t) > 0$ for all $t > 0$ and $x \in \bar{\Omega}$.

Proof. Due to the local existence theory of H. Amann (1997) [459], we can easily check that there exists a $T_{\max}$ depending only on u_0 and e_0 such that (11.15) has a unique solution $(u(\cdot,t),(e(\cdot,t))$ in $C\left([0,T_{\max}),\mathbb{X}\right)$. Moreover, if $T_{\max} < \infty$ then

$$\lim_{t \to T_{\max}} \sup_{\bar{\Omega}} \{|u(x,t)| + |e(x,t)|\} = +\infty.$$

To prove the nonnegativity of solutions, let $u^- = \min\{0,u\}$. Multiplying the first equation of (11.15) by u^- and integrating over Ω, we will get

$$\frac{1}{2}\frac{d}{dt}\int_\Omega (u^-)^2\, dx + d_1 \int_\Omega |\nabla u^-|^2\, dx = \int_\Omega (r(1-u) - \chi(x)e)uu^-\, dx$$

$$\leq \|r - ru - \chi(x)e\|_{L^\infty(\Omega)} \int_\Omega |u^-|^2\, dx$$

$$= h(t)\int_\Omega |u^-|^2\, dx.$$

Since $u^-(x,0) = 0$ for all $x \in \Omega$, by Gronwall inequality [460], it follows that

$$\int_\Omega |u^-|^2\, dx = 0.$$

Hence,

$$u^-(x,t) = 0 \text{ for all } x \in \Omega \text{ and all } t \geq 0.$$

This means that

$$u(x,t) \geq 0 \text{ for all } x \in \Omega \text{ and } 0 < t < T_{\max}.$$

To prove $e(x,t) \geq 0$, we compare it with $v \equiv 0$. From the inequality

$$\frac{\partial v}{\partial t} - d_2\Delta v - F_2(x,v) = 0 \leq \frac{\partial e}{\partial t} - d_2\Delta e - F_2(x,e).$$

Then from comparison principle of the parabolic equations [461], we conclude that $e(x,t) \geq 0$ for all $x \in \Omega$ and $0 < t < T_{\max}$.

In order to show the global existence of solutions of (11.15) (i.e. $T_{\max} = \infty$), we only need to show that solutions are bounded.

It follows from the maximum principle and from the nonnegativity of (u,e) that the u component satisfies

$$0 \leq u(x,t) \leq 1, \quad x \in \Omega, 0 < t < T_{\max}.$$

For $0 < t < T_{\max}$, an elementary algebraic manipulations yield

$$\frac{d}{dt}\int_{\Omega_*} e(x,t)dx + p\frac{d}{dt}\int_{\Omega} u(x,t)dx = -c\int_{\Omega_*} e(x,t)dx$$
$$+ pr\int_{\Omega} u(x,t)dx - pr\int_{\Omega} u^2(x,t)dx.$$

Following a standard methodology, either

$$\frac{d}{dt}\int_{\Omega_*} e(x,t)dx + p\frac{d}{dt}\int_{\Omega} u(x,t)dx \leq 0, \quad 0 < t < T_{\max},$$

that implies

$$0 \leq \|e(t)\|_{L^1(\Omega_*)} \leq \|e_0\|_{L^1(\Omega_*)} + p\|u_0\|_{L^1(\Omega)}, \quad 0 < t < T_{\max},$$

or for some $t_0 \in [0, T_{\max})$ one has

$$\frac{d}{dt}\int_{\Omega_*} e\,(x,t_0)\,dx + p\frac{d}{dt}\int_{\Omega} u\,(x,t_0)\,dx > 0,$$

in which case

$$c\int_{\Omega_*} e\,(x,t_0)\,dx < pr\int_{\Omega} u\,(x,t_0)\,dx - pr\int_{\Omega} u^2\,(x,t_0)\,dx,$$

that implies

$$0 \leq \|e(t)\|_{L^1(\Omega_*)} \leq \frac{pr}{c}|\Omega|.$$

wherein $|\Omega|$ is the N-dimensional Lebesgue measure of $\Omega \subset \mathbb{R}^N$. One may conclude to

$$0 \leq \|e(t)\|_{L^1(\Omega_*)} \leq M$$
$$:= \max\left(\|e_0\|_{L^1(\Omega_*)} + p\|u_0\|_{L^1(\Omega)}, \frac{pr}{c}|\Omega|\right), \text{ for } 0 < t < T_{\max}.$$

Therefore, $\|e(t)\|_{L^1(\Omega_*)}$ is bounded in $[0, T_{\max})$. In view of Theorem 3.1 in [462], there exists a constant M^* depending on M and on $\|e_0\|_{L^\infty(\Omega)}$ such that

$$\sup_{t\geq 0} \|e(t)\|_{L^\infty(\Omega_*)} \leq M^*.$$

in $\bar{\Omega} \times [0, T_{\max})$. Hence, it follows from the standard theory for semilinear parabolic systems that $T_{\max} = \infty$. The last part of the Theorem on the strict positivity of solutions is a consequence of the Maximum Principle. □

Theorem 11.7. *The set* $D := [0,1] \times [0, M^*]$ *is a global attractor for all solutions of (11.15) in the sense that any non-negative solution* $(u(x,t), e(x,t))$ *of (11.15) lies in* D *as* $t \to \infty$ *for all* $x \in \Omega$.

11.2.3 Stationary Solution

The main aim of this section is to study the existence and non-existence of non-constant positive steady-states of (11.15) by using bifurcation theorem of Crandall-Rabinowitz [463]. The corresponding steady-state problem of (11.15) is the coupled elliptic system

$$\begin{cases} -d_1\Delta u = ru(1-u) - \chi(x)ue, & x \in \Omega, \\ -d_2\Delta e = -ce + p\chi(x)ue, & x \in \Omega_*, \\ \left.\frac{\partial u}{\partial \eta}\right|_{\partial\Omega} = 0, & \frac{\partial e}{\partial \eta} \mid \partial\Omega_* = 0. \end{cases} \tag{11.17}$$

Denote by $\lambda_1(q,\Omega)$ the first eigenvalue of $-\Delta + q$ over Ω under homogeneous Neumann boundary conditions with $q = q(x) \in L^\infty(\Omega)$ and $0 = \mu_0(\Omega) < \mu_1(\Omega) < \mu_2(\Omega) < \dots$ are the eigenvalues of $-\Delta$ in Ω under homogeneous Neumann boundary condition. We recall some well-known properties of $\lambda_1(\omega,\Omega)$ (see for example [464]):

(a) $\lambda_1(0,\Omega) = \mu_0(\Omega) = 0$;
(b) $\lambda(q_1,\Omega) > \lambda(q_2,\Omega)$ if $q_1 \geq q_2$ and $q_1 \not\equiv w_2$;
(c) $\lambda_1(q,\Omega)$ is continuous with respect to $q \in L^\infty(\Omega)$.

11.2.3.1 Nonexistence of non-constant positive solutions

In this subsection we shall give conditions for the non-existence of non-constant positive solutions to (11.17). The steady state equation (11.17) has two non-negative constant solutions: the trivial solution $(0,0)$, and a semi-trivia solution $(1,0)$. The local stability of the last one solution can be determined through linear stability as follows.

Proposition 11.1. *Suppose that the parameters $r, p, c > 0$ and $\chi(x)$ satisfy (1.2). Then $(1,0)$ is locally asymptotically stable when $c > c^*$ and it is unstable for $c \leq c^*$, where*

$$c^* = -d_2\lambda_1\left(-\frac{pq(x)}{d_2}, \Omega_*\right) > 0, \tag{11.18}$$

Proof. The linearized problem of (11.17) at $(1,0)$ is

$$\begin{cases} d_1\Delta v - rv - \chi(x)w - \mu v = 0, & x \in \Omega, \\ d_2\Delta w + (pq(x) - c)w - \mu w = 0, & x \in \Omega_*, \\ \frac{\partial v}{\partial \eta} = 0 & x \in \partial\Omega, \\ \frac{\partial w}{\partial \eta} = 0 & x \in \partial\Omega_*. \end{cases} \tag{11.19}$$

which has a sequence of real eigenvalues $\mu_1 < \mu_2 < \ldots < \mu_n \leq \ldots \to \infty$, as μ_1 is determined by the equation of w only. The solution $(1,0)$ is stable when $\mu_1 > 0$, that is

$$\frac{c}{d_2} + \lambda_1\left(-\frac{pq(x)}{d_2}, \Omega_*\right) = \lambda_1\left(\frac{c}{d_2} - \frac{pq(x)}{d_2}, \Omega_*\right) > 0. \qquad \Box$$

Denote by $q^* = \max_{\Omega_*} q(x)$. We observe that c^* defined in (11.18) satisfies

$$c^* = -d_2\lambda\left(\frac{-pq(x)}{d_2}, \Omega_*\right) \leq -d_2\lambda\left(\frac{-pq^*}{d_2}, \Omega_*\right) = pq^* = \tilde{c},$$

which is the threshold value for c.

Proposition 11.2. *Suppose $c > \tilde{c}$, hence (11.17) has no positive solution.*

Proof. First, it is easy to see that (11.17) also has no constant positive solution. Next, we suppose on the contrary that (11.17) has a non-constant positive (u, e). Integrating the second equation of (11.17) over Ω_*, we have

$$0 = \int_{\Omega_*} e(-c + pq(x)u)dx.$$

Since $e > 0$, it follows that there exists $x_0 \in \bar{\Omega}_*$ such that

$$-c + pq(x_0)u(x_0) = 0.$$

While $\frac{c}{pq(x_0)u(x_0)} > \frac{c}{pq^*} > 1$, it is a contradiction. $\qquad \Box$

Theorem 11.8. *If $c > \tilde{c}$, then $(1,0)$ is globally asymptotically stable, that is to say, $(1,0)$ attracts every positive solution of (11.15).*

Proof. From Theorem 11.6, for a sufficiently small $\varepsilon > 0$ there exist $T > 0$ such that

$$u(x, t) \leq 1 + \varepsilon \text{ in } [T, \infty) \times \bar{\Omega}, \qquad (11.20)$$

and thus

$$\begin{cases} \frac{de}{dt} - d_2\Delta e = e(-c + p\chi(x)u), \\ \qquad\qquad \leq e\left(-c + pq^*(1 + \varepsilon)\right) \text{ in } [T, \infty) \times \Omega_*, \\ \frac{\partial e}{\partial \eta} = 0 \qquad \text{ on } [T, \infty) \times \partial\Omega_*, \\ e(x, T) > 0 \qquad \text{ in } \Omega_*. \end{cases} \qquad (11.21)$$

The comparison argument in (11.21) yields

$$\lim_{t \to \infty} e(x, t) = 0 \text{ in } \bar{\Omega}_*, \qquad (11.22)$$

so that the existence of $T_1 \geq T$ such that $e(x, t) \leq \varepsilon$ in $[T_1, \infty) \times \bar{\Omega}_*$. Therefore, we have

$$\begin{cases} \frac{du}{dt} - d_1 \Delta u &= ru(1 - u) - \chi(x)ue, \\ &\geq u\left[r - q^*\varepsilon - ru\right] \qquad \text{in } [T_1, \infty) \times \Omega, \\ \frac{\partial u}{\partial \eta} = 0 &\quad \text{on } [T_1, \infty) \times \partial\Omega, \\ e\left(x, T_1\right) > 0 &\quad \text{in } \Omega. \end{cases} \tag{11.23}$$

By again applying the comparison argument, we see that

$$u(x, t) \geq \frac{1}{r}\left(r - q^*\varepsilon\right). \tag{11.24}$$

From Eqs. (11.20) and (11.24), we conclude that $\lim_{t \to \infty} u(x, t) = 1$ in $\bar{\Omega}$ by using the continuity as $\varepsilon \to 0$, which implies

$$\|(u(x, t), e(x, t)) - (1, 0)\|_{C(\bar{\Omega} \times \bar{\Omega}_*)} \to 0$$

as $t \to \infty$ together with (11.22). $\qquad \square$

Remark 11.4. Theorem 11.8 shows that when the cost is too high, the fishery is not viable and fishing effort disappears.

In the case $c < \tilde{c}$, the following results show the non-existence of non-constant positive steady-state solutions for certain ranges of diffusion coefficients, when the other parameters r, c, p are fixed. Simple computations give

$$\frac{\partial f}{\partial u} = r - 2ru - \chi(x)e \leq r, \qquad \frac{\partial g}{\partial e} = -c + p\chi(x)u \leq pq^* - c.$$

Theorem 11.9. *Suppose that $c < \tilde{c}$, then the following hold:*

(i) *There exists a positive constant $\tilde{d}_1$ such that (11.17) has no positive non-constant solution, provided that $d_1 > \tilde{d}_1$ and $d_2\mu_1\left(\Omega_*\right) > pq^* - c$.*

(ii) *There exists a positive constant $\tilde{d}_2$ such that (11.17) has no positive non-constant solution, provided that $d_2 > \tilde{d}_2$ and $d_1\mu_1(\Omega) > r$.*

Proof. We only prove (i) and the proof of (ii) can be accomplished similarly. Let (u, e) be a positive solution of (11.17) and write

$$\bar{u} = \frac{1}{|\Omega_*|} \int_{\Omega_*} u(x)dx, \quad \tilde{u} = \frac{1}{|\Omega_0|} \int_{\Omega_0} u(x)dx, \quad \bar{e} = \frac{1}{|\Omega_*|} \int_{\Omega_*} e(x)dx.$$

Then,

$$\int_{\Omega\Omega_*} (u - \bar{u})dx = 0, \quad \int_{\Omega\Omega_0} (u - \tilde{u})dx = 0, \quad \int_{\Omega_*} (e - \bar{e})dx = 0.$$

Multiplying the first equation in (11.17) by $(u - \bar{u})$, and integrating over Ω_* by parts, we get

$$d_1 \int_{\Omega_*} |\nabla(u - \bar{u})|^2 dx = \int_{\Omega_*} f(u, e)(u - \bar{u}) dx$$

$$= \int_{\Omega_*} [f(u, e) - f(\bar{u}, \bar{e})](u - \bar{u}) dx$$

$$= \int_{\Omega_*} \left[\frac{\partial f}{\partial u}(\xi, \mu)(u - \bar{u})^2 + \frac{\partial f}{\partial e}(\xi, \mu)(u - \bar{u})(e - \bar{e}) \right] dx$$

$$\leq \int_{\Omega_*} \left[C_1(u - \bar{u})^2 + \varepsilon_1(e - \bar{e})^2 \right] dx,$$

for some constant C_1 and $\varepsilon_1 < 1$ obtained from Young inequality, where ξ and μ lie between u and $\bar{u}$, and e and $\bar{e}$ respectively. Hence by multiplying the first equation in (11.15) by $(u - \tilde{u})$, and integrating over Ω_0 by parts, we obtain the same expression

$$d_1 \int_{\Omega_0} |\nabla(u - \tilde{u})|^2 dx \leq \int_{\Omega_0} r(u - \tilde{u})^2.$$

Similarly, from the second equation in (11.17), it follows that

$$d_2 \int_{\Omega_*} |\nabla(e - \bar{e})|^2 dx$$

$$= \int_{\Omega_*} g(u, e)(e - \bar{e}) dx$$

$$= \int_{\Omega_*} [g(u, e) - g(\bar{u}, \bar{e})](e - \bar{e}) dx$$

$$= \int_{\Omega_*} \left[\frac{\partial g}{\partial e}(\xi_1, \mu_1)(e - \bar{e})^2 + \frac{\partial g}{\partial u}(\xi_1, \mu_1)(u - \bar{u})(e - \bar{e}) \right] dx$$

$$\leq \int_{\Omega_*} \left[(pq^* - c + \varepsilon_2)(e - \bar{e})^2 + C_2(u - \bar{u})^2 \right] dx,$$

for some constant C_2 and $\varepsilon_2 < 1$, where ξ_1 and μ_1 lie between u and $\bar{u}$, and e and $\bar{e}$ respectively.

From the calculations above, we obtain that

$$d_1 \int_{\Omega_*} |\nabla(u - \bar{u})|^2 dx + d_1 \int_{\Omega_0} |\nabla(u - \tilde{u})|^2 dx + d_2 \int_{\Omega_*} |\nabla(e - \bar{e})|^2 dx$$

$$\leq \int_{\Omega_*} (C_1 + C_2)(u - \bar{u})^2 dx$$

$$+ r \int_{\Omega_0} (u - \tilde{u})^2 dx + \int_{\Omega_*} (pq^* K - c + \varepsilon_1 + \varepsilon_2)(e - \bar{e})^2 dx.$$

Hence, applying Poincarés inequality [465], we have that

$$d_1\mu_1\left(\Omega_*\right)\int_{\Omega_*}(u-\bar u)^2dx+d_1\mu_1\left(\Omega_0\right)\int_{\Omega_0}(u-\tilde u)^2dx+d_2\mu_1\left(\Omega_*\right)\int_{\Omega}(e-\bar e)^2dx$$

$$\leq\int_{\Omega_*}(C_1+C_2)\,(u-\bar u)^2dx$$

$$+r\int_{\Omega_0}(u-\tilde u)^2dx+\int_{\Omega_*}(pq^*-c+\varepsilon_1+\varepsilon_2)\,(e-\bar e)^2dx.$$

Since $d_2\mu_1\left(\Omega_*\right)>pq^*-c$, we may choose ε_1 and ε_2 to be sufficiently small so that $d_2\mu_1\left(\Omega_*\right)>pq^*-c+\varepsilon_1+\varepsilon_2$. Consequently,

$$d_1\mu_1\left(\Omega_*\right)\int_{\Omega_*}(u-\bar u)^2dx+d_1\mu_1\left(\Omega_0\right)\int_{\Omega_0}(u-\tilde u)^2dx$$

$$\leq(C_1+C_2)\int_{\Omega_*}(u-\bar u)^2dx+r\int_{\Omega_0}(u-\tilde u)^2dx,$$

which implies that $u\equiv\bar u=$ constant on Ω_* and $u\equiv\tilde u=$ constant on Ω_0 and hence $e\equiv\bar e=$ constant on Ω_* if d_1 is large. $\qquad\square$

11.2.3.2 *Bifurcation from semi-trivial solutions*

In this section, we prove the existence of non-constant steady state solutions of (11.15) using bifurcation theory. We fix r,p and take c as the bifurcation parameter. From the strong maximum principle, any non-negative solution (u,e) of (11.17) is either the trivial one $(0,0)$, or a semi-trivial solution $(1,0)$, or a positive one. We will apply the local bifurcation theorem of Crandall and Rabinowitz [463] in order to obtain a branch of positive solutions of (11.17) which bifurcates from the line of semi-trivial solution: $\Gamma=\{(c,1,0):0<c<\infty\}$. We now set up the abstract framework for our bifurcation analysis. For $p>N$, we define

$$\mathbb{X}_1=\left\{u\in W^{2,p}(\Omega):\frac{\partial u}{\partial n}=0\text{ on }\partial\Omega\right\},\quad\mathbb{Y}_1=L^p(\Omega),$$

and

$$\mathbb{X}_2=\left\{e\in W^{2,p}\left(\Omega_*\right):\frac{\partial e}{\partial n}=0\text{ on }\partial\Omega_*\right\},\quad\mathbb{Y}_2=L^p\left(\Omega_*\right).$$

Theorem 11.10. *Suppose $r,p>0$, and $\chi(x)\in C(\Omega)$. Then there are positive solutions of (11.17) bifurcating from Γ if and only if*

$$c=c^*=-d_2\lambda_1\left(-\frac{pq(x)}{d_2},\Omega_*\right).$$

Moreover, All positive solutions of (11.17) *near* $(c^*, 1, 0) \in \mathbb{R}_+ \times \mathbb{X}_1 \times \mathbb{X}_2$ *can be parameterized as* $\Gamma_1 = \{(c^*(s), 1 + u_1(s), e_1(s)) : s \in s \in (0, \sigma)\}$ *for some* $\sigma > 0$, $(c(s), u(s), e(s))$ *is a smooth function with respect to s and satisfies* $c^*(0) = c^*, u_1(0) = e_1(0) = 0, u_1'(0) < 0, e_1'(0) > 0,$ *and* $c^{*'}(0) < 0$.

In terms of the model, this means that when the cost reaches some threshold value, a non trivial solution bifurcates from $(c, 1, 0)$.

Theorem 11.11 ([463]). *Let* $\mathbb{X}, \mathbb{Y}$ *be Banach spaces and* $\mathcal{F} : \mathbb{R} \times \mathbb{X} \to \mathbb{Y}$. *Assume that there is* $(\lambda_0, U_0) \in \mathbb{R} \times \mathbb{X}$ *such that* $\mathcal{F}(\lambda_0, U_0) = 0$ *and* $\mathcal{F}$ *is a continuously differentiable function in an open neighborhood of* (λ_0, U_0). *Suppose the following conditions hold:*

(11.26.1) $\dim \mathcal{N}(\mathcal{F}_U(\lambda_0, U_0)) = \operatorname{codim} \mathcal{R}(\mathcal{F}_U(\lambda_0, U_0)) = 1$ *and* $\mathcal{N}(\mathcal{F}_U(\lambda_0, U_0))) = \operatorname{span}\{\Phi_0\};$
(11.26.2) $\mathcal{F}_\lambda(\lambda_0, U_0) \notin \mathcal{R}(\mathcal{F}_U(\lambda_0, U_0))$.

Let $\mathbb{Z}$ *be the complement of* $\operatorname{span}\{\Phi_0\}$ *in* $\mathbb{X}$. *Then the solutions of* $\mathcal{F}(\lambda, U) = 0$ *near* (λ_0, U_0) *form a curve* $(\lambda(s), U(s)) = (\lambda_0 + \tau(s), U_0 + s\Phi_0 + sz(s))$, *where* $s \to (\tau(s), z(0)) \in \mathbb{R} \times \mathbb{Z}$ *is a continuously differentiable function near* $s = 0$ *satisfying* $\tau(0) = \tau'(0) = 0, z(0) = z(0) = 0$. *Furthermore, if* $\mathcal{F}$ *is k-times continuously differentiable, so are* $\tau(s)$ *and* $z(s)$.

Proof. We define a mapping $F : \mathbb{R} \times \mathbb{X}_1 \times \mathbb{X}_2 \to \mathbb{Y}_1 \times \mathbb{Y}_2$ by

$$F(c, u, e) = \begin{pmatrix} d_1 \Delta u + f(u) - \chi(x)ue \\ d_2 \Delta e - ce + p\chi(x)ue \end{pmatrix},$$

where $f(u) = ru(1 - u)$ By a simple calculation, the Fréchet derivatives of F at (u, e) are given by

$$F_{(u,e)}(c, u, e)[\phi, \psi] = \begin{pmatrix} d_1 \Delta\phi + f'(u)\phi - \chi(x)e\phi - \chi(x)u\psi \\ d_2 \Delta\psi - c\psi + p\chi(x)e\phi + p\chi(x)u\psi \end{pmatrix},$$

$$F_c(u, e) = \begin{pmatrix} 0 \\ -e \end{pmatrix},$$

$$F_{c(u,e)}(\phi, \psi) = \begin{pmatrix} 0 \\ -\psi \end{pmatrix},$$

$$F_{(u,e)(u,e)}(c, u, e)[\phi, \psi]^2 = \begin{pmatrix} -2\phi^2 - 2\chi(x)\phi\psi \\ 2pq(x)\phi\psi \end{pmatrix}.$$

The equation $F_{(u,w)}(c,1,0)[\phi,\psi] = 0$ is equivalent to

$$d_1\Delta\phi^* - r\phi^* - \chi(x)\psi^* = 0 \quad \text{in } \Omega, \quad \frac{\partial\phi^*}{\partial\eta} = 0,$$

$$d_2\Delta\psi^* - c\psi^* + p\chi(x)\psi = 0 \quad \text{in } \Omega_*, \quad \frac{\partial\psi^*}{\partial\eta} = 0,$$
(11.25)

which has a non trivial solution with $\psi > 0$ if and only

$$c = c^* = -d_2\lambda_1\left(-\frac{pq(x)}{d_2}, \Omega_*\right).$$

Hence $c = c^*$ is the only possible bifurcation point along Γ where positive solution of (11.25) bifurcate. A direct calculation shows $\text{Ker } F_{(u,e)}(c^*,1,0) = \text{span}\{(\phi_1,\phi_2)\}$ where (ϕ_1,ϕ_2) (with $\phi_2 > 0$) satisfies (11.25) with $c = c^*$. The uniqueness (up to a constant scale) of (ϕ_1,ϕ_2) follows from the fact that c^* is a principal eigenvalue. Since $\phi_2 > 0$, then

$$\phi_1 = (-d_1\Delta + r)^{-1}[-\chi(x)\phi_2] < 0.$$

The range of the operator is given by

$$\mathcal{R}\left(F_{(u,e)}(c,1,0)\right) = \left\{f,g) \in \mathbb{Y} : \int_{\Omega_*} g\phi_2 dx = 0\right\},$$

which is of co-dimension one, and

$$F_{c(u,e)}(c^*,1,0)[(\phi_1,\phi_2)] = (0,-\phi_2) \notin \mathcal{R}\left(F_{(u,e)}(c,1,0)\right),$$

since $\int_{\Omega_*}\phi_2^2 > 0$. Thus we can apply the result of [463] to conclude that the set of positive solutions to (11.17) near $(c^*,1,0)$ is a smooth curve

$$\Gamma = \{(c(s),1+u(s),e_1(s)) : s \in [0,\sigma)\},$$

such that $c^*(0) = c^*, u_1(s) = s\phi_1 + o(|s|), v_1(s) = s\phi_2 + o(|s|)$. Moreover, $c'(0)$ can be calculated as

$$c'(0) = -\frac{< l_1, F_{(u,v)(u,v)}(c^*,1,0)[\phi_1,\phi_2]^2 >}{2 < l_1, F_{d(u,v)}(c^*,1,0)[\phi_1,\phi_2] >} = \frac{p\int_{\Omega_*}q(x)\phi_1\phi_2^2 dx}{\int_{\Omega_*}\phi_2^2 dx} < 0,$$

where l_1 is the linear functional on $\mathbb{Y}$ defined by $< l_1,[f,g] >= \int_{\Omega_*}g\phi_2 dx$. $\qquad\square$

Remark 11.5. Theorem 11.10 shows that when the cost of fishing is too large, it is preferable to decrease the size of reserve until we arrive to $c < c^*$. This favors the fishing activity and ensures the durability of the fishery.

11.3 Notes and Remarks

The results of in this chapter are taken from the papers [466, 467]. Form more details and results see the publications [438–465].

Bibliography

[1] N. U. Ahmed, *Semigroup Theory with Applications to Systems and Control*, Harlow John Wiley & Sons, Inc., New York, 1991.

[2] J. Wu, *Theory and Application of Partial Functional Differential Equations*, Springer-Verlag, New York, 1996.

[3] K. Balachandran, D.G. Park and S.M. Anthoni, Existence of solutions of abstract nonlinear second-order neutral functional integrodifferential equations, *Comput. Math. Appl.* **46** (2003), 1313–1324.

[4] M. Benchohra, J. Henderson and N. Rezoug, Global existence results for second order evolution equations, *Comm. Appl. Nonlinear Anal.* **23** (2016), 57–67.

[5] M. Benchohra and S. K. Ntouyas, Existence of mild solutions on noncompact intervals to second order initial value problems for a class of differential inclusions with nonlocal conditions, *Comput. Math. Appl.* **39** (2000), 11–18.

[6] H. O. Fattorini, *Second Order Linear Differential Equations in Banach Spaces*, North-Holland Mathematics Studies, Vol. **108**, North-Holland, Amsterdam, 1985.

[7] H. Mönch, Boundary value problems for nonlinear ordinary differential equations of second order in Banach spaces, *Nonlinear Anal.* **4**(5) (1980), 985–999.

[8] C. C. Travis and G.F. Webb, Second order differential equations in Banach spaces, in: Nonlinear Equations in Abstract Spaces, *Proc. Internat. Sympos. (Univ. Texas, Arlington, TX, 1977), Academic Press, New-York*, 1978, 331–361.

[9] H. L. Tidke, M. B. Dhakne, Existence and uniqueness of solutions of certain second order nonlinear equations. *Note Mat.* **30** (2010), no. 2, 73–81.

[10] N. Abada, M. Benchohra, H. Hammouche and A. Ouahab, Controllability of impulsive semilinear functional differential inclusions with finite delay in Fréchet spaces, *Discuss. Math. Differ. Incl. Control Optim.* **27** (2) (2007), 329–347.

[11] A. Pazy, *Semigroups of Linear Operators and Applications to Partial Differential Equations*, Springer-Verlag, New York, 1983.

[12] M. Frigon and A. Granas, Résultats de type Leray-Schauder pour des contractions sur des espaces de Fréchet, *Ann. Sci. Math. Québec* **22** (2) (1998), 161–168.

[13] A. Baliki and M. Benchohra, Global existence and asymptotic behaviour for functional evolution equations, *J. Appl. Anal. Comput.* **4** (2) (2014), 129–138.

[14] J.H. Liu, Periodic solutions of infinite delay evolution equations, *J. Math. Anal. Appl.*, **247** (2000), 627–644.

[15] J. Liang, J.H. Liu, M.V. Nguyen and T.J. Xiao, Periodic mild solutions of impulsive differential equations with infinite delay in Banach spaces, *J. Nonlinear Funct. Anal.*, **2019** (2019), Article ID 18, 1–10.

[16] M. Büger and M. Marcus, The escaping disaster: a problem related to state-dependent delays, *Z. Angew. Math. Phys.* **55** (2004), no. 4, 547–574.

[17] S. Abbas, M. Benchohra and G. N'Guérékata, Periodic mild solutions of infinite delay not instantaneous impulsive evolution inclusions, *Vietnam J. Math.* **50** (2022), 287–299.

[18] S. Djebali, L. Gorniewicz, and A. Ouahab, First-order periodic impulsive semilinear differential inclusions: existence and structure of solution sets, *Math. Comput. Model.* **52** (5–6) (2010), 683–714.

[19] J.H. Liu, T. Naito and N.V. Minh, Bounded and periodic solutions of infinite delay evolution equations, *J. Math. Anal. Appl.*, **286** (2003), 705–712.

[20] X. Yu, J. Wang, Periodic boundary value problems for nonlinear impulsive evolution equations on Banach spaces, *Commun. Nonlinear Sci. Numer. Simulat.* **22** (2015), 980–989.

[21] A. Benchaib, A. Salim, S. Abbas and M. Benchohra, New stability results for abstract fractional differential equations with delay and non-instantaneous impulses. *Mathematics.* **11** (16) (2023), 3490. https://doi.org/10.3390/math11163490

[22] M. Fečkan, J. Wang, Y. Zhou. Periodic solutions for nonlinear evolution equations with non-instantaneous impulses, *Nonauton. Dyn. Syst.*, 1(1) (2014), 93–101.

[23] D. Guo, V. Lakshmikantham and X. Liu, *Nonlinear Integral Equations in Abstract Spaces*, Kluwer Academic Publishers Group, Dordrecht, 1996.

[24] H. O. Walther, A periodic solution of a differential equation with state-dependent delay, *J. Differential Equations.* **244** (2008), 1910–1945.

[25] G. M. N'Guérékata, *Almost Automorphic and Almost Periodic Functions in Abstract Spaces*, Kluwer Academic, New York, London, Moscow, 2001.

[26] D. Benzenati, S. Bouriah, A. Salim and M. Benchohra, On periodic solutions for some nonlinear fractional pantograph problems with Ψ-Hilfer derivative, *Lobachevskii J Math.* **44** (2023), 1264–1279. https://doi.org/10.1134/S1995080223040054

[27] B.C. Dhage, V. Lakshmikantham, On global existence and attractivity results for nonlinear functional integral equations, *Nonlinear Anal.* **72** (2010), 2219–2227.

[28] M. Benchohra, N. Rezoug, Existence and Attractivity of Solutions of Semilinear Volterra Type Integro-Differential Evolution Equations, *Surv. Math. Appl.* **13** (2018), 215–235.

[29] J. Blot, C. Buse, P. Cieutat, Local attractivity in nonautonomous semilinear evolution equations. *Nonauton. Dyn. Syst.* **1** (2014), 72–82.

[30] A. Diop, M. A. Diop, O. Diallo, and M. B. Traoré Local attractivity for integro-differential equations with noncompact semigroups, *Nonauton. Dyn. Syst.* **7** (2020), 102–117.

[31] K. Aissani, M. Benchohra, J.J. Nieto, Controllability for impulsive fractional evolution inclusions with state-dependent delay. *Adv. Theory Nonlinear Anal. Appl.* **3** (2019), 18–34.

[32] M. Benchohra, L. Gorniewicz, and S.K. Ntouyas, Controllability on infinite time horizon for first and second order functional differential inclusions in Banach spaces. *Discuss. Math. Differ. Incl. Control Optim.* **21** (2001), 261–282.

[33] M. Benchohra and S.K. Ntouyas, Existence and controllability results for multivalued semilinear differential equations with nonlocal conditions, *Soochow J. Math.* **29** (2003), 157–170.

[34] N. I. Mahmudov, Approximate controllability of semilinear deterministic and stochastic evolution equations in abstract spaces, *SIAM J. Control Optim.* **42** (2003), 1604–1622.

[35] N. I. Mahmudov, Approximate controllability of evolution systems with nonlocal conditions, *Nonlinear Anal.* **68** (2008), 536–546.

[36] F. Z. Mokkedem and X. Fu, Approximate controllability of semi-linear neutral integro-differential systems with finite delay. *Appl. Math. Comput.* **242** (2014), 202–215.

[37] F. Z. Mokkedem and X. Fu, Approximate controllability of a semi-linear neutral evolution system with infinite delay, *Int. J. Robust Nonlinear Control,* **27** (2017), 1122–1146.

[38] M. Benchohra, F. Bouazzaoui, E. Karapınar and A. Salim, Controllability of second order functional random differential equations with delay. *Mathematics.* **10** (2022), 16pp. `https://doi.org/10.3390/math10071120`

[39] N. Benkhettou, K. Aissani, A. Salim, M. Benchohra and C. Tunc, Controllability of fractional integro-differential equations with infinite delay and non-instantaneous impulses, *Appl. Anal. Optim.* **6** (2022), 79–94.

[40] A. Bensoussan, G. Da Prato, M.C. Delfour and S.K. Mitter, *Representation and Control of Infinite Dimension Systems.* Vol. **II**. Systems and Control: Foundations and Applications, Birkhauser, Boston, Inc., Boston, MA, 1993.

[41] A. Baliki and M. Benchohra, Global existence and stability for neutral functional evolution equations. *Rev. Roumaine Math. Pures Appl.* **LX** (1) (2015), 71–82.

[42] W. Desch, R.C. Grimmer and W. Schappacher, Some considerations for linear integrodifferential equations, *J. Math. Anal. Appl.* **104** (1984), 219–234.

[43] X. Hao, L. Liu, Mild solution of semilinear impulsive integro-differential evolution equation in Banach spaces. *Math. Methods Appl. Sci.* **40** (13) (2017), 4832–4841.

[44] A. Heris, A. Salim, M. Benchohra and E. Karapınar, Fractional partial random differential equations with infinite delay. *Results in Physics.* (2022). `https://doi.org/10.1016/j.rinp.2022.105557`

[45] A. Heris, A. Salim and M. Benchohra, Some new existence results for fractional partial random nonlocal differential equations with delay. *Ann. Univ. Paedagog. Crac. Stud. Math.* **22** (2023), 135–148. http://dx.doi.org/10.2478/aupcsm-2023-0011

[46] L. Lasiecka and R. Triggiani, Exact controllability of semilinear abstract systems with application to waves and plates boundary control problems, *Appl. Math. Optim.* **23** (1991), 109–154.

[47] K. Naito, On controllability for a nonlinear Volterra equation, *Nonlinear Anal.* **18**(1) (1992), 99–108.

[48] S. Nakagiri and R. Yamamoto, Controllability and observability for linear retarded systems in Banach space, *Inter. J. Control.* **49** (5) (1989), 1489–1504.

[49] X. Fu, Controllability of neutral functional differential systems in abstract space, *Appl. Math. Comput.* **141** (2003), 281–296.

[50] X. Fu and K. Ezzinbi, Existence of solutions for neutral functional differential evolution equations with nonlocal conditions, *Nonlinear Anal.* **54** (2003), 215- 227.

[51] Y. C. Kwun, J. Y. Park and J. W. Ryu, Approximate controllability and controllability for delay Volterra system, *Bull. Korean Math. Soc.* **28**(2) (1991), 131–145.

[52] K. Balachandran and J. P. Dauer, Controllability of nonlinear systems in Banach spaces: A survey. Dedicated to Professor Wolfram Stadler, *J. Optim. Theory Appl.* **115** (2002), 7–28.

[53] A. Arara, M. Benchohra, L. Gorniewicz and A. Ouahab, Controllability results for semilinear functional differential inclusions with unbounded delay, *Math. Bulletin.* **3** (2006), 157–183.

[54] M. Benchohra, E. P. Gatsori, L. Gorniewicz and S. K. Ntouyas, Controllability results for evolution inclusions with non-local conditions, *Z. Anal. Anwendungen* **22** (2) (2003), 411–431.

[55] M. Benchohra, L. Gorniewicz and S. K. Ntouyas, Controllability of neutral functional differential and integrodifferential inclusions in Banach spaces with nonlocal conditions, *Nonlinear Anal. Forum* **7** (2002), 39–54.

[56] M. Benchohra, L. Gorniewicz and S. K. Ntouyas, Controllability results for multivalued semilinear differential equations with nonlocal conditions, *Dynam. Syst. Appl.* **11** (2002), 403–414.

[57] M. Benchohra and S. K. Ntouyas, Controllability results for multivalued semilinear neutral functional equations, *Math. Sci. Res. J.* **6** (2002), 65–77.

[58] K. Aissani, M. Benchohra, Controllability of impulsive fractional differential equations with infinite delay. *Lib. Math. (N.S.)* **2013**, 33, 47–64.

[59] K. Aissani, M. Benchohra, Controllability of fractional integrodifferential equations with state-dependent delay. *J. Integral Equations Appl.* **2016**, 28, 149–167.

[60] K. Aissani, M. Benchohra, M. A. Darwish, Controllability of fractional order integro-differential inclusions with infinite delay. *Electron. J. Qual. Theory Differ. Equ.* **2014**, 1–18.

[61] K. Aissani, M. Benchohra, J. J. Nieto, Controllability of fractional order integro-differential inclusions with infinite delay. *Adv. Theory Nonlinear Anal. Appl.* **2019**, 3, 18–34.

[62] K. Balachandran, J.Y. Park, Controllability of fractional integrodifferential systems in Banach spaces, *Nonlinear Anal. Hybr. Syst.* **2009**, 3, 363–367.

[63] G. M. Mophou, G.M. N'Guérékata, Controllability of semilinear neutral fractional functional evolution equations with infinite delay. *Nonlinear stud.* **2011**, 18, 149–165.

[64] R. F. Curtain and H. J. Zwart, *An Introduction to Infinite Dimensional Linear Systems Theory.* Springer Verlag, New York, 1995.

[65] X. Li and J. Yong, *Optimal Control Theory for Infinite Dimensional Systems*, Birkhauser, Berlin, 1995.

[66] J. Zabczyk, *Mathematical Control Theory*, Birkhauser, Berlin, 1992.

[67] S. Bochner, A new approach in almost-periodicity. *Proc. Nat. Acad. Sci. USA,* **48** (1962), 2039–2043.

[68] T. Dianaga, G. M. N'Guérékata, Almost automorphic solutions to some classes of partial evolution equations, *Appl. Math. Lett.* **20** (2007), 462–466.

[69] J. A. Goldstein, G.M. N'Guérékata, Almost automorphic solutions of semilinear evolution equations, *Proc. Amer. Math. Soc.* **133** (2005), 2401–2408.

[70] K. Ezzinbi, and G. M. N'Guérékata, Almost automorphic solutions for some partial functional differential equations. *J. Math. Anal. Appl.* **328** (1) (2007), 344–358.

[71] I. Mishra, D. Bahuguna, S. Abbas, Existence of almost automorphic solutions of neutral functional differential equation, *Nonlinear Dyn. Syst. Theory* **11** no. 2 (2011), 165–172.

[72] D. Araya, C. Lizama, Almost automorphic mild solutions to fractional differential equations, *Nonlinear Anal.* **69** (2008), 3692–3705.

[73] G. M. N'Guérékata, *Topics in Almost Automorphy*, Springer, New York, Boston, Dordrecht, London, Moscow, 2005.

[74] G. M. N'Guérékata, Sur les solutions presque automorphes d'équations différentielles abstraites, *Ann. Sci. Math. Québec*, **1**(1981), 69–79.

[75] R. Xie and C. Zhang. Criteria of asymptotic ω-periodicity and their applications in a class of fractional differential equations. *Adv. Difference Equ.*, 2015, 2015:68, 20 pp.

[76] J. Cao, Z. Huang, G. M. N'Guérékata, Existence of asymptotically almost automorphic mild solutions for nonautonomous semilinear evolution equations. *Electron. J. Differential Equations* **2018**, Paper No. 37, 16 pp.

[77] V. Kavitha, S. Abbas, R. Murugesu, Asymptotically almost automorphic solutions of fractional order neutral integro-differential equations. *Bull. Malays. Math. Sci. Soc.* **39** (2016), no. 3, 1075–1088

[78] H. S. Ding, T. J. Xiao, J. Liang. Asymptotically almost automorphic solutions for some integrodifferential equations with nonlocal initial conditions. *J. Math. Anal. Appl.* **338** (2008), no. 1, 141–151.

[79] T. Diagana, Almost periodic solutions to some second-order nonautonomous differential equations. *Proc. Am. Math. Soc.* **140** (2012), 279–289.

[80] J.H. Liu, Bounded and periodic solutions of semi-linear evolution equations. *Dyn. Syst. Appl.* **4**(1995), 341–350.

[81] M. Pierri. On S-Asymptotically ω-periodic functions and applications. *Nonlinear Anal.* **75** (2012), no. 2, 651–661.

[82] H. R. Henríquez, M. Pierri, P. Taboas, Existence of S-asymptotically ω-periodic solutions for abstract neutral equations, *Bull. Aust. Math. Soc.*, **78** (2008), 365–382.

[83] H. R. Henríquez, M. Pierri, P. Taboas, On S-asymptotically ω-periodic functions on Banach spaces and applications, *J. Math. Anal. Appl.*, **343** (2008), 1119–1130.

[84] J. Blot, P. Cieutat, G. M. N'Guérékata, S-asymptotically ω-periodic functions applications to evolution equations, *African Diaspora J. Math.*, **12** (2011), 113–121.

[85] C. Cuevas and J.C. Souza, S-asymptotically ω-periodic solutions of semi-linear fractional integro-differential equations. *Appl. Math. Lett.* **22** (2009), 865–870.

[86] C. Cuevas and C. Lizanna. Existence of S-asymptotically ω-periodic solutions for two-times fractional order differential equations. *Southeast Asian Bull. Math.* **37** (2013), no. 5, 683–690.

[87] C. Cuevas and C. Lizanna. S-asymptotically ω-periodic solutions of semilinear Volterra equations. *Math. Meth. Appl. Sci.* **33** (2010), 1628–1636.

[88] W. Dimbour, J-C. Mado. S-asymptotically ω-periodic solution for a nonlinear differential equation with piecewise constant argument in a Banach space. *Cubo* **16** (2014), no. 3, 55–65.

[89] W. Dimbour, G.M. N'Guérékata, S-asymptotically ω-periodic solutions to some classes of partial evolution equations. *Appl. Math. Comput.* **218** (2012), 7622–7628.

[90] F. Li, J. Liang, H. Wang, S-asymptotically ω-periodic solution for fractional differential equations of order $q \in (0,1)$ with finite delay. *Adv. Difference Equ.* (2017), Paper No. 83, 14 pp.

[91] F. Li, H. Wang, S-asymptotically ω-periodic mild solutions of neutral fractional differential equations with finite delay in Banach space. *Mediterr. J. Math.* **14** (2017), no. 2, Art. 57, 16 pp.

[92] E. Hernandez, M. Pierri, S-asymptotically ω-periodic solutions for abstract equations with statedependent delay, *Bull. Aust. Math. Soc.* **98** (2018), 456–464.

[93] H. R. Henrique, M. Pierri, P. Taboas, On S-asymptotically ω-periodic functions on Banach spaces and applications, *J. Math. Anal. Appl.* **343** (2008), 1119–1130.

[94] M. Pierri, D. O'Regan, S-asymptotically ω-periodic solutions for abstract neutral differential equations, *Electron. J. Differential Equations* **2015** (2015), 1–14.

[95] J. Zhu and X. Fu, Existence and asymptotic periodicity of solutions for neutral integro-differential evolution equations with infinite delay, *Math. Slovaca* **72** (1) (2022), 121–140.

[96] S. Bochner, Uniform convergence of monotone sequences of functions, *Proc. Natl. Acad. Sci. USA* **47** (1961), 582–585.

[97] A.M. Fink, Almost Periodic Differential Equations, Springer-Verlag, Berlin, 1974.

[98] P. Chen, Y. Li, Monotone iterative technique for a class of semilinear evolution equations with nonlocal conditions, *Results Math.* **63** (2013), 731–744.

[99] R. Ortega, M. Tarallo, Almost periodic linear differential equations with non-separated solutions, *J. Funct. Anal.* **237** (2006), 402–426.

[100] T. Caraballo, D. Chaban, Almost periodic and almost automorphic solutions of linear differential/difference equations without Favard's separation condition II, *J. Differential Equations* **246** (2009), 1164–1186.

[101] L. Arnold, C. Tudor, Stationary and almost periodic solutions of almost periodic affine stochastic differential equations, *Stoch. Stoch. Rep.* **64** (1998), 177–193.

[102] D.X. Piao, Pseudo almost periodic solutions for the systems of differential equations with piecewise constant argument $[t + \frac{1}{2}]$, *Sci. China Ser. A* **1** (2004), 31–38.

[103] C.A. Tudor, M. Tudor, Pseudo almost periodic solutions of some stochastic differential equations, *Math. Rep. (Bucur.)* 1 **51** (1999), 305–314.

[104] C. Tudor, Almost periodic solutions of affine stochastic evolution equations, *Stoch. Stoch. Rep.* **38** (1992), 251–266.

[105] M.M. Fu, Z.X. Liu, Square-mean almost automorphic solutions for some stochastic differential equations, *Proc. Amer. Math. Soc.* **138** (2010), 3689–3701.

[106] A. Baliki, M. Benchohra and J. Graef, Global existence and stability for second order functional evolution equations with infinite delay. *Electron. J. Qual. Theory Differ. Equ.* vol. 2016 no. 23 (2016), pp. 1–10.

[107] A. Bensalem, A. Salim and M. Benchohra, A. Bensalem, A. Salim, M. Benchohra and G. N'Guérékata, Functional integro-differential equations with state-dependent delay and non-instantaneous impulsions: existence and qualitative results, *Fractal Fract.* **6** (2022), 1–27. https://doi.org/10.3390/fractalfract6100615

[108] J. P. C. dos Santos, On state-dependent delay partial neutral functional integrodifferential equations, *Appl. Math. Comput.* **100** (2010), 1637–1644.

[109] N. Benkhettou, A. Salim, K. Aissani, M. Benchohra and E. Karapınar, Non-instantaneous impulsive fractional integro-differential equations with state-dependent delay, *Sahand Commun. Math. Anal.* **19** (2022), 93–109. https://doi.org/10.22130/scma.2022.542200.1014

[110] S. Krim, A. Salim and M. Benchohra, On implicit Caputo tempered fractional boundary value problems with delay. *Lett. Nonlinear Anal. Appl.* **1** (2023), 12–29.

[111] A. Salim, M. Benchohra, J. E. Lazreg and J. Henderson, On k-generalized ψ-Hilfer boundary value problems with retardation and anticipation. *ATNAA.* **6** (2022), 173–190. https://doi.org/10.31197/atnaa.973992

[112] A. Salim, M. Benchohra, J. E. Lazreg and E. Karapınar, On k-generalized ψ-Hilfer impulsive boundary value problem with retarded and advanced arguments. *J. Math. Ext.* **15** (2021), 1–39. https://doi.org/10.30495/JME.SI.2021.2187

[113] A. Salim, M. Benchohra, J. E. Lazreg and Y. Zhou, On k-generalized ψ-Hilfer impulsive boundary value problem with retarded and advanced arguments in Banach spaces. *J. Nonl. Evol. Equ. Appl.* **2022** (2023), 105–126.

[114] F. Hartung, T. Krisztin, H. O. Walther, J. Wu, Functional differential equations with state-dependent delays: theory and applications, in: Handbook of differential equations: ordinary differential equations, Vol. III, Elsevier/North-Holland, Amsterdam, 2006, 435–545.

[115] H. O. Walther, Differential equations with locally bounded delay, *J. Differential Equations.* **252** (2012), 3001–3039.

[116] J. D. Murray, *Mathematical biology I. An introduction*, 3rd ed, Springer-Verlag, Berlin Heidelberg, 2002.

[117] R. Gambell, Birds and mammals: Antarctic whales, in "Key Environments Antarctica" (eds. W. N. Bonner and D. W. H. Walton), Pergamon Press, New York, (1985), 223–241.

[118] W. G. Aiello, H. I. Freedman, J. Wu, Analysis of a model representing stage-structured population growth with statedependent time delay, *SIAM J Appl Math*, **52** (1992), 855–869.

[119] A. T. Bharucha-Reid, *Random Integral Equations*, Academic Press, New York, 1972.

[120] G.S. Ladde and V. Lakshmikantham, *Random Differential Inequalities*, Academic Press, New York, 1980.

[121] C. P. Tsokos and W. J. Padgett, *Random Integral Equations with Applications to Life Sciences and Engineering*, Academic Press, New York, 1974.

[122] C. P. Tsokos, and W. J. Padgett, *Random Integral Equations with Applications to Life Sciences and Engineering*, Academic Press, New York, 1974.

[123] T. T. Soong, *Random Differential Equations in Science and Engineering*, Academic Press, New York, 1973.

[124] A.I. Perov On the Cauchy problem for a system of ordinary differential equations, *Pviblizhen. Met. Reshen. Differ. Uvavn.* **2** (1964), 115–134.

[125] J.R. Graef, J. Henderson, A. Ouahab, *Topological Methods for Differential Equations and Inclusions*, CRC Press: Boca Raton, FL, 2018.

[126] O. Bolojan-Nica, G. Infante, R. Precup, Existence results for systems with coupled nonlocal initial conditions, *Nonlinear Anal.* **4** (2014), 231–242.

[127] C. Derbazi, Z. Baitiche, M. Benchohra, J.R. Graef, Extremal solutions to a coupled system of nonlinear fractional differentialequations with Caputo fractional derivative, *J. Math. Appl.* **44** (2021), 19–34.

[128] O. Nica, Existence results for second order three point boundary value problems, *Differ. Equ. Appl.* **4** (2012), 547–570.

[129] O. Nica, R. Precup, On the nonlocal initial value problem for first order differential systems, *Stud. Univ. Babes Bolyai Math.* **56** (2011), 125–137.

[130] R. Precup, A. Viorel, Existence results for systems of nonlinear evolution equations, *Int. J. Pure Appl. Math.* **47** (2008), 199–206.

[131] C. Derbazi, H. Hammouche, A. Salim and M. Benchohra, Weak solutions for fractional Langevin equations involving two fractional orders in Banach spaces. *Afr. Mat.* **34** (2023), 10 pages. `https://doi.org/10.1007/s13370-022-01035-3`

[132] I. Hammoumi, H. Hammouche, A. Salim and M. Benchohra, Mild solutions for impulsive fractional differential inclusions with Hilfer derivative in Banach spaces. *Rend. Circ. Mat. Palermo (2).* (2023), 1–14. `https://doi.org/10.1007/s12215-023-00944-x`

[133] S. Krim, A. Salim, S. Abbas and M. Benchohra, Functional k-generalized ψ-Hilfer fractional differential equations in b-metric spaces. *Pan-Amer. J. Math.* **2** (2023), 10 pages. `https://doi.org/10.28919/cpr-pajm/2-5`

[134] S. Krim, A. Salim and M. Benchohra, Nonlinear contractions and Caputo tempered implicit fractional differential equations in b-metric spaces with infinite delay. *Filomat.* **37** (22) (2023), 7491–7503. `https://doi.org/10.2298/FIL2322491K`

[135] S. Krim, A. Salim and M. Benchohra, Nonlinear contractions and Caputo tempered impulsive implicit fractional differential equations in b-metric spaces. *Math. Morav.* **27** (2) (2023), 1–24.

[136] J. E. Lazreg, M. Benchohra and A. Salim, Existence and Ulam stability of k-generalized ψ-Hilfer fractional problem. *J. Innov. Appl. Math. Comput. Sci.* **2** (2022), 01–13. `https://doi.org/10.58205/jiamcs.v2i2.19`

[137] A. Salim, S. Abbas, M. Benchohra and J. E. Lazreg, Caputo fractional q-difference equations in Banach spaces. *J. Innov. Appl. Math. Comput. Sci.* **3** (1) (2023), 1–14. `https://doi.org/10.58205/jiamcs.v3i1.67`

[138] A. Salim, B. Ahmad, M. Benchohra and J. E. Lazreg, Boundary value problem for hybrid generalized Hilfer fractional differential equations, *Differ. Equ. Appl.* **14** (2022), 379–391. `http://dx.doi.org/10.7153/dea-2022-14-27`

[139] A. Salim and M. Benchohra, Existence and Uniqueness Results for Generalized Caputo Iterative Fractional Boundary Value Problems. *Fract. Differ. Calc.* **12** (2022), 197–208. `http://dx.doi.org/10.7153/fdc-2022-12-12`

[140] A. Salim, M. Benchohra and J. E. Lazreg, Implicit coupled k-generalized ψ-Hilfer fractional differential systems with terminal conditions in Banach spaces. In: *Candela, A.M., Cappelletti Montano, M., Mangino, E. (eds) Recent Adv. Math. Anal. Trends in Mathematics.* Birkhäuser, Cham, 2023. `https://doi.org/10.1007/978-3-031-20021-2_22`

[141] A. Salim, S. Krim, S. Abbas and M. Benchohra, On deformable implicit fractional differential equations in b-metric spaces. *J. Math. Ext.* **17** (2023), 1–17. `https://doi.org/10.30495/JME.2023.2468`

[142] A. Salim, S. Krim and M. Benchohra, On implicit boundary value problems with deformable fractional derivative and delay in b-metric spaces. *Appl. Anal. Optim.* **7** (2023), 1–16.

[143] A. Salim, S. Krim and M. Benchohra, Three-point boundary value problems for implicit Caputo tempered fractional differential equations in b-metric

spaces. *Eur. J. Math. Appl.* **3** (2023), Article ID 16. https://doi.org/10.28919/ejma.2023.3.16

[144] A. Salim, S. Krim, J. E. Lazreg and M. Benchohra, On Caputo tempered implicit fractional differential equations in *b*-metric spaces. *Analysis.* **43** (2) (2023), 129–139. https://doi.org/10.1515/anly-2022-1114

[145] A. Salim, J. E. Lazreg, S. Abbas, M. Benchohra and Y. Zhou, Initial value problems for hybrid generalized Hilfer fractional differential equations, *DNC.* **12** (2) (2023), 287–298. http://dx.doi.org/10.5890/DNC.2023.06.005

[146] N. Benkhettou, A. Salim, J. E. Lazreg, S. Abbas and M. Benchohra, Lakshmikantham monotone iterative principle for hybrid Atangana-Baleanu-Caputo fractional differential equations, *An. Univ. Vest Timiş. Ser. Mat.-Inform.* **59** (1) (2023), 79–91. https://doi.org/10.2478/awutm-2023-0007

[147] N. Benkhettou, A. Salim, J. E. Lazreg, S. Abbas and M. Benchohra, Caputo-Fabrizio fractional hybrid differential equations via new Dhage iteration method, *J. Appl. & Pure Math.* **5** (2023), 211–222. https://doi.org/10.23091/japm.2023.211

[148] M. Chohri, S. Bouriah, A. Salim and M. Benchohra, On nonlinear periodic problems with Caputo's exponential fractional derivative, *ATNAA.* **7** (2023), 103–120. https://doi.org/10.31197/atnaa.1130743

[149] A. Salim, M. Benchohra and J. E. Lazreg, On implicit *k*-generalized ψ-Hilfer fractional differential coupled systems with periodic conditions, *Qual. Theory Dyn. Syst.* **22** (2023), 46 pages. https://doi.org/10.1007/s12346-023-00776-1

[150] A. Salim, S. Bouriah, M. Benchohra, J. E. Lazreg and E. Karapınar, A study on *k*-generalized ψ-Hilfer fractional differential equations with periodic integral conditions. *Math. Methods Appl. Sci.* (2023), 1–18. https://doi.org/10.1002/mma.9056

[151] A. Salim and M. Benchohra, A study on tempered (k, ψ)-Hilfer fractional operator, *Lett. Nonlinear Anal. Appl.* **1** (3) (2023), 101–121. https://doi.org/10.5281/zenodo.8361961

[152] A. Salim, J. E. Lazreg and M. Benchohra, Existence, uniqueness and Ulam-Hyers-Rassias stability of differential coupled systems with Riesz-Caputo fractional derivative, *Tatra Mt. Math. Publ.* **84** (2023), 111–138. https://doi.org/10.2478/10.2478/tmmp-2023-0019

[153] A. Salim, F. Mesri, M. Benchohra and C. Tunç, Controllability of second order semilinear random differential equations in Fréchet spaces. *Mediterr. J. Math.* **20** (84) (2023), 1–12. https://doi.org/10.1007/s00009-023-02299-0

[154] M. Benchohra, J. Henderson and S. K. Ntouyas, *Impulsive Differential Equations and inclusions*, Hindawi Publishing Corporation, New York, 2006.

[155] P. Egbunonu, M. Guay, Identification of switched linear systems using subspace and integer programming techniques. *Nonlinear Anal. Hybrid Syst.* **1** (2007), 577–592.

[156] F. Vaadrager, J. Van Schuppen, *Hybrid Systems, Computation and Control. Lecture Notes in Computer Sciences*, vol. 1569. Springer, New York, 1999.

[157] S. Abbas, M. Benchohra, Uniqueness and Ulam stabilities results for partial fractional differential equations with not instantaneous impulses. *Appl. Math. Comput.* **257** (2015), 190–198.

[158] S. Abbas, M. Benchohra, M. A. Darwish, New stability results for partial fractional differential inclusions with not instantaneous impulses. *Frac. Calc. Appl. Anal.* **18** (2015), 172–191.

[159] R. P. Agarwal, S. Hristova, D. O'Regan, *Non-Instantaneous Impulses in Differential Equations*, Springer, New York, 2017.

[160] S. Abbas, M. Benchohra, J. R. Graef and J. Henderson, *Implicit Differential and Integral Equations: Existence and stability*, Walter de Gruyter, London, 2018.

[161] S. Abbas, W. Albarakati, M. Benchohra and J. J. Nieto, Existence and stability results for partial implicit fractional differential equations with not instantaneous impulses. *Novi Sad J. Math.* **47** (2017), 157–171.

[162] S. Abbas and M. Benchohra, Stability results for fractional differential equations with state-dependent delay and not instantaneous impulses, *Math. Slovaca* **67**(4) (2017), 875–894.

[163] S. Abbas, M. Benchohra, A. Alsaedi and Y. Zhou, Some stability concepts for abstract fractional differential equations with not instantaneous impulses. *Fixed Point Theory.* **18** (2017), 3–16.

[164] L. Bai, J. J. Nieto and J. M. Uzal, On a delayed epidemic model with non-instantaneous impulses. *Commun. Pure Appl. Anal.* **19** (2020), no. 4, 1915–1930.

[165] E. Hernández, K. A. G. Azevedo and M. C. Gadotti, Existence and uniqueness of solution for abstract differential equations with state-dependent delayed impulses. *J. Fixed Point Theory Appl.* **21** (1) (2019), 17 pp.

[166] F. Kong and J. J. Nieto, Control of bounded solutions for first-order singular differential equations with impulses. *IMA J. Math. Control Inform.* **37** (3) (2020), 877–893.

[167] J.R. Wang, M. Feckan, *Non-Instantaneous Impulsive Differential Equations*. Basic Theory And Computation, IOP Publishing Ltd. Bristol, UK, 2018.

[168] J. Wang, A. G. Ibrahim and D. O'Regan, Nonemptyness and compactness of the solution set for fractional evolution inclusions with non-instantaneous impulses. *Electron. J. Differential Equations.* **2019** (2019), 1–17.

[169] S. Abbas, M. Benchohra and G. M. N'Guérékata, *Topics in Fractional Differential Equations*, Springer, New York, 2012.

[170] M. Benchohra, J. Henderson and S. K. Ntouyas, *Impulsive Differential Equations and Inclusions*, Hindawi Publishing Corporation, Vol 2, New York, 2006.

[171] J. R. Graef, J. Henderson and A. Ouahab, *Impulsive Differential Inclusions. A Fixed Point Approch*, De Gruyter, Berlin/Boston, 2013.

[172] S. Abbas and M. Benchohra, Existence and Ulam stability for impulsive discontinuous fractional differential inclusions in Banach Algebras, *Mediter. J. Math.* **12** (4) (2015), 1245–1264.

[173] S. Abbas, M. Benchohra and N. Hamidi, Stability for impulsive fractional differential inclusions via Picard operators in Banach spaces, *J. Nonlinear Funct. Anal.* Vol. 2018 (2018), Article ID 44, pp. 1–17.

[174] S. Abbas, M. Benchohra and S. Sivasundaram, Ulam stability for partial fractional differential inclusions with multiple delay and impulses via Picard operators, *Nonlinear Stud.* **20** (4) (2013), 623–641.

[175] D. D. Bainov and P. S. Simeonov, *Systems with Impulse Effect: Theory and Applications*, Ellis Horwood, Chichister, 1989.

[176] V. Lakshmikantham, D. D. Bainov, and P. S. Simeonov, *Theory of Impulsive Differential Equations*, Series in Modern Applied Mathematics, **6**, World Scientific, New Jersey, 1989.

[177] E. Hernández and D. O'Regan, On a new class of abstract impulsive differential equations, *Proc. Amer. Math. Soc.* **141** (2013), 1641–1649.

[178] N. N. Krylov and N. N. Bogolyubov, *Introduction to Nonlinear Mechanics*, Izd. Acad. Sci. Ukr. SSR, Kiev, 1937.

[179] B. Ahmad, J. Henderson, and R. Luca, *Boundary Value Problems for Fractional Differential Equations and Systems*, World Scientific, USA, 2021.

[180] B. Ahmad, A. Alsaedi, S.K. Ntouyas and J. Tariboon, *Hadamard-type Fractional Differential Equations, Inclusions and Inequalities.* Springer, Cham, 2017.

[181] B. Ahmad, A. Almalki, S. K. Ntouyas, A. Amlsaedi, Existence results for a self-adjoint coupled system of three nonlinear ordinary differential equations with cyclic boundary conditions, *Qual. Theory Dyn. Syst.* **21** (2022). https://doi.org/10.1007/s12346-022-00616-8

[182] B. Ahmad, A. Almalki, S. K. Ntouyas, A. Alsaedi, Existence results for a self-adjoint coupled system of nonlinear second-order ordinary differential inclusions with nonlocal integral boundary conditions, *J. Inequal Appl.* **2022** (2022). https://doi.org/10.1186/s13660-022-02846-5

[183] M. Pierri, D. O'Regan, V. Rolnik, Existence of solutions for semi-linear abstract differential equations with non instantaneous impulses, *Appl. Math. Comput.* **219** (2013), 6743–6749.

[184] S. Abbas, M. Benchohra, G. M. N'Guérékata, Instantaneous and noninstantaneous impulsive integrodifferential equations in Banach spaces, *J. Anal. Appl.* **38** (2) (2020), 143–156.

[185] L. Bai, J. J. Nieto, Variational approach to differential equations with not instantaneous impulses, *Appl. Math. Lett.* **73** (2017), 44–48.

[186] P. Chen, X. Zhang, Y. Li, Existence of mild solutions to partial differential equations with noninstantaneous impulses, *Electron. J. Differential Equations* **241** (2016), 1–11.

[187] D. Yang, J. Wang, Integral boundary value problems fornonlinear noninstataneous impulsive differential equations, *J. Appl. Math. Comput.* **55** (2017), 59–78.

[188] J. C. Alvàrez, Measure of noncompactness and fixed points of nonexpansive condensing mappings in locally convex spaces. *Rev. Real. Acad. Cienc. Exact. Fis. Natur. Madrid.* **79** (1985), 53–66.

[189] J. Banaś and K. Goebel, *Measures of Noncompactness in Banach Spaces*, Marcel Dekker, New York, 1980.

[190] R. Agarwal, Certain fractional q-integrals and q-derivatives. *Proc. Camb. Philos. Soc.* **66** (1969), 365–370.

[191] D. O'Regan, Fixed point theory for weakly sequentially continuous mapping. *Math. Comput. Model.* **27** (1998), 1–14.

[192] S. Abbas, M. Benchohra and G. M. N'Guérékata, *Advanced Fractional Differential and Integral Equations*, Nova Science Publishers, New York, 2014.

[193] J. M. Ayerbee Toledano, T. Dominguez Benavides and G. Lopez Acedo, *Measures of Noncompactness in Metric Fixed Point Theory*, Operator Theory, Advances and Applications, vol 99, Birkhäuser, Basel, Boston, Berlin, 1997.

[194] K. Aissani, M. Benchohra, Global existence results for fractional integrodifferential equations with state-dependent delay. *An. Ştiinţ. Univ. Al. I. Cuza Iaşi. Mat. (N.S.)* **62** (2016), 411–422.

[195] R. R. Akhmerov. M. I. Kamenskii, A. S. Patapov, A. E. Rodkina and B. N. Sadovskii, *Measures of Noncompactness an Condensing Operators*, Birkhauser Verlag, Basel, 1992.

[196] J. C. Alvárez, Measure of Noncompactness and fixed points of nonexpansive condensing mappings in locally convex spaces, *Rev. Real. Acad. Cienc. Exact. Fis. Natur. Madrid* **79** (1985), 53–66.

[197] J. Banaś and K. Goebel, *Measures of Noncompactness in Banach Spaces*. Lecture Note in Pure App. Math. 60, Dekker, New York, 1980.

[198] L. Olszowy. Existence of Mild solutions for semilinear nonlocal cauchy problems in separable Banach spaces. *Z. Anal. Anwend.* **32** (2) (2013), 215–232.

[199] L. Olszowy. Existence of mild solutions for semilinear nonlocal Cauchy problems in separable Banach spaces. *Z Anal. Anwend.* **32** (2) (2013), 215–232.

[200] L. Byszewski and V. Lakshmikantham, Theorem about the existence and uniqueness of a solution of a nonlocal abstract Cauchy problem in a Banach space, *Appl. Anal.* **40** (1990), 11–19.

[201] S. A. Abd-Salam, A. M. A. El-Sayed, On the stability of a fractional-order differential equation with nonlocal initial condition. *Electron. J. Qual. Theory Differ. Equ.* **29** (2008), 1–8.

[202] B. Ahmad, J.R. Graef, Coupled systems of nonlinear fractional differential equations with nonlocal boundary conditions. *Panamer. Math. J.* **19** (2009), 29–39.

[203] G M. N'Guérékata, A Cauchy problem for some fractional abstract differential equation with non local conditions. *Nonlinear Anal.* **70** (2009), 1873–1876.

[204] L. Byszewski, Theorems about the existence and uniqueness of solutions of a semilinear evolution nonlocal Cauchy problem, *J. Math. Anal. Appl.* **162** (1991), 494–505.

[205] L. Byszewski, Existence and uniqueness of mild and classical solutions of semilinear functional-differential evolution nonlocal Cauchy problem, *Selected problems of mathematics, 50th Anniv. Cracow Univ. Technol. Anniv. Issue*, **6**, Cracow Univ. Technol., Krakw, (1995), 25–33.

[206] L. Byszewski and H. Akca, On a mild solution of a semilinear functional-differential evolution nonlocal problem, *J. Appl. Math. Stoch. Anal.* **10** (1997), 265–271.

[207] K. Balachandran and R. R. Kumar, Existence of solutions of integrodifferential evolution equations with time varying delays, *Appl. Math.* **7** (2007), 1–8.

[208] M. Benchohra and S. K. Ntouyas, Existence of mild solutions on semiinfinite interval for first order differential equation with nonlocal conditions, *Comment. Math. Univ. Corolin.* **41** (2000), 485–491.

[209] M. Benchohra and S. K. Ntouyas, Existence and controllability results for nonlinear differential inclusions with nonlocal conditions, *Appl. Anal.* **8** (2002), 31–46.

[210] A. Salim, M. Benchohra, J. E. Lazreg, J. J. Nieto and Y. Zhou, Nonlocal initial value problem for hybrid generalized Hilfer-type fractional implicit differential equations. *Nonauton. Dyn. Syst.* **8** (2021), 87–100. `https://doi.org/10.1515/msds-2020-0127`

[211] L. Byszewski, Theorems about the existence and uniquenessof solutions of a semilinear evolution nonlocal Cauchy problem, *J. Math. Anal. Appl.* **162** (1991), 494–505.

[212] K. Deng, Exponential decay of solutions of semilinearparabolic equations with nonlocal initial conditions, *J. Math. Anal. Appl.* **179** (1993), 630–637.

[213] M. Mckibben, *Discovering Evolution Equations with Applications:* Volume 1, Deterministic Models, Chapman and Hall/CRC Appl. Math. Nonlinear Sci. Ser., 2011.

[214] A. Granas, and J. Dugundji, *Fixed Point Theory.* Springer-Verlag, New York, 2003.

[215] R. S. Varga, *Matrix Iterative Analysis,* Springer Series in Computational Mathematics. **27**. Springer-Verlag, Berlin, 2000.

[216] O. Bolojan and R. Precup, Implicit frst order differential systems with nonlocal conditions, *Electron. J. Qual. Theory Differ. Equ.* **69** (2014), 1–13.

[217] R. Precup, The role of matrices that are convergent to zero in the study of semilinear operator systems, *Math. Comp. Model.* **49** (2009), 703–708.

[218] J. Hale and J. Kato, Phase space for retarded equations with infinite delay, *Funkcial. Ekvac.* **21** (1978), 11–41.

[219] Y. Hino, S. Murakami, and T. Naito, *Functional Differential Equations with Unbounded Delay*, Springer-Verlag, Berlin, 1991.

[220] S. Dudek and L. Olszowy, Continuous dependence of the solutions of nonlinear integral quadratic Volterra equation on the parameter, *J. Funct. Spaces*,V. 2015, Article ID 471235, 9 pages.

[221] D. Bothe, Multivalued perturbation of m-accretive differential inclusions, *Isr. J. Math.* **108** (1998), 109–138.

[222] H.P. Heinz, On the behaviour of measure of noncompactness with respect to differentiation and integration of rector-valued functions, *Nonlinear Anal.* **7** (1983), 1351–1371.

[223] J. Banaś, Measures of noncompactness in the space of continuous tempered functions, *Demonstr. Math.* **14** (1981), 127–133.

[224] L. Liu, F. Guo, C. Wu, Y. Wu, Existence theorems of global solutions for nonlinear Volterra type integral equations in Banach spaces. *J. Math. Anal. Appl.* **309** (2005), 638–649.

[225] Guo, D. J.; Lakshmikantham, V.; Liu, X. *Nonlinear Integral Equations in Abstract Spaces*, Kluwer Academic Publishers: Dordrecht, 1996.

[226] N. U. Ahmed, *Semigroup Theory with Applications to Systems and Control*, Harlow John Wiley & Sons, Inc., New York, 1991.

[227] C. Lizama, Regularized solutions for abstract Volterra equations, *J. Math. Anal. Appl.* **243**, (2000), 278–292.

[228] J. Prüss, *Evolutionary Integral Equations and Applications* Monographs Math. **87**, Bikhäuser Verlag, 1993.

[229] K.J. Engel and R. Nagel, *One-Parameter Semigroups for Linear Evolution Equations*, Springer-Verlag, New York, 2000.

[230] W. Arendt, C. Batty, M., Hieber, and F. Neubrander, *Vector-Valued Laplace Transforms and Cauchy Problems*, Monographs in Mathematics, **96** Birkhauser, Basel, 2001.

[231] A. Pazy, *Semigroups of Linear Operators and Applications to Partial Differential Equations*, Springer-Verlag, New York, 1983.

[232] W. Arendt, Vector valued Laplace transorms and Cauchy problems, *Israel J. Math.* **59** (1987), 327–352.

[233] H. Kellermann, Integrated Semigroups, Thesis, Tubingen, 1986.

[234] P. Magal, S. Ruan, On integrated semigroups and age structured models in L^p spaces, *Differential Integral Equations* **20** (2007), 197–239.

[235] H.R. Thieme, Differentiability of convolutions, integrated semigroups of bounded semi-variation, and the inhomogeneous Cauchy problem, *J. Evol. Equ.* **8** (2008), 283–305.

[236] T. J. Xiao, X. X. Zhu, J. Liang. Pseudo-almost automorphic mild solutions to nonautonomous differential equations and applications. *Nonlinear Anal.* **70** (2009), no. 11, 4079–4085.

[237] Z. Zhao, Y. Chang, J. Nieto, Almost automorphic and pseudo-almost automorphic mild solutions to an abstract differential equation in Banach spaces. *Nonlinear Anal. Theory Methods Appl.* **72** (2010), 1886–1894.

[238] G. M. N'Guérékata, *Spectral Theory for Bounded Functions and Applications to Evolution Equations*. Mathematics Research Developments. Nova Science Publishers, Inc., New York, 2017.

[239] J. Liang, J. Zhang, T. Xiao. Composition of pseudo almost automorphic and asymptotically almost automorphic functions. *J. Math. Anal. Appl.* **340** (2008), no. 2, 1493–1499.

[240] B. G. Pachpatte: A note on Gronwall-Bellman inequality, *J. Math. Anal. Appl.* **44** (1973), 758–762.

[241] A.M. Samoilenko, N. Perestyuk, *Differential Equations with Impulse Effect*, Visha Shkola, Kyiv, 1987.

[242] C. Corduneanu, *Integral Equations and Stability of Feedback Systems*, Acadimic Press, New York, 1973.

[243] S. Itoh, Random fixed point theorems with applications to random differential equations in Banach spaces, *J. Math. Anal. Appl.*, **67** (1979), 261–273.

[244] H. W. Engl, A general stochastic fixed-point theorem for continuous random operators on stochastic domains, *J. Math. Anal. Appl.* **66** (1978), 220–231.

[245] H. W. Engl, A general stochastic fixed-point theorem for continuos random operators on stochastic domains. *Anal. Appl.* **66** (1978), 220–231.

[246] J. Liouville, Second mémoire sur le développement des fonctions ou parties de fonctions en séries dont divers termes sont assujettis á satisfaire à une même équation différentielle du second ordre contenant un paramétre variable. *J. Math. Pure et Appl.* **2** (1837), 16–35.

[247] E. Picard, Mémoire sur la théorie des équations aux derivées partielles et la méthode des approximations successives. *J. Math. Pures et Appl.* **6** (1890), 145–210.

[248] H. Poincaré, Sur les courbes definies par les équations différentielles. *J. Math.* **2** (1886), 54–65.

[249] S. Banach, Sur les opérations dans les ensembles abstraits et leur application aux équations intégrales, *Fund. Math.* **3** (1922), 133–181.

[250] G. Darbo, Punti uniti in transformazioni a condominio non compatto. *Rend. Sem. Math. Univ. Padova*, **24** (1955), 84–92.

[251] J. Dugundji, A. Granas, *Fixed Point Theory*, Springer-Verlag, New York, 2003.

[252] A. Aghajani, J. Banaś, N. Sabzali, Some generalizations of Darbo fixed point theorem and applications. *Bull. Belg. Math. Soc. Simon Stevin* **20** (2013), no. 2, 345–358.

[253] R. Agarwal, M. Meehan and D. O'Regan, *Fixed point theory and applications*, in: Cambridge Tracts in Mathematics, Cambridge University Press, New York, 2001.

[254] K. Deimling, *Nonlinear Functional Analysis,* Springer-Verlag, New York, 1985.

[255] S. Dudek, Fixed point theorems in Fréchet algebras and Fréchet spaces and applications to nonlinear integral equations, *Appl. Anal. Discrete Math.* **11**(2017), 340–357.

[256] E. Zeidler, *Nonlinear Functionnal Analysis and its Applications*, Fixed Point Theorems, Springer-Verlag, New York, 1990.

[257] N. Laksaci, A. Boudaoui, W. Shatanawi, T. Shatnawi. Existence results of global solutions for a coupled implicit Riemann-Liouville fractional integral equation via the vector Kuratowski measure of noncompactness. *Fractal Fract.* **6** (2022), 130.

[258] L. Cai, J. Liang and J. Zhang, Generalizations of Darbo's fixed point theorem and solvability of integral and differential systems, *J. Fixed Point Theory Appl.* (2018), 1–20.

[259] B.C. Dhage, Some generalizations of multivalued version of Schauder's fixed point theorem with applications. *Cubo* **12** (2010), 139–151.

[260] C. J. K. Batty, R. Chill, S. Srivastava, Maximal regularity for second order non-autonomous Cauchy problems, *Studia Math.* **189** (2008), 205–223.

[261] F. Faraci, A. Iannizzotto, A multiplicity theorem for a perturbed second-order non-autonomous system, *Proc. Edinb. Math. Soc.* **49** (2006), 267–275.

[262] H. Henríquez, V. Poblete, J. Pozo, Mild solutions of non-autonomous second order problems with nonlocal initial conditions. *J. Math. Anal. Appl.* **412** (2014), no. 2, 1064–1083.

[263] T. Winiarska, Evolution equations of second order with operator dependent on t, *Selected Problems of Math. Cracow Univ. Tech.* **6** (1995), 299–314.

[264] M. Benchohra and N. Rezoug, Measure of noncompactness and second order evolution equations. *Gulf J. Math.* **4** (2016), no. 2, 71–79.

[265] N. I. Mahmudov, V. Vijayakumar, R. Murugesu, Approximate controllability of second-order evolution differential inclusions in Hilbert spaces. *Mediterr. J. Math.* **13** (2016), 3433–3454.

[266] M. Kozak, A fundamental solution of a second-order differential equation in a Banach space, *Univ. Iagel. Acta Math.* **32** (1995), 275–289.

[267] M. Benchohra, G. M. N'Guérékata and N. Rezoug, Asymptotically almost automorphic mild solutions for second-order nonautonomous semilinear evolution equations, *Journal of Computational Analysis and Applications*, **29** (3) (2021), 468–493.

[268] M. Benchohra, N. Rezoug and Y. Zhou, Semilinear integro-differential evolution equations via Kuratowski measure of noncompactness. *Z. Anal. Anwend.* **38** (2) (2019), 143–156.

[269] Y. Zhou, R.N. Wang and L. Peng, *Topological Structure of the Solution Set for Evolution Inclusions*, Springer-Nature, Singapore, 2017.

[270] X. Zhang, P. Chen, Fractional evolution equation nonlocal problems with noncompact semigroups. *Opuscula Math.* **36** (2016), no. 1, 123–137.

[271] L. Byszewski, Existence, uniqueness and asymptotic stability of solutions of abstract nonlocal Cauchy problems, *Dynam. Systems Appl.* **5** (1996), 595–605.

[272] L. Byszewski, H. Akca, Existence of solutions of a semilinear functional differential evolution nonlocal problem, *Nonlinear Anal.* **34**(1998), 65–72.

[273] E. Hernandez and D. O'Regan, On state dependent non-local conditions, *Appl. Math. Lett.* **83** (2018), 103–109.

[274] E. Hernandez, On abstract differential equations with state dependent non-local conditions, *J. Math. Anal. Appl.* **466** (2018), 408–425.

[275] S. Bouriah, A. Salim and M. Benchohra, On nonlinear implicit neutral generalized Hilfer fractional differential equations with terminal conditions and delay, *Topol. Algebra Appl.* **10** (2022), 77–93. https://doi.org/10.1515/taa-2022-0115

[276] C. Derbazi, H. Hammouche, A. Salim and M. Benchohra, Measure of noncompactness and fractional hybrid differential equations with hybrid conditions. *Differ. Equ. Appl.* **14** (2022), 145–161. http://dx.doi.org/10.7153/dea-2022-14-09

[277] S. Krim, A. Salim, S. Abbas and M. Benchohra, On implicit impulsive conformable fractional differential equations with infinite delay in b-metric spaces. *Rend. Circ. Mat. Palermo (2)*. (2022), 1–14. https://doi.org/10.1007/s12215-022-00818-8

[278] N. Laledj, A. Salim, J. E. Lazreg, S. Abbas, B. Ahmad and M. Benchohra, On implicit fractional q-difference equations: Analysis and stability. *Math. Methods Appl. Sci.* **45** (17) (2022), 10775–10797. https://doi.org/10.1002/mma.8417

[279] A. Salim, M. Benchohra, J. R. Graef and J. E. Lazreg, Initial value problem for hybrid ψ-Hilfer fractional implicit differential equations. *J. Fixed Point Theory Appl.* **24** (2022), 14 pp. https://doi.org/10.1007/s11784-021-00920-x

[280] A. Salim, J. E. Lazreg, B. Ahmad, M. Benchohra and J. J. Nieto, A study on k-generalized ψ-Hilfer derivative operator, *Vietnam J. Math.* (2022). https://doi.org/10.1007/s10013-022-00561-8

[281] A. Salim, M. Benchohra, J. E. Lazreg and G. N'Guérékata, Existence and k-Mittag-Leffler-Ulam-Hyers stability results of k-generalized ψ-Hilfer boundary value problem. *Nonlinear Studies.* **29** (2022), 359–379.

[282] A. A. Kilbas, Hari M. Srivastava, and Juan J. Trujillo, *Theory and Applications of Fractional Differential Equations*. North-Holland Mathematics Studies, 204. Elsevier, Amsterdam, 2006.

[283] V. Lakshmikantham, S. Leela and J. Vasundhara, *Theory of Fractional Dynamic Systems*, Cambridge Academic Publishers, Cambridge, 2009.

[284] R. P. Agarwal, B. Andradec, G. Siracusa, On fractional integro-differential equations with state-dependent delay, *Comput. Math. Appl.* **63** (3) (2011), 1142–1149.

[285] A. Anguraj, P. Karthikeyan, J. J. Trujillo, Existence of solutions to fractional mixed integro-differential equations with nonlocal initial condition. *Adv. Difference Equ.* 2011, Art. ID 690653, 12 pp.

[286] K. Balachandran, S. Kiruthika, J. J. Trujillo, Existence results for fractional impulsive integro-differential equations in Banach spaces. *Commun. Nonlinear Sci. Numer. Simul.* **16** (2011), 1970–1977.

[287] C. Cuevas, J. -C. de Souza, S-asymptotically w-periodic solutions of semilinear fractional integro-differential equations, *Appl. Math. Lett.* **22** (2009), 865–870.

[288] A. Ouahab, Some uniqueness results for functional damped semilinear differential equations in Fréchet spaces, *Acta Math. Sinica* **24** (1) (2008), 95–106.

[289] A. Belarbi, M. Benchohra and A. Ouahab, Uniqueness results for fractional functional differential equations with infinite delay in Fréchet spaces, *Applicable Analysis.* **85** (12) (2006), 1459–1470.

[290] A. Ouahab, Local and global existence and uniqueness results for impulsive functional differential equations with multiple delay, *J. Math. Anal. Appl.* **323** (2006), 456–472.

[291] S. Abbas, A. Arara, M.Benchohra, F. Mesri, Functional random evolution equations in Fréchet spaces, *Adv. Theory Nonlinear Anal. Appl.* **2** (2018), 128–137.

[292] I. Chueshov and A. Rezounenko, Dynamics of second order in time evolution equations with state-dependent delay, *Nonlinear Anal.* **123** (2015), 126–149.

[293] S. Das, D. N. Pandey, and N. Sukavanam, Existence of solution and approximate controllability of a second-order neutral stochastic differential equation with state dependent delay, *Acta Math. Sci. Ser. B Engl. Ed.* **36** (2016), 1509–1523.

[294] E. Hernandez, K. Azevedo and V. Rolnik, Wellposedness of abstract differential equations with state-dependent delay, *Math. Nachr.* **291** (2018), 2045–2056.

[295] R. Grimmer, Resolvent opeators for integral equations in a Banach space, *Trans. Amer. Math. Soc.* **273** (1982), 333–349.

[296] R. Grimmer and A.J. Pritchard, Analytic resolvent operators for integral equations in a Banach space, *J. Differ. Equ.* **50** (1983), 234–259.

[297] M. Dieye, M. A. Diop, K. Ezzinbi, H. Hmoyed, On the existence of mild solutions for nonlocal impulsive integrodifferential equations in banach spaces, *Matematiche*, **74**(1) (2019), 13–34.

[298] M.A. Diop, K. Ezzinbi, M.P. Ly, Nonlocal problems for integrodifferential equation via resolvent operators and optimal control, *Differ. Incl. Control Optim.* **42** (2022), 5–25.

[299] M. Mohan Raja, A. Shukla, J.J. Nieto, V. Vijayakumar, K.S. Nisar, A note on the existence and controllability results for fractional integrodifferential inclusions of order $r \in (1,2]$ with impulses, *Qualitative Theory Dyn. Syst.* (2022) 21(4), 150.

[300] S. Rezapour, K.S. Vijayakumar, H.R. Henriquez, V. Nisar, A. Shukla, A Note on existence of mild solutions for second-order neutral integrodifferential evolution equations with state-dependent delay. *Fractal Fractional* **5** (2021), 1–17.

[301] E. Hernandez, Existence and uniqueness of global solution for abstract second order differential equations with state-dependent delay, *Math. Nachr.* **295** (2022), 124–139.

[302] M. A. Diop, K. H. Bete, R. Kakpo and C. Ogouyandjou, Existence results for some integro-differential equations with state-dependent nonlocal conditions in Fréchet Spaces, *Nonauton. Dyn. Syst.* **7** (2020), 272–280.

[303] J.-G. Si and X. P. Wang, Analytic solutions of a second-order functional-differential equation with a state derivative dependent delay, *Colloq. Math.* **79** (1999), no. 2, 273–281.

[304] J. Kisynski, On cosine operator functions and one parameter group of operators, *Studia Math.* **44** (1972), 93–105.

[305] R. R. Akhmerov, M. I. Kamenskii, A. S. SPatapov, A. E. Rodkina, B. N. Sadovskii, *Measures of noncompactness and condensing operators*, Birkhauser Verlag, Basel, 1992.

[306] J. C. Alvàrez, Measure of noncompactness and fixed points of nonexpansive condensingmappings in locally convex spaces, *Rev. Real. Acad. Cienc. Exact. Fis. Natur., Madrid*, **79** (1985), 53–66.

[307] H.R. Henríquez, J.C. Pozo, Existence of solutions of abstract nonautonomous second order integro-differential equations. *Bound. Value Probl.* **168** (2016), 1–24.

[308] J. Hale and J. Kato, Phase space for retarded equations with infinite delay, *Funkcial. Ekvac.* **21** (1978), 11–41.

[309] Y. Hino, S.Murakami, T. Naito, *Functional-differential equations with infinite delay*. In: Stahy, S. (ed.) Lecture Notes in Mathematics, **1473**. Springer, Berlin (1991).

[310] E. Hernandez, R. Sakthivel, and A. Tanaka, Existence results for impulsive evolution differential equations with state-dependent delay, *Electron. J. Differential Equations*, **2008** (2008), 1–11.

[311] A.E. Bashirov, N.I. Mahmudov, On concepts of controllability for linear deterministic and stochastic systems, *SIAM J. Control Optim.* **37** (1999), 1808–1821.

[312] M. Fall, A. Mane, B. Dehigbe, M.A. Diop, Some results on the approximate controllability of impulsive stochastic integro-differential equations with nonlocal conditions and state-dependent delay. *J. Nonlinear Sci. Appl.* **15** (2022), 284–300.

[313] V. Singh, R. Chaudhary, D. N. Pandey Approximate controllability of second-order non-autonomous stochastic impulsive differential systems. *Stoch. Anal. Appl.* **39** (2020), 339–356.

[314] E. Lakhel, Controllability of neutral stochastic functional integrodifferential equations driven by fractional Brownian motion, *Stoch. Anal. Appl.* **34** (2016), 427–440.

[315] Z. Bouteffal, A. Salim, S. Litimein and M. Benchohra, Uniqueness results for fractional integro-differential equations with state-dependent nonlocal conditions in Fréchet spaces, *An. Univ. Vest Timiş. Ser. Mat.-Inform.* **59** (1) (2023), 35–44. https://doi.org/10.2478/awutm-2023-0004

[316] Z. Bouteffal, A. Salim, S. Litimein and M. Benchohra, Abstract differential equations with state-dependent delay on the half line. (Submitted).

[317] A. Bensalem, A. Salim, M. Benchohra and J. J. Nieto, Controllability results for second-order integro-differential equations with state-dependent delay. *Evol. Equ. Control Theory.* **12** (6) (2023), 1559–1576. http://dx.doi.org/10.3934/eect.2023026

[318] C. C. Travis, G. F. Webb, Cosine families and abstract nonlinear second order differential equations. *Acta Math. Acad. Sci. Hung.* **32** (1978), 75–96.

[319] K. Balachandran, S. M. Anthoni, Controllability of second-order semilinear neutral functional differential systems in Banach spaces. *Comput. Math. Appl.* **41** (2001), 1223–1235.

[320] H. R. Henríquez, E. M. Hernández, Approximate controllability of second-order distributed implicit functional systems. *Nonlinear Anal.* **70** (2009), 1023–1039.

[321] W. Arendt, Vector valued Laplace transforms and Cauchy problems, *Israel J. Math.* **59** (1987), 327–352.

[322] H. Kellermann and M. Hieber, Integrated semigroup, *J. Funct. Anal.* **84** (1989), 160–180.

[323] G. Da Prato and E. Sinestrari, Differential operators with non-dense domains, *Ann. Scuola. Norm. Sup. Pisa Sci.* **14** (1987), 285–344.

[324] A. Salim, S. Abbas, M. Benchohra, Semilinear random differential evolution equations in Fréchet spaces. (Submitted).

[325] A. Salim, F. Mesri, M. Benchohra and C. Tunç, Controllability of second order semilinear random differential equations in Fréchet spaces. *Mediterr. J. Math.* **20** (84) (2023), 1–12. https://doi.org/10.1007/s00009-023-02299-0

[326] A. Benaissa, M. Benchohra and J. R. Graef, Functional differential equations with delay and random effects, *Stoc. Anal. Appl* **33** (6) (2015), 1083–1091.

[327] A. Salim, S. Abbas, M. Benchohra and E. Karapınar, Global stability results for Volterra-Hadamard random partial fractional integral equations. *Rend. Circ. Mat. Palermo (2).* (2022), 1–13. https://doi.org/10.1007/s12215-022-00770-7

[328] A. Diop, M. A. Diop and K. Ezzinbi, Existence results for a class of random delay integrodifferential equations, *Random Oper. Stoch. Equ.* **29** (2021), 79–86.

[329] V. Lupulescu, D. O'Regan, Ghaus ur Rahman, Existence results for random fractional differential equations, *Opuscula Math.* **34** (2014), 813–825.

[330] V. Lupulescu, S.K. Ntouyas, Random fractional differential equations, *Int. Electron. J. Pure Appl. Math.* **4** (2012), 119–136.

[331] A. Salim, M. Boumaaza and M. Benchohra, Random solutions for mixed fractional differential equations with retarded and advanced arguments. *J. Nonlinear Convex Anal.* **23** (2022), 1361–1375.

[332] B. Ahmad, M. Boumaaza, A. Salim and M. Benchohra, Random solutions for generalized Caputo periodic and non-local boundary value problems. *Foundations.* **3** (2) (2023), 275–289. https://doi.org/10.3390/foundations3020022

[333] A. Benaissa, A. Salim, M. Benchohra and E. Karapınar, Functional delay random semilinear differential equations. *J. Anal.* (2023), 1–12. https://doi.org/10.1007/s41478-023-00592-5

[334] H. Smith, *An Introduction to Delay Differential Equations with Applications to the Life Sciences*, Springer, New York, 2011.

[335] Triggiani, R. On the stabilizability problem in Banach space. *J. Math. Anal. Appl.* **1975**, 52, 383–403.

[336] M. D. Quinn, N. Carmichael, An approach to nonlinear control problems using the fixed point methods, degree theory and pseudo-inverses. *Numer. Funct. Anal. Optim.* **1984**, 7, 197–219.

[337] M. A. Diallo, K. Ezzinbi, A. Sene, Controllability for some integrodiffer- ential evolution equations in Banach spaces. *Discuss. Math. Differ. Incl. Control Optim.* **2017**, 37, 69–81.

[338] R. S. Adiguzel, U. Aksoy, E. Karapinar, I. M. Erhan, On the solution of a boundary value problem associated with a fractional differential equation, *Mathematical Methods in the Applied Sciences.* (2020). https://doi.org/ 10.1002/mma.6652

[339] R. S. Adiguzel, U. Aksoy, E. Karapinar, I. M. Erhan, Uniqueness of solution for higher-order nonlinear fractional differential equations with multi-roint and integral boundary conditions, *RACSAM.* (2021). https://doi.org/ 10.1007/s13398-021-01095-3

[340] R. S. Adiguzel, U. Aksoy, E. Karapinar, I. M. Erhan, On The Solutions Of Fractional Differential Equations Via Geraghty Type Hybrid Contractions, *Appl. Comput. Math.* **20** (2021), 313–333.

[341] H. Afshari and E. Karapinar, A discussion on the existence of positive solutions of the boundary value problems via ψ-Hilfer fractional derivative on b-metric spaces, *Adv Differ Equ.* **2020** (2020), 1–11.

[342] K. Balachandran, R. Sakthivel, Controllability of integrodifferential sys- tems in Banach spaces. *Appl. Math. Comput.* **2001**, 118, 63–71.

[343] Yan, Z. Controllability of fractional-order partial neutral functional inte- grodifferential inclusions with infinite delay. *J. Franklin Inst.* **2011**, 348, 2156–2173.

[344] G. Da Prato, E. Sinestrary, Differential operators with non-dense domains, *Ann. Sc. Pisa Cl. Sci.* **14** (1987), 285–344.

[345] V. V. Vasil'ev and S. I. Piskarev, Differential equations in Banach spaces. II. Theory of cosine operator functions. Functional analysis, *J. Math. Sci.* (N. Y.) **122** (2004), no. 2, 3055–3174.

[346] C. C. Travis and G. F. Webb, Compactness, regularity, and uniform con- tinuity properties of strongly continuous cosine families, Houston *J. Math.* **3** (1977), 555–567.

[347] E. Hernandez, R. Sakthivel, A. Tanaka, Existence results for impulsive evo- lution differential equations with state-dependent delay. *Electron. J. Dif- ferential Equations* **2008**, 2008, 1–11.

[348] R. P. Agarwal, M. Meechan, and D. O'Regan, *Fixed Point Theory and Applications*, Cambridge University Press, Cambridge, 2001.

[349] A. Heris, Z. Bouteffal, A. Salim and M. Benchohra, Abstract random dif- ferential equations with state-dependent delay using measures of noncom- pactness. (Submitted).

[350] M. Benchohra and S. Abbas, *Advanced Functional Evolution Equations and Inclusions*, Springer, Cham, 2015.

[351] J. Liang, J.H. Liu and T.J. Xiao, Nonlocal problems for integrodifferential equations. *Dyn. Cont. Disc. Impul. Syst.* **15** (2008), 815–824.

[352] A. Freidman, *Partial Differential Equations*, Holt, Rinehat and Winston, New York, 1969.

[353] S.G. Krein, *Linear Differential Equation in Banach Spaces*, Amer. Math. Soc.,Providence, 1971.

[354] M. Benchohra, G.M. N'Guérékata and N. Rezoug, S-asymptotically ω-periodic mild solutions for second order evolution equations. *Evolutionary processes and applications*, 151–174, Adv. Evol. Equ., Nova Sci. Publ., New York, 2019.

[355] M. Benchohra, A. Bensalem and A. Salim, S-asymptotic ω-periodic solution for a coupled integro-differential equations with nonlocal conditions. *Novi Sad J. Math.* (to appear).

[356] A. Benaissa, F. Bouazzaoui, A. Salim and M. Benchohra, A random S-asymptotically ω-periodic mild solutions for functional evolution equations with delay and random effect. (Submitted).

[357] Y. Hino and S. Murakami, Total stability in abstract functional differential equations with infinite delay, *Electron. J. Qual. Theory Differ. Equ.* Lecture Notes in Mathematics, **1473**, Springer-Verlag, Berlin, 1991.

[358] M. Kamenskii, V. Obukhovskii and P. Zecca, *Condensing Multivalued Maps and Semilinear Differential Inclusions in Banach Spaces*, de Gruyter Series in Nonlinear Analysis and Applications, Berlin, 2001.

[359] Z. Fan, G. Li, Existence results for semilinear differential equations with nonlocal and implusive conditions, *J. Funct. Anal.* **258** (2010), 1709–1727.

[360] X. Fu, K. Ezzinbi, Existence of solutions of a semilinear functional-differential evolution equations with nonlocal conditions, *Nonlinear Anal.* **54** (2003), 215–227.

[361] S. Abbas, A. Arara and M. Benchohra, Global convergence of successive approximations for abstract semilinear differential equations, *PanAmer. Math. J.* **29** (1) (2019), 17–31.

[362] S. Baghli and M. Benchohra, Global uniqueness results for partial functional and neutral functional evolution equations with infinite delay, *Differential Integral Equations* **23** (2010), 31–50.

[363] A. Baliki and M. Benchohra, Global existence and stability for neutral functional evolution equations, *Rev. Roumaine Math. Pures Appl.* **LX** (1) (2015), 71–82.

[364] M. Dieye, M.A. Diop, K. Ezzinbi and H. Hmoyed, On the existence of mild solutions for nonlocal impulsive integro-differential equations in Banach spaces, *Le Matematiche* **LXXIV** (1) (2019), 13–34.

[365] M. Yang, Q. Wang, Existence of mild solutions for a class of Hilfer fractional evolution equations with nonlocal conditions. *Fract. Calc. Appl. Anal.* **20** (2017), 679–705.

[366] X. Xue, Semilinear nonlocal differential equations with measure of noncompactness in Banach spaces, *J. Nanjing. Univ. Math.* **24**(2007), 264–276.

[367] X. Xue, Existence of semilinear differential equations with nonlocal initial conditions. *Acta. Math. Sini.* **23**(2007), 983–988.

[368] S. Abbas, W. Albarakati, M. Benchohra and J.J. Nieto, Existence and stability results for partial implicit fractional differential equations with not instantaneous impulses, *Novi Sad J. Math.* **47** (2) (2017), 157–171.

[369] S. Abbas, M. Benchohra, A. Alsaedi and Y. Zhou, Some stability concepts for abstract fractional differential equations with not instantaneous impulses, *Fixed Point Theory* **18** (1) (2017), 3–16.

[370] S. Abbas, M. Benchohra and M. A. Darwish, Some existence and stability results for abstract fractional differential inclusions with not Instantaneous impulses, *Math. Reports* **19** (69) (2017), 245–262.

[371] J.M.A. Toledano, T.D. Benavides and G.L. Acedo, *Measures of Noncompactness in Metric Fixed Point Theory*, Birkhauser, Basel, 1997.

[372] R. R. Kumar, Regularity of solutions of evolution integrodifferential equations with deviating argument, *Appl. Math. Comput.* **217** (2011), 9111–9121.

[373] R. R. Kumar, Nonlocal Cauchy problem for analytic resolvent operator integrodifferential equations in Banach spaces, *Appl. Math. Comput.* **204** (2008), 352–362.

[374] R. P. Agarwal and D. O'Regan, *Infinite Interval Problems for Differntial, Difference and Integral Equation*, Academic Publishers, Dordrecht, 2001.

[375] R. P. Agarwal and D. O'Regan, Infinite interval problems modeling phenomena which arise in the theory of plasma and electrical potential theory, *Stud. Appl. Math.* **111** (2003), 339–358.

[376] A. Belleni-Morante, An integrodifferential equation arising from the theory of heat conduction in rigid material with memory, *Boll. Un. Mat. Ital.* **15** (1978), 470–482.

[377] A. Jawahdou, Mild solutions of functional semilinear evolution Volterra integrodifferential equations on an unbounded interval,*Nonlinear Anal.* **74** (2011), 7325–7332.

[378] R. S. Adiguzel, U. Aksoy, E. Karapinar, I. M. Erhan, On the solution of a boundary value problem associated with a fractional differential equation. *Math. Meth. Appl. Sci.* (2020), 1–12.

[379] R. S. Adiguzel, U. Aksoy, E. Karapinar, I. M. Erhan, Uniqueness of solution for higher-order nonlinear fractional differential equations with multi-point and integral boundary conditions. *Rev. R. Acad. Cienc. Exactas Fis. Nat. Ser. A Mat. RACSAM* **115** (2021), 115–155.

[380] R. S. Adiguzel, U. Aksoy, E. Karapinar, I. M. Erhan, On the solutions of fractional differential equations via Geraghty type hybrid contractions. *Appl. Comput. Math.* **20** (2021), 313–333.

[381] H. Afshari and E. Karapinar, A solution of the fractional differential equations in the setting of *b*-metric space. *Carpathian Math. Publ.* **13** (2021), 764–774. https://doi.org/10.15330/cmp.13.3.764-774

[382] H. Afshari and E. Karapinar, A discussion on the existence of positive solutions of the boundary value problems via ψ-Hilfer fractional derivative on *b*-metric spaces. *Adv. Difference Equ.* **2020** (2020), 616. https://doi.org/10.1186/s13662-020-03076-z

[383] L. Byszewski, Application of properties of the right-hand sides of evolution equations to an investigation of nonlocal evolution problems, *Nonl. Anal.* **33** (1998), 413–426.

[384] L. Byszewski, Existence and uniqueness of a classical solution to a functional-differential abstract nonlocal Cauchy problem. *Appl. Math. Stoch. Anal.* **12** (1999), 91–97.

[385] E. Hernández, D. O'Regan. On a new class of abstract impulsive differential equations, *Proc. Amer. Math. Soc.*, 141(5) (2013), 1641–1649.

[386] S. Abbas and M. Benchohra, M. Darwish, Some existence and stability results for abstract fractional differential inclusions with not instantaneous impulses, *Math. Rep. (Bucur.)*, 19(2) (2017), 245–262.

[387] A. Anguraj, S. Kanjanadevi, Existence results for fractional non-instantaneous impulsive integro-differential equations with nonlocal conditions. Dyn. Contin. Discrete Impuls, *Syst. Ser. A Math. Anal.*, 23(6) (2016), 429–445.

[388] M. Benchohra, S. Litimein. Existence results for a new class of fractional integro-differential equations with state dependent delay, *Mem. Differ. Equ. Math. Phys.*, 74 (2018), 27–38.

[389] R. Ganga, D. Jaydev, Existence result of fractional functional integro-differential equation with not instantaneous impulse, *Int. J. Adv. Appl. Math. and Mech.*, 1(3) (2014), 11–21.

[390] G. Gautam, J. Dabas. Mild solution for nonlocal fractional functional differential equation with not instantaneous impulse, *Int. J. Nonlinear Sci.*, 21(3) (2016), 151–160.

[391] M. Muslim, A. Kumar. Controllability of fractional differential equation of order $\alpha \in (1,2]$ with non-instantaneous impulses, *Asian J. Control*, **20** (2) (2018), 935–942.

[392] R. Saadati, E. Pourhadi, B. Samet, On the $\mathcal{PC}$-mild solutions of abstract fractional evolution equations with non-instantaneous impulses via the measure of noncompactness, *Bound. Value Probl.* (2019), No 19.

[393] J. Wang, Y. Zhou, Z. Lin, On a new class of impulsive fractional differential equations, *Appl. Math. and Comput.* **242** (2014), 649–657.

[394] R.C. Grimmer, Resolvent opeators for integral equations in a Banach space, *Trans. Amer. Math. Soc.* **273** (1982), 333–349.

[395] J. Simon, *Banach, Fréchet, Hilbert and Neumann spaces, Analysis for PDEs set.* **1** Mathematics and Statistics Series. Wiley, Hoboken, NJ 2017.

[396] L. Olszowy and S. Wedrychowicz, Mild solutions of semilinear evolution equation on an unbounded interval and their applications. *Nonlinear Anal.* **72** (2010), no. 3–4, 2119–2126.

[397] L. Olszowy, Fixed point theorems in the Fréchet space $C(R+)$ and functional integral equations on an unbounded interval, *Appl. Math. Comput.* **218** (18) (2012), 9066–9074.

[398] L. Olszowy and S. Wędrychowicz, On the existence and asymptotic behaviour of solution of an evolution equation and an application to the Feynman-Kac theorem *Nonlinear Anal.* **72** (2011), 6758–6769.

[399] H. Henríquez. Existence of solutions of non-autonomous second order functional differential equations with infinite delay, *Nonlinear Anal.* **74** (10) (2011), 3333–3352.

[400] S. Abbas, M. Benchohra and G. N'Guérékata, Instantaneous and noninstantaneous impulsive integro-differential equations in Banach spaces, *Abstract and Applied Analysis.* **2020** (2020), Art. ID 2690125, 8 pages.

[401] A. Bensalem, A. Salim, M. Benchohra and E. Karapınar, Existence and attractivity results on semi-infinite intervals for integrodifferential equations with non-instantaneous impulsions in Banach spaces, *An. Ştiinţ. Univ. "Ovidius" Constanţa Ser. Mat.* **32** (1) (2024), 65–84. http://dx.doi.org/10.2478/auom-2024-0004

[402] A. Bensalem, A. Salim, B. Ahmad and M. Benchohra, Existence and controllability of integrodifferential equations with non-instantaneous impulses in Fréchet spaces, *CUBO.* **25** (2) (2023), 231–250. https://doi.org/10.56754/0719-0646.2502.231

[403] M. Benchohra, N. Rezoug, B. Samet, Y. Zhou, Second order semilinear Volterra-type integro-differential equations with non-instantaneous impulses, *Mathematics.* **2019** (7) (2019), 1134.

[404] M. Benchohra, E. Karapınar, J. E. Lazreg and A. Salim, *Advanced Topics in Fractional Differential Equations: A Fixed Point Approach*, Springer, Cham, 2023.

[405] M. Benchohra, E. Karapınar, J. E. Lazreg and A. Salim, *Fractional Differential Equations: New Advancements for Generalized Fractional Derivatives*, Springer, Cham, 2023.

[406] W. Rahou, A. Salim, J. E. Lazreg and M. Benchohra, Existence and stability results for impulsive implicit fractional differential equations with delay and Riesz–Caputo derivative. *Mediterr. J. Math.* **20** (2023), 143. https://doi.org/10.1007/s00009-023-02356-8

[407] W. Rahou, A. Salim, J. E. Lazreg and M. Benchohra, On fractional differential equations with Riesz-Caputo derivative and non-instantaneous impulses. *Sahand Commun. Math. Anal.* **20** (3) (2023), 109–132. https://doi.org/10.22130/scma.2023.563452.1186

[408] A. Bensalem, A. Salim and M. Benchohra, Ulam-Hyers-Rassias stability of neutral functional integrodifferential evolution equations with non-instantaneous impulses on an unbounded interval, *Qual. Theory Dyn. Syst.* **22** (2023), 29 pages. https://doi.org/10.1007/s12346-023-00787-y

[409] A. Bensalem, A. Salim, B. Ahmad and M. Benchohra, Existence and controllability of integrodifferential equations with non-instantaneous impulses in Fréchet spaces, *CUBO.* **25** (2) (2023), 231–250. https://doi.org/10.56754/0719-0646.2502.231

[410] A. Bensalem, A. Salim, M. Benchohra and M. Fečkan, Approximate controllability of neutral functional integro-differential equations with state-dependent delay and non-instantaneous impulses, *Mathematics.* **11** (2023), 1–17. https://doi.org/10.3390/math11071667

[411] S. Abbas, W. Albarakati and M. Benchohra, Successive approximations for functional evolution equations and inclusions, *J. Nonlinear Funct. Anal.*, Vol. 2017 (2017), Article ID 39, pp. 1–13.

[412] S. Heikkila and V. Lakshmikantham, *Monotone Iterative Technique for Nonlinear Discontinuous Differential Equations*, Marcel Dekker Inc., New York, 1994.

[413] S. Baghli, M. Benchohra, Uniqueness results for partial functional differential equations in Fréchet spaces, *Fixed Point Theory*, **9** (2008), 395–406.

[414] R.P. Agarwal, M. Benchohra and B.A. Slimani, Existence results for differential equations with fractional order impulses, *Mem. Differ. Equs. Math. Phys.*, **44** (2008), 1–21.

[415] K. Balachandran and S. Kiruthika, Existence of solutions of abstract fractional impulsive semilinear evolution equations, *Electron. J. Qual. Theor. Differ. Equat.*, **2010**(4) (2010), 1–12.

[416] V. Lakshmikantham, D.D. Bainov and P.S. Simeonov, *Theory of Impulsive Differential Equations*, World Scientific, NJ, 1989.

[417] M. Meghnafi, M. Benchohra and K. Aissani, Impulsive fractional evolution equations with state-dependent delay, *Nonlinear Stud.* **22** (4) (2015), 659–671.

[418] R. P. Agarwal, S. Hristova, and D. O'Regan, *Non-instantaneous Impulses in Differential Equations.* Springer, Cham, 2017.

[419] D. N. Pandey, S. Das and N. Sukavanam, Existence of solution for a second-order neutral differential equation with state dependent delay and non-instantaneous impulses, *Int. J. Nonlin. Sci.* **18**(2) (2014), 145–155.

[420] M. Pierri, D. O'Regan and V. Rolnik, Existence of solutions for semilinear abstract differential equations with not instantaneous impulses, *Appl. Math. Comput.* **219** (2013), 6743- 6749.

[421] S. Abbas, N. Al Arifi, M. Benchohra and J. Graef, Periodic mild solutions of infinite delay evolution equations with not instantaneous impulses. *J. Nonl. Funct. Anal.* **2020** (2020), Art. 7, 11 pp.

[422] S. Abbas, M. Benchohra, G.M. N'Guérékata and Y. Zhou, Periodic mild solutions of infinite delay second order evolution equations with impulses, *Electron. J. Math. Anal. Appl.*, **9** (1) (2021), 179–190.

[423] S. Abbas, M. Benchohra, J. E. Lazreg, J. J. Nieto and Y. Zhou, *Fractional Differential Equations and Inclusions: Classical and Advanced Topics*, World Scientific, Hackensack, NJ, 2023.

[424] A. Salim, J. Alzabut, W. Sudsutad and C. Thaiprayoon, On impulsive implicit ψ-Caputo hybrid fractional differential equations with retardation and anticipation. *Mathematics.* **10** (2022), 20 pages. https://doi.org/10.3390/math10244821

[425] A. Salim, M. Benchohra, J. E. Lazreg and G. N'Guérékata, Boundary value problem for nonlinear implicit generalized Hilfer-type fractional differential equations with impulses. *Abstr. Appl. Anal.* **2021** (2021), 17pp. https://doi.org/10.1155/2021/5592010

[426] A. Salim, M. Benchohra, E. Karapınar and J. E. Lazreg, Existence and Ulam stability for impulsive generalized Hilfer-type fractional differential equations. *Adv. Differ. Equ.* **2020** (2020), 21 pp. https://doi.org/10.1186/s13662-020-03063-4

[427] A. Salim, M. Benchohra, J. R. Graef and J. E. Lazreg, Boundary value problem for fractional generalized Hilfer-type fractional derivative with non-instantaneous impulses. *Fractal Fract.* **5** (2021), 1–21. https://dx.doi.org/10.3390/fractalfract5010001

[428] A. Salim, M. Benchohra, J. E. Lazreg and J. Henderson, Nonlinear implicit generalized Hilfer-type fractional differential equations with non-

instantaneous impulses in Banach spaces. *ATNAA.* **4** (2020), 332–348. https://doi.org/10.31197/atnaa.825294

[429]　A. Salim, M. Benchohra and J. E. Lazreg, Nonlocal k-generalized ψ-Hilfer impulsive initial value problem with retarded and advanced arguments, *Appl. Anal. Optim.* **6** (2022), 21–47.

[430]　S. Abbas, M. Benchohra and G. N'Guérékata, Periodic mild solutions of infinite delay not instantaneous impulsive evolution equations, *Vietnam J. Math.* (2020).

[431]　S. Meslem, A. Salim, S. Abbas and M. Benchohra, Periodic mild solutions of infinite delay integro-differential inclusions with non instantaneous impulses, *Appl. Anal. Optim.* **8** (1) (2024), 1–14.

[432]　S. Meslem, A. Salim, S. Abbas, A. Arara, and M. Benchohra, Periodic mild solutions of infinite delay second order evolution inclusions with non instantaneous impulses. (Submitted).

[433]　A. Salim, S. Abbas, M. Benchohra and E. Karapınar, A Filippov's theorem and topological structure of solution sets for fractional q-difference inclusions. *Dynam. Systems Appl.* **31** (2022), 17–34. https://doi.org/10.46719/dsa202231.01.02

[434]　P. Aviles and J. Sandefur: Nolinear second order equations wtih applications to partial differential equations, *J. Differential Equations* **58** (1985), 404–427.

[435]　A. Belleni-Morante, An integrodifferential equation arising from the theory of heat conduction in rigid material with memory, *Boll. Un. Mat. Ital.* **15** (1978), 470–482.

[436]　A. Belleni-Morante and G. F. Roach, A mathematical model for Gamma ray transport in the cardiac region, *J. Math. Anal. Appl.* **244** (2000), 498–514.

[437]　D. Tang and M. S. Rankin III, Peristaltic transport of a heat conducting viscous fluid as an application of abstract differential equations and semigroup of operators, *J. Math. Anal. Appl.* **169** (1992), 391–407.

[438]　R. Anguelov, C. Dufourd, Y. Dumont, Mathematical model for pest-insect control using mating disruption and trapping, *Appl. Math. Model.* **52** (2017), 437–457.

[439]　A. Malik, N. Singh, S. Satya, House fly (Musca domestica): a review of control strategies for a challenging pest, *J. Environ. Sci. Health.* **42** (4) (2007), 453–69.

[440]　P. Bates, P. Fife, X. Ren, X. Wang, Travelling waves in a convolution model for phase transitions, *Arch. Ration. Mech. Anal.* **138** (1997), 105–136.

[441]　F. Andreu-Vaillo, J. M. Mazon, J. D. Rossi, J. Toledo-Melero, *Nonlocal Diffusion Problems, Mathematical Surveys and Monographs*, AMS, Providence, Rhode Island, 2010.

[442]　M. A. Bogoya, nonlocal nonlinear diffusion equation in higher space dimensions, *J. Math. Anal. Appl.* **344** (2008), 601–615.

[443]　H. Kang, S. Ruan, Nonlinear age-structured population models with nonlocal diffusion and nonlocal boundary conditions, J. Differential Equations. **278** (2021), 430–462.

[444] F. Y. Yang, W. T. Li, and S. Ruan, Dynamics of a nonlocal dispersal SIS epidemic model with Neumann boundary conditions, *J. Differential Equations.*, **267** (3) (2019), 2011–2051.

[445] H. Kang, S. Ruan, and X. Yu, Age-structured population dynamics with nonlocal diffusion. *J. Dynam. Differential Equations.*, (2020).

[446] H. Amann, *Ordinary Differential Equations, An introduction to nonlinear Analysis*, De Gruter Studies in Mathematics 13, Berlin, NY, 1990.

[447] G. F. Webb, *Theory of Nonlinear Age-Dependent Populations Dynamics*, Monographs Textbooks in Pure and Applied Math. Dekker, NY 1985.

[448] N. Kato, Positive global solution for a general model of size-dependent populations dynamics, *Abstr. Appl. Anal.*, **5** (3) (2000), 191–206.

[449] B. Ainseba, S. M. Bouguima, An adaptive model for a multistage structured population under fluctuation environment, *Discrete Contin. Dyn. Syst. Ser. B.* **25** (6) (2020), 2331–2349.

[450] S. Anita, *Analysis and Control of Age-Dependent Population Dynamics*, Kluwer, Dordrecht, Netherlands, 2000.

[451] M. T. Agardy, Advances in marine conservation: the role of marine protected areas, *rend. Ecol. Evol.* **9** (1994), 267–270.

[452] G. W. Allison, J. Lubchenco, M.H. Carr, Marine reserves are necessary but not sufficient condition for marine conservation, *Ecol Appl.* **8** (1994), 79–92.

[453] L. Anderson, A bioeconomic analysis of marine reserves, *Nat. Resour. Model.* **15** (2002), 311–334.

[454] T. Luck, C. W. Clark, M. Mangel, G. R. Munro, Implementing the precautionary principles in fisheries management through marine reserves. *Ecol. Appl.* **8** (1998), 72–78.

[455] M. Bensenane, A. Moussaoui. P. Auger, On the Optimal Size of Marine Reserves. *Acta Biotheoretica.* **61** (2013), 109–118.

[456] A Moussaoui. P. Auger. Simple fishery and marine reserve models to study the SLOSS problem. *ESAIM: Proceedings and surveys.* **49** (2015), 78–90.

[457] P. Auger, C. Lett, A. Moussaoui, S. Pioch, Optimal number of sites in artificial pelagic multi-site fisheries. *Can. J. Fish. Aquat. Sci.* **67** (2010), 296–303.

[458] A. Moussaoui, P. Auger P, C. Lett, Optimal number of sites in multi-site fisheries with fish stock dependent migrations. *Math. Biosci. Eng.* **8** (2011), 769–783.

[459] H. Amann, Dynamics theory of quasilinear parabolic equation-I. Abstract evolution equation, *Nonlinear Anal.* **12** (1997), 219–250.

[460] G. R. Sell, Y. You, *Dynamics of Evolutionary Equations*, Applied Mathematical Sciences 143, Springer-Verlag, N.Y, 2002.

[461] C. V. Pao, *Nonlinear Parabolic and Elliptic Equations*, Plenum, 1992.

[462] N. D. Alikakos, An Application of the Invariance Principle to Reaction-Diffusion Equations. *J. Differential Equations.* **33** (1979), 201–225.

[463] M. G. Crandall, P. H. Rabinowitz, Bifurcation from simple eigenvalues, *J. Funct. Anal.* **8** (1971), 321–340.

[464] R.S. Cantrell, C. Cosner, On the dynamics of predator-prey models with the Beddington-DeAngelis functional response, *J. Math. Anal. Appl.* **257** (2001), 206–222.

[465] J. Smoller, *Shock Waves and Reaction Diffusion Equations*, Springer-Verlag, Berlin, 1983.

[466] S. M. Bouguima and K. A. Kada, Analysis and control of physiologically structured models with nonlocal diffusion, *Differ. Equ. Appl.* **15** (1) (2023), 29–60.

[467] S. M. Bouguima and K. A. Kada, Qualitative analysis of a stock-effort parabolic model incorporating marine reserve, *Appl. Anal.* **102** (15) (2023), 4045–4057.

Index